Lecture Notes in Mathematics

Edited by A. Dold and B. Eckmann

792

Geometry and Differential Geometry

Proceedings of a Conference Held at the University of Haifa, Israel, March 18–23, 1979

Edited by
R. Artzy and I. Vaisman

Springer-Verlag
Berlin Heidelberg New York 1980

Geometry, under its various aspects, has been a very exciting intellectual activity during the whole history of civilized mankind. It is an old and always new science which, undoubtedly, has given us much an important part of our knowledge about the world and with many applications.

The present volume includes the text of most of the lectures presented at a Conference on Geometry and Differential Geometry which was held at the University of Haifa, Israel, on March 18-23, 1979. The Conference was attended by some 70 mathematicians from all over the world. The subject matters covered a broad range and many of the aspects of modern research in the field were discussed. This is why we decided to publish the Proceedings of the Conference and it is our hope that they will be of interest to the mathematical community.

The organizers of the Conference are glad to acknowledge here their gratitude to the participants in the Conference, to all those who contributed to its success and to Springer Verlag who kindly accepted this volume in its Lecture Notes Series.

R. Artzy, I. Vaisman

TABLE OF CONTENTS

ON FREE MINKOWSKI PLANES

R. Artzy
University of Haifa
31999 Haifa, Israel

The structure of the automorphism groups of free planes has been determined for various geometries; however, this was successfully done only in cases of minimal generation, that is, when the plane originates by free generation from a minimal fundamental configuration. The results so far have been:

Plane	Fundamental configuration	Automorphism group	Reference
Projective	4 independent points	$\langle S_4 * S_4; D_4 \rangle$	[3]
Affine	3 independent points	$\langle D_3 * D_4; S_2 \rangle$	[1]
Möbius	1 circle c, 3 points 2 of which lie on c	$S_2 * V_4$	[4]
Laguerre	1 circle c, 3 mutually nonparallel points 2 of which lie on c	$V_4 * V_4 * V_4$ (contains an error: one of these products is indeed not free)	[4]

Here S_m, D_n, V_4 are, respectively, the symmetric group on m objects, the dihedral group of order 2n, and Klein's 4-group. By $\langle A*B; C \rangle$ we mean the free product of A and B with C amalgamated. In the Möbius and Laguerre planes, as also in the Minkowski plane in the following, circles do not necessarily intersect (in terms of [6], the circle structures are "affine").

The procedures used in our approach to finding the structure of the automorphism group of the minimally generated free Minkowski plane will be based on previous extensive work by various authors: The theory of free extensions in general was thoroughly investigated by Schleiermacher and Strambach [6], and the case of the Minkowski plane was worked out by Heise and Sörensen [2]. The practice for determining the automorphism groups was first developed by Sandler [5] and later refined by Iden [3,4]. We will follow these sources, their definitions and terminology rather closely.

A few definitions follow: The circle containing the points A, B, C is called ABC . Parallelism of points is denoted by $\parallel_+$ and $\parallel_-$. A couple (P,c) of a point P and a circle c incident with it is called a flag. The set of all points $\parallel_+$ to a given point is called a generator line, and so is the set of all points $\parallel_-$ to the point. Two circles c and d are called tangent or touching if $|c\cap d|=1$.

We now start out from a minimal fundamental configuration consisting of 3 mutually nonparallel points B_{11}, B_{22}, B_{33} in a Minkowski plane. We perform the following operations ("steps") successively:

I. Draw all lines through triples of mutually nonparallel existing points,

II. draw all possible tangents to flags (P,c) already obtained through already obtained points Q such that $P \nparallel Q \notin c$,

III. intersect obtained circles with obtained +generator and -generator lines,

IV. intersect intersecting but nontangent circles once more,

V. draw all genrator lines through given points,

VI. intersect all +generator lines with all -generator lines.

Each full set of these 6 steps will be called a "stage". We start with stage 0: Steps I through VI operating on the fundamental configuration yield 6 points B_{ij} in addition to B_{11}, B_{22}, B_{33}: let $B_{jj} \parallel_+ B_{ij} \parallel_- B_{ii}$; $i,j=1,2,3$. Thus, at the end of stage 0 we have 9 points, that is, 36 ordered point triples which can serve as fundamental triples.

Definition. A point triple which can serve as fundamental triple (that is, which yields by free extension the same set of points as B_{11}, B_{22}, B_{33}) will be called a generating triple (in short, GT). GT's will be designated by brackets.

Proposition 1. Stage 0 provides 36 ordered GT's.

Now we begin stage 1. Step I produces 6 circles $B_{1i}B_{2j}B_{3k}$, where (ijk) is a permutation of (123). By step II we obtain 2 tangents to each of the 3 flags on each circle, that is, 6 distinct tangents to each of the 6 circles, a total of 36 such tangents. Each of these tangents intersects, by step III, two of the generator lines obtained so far. Thus we obtain 72 new points in stage 1.

Proposition 2. After completion of step III, one has in stage 1, $72\cdot 6=432$ ordered GT's.

Proof. Step III yields 2 points on each tangent in addition to the two points used in step II. Thus we have on each tangent 4 triples of mutually nonparallel points. However, such a triple is a GT if and only if two of its points are of the type B_{ij}. Hence each of the 36 tangents produces just two GT's, a total of 72 GT's, or 72·6 ordered GT's. By inspection one sees also that no other combination of points produces GT's.

Step IV requires that two intersecting circles, say d and b, can be shown not to be tangent. This is possible if it can be proved that at their point of intersection, say P, another circle, say c, touches b and intersects d at $Q \neq P$. Then d is not the tangent to b through Q at P, and hence d and b must intersect once more, say at R. The first figure describes this situation for the case where P is the point at infinity in the hyperbola model of the Minkowski plane.

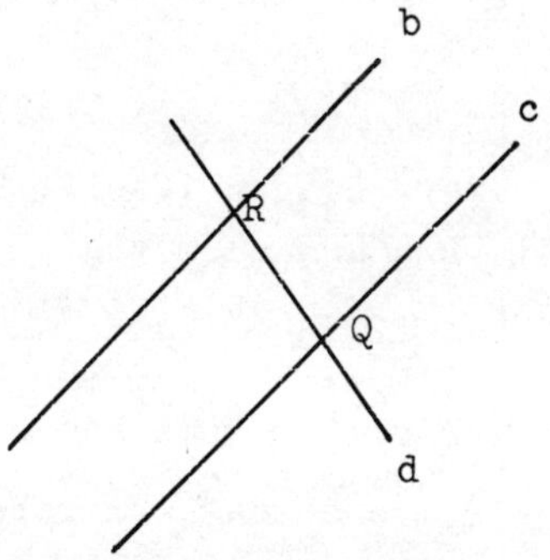

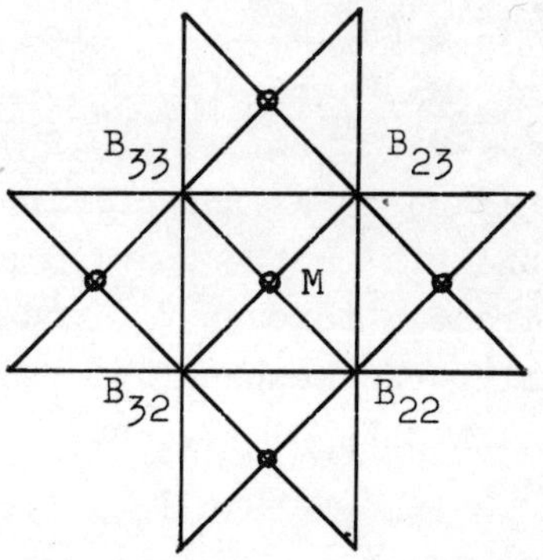

Proposition 3. By step IV, 18 GT's result from each GT.

Proof. Let B_{11} play the role of P, and for the sake of illustration choose P again as the point of infinity in the hyperbola model. Then we get the second figure. For $P=B_{11}$ we obtain the 5 marked new points. Of these, only $(B_{11}B_{23}B_{32} \cap B_{11}B_{22}B_{33}) \setminus B_{11} =: M$ can be used for GT's. There are indeed two GT's involving M, namely $[B_{33}B_{22}M]$ and $[B_{32}B_{23}M]$. For every choice of P among the nine B_{ij}, we would obtain again one new point and two GT's, a total of 18 GT's.

Proposition 4. Stage 1 provides a total of $36 \cdot 9^3$ ordered GT's.

Proof. In view of Proposition 3, we get from the GT's of Proposition 2, 72·18 GT's. By steps V and VI each pair of triples of the type $[B_{22}B_{33}M]$ and $[B_{23}B_{32}M]$ yields 6 GT's. Thus we have, respectively, 18·3 and 72·18·3 GT's, together (54+72·54)6 ordered GT's. From Proposition 2, we have 432 ordered GT's, which after steps V and VI make 432·6 ordered GT's. The total is $(432+54+72 \cdot 54)6 = 36 \cdot 9^3$ ordered

GT's from stage 1.

Remark. A consequence of Proposition 4 is that the number of ordered GT's of stage $k \geq 2$ is that of stage k-1 multiplied by 9^3, minus the number of the GT's where (i) step IV is performed twice in a row on the same point triple and therefore cancels itself out, (ii) a tangent is drawn to a circle in stage k-1, and in stage k the original circle is regained as a tangent to the tangent.

We now introduce collineations induced by permutations of the B_{ij}. Define r, s, u, v as collineations induced, respectively, by the permutations as follows: $[B_{11}B_{22}B_{33}]$ maps under r on $[B_{11}B_{23}B_{32}]$, under s on $[B_{13}B_{21}B_{32}]$, under u on $[B_{11}B_{32}B_{23}]$, and under v on $[B_{31}B_{12}B_{23}]$. By abuse of language we may write, in cycle notation

$$r=(B_{22}B_{23})(B_{33}B_{32})(B_{13}B_{12})$$
$$s=(B_{31}B_{33}B_{32})(B_{21}B_{23}B_{22})(B_{11}B_{13}B_{12})$$
$$u=(B_{22}B_{32})(B_{33}B_{23})(B_{21}B_{31})$$
$$v=(B_{31}B_{21}B_{11})(B_{33}B_{23}B_{13})(B_{32}B_{22}B_{12}).$$

We also define t_1 as the collineation which maps B_{11} on itself, B_{33} on B_{23}, and B_{22} on the point C such that $C\|_+ B_{22}$ and such that C lies on the circle c through B_{23} and tangent to $B_{11}B_{22}B_{33}$ at B_{11}. Similarly, t_2 is defined as mapping B_{11} on itself, B_{22} on B_{23}, and B_{33} on the point D such that $D\|_- B_{33}$ and such that D lies on the same tangent c.

Finally, we remark that in a free plane $C \neq B_{32}$ and that, therefore, the circle $B_{11}B_{32}B_{23}$ is not the tangent to $B_{11}B_{22}B_{33}$ at B_{11} through B_{23}. Hence $B_{11}B_{22}B_{33}$ and $B_{11}B_{32}B_{23}$ have to intersect in a point M distinct from B_{11}. Again, we define a collineation m such that it maps $[B_{11}B_{22}B_{33}]$ on $[MB_{22}B_{33}]$.

Proposition 5. For the collineations defined above, $r^2 = u^2 = s^3 = v^3 = (rs)^2 = (uv)^2 = (rut_1)^2 = (rut_2)^2 = m^2 = (mr)^2 = (mu)^2 = 1$. Moreover, r and s commute with u and v.

Proof. The first 4 relations are immediate, in view of the cycle lengths of the permutations. We get $rs=(B_{11}B_{13})(B_{33}B_{31})(B_{23}B_{21})$ and $uv=(B_{11}B_{31})(B_{33}B_{13})(B_{12}B_{32})$, hence they are involutory. We have $ru=(B_{22}B_{33})(B_{32}B_{23})(B_{13}B_{12})(B_{21}B_{31})=ur$, $rv=(B_{11}B_{31}B_{21})(B_{22}B_{13}B_{32}B_{23}B_{12}B_{33})=vr$, $su=(B_{11}B_{33}B_{22})(B_{32}B_{21}B_{13})(B_{31}B_{23}B_{12})=us$, $sv=(B_{11}B_{13}B_{12})(B_{22}B_{31}B_{23}B_{32}B_{21}B_{33})=vs$. Under rut_1, $[B_{11}B_{22}B_{33}]$ maps on $[B_{11}B_{23}C]$; when rut_1 is applied again, B_{11} maps on itself, and B_{23} on B_{22}. Under ru, C maps on a point $E\|_+ B_{33}$ which lies on the tangent to $B_{11}B_{22}B_{33}$ at B_{11} through B_{32}. Finally, under t_1, E goes into the point $\|_+ B_{23}$ which lies on the tangent to $B_{11}CB_{23}$ at B_{11} through B_{22}. This, however, is the point B_{33}, which proves that $(rut_1)^2=1$. Similarly, rut_2 takes $[B_{11}B_{22}B_{33}]$ into $[B_{11}DB_{23}]$, and $[B_{11}DB_{23}]rut_2=[B_{11}B_{22}B_{33}]$, implying $(rut_2)^2=1$.

By its very definition, m maps M on B_{11}, and hence $m^2=1$. We have $[B_{11}B_{22}B_{33}](mr)^2=[MB_{23}B_{32}]mr=[B_{11}B_{22}B_{33}]$ and $[B_{11}B_{22}B_{33}](mu)^2 = =[MB_{32}B_{23}]mu=[B_{11}B_{22}B_{33}]$, hence $(mr)^2=(mu)^2=1$.

Proposition 6. Every word composed of r, s, u, v, t_1, t_2, m is of the form

$$\left(\prod_{j=0}^{n-1} \beta_{n-j} m^{e_{n-j}} \delta_{n-j} t_{i_{n-j}}^{\varepsilon_{n-j}}\right)\beta_o$$

with $\beta_k \in \langle r,s,u,v\rangle$, $\delta_k \in \langle s,v\rangle$, $i_k \in \{1,2\}$, ε_k and $e_k \in \{0,1\}$.

Proof. In the words we are able to perform the following reduction procedures: No powers of r, u, m will appear because they are involutory. The generators s and v will occur only as s, s^2, v, v^2 since their order is 3. In every product involving r, s, u, v, we shift r and u to the left by means of $sr=rs^2$ and $vu=uv^2$ and in view of r and s commuting with u and v. Negative powers of t_1 and t_2 will be avoided by means of $t_i^{-1}=(ru)^{-1}t_i ru=rut_i ru$. Any factors r, u to the immediate right of m will be shifted to the left by means of mr=rm and mu=um. The assertion is now proven.

Definition. The number of factors $\beta m^e \delta t_i^\varepsilon$ in a collineation is called the stage of the collineation. It is assumed that in each such product at least one of e and ε is 1. Moreover, a product $(\beta_j m\delta_j)(\beta_{j-1}\delta_{j-1}t_i)$ is to be counted as one factor only. A collineation $\beta \in \langle r,s,u,v\rangle$ is defined to be of stage 0.

Proposition 7. Every GT of stage $k \geqq 0$ can be obtained from the fundamental triple by a collineation of stage $\leqq k$.

Proof by induction on k. The 36 ordered GT's of Proposition 1 are exactly those produced by the 36 elements of the group $\langle r,s,u,v\rangle$, the direct product of the two groups $\langle r,s\rangle$ and $\langle u,v\rangle$, both isomorphic to S_3. Steps II and III of stage 1 correspond to the collineations t_1 and t_2: just as each of the new tangents produced just two GT's so do t_1 and t_2. Step IV corresponds to $m\delta$: there are 9 distinct elements $\delta \in \langle s,v\rangle$ and thus 9 collineations $m\delta$. Finally, steps V and VI correspond to the 36 elements $\beta \in \langle r,s,u,v\rangle$. Thus, if in $\beta_1 m^\varepsilon \delta t_i^e \beta_o$ $e=\varepsilon=0$, we get $36\cdot 2\cdot 9\cdot 36$ elements. If $e=1$, $\varepsilon=0$, we have $36\cdot 2\cdot 36$ elements. For $e=0$, $\varepsilon=1$, we have only $9\cdot 36$ collineations because in $\beta_o m\delta\beta_1$ the factors u and r in $\delta\beta_1$ can be shifted to the left of m so that only elements of the type $\beta m\delta$ remain. The total is then

$36(2\cdot 9\cdot 36+2\cdot 36+9) = 36\cdot 9^3$, the same as in Proposition 4. Thus, our Proposition is proven for k=1.

For stage $k\geqq 1$, a collineation of stage k in its reduced form starts with an expression $\beta m^{\varepsilon}\delta t_i^e$. To raise the stage of the triple to k+1 we have to premultiply by another such expression $\beta' m^{\varepsilon'}\delta' t_{i'}^{e'}$. This will yield a collineation of stage k+1, except in two cases: (i) $e'=0$, $\delta'\beta\in\{1,r,u,ru\}$, $\varepsilon=\varepsilon'=1$, (ii) $\varepsilon=0$, $\beta\delta=ru$, $i=i'$, $e=e'=1$. In these cases the resulting collineation is of stage $\leqq k$. The two cases (i) and (ii) correspond exactly to the cases (i) and (ii), respectively, in the Remark after Proposition 4.

<u>Theorem 1</u>. The GT's are obtained from the fundamental triple by collineations whose group G is presented by the generators r, s, u, v, t_1, t_2, m and the relators r^2, s^3, u^2, v^3, $rur^{-1}u^{-1}$, $rvr^{-1}v^{-1}$, $sus^{-1}u^{-1}$, $svs^{-1}v^{-1}$, $(rut_1)^2$, $(rut_2)^2$, m^2, $(rm)^2$, $(um)^2$, $(rs)^2$, $(uv)^2$. Proof follows immediately from Propositions 5, 6, and 7.

<u>Theorem 2</u>. G is the free product of 4 groups G_1, G_2, G_3, G_4, where $G_1 \cong S_3\times S_3$, $G_2 \cong G_3 \cong D_\infty$, $G_4 \cong S_2\times S_2\times S_2$ and such that G_1 and G_4 have an amalgamated subgroup H isomorphic to $S_2\times S_2$ and such that there is an amalgamated subgroup of H of order 2 in all 4 factors.

Proof. From Theorem 1 we have the groups

$$G_1 = \langle r,s,u,v;r^2,u^2,s^3,v^3,(rs)^2,(uv)^2,\text{s and r commute with u and v}\rangle \cong S_3\times S_3$$

$$G_2 = \langle t_1,ru;(rut_1)^2,(ru)^2\rangle \cong D_\infty$$

$$G_3 = \langle t_2,ru;(rut_2)^2,(ru)^2\rangle \cong D_\infty$$

$$G_4 = \langle m,r,u;m^2,r^2,u^2,(mr)^2,(mu)2,(ru)^2\rangle \cong S_2\times S_2\times S_2.$$

G_1 and G_4 have the amalgamated subgroup $H=\langle r,u;r^2,u^2,(ru)^2\rangle \cong S_2\times S_2$. G_1, G_2, G_3, G_4 have the amalgamated subgroup $\langle ru;(ru)^2\rangle < H$.

References

[1] Artzy, R.: Collineation group presentations for some minimally generated planes. Proc. Symposium on Geometry, Silivri, 1978, forthcoming

[2] Heise, W.; Sörensen, K.: Freie Minkowski-Ebenenerweiterungen. J. Geometry 3, 1-4 (1973)

[3] Iden, O.: The collineation group of the free plane F_4. Math. Z. 119, 60-62 (1971)

[4] Iden, O.; Moe, J. G.: Automorphism groups of free Moebius planes and free Laguerre planes. Geometriae dedicata 7, 209-222 (1978)

[5] Sandler, R.: The collineation groups of free planes. Trans. Amer. math. Soc. 107, 129-139 (1963), and Proc. Amer. math. Soc. 16, 181-186 (1965)

[6] Schleiermacher, A.; Strambach, K.: Freie Erweiterungen in der affinen Geometrie und in der Geometrie der Kreise. Abh. math. Sem. Univ. Hamburg 34, 22-37, 209-226 (1969/70)

Rigidity in the geometry of involutory elements of a group

F. Bachmann
Mathematisches Seminar der Universität Kiel
D-2300 Kiel 1
Bundesrepublik Deutschland

1. Postulates of motion

F. Schur, in his book "Grundlagen der Geometrie" (1909), uses so-called projective postulates and postulates of motion. Schur's "projective" postulates correspond to Hilbert's axioms of incidence and order. To his postulates of motion, the following belong:

free mobility
rigidity
angle inversion
segment inversion.

In this lecture we shall restrict ourselves to plane geometry. If A is a point, h a half-line with origin A, and H one of the two half-planes defined by the line which carries h, then A,h,H is called a flag (in the classical sense). Rigidity says that the identity is the only motion which leaves a flag fixed. It implies that for any two flags, there is at most one motion which maps the first onto the second one. Free mobility says that any flag can be moved into any flag. Using the above axioms, we can move a given flag A,h,H into a given flag A*,h*,H* as follows: first by reflection in the midpoint of A,A* the flag A,h,H can be moved into a flag A*,h',H'; then this flag can be moved into a flag A*,h*,H" by reflection in the angle bisector of h',h*; finally if H" $\neq$ H*, we apply the reflection in the line carrying h*.

Remark. A point M is called a midpoint of two given points if the reflection in M interchanges the two points. A line m is called a bisector of a given angle if the reflection in m interchanges the sides of the angle.

In Euclidean and hyperbolic planes over ordered fields we do not always have free mobility, but we do have:

(1) conditional existence of midpoints
(2) uniqueness of midpoints
(3) conditional existence of angle bisectors
(4) rigidity.

Two points (lines, half-lines etc.) are called congruent if they can be moved into each other. Conditional existence of midpoints means: any two congruent points have a midpoint. Conditional existence of angle bisectors means: each angle whose sides are congruent has a bisector.

If we set aside order and wish to include, for instance, Euclidean planes over finite fields (with characteristic $\neq 2$), we shall modify some of the classical definitions as follows: an angle is a pair of intersecting lines; the two lines are the sides; a flag is an incident point-line-pair; rigidity means that the stabilizer of a flag $\{A,b\}$ is a Klein 4-group: the only motions fixing a point A and a line b incident with it are the identity, the reflection in the point A, the reflection in the line b, and the reflection in the line erected perpendicular to b through A.

If we adopt these definitions, conditions (1) - (4) are satisfied in many metric planes. Moreover, in many metric planes the motion group contains a subgroup of index 2, and the point reflections are the involutory elements of the subgroup while the line reflections are the involutory elements of the other coset.

2. Geometry of involutory elements of a group

Turning now to the geometry of involutory elements of a group, we start with the

Basic assumption. Let G be a group and S,P sets of involutory elements of G which are both invariant under inner automorphisms of G. Each involutory element of G lies in either S or P. Let $S,P \neq \emptyset$.

Notations. If $\alpha,\beta \in G$ we write β^{α} instead of $\alpha^{-1}\beta\alpha$. For involutory elements α,β the statement "$\alpha\beta$ is involutory" is abbreviated by $\alpha|\beta$; this stroke relation is a symmetric, irreflexive binary relation on the set of involutory elements of G. We denote elements of S by a,b,m,n,x, elements of P by A,B,M,N,X.

To a triple (G,S,P) which satisfies the basic assumption we assign a geometric structure, the group plane of (G,S,P), as follows:

We call the elements of S lines, the elements of P points. We define orthogonality of lines a,b by $a|b$, and incidence of a point A and a line b by $A|b$. By these definitions, the group plane is an incidence structure with orthogonality.

Moreover, the inner automorphisms of G restricted to S and P are called motions: For $\alpha \in G$, we call the mapping of S onto S and P onto P defined by

$$x \longmapsto x^{\alpha},\ X \longmapsto X^{\alpha} \qquad (\text{for } x \in S,\ X \in P)\ ,$$

the motion of the group plane induced by α. Especially, the motion induced by $a \in S$ is called the reflection in the line a, the motion induced by $A \in P$ is called the reflection in the point A. Thus G acts by conjugation as the motion group of the group plane[1].

3. Postulates of motion in the geometry of involutory elements of a group

The geometry of arbitrary triples which satisfy the basic assumption is merely a general framework in which we single out richer geometries by axioms. We are interested de facto in triples (G,S,P) where the group G has some properties of motion groups of classical metric planes and S and P have properties of the set of line reflections and the set of point reflections in such planes. Axioms concerning (G,S,P) can be chosen which differ both in form and content. The book "Aufbau der Geometrie aus dem Spiegelungsbegriff" is based on the observation that rules for involutory elements can have far-reaching consequences; therefore axioms about the elements of S and P were preferred.

It seems sensible to assume the postulates of motion (1) - (4) as axioms and investigate their consequences. In the present

---

[1] In the really interesting cases, G is usually isomorphic to the motion group of the group plane.

situation determined by the basic assumption and the definition of the group plane, we formulate the postulates as follows:

(1') For $A \in P$, $\alpha \in G$ there exists an $M \in P$ such that $A^M = A^\alpha$;
(2') $A^M = A^N$ implies $M = N$, for all $A,M,N \in P$;
(3') For all $b \in S$ and $\alpha \in G$, if there exists an $A \in P$ with $A|b,b^\alpha$, then there exists an $m \in S$ with $b^m = b^\alpha$;
(4') $A|b$ implies $C_G(A,b) = \{1,A,b,Ab\}$, for all $A \in P$, $b \in S$ [1].

These postulates say, in other words: any two conjugate points have a midpoint; two points have at most one midpoint; any two conjugate intersecting lines have an angle bisector; the centralizer of a flag is the Klein 4-group generated by the flag.

Now let G be a group, H a subgroup of index 2. Define P as I(H), S as I(G-H) [2]. Assume that P and S are non-empty[3] and that (G,S,P) satisfies conditions (1') - (4').

I think that these triples (G,S,P) deserve interest. They are the Hjelmslev groups[4], and several hundred theorems have been proved about them during the last ten years.

<u>Theorem 1</u>. Let (G,H) be a pair consisting of a group G and a subgroup H of index 2. Let $P := I(H)$, $S := I(G-H)$, and $P,S \neq \emptyset$. Let (G,S,P) satisfy conditions (1') - (4'). Then (G,S,P) is a Hjelmslev group. - Conversely, in an arbitrary Hjelmslev group (G,S,P), the pair (G,S^{even}) has all these properties.

---

[1] $C_G(A,b)$ denotes the centralizer of $\{A,b\}$ in G.

[2] For a subset T of a group, we denote by I(T) the set of involutory elements of T.

[3] Note that (G,S,P) satisfies the basic assumption.

[4] By a Hjelmslev group we always understand in this lecture a non-elliptic Hjelmslev group.

Hjelmslev groups are defined in AGS §20. By definition, a Hjelmslev group is a pair (G,S) consisting of a group G and a set S of involutory elements of G which is invariant under inner automorphisms of G and generates G. The elements of S and the involutory products of two elements of S satisfy certain axioms.

By the statement "(G,S,P) is a Hjelmslev group" we summarize that (G,S) is a Hjelmslev group and P the set of involutory products of two elements of S.

In a (non-elliptic) Hjelmslev group (G,S,P), the set S^{even} of all products of an even number of elements of S is a subgroup of index 2 of G.

It is remarkable that the conditions (1') - (4') say nothing about the group plane as an incidence structure with orthogonality. But from the basic assumption and (1'),(2') we easily deduce that in the group plane, for any point A and any line b, there exists a line through A perpendicular to b; using (4') in addition, we can also prove that this perpendicular is unique for given A,b. The conditions (1') - (4') do not exclude group planes which contain points with several joining lines or points which have no joining line.

I value theorem 1 as an argument in favour of the notion of a Hjelmslev group. In Hjelmslev groups and their group planes, some curious phenomena can occur. I feel we have to face these phenomena.

4. A characterisation of finite Hjelmslev groups

The rigidity condition (4') is (under the basic assumption and conditions (1'),(2')) equivalent to

(4*) For any Klein 4-group V with $V \leq G$, $C_G(V) = V$
(Any Klein 4-group contained in G is its own centralizer).

The finite Hjelmslev groups can be characterised as finite triples (G,S,P) which satisfy the basic assumption, condition (4*) and two further simple conditions:

Theorem 2 (F.Knüppel). Let (G,S,P) be a triple satisfying the basic assumption and let G be finite. Let the following conditions hold:

(4*) any Klein 4-group contained in G is its own centralizer,
(2*) the product of two elements of P is never involutory,
(3*) S is not a conjugacy class.

Then (G,S,P) is a Hjelmslev group. - Any finite Hjelmslev group (G,S,P) has all these properties.

The conditions of theorem 2 are purely group theoretic and can be understood without reference to a geometric structure. For the proof of the theorem we use the following lemmas about finite groups:

Lemma 1 (K.Johnsen). Let G be a finite group and P a set of involutory elements of G which is invariant under inner automorphisms of G. The elements of P are called points. Then the following conditions are equivalent:

(2*) the product of two elements of P is never involutory,

(1*) existence of midpoints: for all $A,B \in P$ there exists an $M \in P$ with $A^M = B$,

(2') uniqueness of midpoints: $A^M = A^N$ implies $M = N$ for all $A,M,N \in P$.

Lemma 2. If all the assumptions of the first half of theorem 2, except perhaps the last one, are fulfilled, the conditional existence of angle bisectors (3') is equivalent to the condition that S is not a conjugacy class.

The geometry of reflections is the geometry of motion groups with special regard to the reflections contained in them. At any rate our theorems give a partially new aspect to a major part of the geometry of reflections, namely the geometry of Hjelmslev groups. Theorem 1 emphasizes the purely metric standpoint of the geometry of Hjelmslev groups which is abstaining from all axioms of projective geometry.

References

1. Bachmann, F.: Aufbau der Geometrie aus dem Spiegelungsbegriff. Zweite ergänzte Auflage. Springer 1973. - Quoted as AGS.
2. - Hjelmslev-Gruppen. Mathematisches Seminar der Universität Kiel 1970/71. Zweiter Neudruck 1976.
3. - Hjelmslev groups. Atti del Colloquio Internazionale sulle Teorie Combinatorie (Roma 1973) 1, 469 - 479. Roma 1976.
4. Bachmann, F., und Knüppel, F.: Starrheit in der Geometrie involutorischer Gruppenelemente. To appear.
5. Bachmann, F., und Schnabel, R.: Kriterien für Hjelmslevgruppen. Abh. Math. Sem. Univ. Hamburg 44, 222 - 232 (1976).
6. Dennert, M.: Zum Schurschen Vorschlag für die Behandlung von Geraden und Ebenen. Wiss. Z. Humboldt-Univ. Berlin Math.-Natur. Reihe 23, 487 - 499 (1974).
7. Johnsen, K.: Endliche Gruppen mit nicht-elliptischer Spiegelungsgeometrie. Geometriae Dedicata 2, 51 - 56 (1973).
8. Schur, F.: Grundlagen der Geometrie. Leipzig 1909.
9. Stölting, R.: Über endliche Hjelmslev-Gruppen. Math. Z. 135, 249 - 255 (1974).

Ein Trennungsaxiom in der Orthogonalgeometrie und eine Charakterisierung der reellen Ebene.

Walter Benz
Mathematisches Seminar
der Universität Hamburg
2000 Hamburg 13
Bundesrepublik Deutschland

1. Eine präeuklidische Ebene (s. W.Degen, L.Profke [3])ist eine Translationsebene Σ, die dem Axiom von Fano genügt und in der es eine anisotrope Orthogonalitätsrelation $\perp$ gibt mit Gültigkeit des Höhenschnittpunktsatzes.
Nach R.Baer [2] ist dann Σ eine pappus'sche Ebene. Also ist der Koordinatenkörper von Σ ein kommutativer, quadratisch nicht abgeschlossener Körper mit einer von 2 verschiedenen Charakteristik. Umgekehrt gehört zu jedem kommutativen Körper K, char $K \neq 2$, und zu einem Nichtquadrat γ (der"Orthogonalitätskonstanten")eine präeuklidische Ebene. Sind A,B verschiedene Punkte von Σ, ist $g := A+B$ die Verbindungsgerade von A,B, so heiße

$$k := \{A,B\} \cup \{X \notin g \mid X+A \perp X+B\}$$

ein Kreis. Eine Gerade t werde Tangente von k genannt, wenn $|k \cap t| = 1$ ist. Der Punkt $P \notin k$ heiße äußerer Punkt von k, wenn durch P eine Tangente von k geht. $\Phi(k)$ sei die Menge der äußeren Punkte von k. Wir schreiben ($\mathbb{P}$ sei die Menge aller Punkte von Σ)

$$\overline{I}(k) := \mathbb{P} \setminus \Phi(k) \quad \text{und}$$
$$I(k) := \overline{I}(k) \setminus k \quad .$$

Genau die Punkte von $I(k)$ heißen innere Punkte von k. Ein innerer Punkt von k ist also einfach ein Punkt $P \notin k$, durch den keine Tangente von k geht. Wir richten nunmehr unser Augenmerk auf die beiden folgenden Eigenschaften

(TA) Ist k Kreis, ist P innerer und Q äußerer Punkt von k, so gilt $(P+Q) \cap k \neq \emptyset$.

(AKK) Ist k Kreis, ist $\mathbb{M}$ eine wenigstens zweielementige Menge von Punkten mit $\mathbb{M} \subset \overline{I}(k)$, so gibt es einen minimalen, zu k konzentrischen Kreis k_0 mit $\mathbb{M} \subset \overline{I}(k_0)$.

Hier muß zunächst der Mittelpunkt eines Kreises erklärt werden: Dies sei der Punkt, der dem Lot in P auf die Tangente von k durch P für alle $P \in k$ angehört. Mit "minimalem" Kreis k_o ist in (A K K) ein Kreis gemeint, für den $\bar{I}(k_o) \subset \bar{I}(k_1)$ für alle zu k_o konzentrischen Kreise k_1 gilt, die $\mathbb{M} \subset \bar{I}(k_1)$ genügen.

Wir nennen (T A) das Trennungsaxiom und (A K K) das Axiom der konzentrischen Kreise. Wir zeigen, daß (T A) in Σ genau dann gilt, wenn der Koordinatenkörper K euklidisch ist, und daß (T A), (A K K) in Σ genau dann gleichzeitig gelten, wenn K (natürlich bis auf Isomorphie) der Körper $\mathbb{R}$ ist. Auf zur Trennungseigenschaft (T A) verwandte Eigenschaften in angeordneten Ebenen (wie: "Eine Gerade g schneidet einen Kreis k, wenn sie einen inneren Punkt von k enthält") und ihren Zusammenhang mit euklidischen Körpern wird auch schon an anderer Stelle hingewiesen (s. F.Bachmann [1], S. 281, wo eine Ebene über einem pythagoreischen - also angeordneten - Körper zugrundegelegt ist). Das vielleicht Bemerkenswerte bei unserem Vorgehen ist, daß innere und äußere Punkte eines Kreises ohne Vorhandensein von Anordnungseigenschaften definiert werden, so daß die Formulierung (T A) Sinn bekommt und ihre Gültigkeit eben zur euklidischen Anordnung des Koordinatenkörpers führt. Hinsichtlich Untersuchungen zur Orthogonalgeometrie verweisen wir neben der schon zitierten Arbeit [2] von R.Baer auf die Arbeit [4] von Rautenberg und Quaisser, wo auch der Fall isotroper Geraden mit in die Betrachtungen einbezogen ist.

2. Unter einer Orthogonalitätsrelation einer affinen Ebene Σ verstehen wir eine binäre Relation $\perp$ auf der Menge $\mathbb{L}$ der Geraden von Σ, die den Eigenschaften (O1), (O2), (O3) genügt:

(O1) $g \perp h \Rightarrow h \perp g$

(O2) $g \perp h \parallel h' \Rightarrow g \perp h'$

(O3) (Lotaxiom) Ist P Punkt, g Gerade, so gibt es genau eine Gerade h mit $P \in h \perp g$.

Eine Gerade g heißt isotrop, wenn $g \perp g$ gilt. Die Orthogonalitätsrelation $\perp$ heiße anisotrop, wenn kein $g \in \mathbb{L}$ isotrop ist.

Sei von nun ab Σ präeuklidische Ebene. Für die Vektoren $\mathfrak{a} \neq o$, $\mathfrak{b} \neq o$ schreiben wir $\mathfrak{a} \perp \mathfrak{b}$, wenn es Geraden g,h gibt mit $g \perp h$ so, daß $\mathfrak{a}$(bzw. $\mathfrak{b}$) Richtungsvektor von g(bzw. h) ist. Nimmt man Vektoren $\mathfrak{a}, \mathfrak{b}$ mit $\mathfrak{a} \perp \mathfrak{b}$, so schreibt sich die Orthogonalitätsrelation mit einer geeigneten Orthogonalitätskonstanten $\gamma \neq 0$ in der Form

(1) $\mathfrak{a}x + \mathfrak{b}y \perp \mathfrak{a}\gamma y + \mathfrak{b}x$

für alle $x,y \in K$, die nicht beide 0 sind.

Hier ist $\gamma \notin K^2 := \{r^2 \mid 0 \neq r \in K\}$. Sei

$$W_\gamma := \{p^2 - \gamma q^2 \neq 0 \mid p,q \in K\} .$$

Sei mit einem festen Punkt U (der "Koordinatenursprung") das Koordinatensystem $U, \mathfrak{a}, \mathfrak{b}$ betrachtet, in dem $X \in \mathbb{P}$ die Koordinaten (x,y) hat, wenn gilt

$$\overrightarrow{UX} = \mathfrak{a}x + \mathfrak{b}y .$$

Sind a,b,c beliebige Elemente aus K mit $c \in W_\gamma$, so stellt

(2) $(x-a)^2 - \gamma(y-b)^2 = c$

die Gleichung eines Kreises dar, wobei kein Kreis von Σ vergessen ist. Der Mittelpunkt von (2) hat die Koordinaten (a,b).

Satz 1: In Σ gilt (TA) genau dann, wenn K euklidisch ist.

Beweis: Wir übergehen den trivialen Nachweis dafür, daß (TA) gilt, sofern K euklidisch ist. Gelte also umgekehrt (TA). Wir betrachten den Kreis k der Gleichung

(3) $x^2 - \gamma y^2 = 1$.

Hier ist U innerer Punkt, da eine Gerade durch U, die k in (x,y) trifft, k auch in (-x,-y) schneidet, also keine Tangente an k sein kann. Die Punkte (1,p), $p \neq 0$, sind äußere Punkte von k, da sie auf der Tangente $\{(1,\lambda) \mid \lambda \in K\}$ von k liegen. Also wird k von der Verbindungsgeraden von U und (1,p), $p \neq 0$, geschnitten. Dies bedeutet $1 - \gamma p^2 \in K^2$ für alle $p \neq 0$, d.h. also $1 - \gamma p^2 \in K^2$ für alle $p \in K$.

Wir wollen jetzt $-\gamma \in K^2$ nachweisen. Sei ohne Einschränkung $\gamma \neq -1$. Dann gilt

$$(0,q) \notin k \quad \text{mit} \quad q = \frac{1}{2}\left(1 - \frac{1}{\gamma}\right) .$$

Die Gerade (sei $v := 1 - q$)

$$t = \{(\lambda, q + \lambda v) \mid \lambda \in K\}$$

durch $(0,q)$ ist Tangente von k, da

$$\lambda^2 - \gamma(q + \lambda v)^2 = 1$$

in $\left(\lambda z - \frac{\gamma q v}{z}\right)^2 = 0$ umgeschrieben werden kann, wobei z ein Element aus K mit $z^2 = 1 - \gamma v^2 \in K^2$ ist.
Also ist $(0,q)$ äußerer Punkt von k, und die Gerade durch U, $(0,q)$ schneidet k nach (TA). Dies bedeutet $-\gamma \in K^2$.
Die Aussagen $-\gamma \in K^2$ und $1 - \gamma p^2 \in K^2$ für alle $p \in K$ können wir zu $W_\gamma = K^2$ zusammenfassen. Um sicher zu sein, daß K euklidisch ist, haben wir noch die folgende Aussage zu beweisen:
Gilt $0 \neq z \notin K^2$, fo folgt $-z \in K^2$. Sei o.B.d.A. $z \neq -1$ und sei k der Kreis der Gleichung

$$(4) \qquad x^2 - \gamma y^2 = \left(\frac{1-z}{2}\right)^2 .$$

Ist s ein Element aus K mit $s^2 = -\gamma$, so sei gesetzt

$$A = \left(\frac{1+z}{2}, 0\right), \quad B = \left(\frac{1+z}{2}, \frac{1-z}{2s}\right) .$$

Es ist A innerer Punkt von k: Wegen $z \neq 0$ ist $A \notin k$.
Wäre

$$t : (p,0) + \lambda(u,v), \quad p := \frac{1+z}{2}, \quad (u,v) \neq (0,0) ,$$

Tangente an k, so müßte

$$\lambda^2(u^2 - \gamma v^2) + 2\lambda p u + z = 0$$

genau eine Lösung λ haben. Mit dem Satz von Vieta würde dies aber

$$\lambda^2 = \frac{z}{u^2 - \gamma v^2} ,$$

d.h. $z \in K^2$, bedeuten unter Berücksichtigung von $u^2 - \gamma v^2 \in W_\gamma = K^2$.
Es ist B äußerer Punkt von k: Wegen $z \neq -1$ ist $B \notin k$.
Außerdem ist

$$t : \left(\frac{1+z}{2}, \frac{1-z}{2s}\right) + \lambda\left(1, \frac{1-z^2}{2sz}\right)$$

Tangente durch B an k .
Damit ist $(A+B) \cap k \neq \emptyset$, d.h. die Gerade

$$g : \left(\frac{1+z}{2}, 0\right) + \lambda\left(0, \frac{1-z}{2s}\right)$$

schneidet k . Ist λ_o ein Parameterwert für einen solchen Schnittpunkt, so folgt

$$-z = \lambda_o^2 \left(\frac{1-z}{2}\right)^2 \in K^2$$

wegen $\lambda_o \neq 0$.

Setzt man $W_\gamma = K^2$ voraus, so läßt sich ohne Forderung von (TA) leicht zeigen, daß $P(p,q)$ genau dann äußerer (bzw. innerer) Punkt von

$$k : (x-a)^2 - \gamma(y-b)^2 = c$$

ist, wenn $N_k(P) \in K^2$ (bzw. $0 \neq N_k(P) \notin K^2$) gilt, wo

$$N_k(P) = (p-a)^2 - \gamma(q-b)^2 - c$$

gesetzt ist ("Potenz" in $U, \mathfrak{v}, \mathfrak{b}$ von P bzgl. k) .

Satz 2: In Σ gilt (TA) und gleichzeitig (AKK) genau dann, wenn $K \cong \mathbb{R}$ ist.

Beweis: Bei einem euklidischen Körper ist bekanntlich $P = K^2$ Positivitätsbereich, d.h. K ist ein angeordneter Körper, in dem jedes positive Element Quadrat ist. Die Gerade

$$\{(\lambda,0) \mid \lambda \in K\}$$

heiße x-Achse. Gegeben sei nun eine nach oben beschränkte nichtleere Teilmenge T von K . Wir identifizieren die Elemente $\tau \in T$ mit den Punkten $(\tau,0)$ der x-Achse. Wir haben zu zeigen, daß T eine obere Grenze besitzt.
Sei $\tau_o \in T$. Ist $\{\tau \in T \mid \tau > \tau_o\}$ leer, so ist nichts zu zeigen. Sei also diese Menge nichtleer und sei $a \in K$ obere Schranke von T . Offenbar ist $a > \tau_o$. Wir betrachten den Kreis k der Gleichung

$$(x-\tau_o)^2 - \gamma y^2 = (a-\tau_o)^2 \tag{5}$$

und

$$\mathbb{M} := \{\tau \in T \mid \tau \geq \tau_o\}.$$

Sei nun

$$k_o : (x-\tau_o)^2 - \gamma y^2 = r^2 , \quad r > 0 ,$$

ein nach (A K K) existierender minimaler Kreis mit Mittelpunkt $(\tau_o, 0)$ und

$$\mathbb{M} \subset \overline{I}(k_o) .$$

Dann ist $\tau_o + r$ obere Grenze von T: Ist $\tau \in T$, so gilt sicherlich $\tau \leq \tau_o + r$ für $\tau \leq \tau_o$. Ist $\tau > \tau_o$, so folgt $(\tau, 0) \in \mathbb{M} \subset \overline{I}(k_o)$, d.h. $\tau - \tau_o \leq r$, d.h. $\tau \leq \tau_o + r$.

Gäbe es nun eine obere Schranke b von T mit $b < \tau_o + r$, so wäre $\tau_o < b$ und würde auch in $\overline{I}(k_1)$ mit

$$k_1 : (x-\tau_o)^2 - \gamma y^2 = (b-\tau_o)^2$$

liegen, was (A K K) widerspricht.

Literatur

[1] F. Bachmann, Aufbau der Geometrie aus dem Spiegelungsbegriff. Springer-Verlag, Berlin 1959.

[2] R. Baer, The fundamental theorems of elementary geometry. An axiomatic analysis. Trans. Amer. math. Soc. 56 (1944), 94 - 129.

[3] W. Degen, L. Profke, Grundlagen der affinen und euklidischen Geometrie. B.G.Teubner, Stuttgart 1976.

[4] E. Quaisser, W. Rautenberg, Orthogonalitätsrelationen in der affinen Geometrie. Zeitschr. math. Logik und Grundl. d.Math. 15 (1969), 19 - 24.

APPLICATIONS OF THE THEORY OF SURFACES TO THE THEORY OF GRAPHS

L. Ja. Beresina

Department of Mathematics
University of Haifa
31999 Haifa, Israel

1. For each simple graph with n vertices and its adjacency matrix G there is a well-defined cone

$$x^T Gx = 0 \tag{1}$$

in n-space.

Various problems in graph theory can be formulated in terms of the geometry of the cone. Thus, deleting a vertex numbered m corresponds to substituting $x_m = 0$ in (1). If, after deleting k vertices numbered $m_1, m_2, \ldots, m_k$ the graph consists of isolated points only, then the coordinate plane $x_{m_i} = 0$ $(i = 1,2,\ldots,k)$ lies on the cone (1).

Using the notations p and q for the inertial indices of the form $x^T Gx$, where, without loss of generality, we may assume $p \geq q$, we remark that the planes forming the surface (1) have dimension $n-p$.

Consequently, we have the following bound for the chromatic number

$$\nu(G) \geq \frac{n}{n-p} \; . \tag{2}$$

If $rk(G) < n$, then

$$\nu(G) \geq \frac{r}{q} \tag{3}$$

2. Another problem, which can be solved geometrically, is the problem of the isomorphism.

Let us have two isomorphic graphs $G = \{g_{ik}\}$ and $\overline{G} = \{\overline{g}_{ik}\}$. Since we refer to a graph as a cone, g_{ik} and $\overline{g}_{ik}$ should be the coordinates of the same tensor in two distinct coordinate systems.

EXAMPLE 1.

We have

$$G = \begin{pmatrix} 0 & 1 & 1 \\ 1 & 0 & 0 \\ 1 & 0 & 0 \end{pmatrix} \quad , \quad \overline{G} = \begin{pmatrix} 0 & 0 & 1 \\ 0 & 0 & 1 \\ 1 & 1 & 0 \end{pmatrix} \tag{4}$$

and consequently,

$$\begin{aligned} g_{12} &= 1 \quad , \quad g_{13} = 1 \quad , \quad g_{23} = 0 \\ \overline{g}_{12} &= 0 \quad , \quad \overline{g}_{13} = 1 \quad , \quad \overline{g}_{23} = 1 \end{aligned} \tag{5}$$

If the graphs are isomorphic, there must exist a transformation $C = \{c_{ik}\}$, such that

$$\begin{aligned}\overline{g}_{ij} &= (c_{i1}c_{j2} + c_{i2}c_{j1})g_{12} + \\ &+ (c_{i1}c_{j3} + c_{i3}c_{j1})g_{13} + \\ &+ (c_{i2}c_{j3} + c_{i3}c_{j2})g_{23}\end{aligned} \tag{6}$$

This is the well-known formula for the transformation of a tensor.

By inserting the values (5) of g_{ik} and $\overline{g}_{ik}$ into (6), we get the following system for the c_{ik}.

$$\begin{aligned}c_{11}c_{22} + c_{12}c_{21} + c_{11}c_{23} + c_{13}c_{21} &= 0 \,, \\ c_{11}c_{32} + c_{12}c_{31} + c_{11}c_{33} + c_{13}c_{31} &= 1 \,, \\ c_{21}c_{32} + c_{22}c_{31} + c_{21}c_{33} + c_{23}c_{31} &= 1 \,.\end{aligned} \tag{7}$$

Since we are only interested in permutations, each c_{ik} has to equal 1 or 0 and

$$c_{ik} = 1 \Rightarrow c_{ij} = 0 \,, \quad c_{jk} = 0 \tag{8}$$

Thus we have to solve the system (7) as a system of boolean equations with the requirement (8).

If the graphs are not isomorphic, there should not be any solution.

An easy calculation shows that the system (7) has the solutions

$$c_{12} = 1 \,, \quad c_{31} = 1 \,, \quad c_{23} = 1 \tag{9}$$

and

$$c_{13} = 1 \,, \quad c_{31} = 1 \,, \quad c_{22} = 1 \tag{10}$$

Thus, our graphs are isomorphic and $\begin{pmatrix}1 & 2 & 3\\ 2 & 3 & 1\end{pmatrix}$, $\begin{pmatrix}1 & 2 & 3\\ 3 & 2 & 1\end{pmatrix}$ are the permutations we have looked for.

In a similar fashion we can obtain the automorphisms of a graph. We shall require only

$$\overline{g}_{ik} = g_{ik} \tag{11}$$

EXAMPLE 2. For the graph

$$G = \begin{pmatrix}0 & 1 & 1\\ 1 & 0 & 0\\ 1 & 0 & 0\end{pmatrix} \tag{12}$$

the system (6) gives

$$\begin{aligned}c_{11}c_{22} + c_{12}c_{21} + c_{11}c_{23} + c_{13}c_{21} &= 1 \,, \\ c_{11}c_{32} + c_{12}c_{31} + c_{11}c_{33} + c_{13}c_{31} &= 1 \,, \\ c_{21}c_{32} + c_{22}c_{31} + c_{21}c_{33} + c_{23}c_{31} &= 0 \,.\end{aligned} \tag{13}$$

The solutions of (13) are

$$c_{11} = 1 \quad , \quad c_{22} = 1 \quad , \quad c_{33} = 1 \tag{14}$$

this is the trivial solution $C = E$, and

$$c_{11} = 1 \quad , \quad c_{23} = 1 \quad , \quad c_{32} = 1 \tag{15}$$

Thus, $\begin{pmatrix} 1 & 2 & 3 \\ 1 & 2 & 3 \end{pmatrix}$ and $\begin{pmatrix} 1 & 2 & 3 \\ 1 & 3 & 2 \end{pmatrix}$ is the group of permutations we sought.

3. We can use the differential geometry of surfaces for the graph. In this case it is easier to introduce the surface

$$2z = x^T Gx \tag{16}$$

in the $(n+1)$-dimension Euclidean space. The vertex v_i corresponds to the point

$$M_i : z = 0 \quad , \quad x_i = 1 \quad , \quad x_j = 0 \quad (j \neq i) \tag{17}$$

on the surface (16).

For a surface $z = f(x_i)$ we have the first quadratic form

$$a_{ii} = 1 + (f_i)^2 \quad , \quad a_{ij} = f_i f_j \qquad (i \neq j) \tag{18}$$

and the second quadratic form

$$b_{ij} = \frac{f_{ij}}{\Delta} \quad , \quad \Delta^2 = 1 + (f_1)^2 + (f_2)^2 + \ldots + (f_n)^2 \tag{19}$$

where

$$f_i = \frac{\partial z}{\partial x_i} \quad , \quad f_{ij} = \frac{\partial^2 z}{\partial x_i \, \partial x_j}$$

For the surface (16) we have

$$f_i = \sum_{j=1}^{n} g_{ij} x_j \quad , \quad f_{ij} = g_{ij} \tag{20}$$

and, consequently

$$A = E + Gxx^T G$$
$$B = \frac{G}{\Delta} \tag{21}$$

The normal curvatures of a surface are given by the following equation

$$\det(B - kA) = 0 \tag{22}$$

Thus the coefficients of (22) form a set of Euclidean invariants for the point M_i and, consequently, a set of arithmetical invariants for each vertex v_i. For corresponding vertices of isomorphic graphs these invariants should be equal.

It is easier to consider the equation

$$\det(G - kA) = 0 \tag{23}$$

whose coefficients are those of (22) multiplied with a power of $\Delta = 1 + \delta$, where δ is the valency of v_i .

An easy calculation shows that the equation (23) for the point M_i can be written in the following way

$$\chi(k) - k^2\{\chi(k) + k\chi_i(k)\} = 0 \quad , \tag{24}$$

where $\chi(k)$ and $\chi_i(k)$ are, respectively, the characteristic polynomials of G and $G_i = G - v_i$.

The most important part of (24) is the polynomial

$$\Delta_i(k) = \chi(k) + k\chi_i(k) \tag{25}$$

The leading coefficient of $\Delta_i(k)$ is the valency of v_i , the second is twice the number of triangles through v_i and so on. The absolute term is $\det G$ for each $\Delta_i(k)$.

CENTRAL TRANSLATION S-SPACE

Flavio Bonetti
Dipartimento di Matematica
Cosenza-Italy

Guglielmo Lunardon
Istituto di Geometria dell'Università
Napoli-Italy

1. Introduction

A Sperner space, or an S-space S, consists of a set P of elements called points and a set L of elements called lines, with an incidence relation I defined in $P \times L$ and a relation $||$ (parallelism) defined in $L \times L$, satisfying the following axioms:

(S1) Any two distinct points are incident with exactly one line;

(S2) Any line is incident with the same cardinal number $\beta(>1)$ of points, called the order of S;

(S3) Parallelism is an equivalence relation;

(S4) For any point-line pair A,a there exist exactly one line incident with A and parallel to a.

An S-space is said *regular* if any point is incident with the same cardinal number of lines. Finite regular S-spaces, that is regular S-spaces with a finite number of points, are exactly those structures called resolvable 2-(v,k,1) designs ([2],2).

In [4], A.Barlotti and J.Cofman generalizing the techniques developed by Andrè in [1] for the construction of translation plane, shown a technique for the construction of finite S-spaces from finite projective spaces with a t-spread of one hyperplane. In [5] such S-spaces are characterized and in [6] an example of such S-spaces is given with a technique that generalizes that for the construction of Hall-plane (see [8]).

In this note after some remarks on the results of [5] and [6] we determine the substructure of such S-spaces.

2. Barlotti-Cofman construction

Let Σ be a projective space. A *spread* of Σ is a family F of non-trivial flats of Σ pairwise isomorphic such that every point of Σ lies in exactly one element of F.

Let Σ' be a projective space and let Σ be an hyperplane of Σ'. Let F be a spread of Σ; then we can define an incidence structure $S=S(\Sigma',\Sigma,F)$ in the following way: the points of S are the points of Σ' not belonging to Σ; the lines of S are the flats of Σ' which intersection with Σ is an element of F; the incidence relation in S is defined by the usual set-inclusion relation; furthemore, two lines in S are said to be parallel if and only if they contain the same element of F.

It has been shown ([4], [5]) that $S=S(\Sigma',\Sigma,F)$ is an S-space. The S-space S is an affine space if and only if F is a *normal* spread of Σ (that is, a spread such that, if $<\alpha,\beta>$ is the flat of Σ spanned by two distinct elements α,β of F, then $F(<\alpha,\beta>)= =\{A\cap<\alpha,\beta>:A\in F\}$ is a spread of $<\alpha,\beta>$); (see [4]).

A *collineation* of an S-space S is a bijection of S preserving incidence and parallelism. A collineation τ of S is called a *central translation* if and only if τ has not fixed points and $a\tau \| a$ for every line a of S.

An S-space S is said to be a *central translation S-space* if and only if the central translations of S form an abelian group which is transitive on the points of S.

It can be shown (see [4],[5]) that $S(\Sigma',\Sigma,F)$ is a central translation S-space.

If S is a central translation S-space and O is a fixed point of S, the group of all the collineations of S fixing O is called *complement of central translations*.

Obviously, if T is the group of central translations and C is the complement of central translations of S, then the group $G:=T\times C$ is the collineation group of S.

It has been shown (see [5]) that the complement of central translations of $S= =S(\Sigma',\Sigma,F)$ consists of all the collineations α of Σ' fixing the point O ($\notin \Sigma$) such that $F\alpha=F$.

3. Construction of a class of translation S-space

A spread F of Σ such that every element of F has dimension t is called a *t-spread*

Let $\Sigma= PG(n^2-1,q)$ and let R,R' two no-empty families of flats of Σ. R is said to be an *(n-1)-regulus* if the following conditions are satisfied:

(R1) The elements of R and R' are of dimension n-1;

(R2) The flats of R are pairwise skew;

(R3) If A R, then every point of A belongs to exactly one element of R';

(R4) If A R', then every element of R intersects A in exactly one points.

The elements of R' are called *transversal flats* of R. Note that an (n-1)-regulus has q^{n-1}+------+q+1 elements.

It has been shown (see [6]) that, given n+1 flats, of dimension n-1, of Σ, $A_1,\ldots,A_{n+1}$, pairwise skew, if, for every j=1,...,n+1, $<A_1,\ldots,\hat{A}_j,\ldots,A_{n+1}> =\Sigma$, then there exists exactly one (n-1)-regulus, $R(A_1,\ldots,A_{n+1})$ containing $A_1,\ldots,A_{n+1}$. Furthemore, it follows from the construction of $R(A_1,\ldots,A_{n+1})$ that the transversal flats of an (n-1)-regulus form another (n-1)-regulus which is called the *opposite regulus* of R, and that the opposite (n-1)-regulus is unique (see [6]). Note that an (n-1)-regulus is a Segre manifold $S_{n-1,n-1}$ of $PG(n^2-1,q)$ (see [7]).

If we identify Σ with the lattice of flats of a vector space V of dimension n^2 over $K:=GF(q)$, there exist a basis of V, $\{t_{11},\dots,t_{1n},t_{21},\dots,t_{2n},\dots,t_{nn}\}$, such that

$$A_i = \langle t_{i1},\dots,t_{in}\rangle, \quad i=1,2,\dots,n$$

$$A_{n+1} = \langle t_{11}+\dots+t_{n+1},\dots,t_{n1}+\dots+t_{nn}\rangle.$$

For $i=2,3,\dots,n$, let α_i the linear regular map of A_1 in A_i defined as follows: if $v=\sum a_j t_{1j}$ then $\alpha_i(v)=\sum a_j t_{ij}$. From this the elements of the (n-1)-regulus are given by the flats of V: $J(a_1,\dots,a_n)=\{a_1v_1+a_2\alpha_2(v)+\dots+a_n\alpha_n(v):\ v\in A_1\}$ with $(a_1,\dots,a_n)\in$ $\in K^n-\{(0,\dots,0)\}$.

The transversal flats of $R(A_1,\dots,A_n)$ are determined by the flats of V of the kind $\langle v,\alpha_2(v),\dots,\alpha_n(v)\rangle$ with v a non-zero vector of A_1 (see [6]).

Let $F=GF(q^n)$, K the subfield of F of order q, and let V a vector space of dimension n over F.

In the following we shall regard V as a vector space over K.

Let L the set of the flats A of V, with $A=\{at:\ a\in F\}$ and $t\in V$.

Let Σ be the lattice of flats of V; it is immediately seen that L is an (n-1)-spread of $\Sigma=PG(n^2-1,q)$; further, $S=S(\Sigma',\Sigma,L)=A(n,q^n)$, since L is normal.

It has been shown (see [6]) that if $A_1,\dots,A_{n+1}$ are element of L, pairwise different, such that, for every $j<n+2:\Sigma = \langle A_1,\dots,\hat{A}_j,\dots,A_{n+1}\rangle$ then, the (n-1)-regulus $R=R(A_1,\dots,A_{n+1})$ is contained in L.

If R' is the opposite regulus of R, obviously we have that $L' = (L-R)\cup R'$ is an (n-1)-spread of Σ.

If n=2 and q>2, then $S=S(\Sigma',\Sigma,L')$ is a Hall plane (see [8], page 225).

If n>2, then S is a central translation S-space which is not an affine desarguesian space (see [6]). It is easily seen that S is also a resolvable 2-$(q^{n^2},q,1)$-design.

Note that the same considerationn can be repeated also when Σ is not finite (see [6]).

In the following, L' will be called *Hall*-(n-1)-*spread* of Σ. Further, we suppose

$n>2$ and R an $(n-1)$-regulus of $\Sigma=PG(n^2-1,q)$.

Lemma 1: Let $R=R(A_1,\ldots,A_{n+1})$, and, for $i=1,2,\ldots,m$, let B_i be an $(m-1)$-flat of A_i $(m<n)$. Let $\Gamma=<B_1,\ldots,B_m>$; if $\Gamma \; A_{n+1}=B_{m+1}$, with B_{m+1} of dimension $m-1$, then $R(\Gamma)=\{A\cap\Gamma;A\in R\}$ is an $(m-1)$-regulus of Γ. The opposite regulus of $R(\Gamma)$ is $R'(\Gamma)=\{A\cap\Gamma;A\in R'\}$.

Proof:

It is sufficient to choose the basis $\{t_{11},\ldots,t_{1n},\ldots,t_{n1},\ldots,t_{nn}\}$ so that $B_i=$ $=<t_{i1},\ldots,t_{im}>$ for $i=1,\ldots,m$. The elements of $R(\Gamma)$ are obtained by intersecting Γ with the elements of R of the kind $J(a_1,\ldots,a_m,0,\ldots,0)$.

Theorem 1: Let Γ be an (m^2-1)-flat of Σ $(1<m<n)$; if $L'(\Gamma)=\{A\cap\Gamma;A\in L'\}$ is an $(m-1)$-spread of Γ, then one of the following statements holds:

(1) there are no elements of $L-R$ intersecting Γ;

(2) there are no elements of R' intersecting Γ;

(3) $R'(\Gamma)$ is the opposite regulus of $R(\Gamma)$.

Proof:

If $L(\Gamma)=\{A\cap\Gamma;\ A\in L\}$, it follows that $L(\Gamma)$ is a t-spread of Γ (see [4]). If $t<m-1$, obviously there are no elements of $L-R$ intersecting Γ; in this case, an element of R can intersect Γ in a t-flat or in $\{0\}$. Let now $t=m-1$. If A is an element of R' intersecting Γ, since $L'(\Gamma)$ is an $(m-1)$-spread, then A intersects Γ in an $(m-1)$-flat. Since $L(\Gamma)$ is an $(m-1)$-spread and A is a transversal flat of R, it follows that $q^{m-1}+q^{m-2}+\cdots+q+1$ elements of R intersect Γ in an $(m-1)$-flat. Therefore, $R(\Gamma)$ is an $(m-1)$-regulus and $R'(\Gamma)$ is the opposite regulus.

Theorem 2: If L' is an Hall-$(m-1)$-spread of Σ, then there exist two flats of Σ, Γ_1 and Γ_2, such that:

(1) $\Gamma_1=PG(2n-1,q)$ and $L'(\Gamma_1)$ is a normal $(n-1)$-spread of Γ_1;

(2) $\Gamma_2=PG(3,q)$ and $L'(\Gamma_2)$ is a Hall-1-spread.

Proof:

Note that if $\Gamma=PG(n,q^n)$ there exist a bijection π between the points of Γ and the elements of L, such that the elements of R correspond to the points belonging to a subgeometry Ω of Γ, Ω isomorphic to PG(n,q). Hence, there exist a line s in Γ which is skew with Ω. Let $A_1,\dots,A_{q^n+1}$ the flats of L corresponding to the points of s. Evidently, $\{A_1,\dots,A_{q^n+1}\}$ is a spread of $\Gamma_1=\langle A_1,A_2\rangle$. It follows that $S(\Gamma,\Gamma_1,s)$ is a desarguesian plane of order q^n.

Let now A be an element of $L-R$ and B an element of R; let r,t be two lines of Σ belonging respectively to A and B. Then $\Gamma_2=\langle s,t\rangle$ does not verify conditions (1), (2) of Theorem 1. $L(\Gamma_2)=\{A\cap\Gamma_2;\ A\in L\}$ is a normal 1-spread of Γ_2, since L is normal; furthemore, $R'(\Gamma_2)$ is the opposite regulus of $R(\Gamma_2)$; hence $L'(\Gamma_2)=(L(\Gamma_2)-R(\Gamma_2))\cup R'(\Gamma_2)$. It follows that L' is a Hall-1-spread.

If q 2, Theorem 2 implies that $S(\Sigma',\Sigma,L)$ is not an affine desarguesian space.

References

[1] André J.: Über Nicht-Desarguessche Ebene mit Transitiver Translations-gruppe, *Math.Z., 60(1954), 156-180.*

[2] André J.: Über Parallelstrukturen-II: Translatinsstrukturen, *Math.Z., 76(1961) 155-163.*

[3] Barlotti A.: Un nuovo procedimento per la costruzione di spazi affini generalizzati di Sperner, *Le Matematiche 21(1966), 302-312.*

[4] Barlotti A.-Cofman J.: Finite Sperner spaces constructed from Projective and Affine spaces, *Abh.Math.Sem.Univ.Hamburg 40(1974), 230-241.*

[5] Bonetti F.-Lunardon G.: Sugli S-spazi di traslazione, *Boll.U.M.I.(5)14-A(1977) 368-374.*

[6] Bonetti F.-Lunardon G.: Un esempio di S-spazio di traslazione, *to appear Atti Acc.Univ.Modena.*

[7] Burau W.: Mehrdimensionale proiective und höhere Geometrie, *VEB Deutscher Verlag der Wissenschaften Berlin (1961).*

[8] Dembowski P.: Finite Geometries, *Springer-Verlag, New-York (1968).*

[9] Schulz R.H.: Über Blochpläne mit Transitiver Dilatatinsgruppe, *Math.Z., 98 (1967), 60-82.*

[10] Segre B.: Teoria di Galois, fibrazioni proiettive e geometrie non desarguesiane, *Annali Mat.Pura ed Appl., Serie IV, vol.64(1964), 1-76.*

[11] Sperner E.: Affine Räume mit Schwacher Incidenz und Zugerörige Algebraische Strukturen, *J.Reine Angew.Math., 204(1960), 205-215.*

Systems of quadrics through a general variety of Segre and their reduction to irreducible parts [2]

W. Burau
Mathematisches Seminar der
Universität Hamburg
West-Germany

In the following pages we first put together some known facts about the varieties of Segre, the Veronesian V_n^2 and the Grassmannian $G_{n,1}$. Most of these concepts are treated more in detail in my book from 1962 [1].

§1. Some generalities about the varieties, the Veronesian V_n^2 and the Grassmannian $G_{n,1}$.

In suitable coordinates and parameters a Segre-variety $S_{n_1,\dots,n_s}$ is described by

(1) $x_{i_1,\dots,x_s} = x_{i_1}^{(1)}\dots x_{i_s}^{(s)}$ $(s \le i_v \le n_v;\ v=1,\dots,s)$.

The following facts on these are valid:

a) $S_{n_1,\dots,n_s}$ spans a space P_m of dimension
$m = (n_1+1)\dots(n_s+1) - 1$.

b) The points of $S_{n_1,\dots,n_s}$ are bijective maps of the s-tuples of points

(2) $(X_o^1,\dots,X_o^s)$ with $X_o^i \in X_{n_i}^i$ $(i=1,\dots,s)$,

so that $S_{n_1,\dots,n_s}$ also is to be called the Segre-product

(3) $X_{n_1}^1 \times\dots\times X_{n_s}^s$.

Extending this notation we speak of the Segre-product

$$M^1 \times\dots\times M^s \subset S_{n_1,\dots,n_s}$$

of the subsets

$$M^i \subset X_{n_i}^i \quad (i=1,\dots,s) .$$

c) To the s-tuples of hyperplanes

(4) $(X_{n_1-1}^1,\dots,X_{n_s-1}^s)$ with $X_{n_i-1}^i \subset X_{n_i}^i$

we adjoin the hyperplane in P_m, spanned by the s Segre-products:

$\{X^1_{n_1-1} \times \ldots \times X^s_{n_s}, \ldots, X^1_{n_1} \times \ldots \times X^s_{n_s-1}\}$.
All these $P_{m-1} \subset P_m$ define the socalled Dual-Segre $\hat{S}_{n_1,\ldots,n_s}$, connected with $S_{n_1,\ldots,n_s}$.

d) The totality of s autocollineations $\pi_1,\ldots,\pi_s$, given in each of the spaces $X^1_{n_1},\ldots,X^s_{n_s}$ induces an autocollineation π of P_m, transforming $S_{n_1,\ldots,n_s}$ and equally $\hat{S}_{n_1,\ldots,n_s}$, introduced in c) into itself. In this way we get a group Γ of autocollineations of $S_{n_1,\ldots,n_s}$.

e) The totality of s autocorrelations, $K_1,\ldots,K_s$, given in each of the spaces $X^1_{n_1},\ldots,X^s_{n_s}$, induces an autocorrelation K of P_m, transforming $S_{n_1,\ldots,n_s}$ into the dual-Segre of c), connected with it.

f) Among the correlations of $S_{n_1,\ldots,n_s}$ into $\hat{S}_{n_1,\ldots,n_s}$, treated in e), there are also involutory ones, i.e. polarities and nullcorrelations. But only in the special cases of the $S_{1,\ldots,1}$, also written $S_{(1)s}$, there exists such an involutory ρ , exchangeable with each element of the group Γ in d) and induced by the identity in each of the parameter lines.
This ρ is a polarity for s even and a null-correlation for s odd.

The next type of varieties, appearing in the following, the Veronesian V^2_n , are defined from the $S_{n,n}$ in the following way: Given a regular projectivity σ with

(5) $\sigma : (X^1_n \longrightarrow X^2_n)$, i.e. $X^2_n = \sigma(X^1_n)$,

then V^2_n is the map of all pairs

(6) $(X^1_o , \sigma(X^1_o))$ with $X^1_o \in X^1_n$,

so that we have $V^2_n \subset S_{n,n}$.

The following facts and definitions about V^2_n are valid:

α) V^2_n spans a space $<:V^2_n> = A_{\binom{n+2}{2}-1}$ of dimension $\binom{n+2}{2}-1$.

β) Owing to (6) V^2_n is another model of the projective X^1_n , and therefore for each subset $M \subset X^1_n$ there exists a well-defined subset $V^2(M) \subset V^2_n$, called the V^2-image of M .

γ) For the V^2-image of $S_{n_1,\ldots,n_s}$ we have the following

<u>Theorem</u>: It is

$$V^2(S_{n_1,\ldots,n_s}) = V^2(X^1_{n_1} \times \ldots \times V^s_{n_s}) = V^2_{n_1} \times \ldots \times V^2_{n_s} .$$

This important theorem, which can be generalized in a far-reaching manner, expresses the following fact: V^2-transformation and formation of Segre-products are exchangeable operations. The simple inductive proof on s for this fact will be omitted here.

The last classical varieties to be of use for us, are the grassmannian $G_{n,1}$, the point-models of all lines $X_1 \subset X_n^1$. These also may be defined with help of the $S_{n,n}$. For that we put

(7) $\langle S_{n,n}\rangle_{n^2+2n} = A_{\binom{n+2}{2}-1} + B_{\binom{n+1}{2}-1}$

with $A_{\binom{n+2}{2}-1} = \langle V_n^2\rangle$, so that the right spaces are skew to one another. Then by projection of $S_{n,n}$ from the center A onto B, we get the desired

$$G_{n,1} \subset B_{\binom{n+1}{2}-1} \quad .$$

§ 2. Quadrics and dualquadrics in P_m defined by the V_m^2 and the linear systems of them through $S_{n_1,\ldots,n_s}$ and $\hat{S}_{n_1,\ldots,n_s}$.

Now we recall the following well-known facts:

A) By intersection of a hyperplane

(1) $H_{\binom{m+2}{2}-2} \subset \langle V_m^2\rangle$

with $V_m^2 = V^2(P_m)$ we define the V^2-image of a quadric

(2) $Q_{m-1} \subset P_m$.

B) Connected with the correspondence in A) between hyperplanes (1) and quadrics (2) there exists a well-defined correspondence between points of $\langle V_m^2\rangle$ and dual-quadrics

(3) $\hat{Q}_{m-1} \subset P_m$.

C) All hyperplanes (1) passing through a space of codimension $r+1$ in $\langle V_m^2\rangle$ define a linear ∞^r-system of $Q_{m-1} \subset P_m$ and dualy: All points of an r-dimensional subspace in $\langle V_m^2\rangle$ define a linear ∞^r-system of $\hat{Q}_{m-1}$ in P_m .

D) Now it is a well known fact for all classical varieties (i.e. those of Segre, Veronese, Grassmann and so on), that the ideal of all polynomials, which vanish on all points of such a variety, has a quadratical basis. That means for our $S_{n_1,\ldots,n_s}$, spanning P_m: We have

(4) $\langle V^2(S_{n_1,\ldots,n_s})\rangle \cap V_m^2 = V^2(S_{n_1,\ldots,n_s})$

Using the theorem in § 1, γ, we get the following

<u>Theorem</u>: The quadrics $Q_{m-1} \subset P_m$, passing through $S_{n_1,\ldots,n_s}$ form a linear ∞^r-system with

(5) $r = \binom{m+2}{2} - \binom{n_1+2}{2}\binom{n_2+2}{2}\ldots.\binom{n_r+2}{2} - 1$,

which is represented by the hyperplanes in $\langle V_m^2\rangle$ passing through $V^2(S_{n_1,\ldots,n_s})$ and dualy : The $\hat{Q}_{m-1} \subset P_m$ passing through $\hat{S}_{n_1,\ldots,n_s}$ connected with $S_{n_1,\ldots,n_s}$, are mapped onto the points of a well-defined space

(6) $R_r \subset \langle V_m^2\rangle$,

where r is given in (5).

The space R_r in (6), the socalled relationspace

$$R_r(\hat{S}_{n_1,\ldots,n_s})$$

will be the chief subject of this conference. Now at first there exists the following important

<u>Theorem</u>: In projective spaces over fields of $p=0$ the spaces $\langle V^2(S_{n_1,\ldots,n_s})\rangle$ and $R_r(\hat{S}_{n_1,\ldots,n_s})$ are in skew position, so that we have:

(7) $\langle V^2(P_m)\rangle = \langle V^2(S_{n_1,\ldots,n_s})\rangle + R_r(\hat{S}_{n_1,\ldots,n_s})$.

Since the proof of this theorem is not easy, we must be satisfied here with a reference to a proof in a paper of H. Timmermann [3], to whom we owe this theorem.

Now by the group Γ of autocollineations, introduced in § 1, certainly are transformed into itself: a) The system of $Q_{m-1} \supset S_{n_1,\ldots,n_s}$, b) that of $\hat{Q}_{m-1} \supset \hat{S}_{n_1,\ldots,n_s}$. Passing to the space $\langle V_m^2\rangle$, at first we state, that Γ is represented in it, however not irreducibly like in P_m, but so that both parts in (7) are transformed into themselves. Since the representation in $\langle V^2(S_{n_1,\ldots,n_s})\rangle$ is irreducible and therefore of minor interest, our chief task in the following will be the study of the geometric meaning of the irreducible parts, into which the representation in $R(\hat{S}_{n_1,\ldots,n_s})$ will split.

§ 3. Structure of the relation-space $R(\hat{S}_{a,b})$.

The main result about the structure of the relation-space $R(\hat{S}_{a,b})$ is contained in the following

Theorem: The relation-space $R(\hat{S}_{a,b})$ is spanned by the Segre-product of 2 Grassmannians as

(1) $$R(\hat{S}_{a,b}) = \langle G_{a,1} \times G_{b,1} \rangle .$$

The points of the product-variety on the right side in (1) are in 1-1-correspondence to the dual-quadrics

$$\hat{Q}_{a\,b+a+b-1} \supset \hat{S}_{a,b}$$

of lowest rank, i.e. to those, the kernel of which is a regular

$$Q_2 = S_{1,1} \subset S_{a,b} \quad .$$

Dualy: The linear system of all

$$Q_{a\,b+a+b-1} \supset S_{a,b} \qquad (a \geq 2,\ b \geq 1)$$

is spanned by the quadric cones through $S_{a,b}$, the vertex of which is a space

$$\langle S_{a,b-2} \rangle + \langle S_{a-2,b} \rangle$$

for $S_{a,b-2}$, $S_{a-2,b} \subset S_{a,b}$ (if $b=1$ the symbol $S_{a,-1}$ means the void set) .

Demonstration A) At first we treat the case of the $S_{a,1}$. For $a=1$ the space $\langle V_3^2 \rangle_9$ for $V_3^2 = V^2(\langle S_{1,1} \rangle_3)$ is trivially spanned, as:

$$\langle V_3^2 \rangle_9 = \langle V^2(S_{1,1}) \rangle_8 + R_0(\hat{S}_{1,1}) \quad ,$$

where the hyperplane and the point on the right side define the quadric $S_{1,1}$ and the dual-quadric $\hat{S}_{1,1}$ connected with it. Now in case $a > 1$, we conclude about the sub-Segres

(2) $$S_{1,1} \subset S_{a,1}$$

step by step the following facts:

a) They are in (1,1)-correspondence to the lines X_1 in the parameter-space X_a^1 of $S_{a,1}$.

b) They are the kernels of the dual-quadrics $\hat{S}_{1,1}$ of least rank, passing through $\hat{S}_{a,1}$.

From a), b) we conclude:

c) There is a special subset

(3) $$g_{a,1} \subset R(\hat{S}_{a,1}) \quad ,$$

the points of which define the $\hat{S}_{1,1}$ in b) and therefore are in correspondence to the lines in a).

d) For $a=2$ the space $R(\hat{S}_{2,1})$ is a plane, and all

$$Q_4 \supset S_{2,1}$$

are cones, having a line

$$S_{0,1} \subset S_{2,1}$$

as vertex, being already determined by this vertex. Such a cone is generated by the spaces $\langle S_{1,1} \rangle_3$ for all

$$S_{1,1} \supset \text{the vertex } S_{0,1} \quad .$$

Otherwise this cone Q_4 is defined by a hyperplane

$$H_{19}\ (Q_4) \subset \langle V_5^2 \rangle_{20} \quad ,$$

containing $V^2(S_{2,1})$ and therefore because of the splitting

$$\langle V_5^2 \rangle_{20} = \langle V^2(S_{2,1}) \rangle_{17} + R_2(\hat{S}_{2,1})$$

we must have

$$H_{19}\ (Q_4) \cap R_2(\hat{S}_{2,1}) = R_1 \quad .$$

The points of R_1 correspond to all $S_{1,1} \supset S_{0,1}$, conceived as dual quadrics.

e) Returning to the general case of the $S_{a,1}$ $(a \geq 2)$ we conclude from d): The lines of a pencil in X_a^1 correspond to a line on the $g_{a,1}$ of c) (3) .

f) The spaces $\langle S_{1,1} \rangle$ for all $S_{1,1} \subset S_{a,1}$ cover the whole $\langle S_{a,1} \rangle$, as is easily seen, and therefore there exists no quadric, containing all these $\langle S_{1,1} \rangle$, $\Rightarrow$ the set $g_{a,1}$ in c) span the space $R\langle \hat{S}_{a,1} \rangle$. Hence it follows, according to the characterization of the grassmannians, contained in the book [1], that $g_{a,1} = G_{a,1}$.

B) Now let $a > 1, b > 1$. At first, as for $b=1$, the $S_{1,1} \subset S_{a,b}$ define dual quadrics of least rank through $\hat{S}_{a,b}$. On the other hand these $S_{1,1}$ depend on the pairs of lines

(4) $$X_1^1 \subset X_a^1 \ , \quad X_1^2 \subset X_b^2 \quad ,$$

so that we get in $R(\hat{S}_{a,b})$ a distinguished point-model $\mu^{a,b}$ for the totality of pairs (4). Now we assert:

(5) $$\langle \mu^{a,b} \rangle = R(\hat{S}_{a,b}) \quad .$$

If $\langle \mu^{a,b} \rangle$ were a proper subspace of $R(\hat{S}_{a,b})$, then there would exist a quadric Q , containing $S_{a,b}$ and the spaces $\langle S_{1,1} \rangle$ for all $S_{1,1} \subset S_{a,b}$. This Q , according to the results of A) must contain the spaces

$\langle S_{a,1}\rangle$ and $\langle S_{1,b}\rangle$ of all $S_{a,1}, S_{1,b} \subset S_{a,b}$ whence follows immediately, that all points of $S_{a,b}$ are singular for Q, whereas the singular points of a quadric are those of a proper linear subspace. So, the existence of Q is impossible. Consequently $\mu^{a,b}$ spans a space of the same dimension as the Segre-product

(6) $$G_{a,1} \times G_{b,1}$$

and lies in a Segre S_{m_a,m_b}, where $m_a = \dim\langle G_{a,1}\rangle$, $m_b = \dim\langle G_{b,1}\rangle$. Therefore $\mu^{a,b}$ must be the Segre-product (6) .

§ 4. Splitting of the space $R(\hat{S}_{(1)s})$.

Now we'll treat the case of the Segre $S_{(1)s}$ and first we'll put together some facts either known or immediately sensible:

a) $S_{(1)s}$ spans a space P_{2^s-1} and is the Segre-product of the parameter-lines

$$x_1^i \ (i=1,\dots,s) .$$

b) The space $\langle V^2(P_{2^s-1})\rangle$ first splits roughly, as follows:

(1) $$\langle V^2(P_{2^s-1})\rangle = \langle V^2(S_{(1)s})\rangle_{3^s-1} + R_r\langle \hat{S}_{(1)s}\rangle$$

with $r = 2^{2s-1} + 2^{s-1} - 3^s - 1$.

c) The fundamental quadric

$$Q_{2^{2t}-2} \supset S_{(1)^{2t}}$$

and its dual $\hat{Q}_{2^{2t}-2} \supset \hat{S}_{(1)^{2t}}$, introduced in § 1 are to be generalized to F-cones and their dual. We first explain it for the $S_{(1)^{2t+1}}$. We know, that by projection from a space

$$Z = \langle S_{(1)^{2t}0}\rangle \text{ for } S_{(1)^{2t}0} \subset S_{(1)^{2t+1}}$$

our $S_{(1)^{2t+1}}$ is changed into a $S_{(1)^{2t}}$. By joining the points of the fundamental $\overline{Q}_{2^{2t}-2}$, belonging to $\overline{S}_{(1)^{2t}}$ with Z, we get one of the so-called F-cones through $S_{(1)^{2t+1}}$. Dually by taking a $S_{(1)^{2t}0} \subset S_{(1)^{2t+1}}$, we see, that the dual $\hat{Q}_{2^{2t}-2}$ of the fundamental $Q_{2^{2t}-2}$, belonging to $S_{(1)^{2t}0}$, can be conceived as the kernel of a so-called dual F-cone through $\hat{S}_{(1)^{2t+1}}$. Since the $S_{(1)^{2t}0} \subset S_{(1)^{2t+1}}$

depend on the points of $X_1^{(2t+1)}$, we get a ∞^1-set of such dual-F-cones. Replacing in this consideration $X_1^{(2t+1)}$ by another parameter-line, we get altogether 2t+1 sets of different dual F-cones through $\hat{S}_{(1)^{2t+1}}$.

d) Generalizing the consideration in c) we take an even number 2t

$$0 < 2t < s$$

and write:

$$S_{(1)^s} = S_{(1)^{2t}} \times S_{(1)^{s-2t}} ,$$

where the factors on the right shall belong to the parameter-lines

(2) $X_1^1, \ldots, X_1^{2t}$ and $X_1^{2t+1}, \ldots, X_1^s$,

then to

(3) $X_1^1, \ldots, X_1^{2t}$; $X_o^{2t+1} \subset X_1^{2t+1}, \ldots, X_o^s \subset X_1^s$

belongs a sub-Segre

$$S_{(1)^{2t},0,\ldots,0} \subset S_{(1)^s} ,$$

having the fundamental-$Q_{2^{2t}-2}$ with the dual $\hat{Q}_{2^{2t}-2}$. This $\hat{Q}_{2^{2t}-2}$ can be conceived as the kernel of a dual F-cone through $\hat{S}_{(1)^s}$. Thus we get a totality of such dual-F-cones, depending on the (s-2t)-tuples of points $(X_o^{2t+1}, \ldots, X_o^s)$ in (3).

Now we can make the same consideration for any partition of the set $\{1,\ldots,s\}$ into 2 subsets of 2t and s-2t elements. Altogether so we get $\binom{s}{2t}$ different dual-F-quadrics of this kind through $\hat{S}_{(1)^s}$.

e) Before formulating the fundamental split-theorem of this chapter, we have to prove the following

<u>Lemma</u>: Let $S_{a,\underbrace{1,\ldots,1}_{b}}$ be a Segre variety with the parameter-space

$$X_a, Y_1^1, \ldots, Y_1^b$$

and further suppose $Q_{a-1} \subset X_a$ is a regular quadric, which effects a set of projectively equivalent Q_{a-1} in each of the generating

$$S_{a\,o\ldots o} \subset S_{a\,1\ldots 1} .$$

The split

(4) $$\langle V^2(X_a) \rangle = V^2(Q_{a-1}) + R_o ,$$

where R_o corresponds to the dual $\hat{Q}_{a-1}$ of Q_{a-1} leads to a

$$\underbrace{V_1^2 \times \ldots \times V_1^2}_{b} \subset V^2(S_{a,1,\ldots,1})$$

as the "locus" of the point R_o in (4).

<u>Demonstration:</u> We know, that

$$V^2(S_{a,1\dots 1}) = V_a^2 \times V_1^2 \times \dots \times V_1^2$$

is contained in a Segre

$$S_{\binom{a+2}{2}-1,2,\dots,2} .$$

Then from (4) $\langle V^2(S_{a,1,\dots,1}) \rangle$ splits into the parts
$V^2(Q_{a-1}) \times V_1^2 \times \dots \times V_1^2$ and $\langle R_o \times V_1^2 \times \dots \times V_1^2 \rangle = \langle V_1^2 \times \dots \times V_1^2 \rangle$,
where the second member is the asserted "locus" of R_o.

<u>Theorem:</u> By refining the right side in (1), the total space $\langle V^2(P_{2^s-1}) \rangle$ splits in (1) in parts of different types. To each integer t with

$$0 \le 2t \le s$$

is associated one type, each space of which is spanned by a Segre-product

$$(5) \qquad \underbrace{V_1^2 \times \dots \times V_1^2}_{s-2t} = (V_1^2)^{s-2t} .$$

The spaces of this type appear $\binom{s}{2t}$ times. For even s the term $(V_1^2)^0$ means the point corresponding to the dual fundamental quadric, there intervening. In abbreviated form the split of $\langle V^2(P_{2^s-1}) \rangle$ can be written as

$$(6) \quad \langle V^2(P_{2^s-1}) \rangle = \sum_{0 \le t \le [\frac{s}{2}]} \binom{s}{2t} \langle (V_1^2)^{s-2t} \rangle .$$

All the spaces on the right side of (6) are of independent position, i.A. each of them is skew to the union of the other ones.

<u>Demonstration:</u> From the preceding considerations in a) - e) it follows already, that the parts in the right side of (6) appear in the split of the left space. We have only to prove, that these parts span $\langle V^2(P_{2^s-1}) \rangle$ and that they are of independent position. The first of these assertions is trivial for s=2 and may be proved up to s-1. Let there be given an $S_{(1)^s}$ with $s > 2$. Then we take

$$S_{(1)^{s-1}} \subset S_{(1)^s} .$$

Per inductionem the respective formula (6) for s-1 is valid; moreover we find, that each part of the right side for s-1 is contained on one space of (6) in the right. If we suppose, that contrary to our assumption the spaces in (6) on the right side span a proper subspace of $\langle V^2(P_{2^s-1}) \rangle$, then at least one hyperplane H would contain them all. But H defines a quadric

$$Q_{2^s-2} \supset S_{(1)^s} ,$$

containing the $\langle S_{(1)^{s-1}} \rangle$ for all $S_{(1)^{s-1}} \subset S_{(1)^s}$. Then there are the following 2 alternatives:

a) s even: Q_{2^s-2} has in all its regular points, which lie on $S_{(1)^s}$ the tangent-hyperplane common with that of the fundamental quadric, i.e. must coincide with it. But then the point corresponding to the dual fundamental quadric, not belonging to H, also appears in the right of (6), and our assumption was false.

b) s odd. Now in almost all points of $S_{(1)^s}$ the polarity, belonging to Q_{2^s-2} would also coincide with the fundamental-correlation K of $S_{(1)^s}$. But this is impossible, because K for odd s is a null-correlation.

At last the independent position on the right of (6) follows from the formular

$$(2^s+1)2^{s-1} = \sum_{0 \le t \le [\frac{s}{2}]} \binom{s}{2t} 3^{s-2t} .$$

§ 5. Split of the relation-space for a general Segre-variety.

For the general Segre- $S_{n_1,\ldots,n_s}$ we present a pure formula without any prove.

<u>Theorem</u>: The relation-space $R(S_{n_1,\ldots,n_s})$ splits in the following way:

a) s even

(1) $$R(\hat{S}_{n_1,\ldots,n_s}) =$$

$$= \sum_{t=1}^{[\frac{1}{2}(s-1)]+1} \sum_{1 \le i_1 < \ldots < i_{2t} \le s} \left\langle G_{n_1,1} \times \ldots \times G_{n_{i_{2t}},1} \times V^2_{n_{i_{2t+1}}} \times \ldots \times V^2_{i_s} \right\rangle$$

where $\{i_1,\ldots,s_s\} = \{1,\ldots,s\}$.

b) s odd.

The formula in this case differs from that for s even in (1) only in this point, that the first summation is extended only from t = 1 to $t = [\frac{1}{2}(s-1)]$.

<u>References</u>

[1] Burau, W., Mehrdimensionale projektive und höhere Geometrie, Berlin 1962

[2] Obeid, J., doctorial thesis (Hamburg 1977)

[3] Timmermann, H., Dissertation Hamburg 1973.

GENERATORS AND RELATIONS
FOR CLASSICAL GROUPS

E.W. Ellers

Department of Mathematics
University of Toronto
Toronto, Canada
M5S 1A1

§1. Introduction

Many theorems for the real Euclidean plane may be expressed as relations between reflections, and every such relation translates into a geometric statement. Since for every line there is exactly one reflection in this line, we shall identify any line with the reflection in this line. We shall give two examples: First, let h, a, b, c be reflections, then the relation $ahabchbc = 1$ is equivalent with the statement, h is the altitude to the side a in the triangle a, b, c. Second, $dababadbadadab = 1$ implies that a, b, d is an isosceles triangle with base a.

Numerous characterizations of metric geometries are based at least in part on the knowledge of relations between reflections. Clearly, it will be of interest to find a set of short relations between reflections such that every relation between reflections is a consequence of the original ones. This task is known as the relation problem.

Since it is desirable to obtain characterizations of more geometries, it will be necessary to solve the relation problem for a wide variety of groups. We shall state the solution of the relation problem in hyperreflection groups (publication of the proofs is in preparation). This result includes as the most important special case the solution of the relation problem for reflections in the general linear group of a vector space whose field of scalars is commutative.

Finally, we shall give a list of classical groups for which the relation problem has been solved, hoping that it might be helpful for those working in this field and also that we might obtain knowledge of possible omissions.

§2. Hyperreflection Groups

Let V be a vector space over a field K. The dimension of V may be infinite, and the multiplication of K may not be commutative.

With each element π in the general linear group $GL(V)$ we associate two subspaces of V, namely $F(\pi) = \{x \in V;\ x^{\pi} = x\}$, the fix of π, and $B(\pi) = \{x^{\pi} - x;\ x \in V\}$, the path of π. An element $\sigma \in GL(V)$ is simple if $F(\sigma)$ is a hyperplane or, equivalently, if $\dim B(\sigma) = 1$. If $\pi_1, \pi_2 \in GL(V)$, then $F(\pi_1\pi_2) \supset F(\pi_1) \cap F(\pi_2)$ and and $B(\pi_1\pi_2) \subset B(\pi_1) + B(\pi_2)$.

Let $r \in V \setminus \{0\}$ and $\psi \in V^* \setminus \{0\}$, where V^* denotes the dual space of V, such that $r^{\psi} \neq -1$, then $\sigma : x \to x + x^{\psi} r$ is a simple mapping in $GL(V)$. Also, $B(\sigma) = Kr$ and $F(\sigma) = \psi^0 = \{x \in V;\ x^{\psi} = 0\}$. Conversely, if $\sigma \in GL(V)$ is simple, then there are r, ψ with these properties. Clearly, for each $\lambda \in K \setminus \{0\}$, $x \to X + x^{\psi\lambda}\lambda^{-1}r$, also describes σ.

The type of σ is the conjugacy class of $1 + r^{\psi}$: type $\sigma = \overline{1+r^{\psi}}$. Clearly, if K is commutative, then type $\sigma = \det \sigma$.

Simple transformations σ of type $\sigma = 1$ and type $\sigma = -1$ are called transvections and reflections, respectively.

The subgroup of $GL(V)$ that consists of all $\pi \in GL(V)$ with finite codimension of $F(\pi)$, will be denoted by $FL(V)$. For $\pi \in FL(V)$, the dimension of $B(\pi)$ is finite. Therefore, $\det_{B(\pi)} | \pi$ is defined. We put $\det \pi = \det_{B(\pi)} | \pi$. Then $\pi \to \det \pi$ is a homomorphism of $FL(V)$ into $K^*/C(K^*)$, where K^* is the multiplicative group of K and $C(K^*)$ the commutator subgroup of K^*.

If $\sigma \in GL(V)$ is simple and type $\sigma = \bar{\varepsilon}$ for some $\varepsilon \in K^*$, then $\det \sigma = \varepsilon C(K^*)$.

Let Γ be a cyclic subgroup of $K^*/C(K^*)$, γ a generator of Γ and m the order of Γ. We define $G_m = \{\pi \in FL(V);\ \det \pi \in \Gamma\}$ and we say, G_m is a hyperreflection group. The group G_1 is the special linear group $SL(V)$. A simple transformation $\rho \in G_m$ with $\det \rho = \gamma$ will be called a hyperreflection. If K is commutative, then $\det \rho = \text{type } \rho$. Therefore, a hyperreflection is a reflection if $m = 2$, it is a transvection if $m = 1$. This is in general not true if K is not commutative.

Let T be the set of all simple transformations $\rho \in G_m(V)$ with type $\rho = \gamma$ and type $\rho = \gamma^{-1}$. Then T generates G_m, $T^{-1} \subset T$, and T is normal in G_m. If $\sigma_i \in T$, then the n-tuple $(\sigma_1, \ldots, \sigma_n)$ is a word in the free group F generated by the set T. A word $(\sigma_1, \ldots, \sigma_n)$ of length n is an n-relation if $\sigma_1 \ldots \sigma_n = 1$. Let S denote the normal subgroup of F that is generated by all n-relations with $n = 2,4,$ and $n = m$ where $F(\sigma_i) = F(\sigma_j)$ for all i, j.

The word $(\sigma_1', \ldots, \sigma_k') = (\pi')$ in F is derived from the word $(\sigma_1, \ldots, \sigma_n) = (\pi)$ in F if $(\pi') \equiv (\pi) \bmod S$ and if $\bigcap_{i=1}^{k} F(\sigma_1') \supset \bigcap_{i=1}^{n} F(\sigma_i)$.

Now we can prove the following result: Let $\dim V \geq 3$ and $|K| > 3$, or $\dim V = 2$ and $|K| > 5$. Let $(\sigma_1, \ldots, \sigma_n) = (\pi)$ be a relation in F. Then the empty word $\emptyset$ is derived from (π). Thus we have solved the relation problem for hyperreflection groups.

THEOREM. Let G_m be a hyperreflection group and T the set of hyperreflections and their inverses. Then every relation between elements in T is a consequence of relations of length 2, 4 and m.

COROLLARY. If K is a commutative field and V a vector space over K, then every relation between reflections in $GL(V)$ is a consequence of relations of length 2 and 4.

§3. Relation Theorems

We shall now give a list of classical groups, for which the relation problem has been solved. In most cases, but not in all, the relation between generators are consequences of relations of length 4 or less.

Group	Generators	Char K	dim V	Remarks	Date	Reference
GL(V)	simple transformations	arb.	arb.		78	Ellers [6]
SL(V)	transvections	arb.	finite	K commutative	67	Böge-Becken [4]
SL(V)	elementary matrices	arb.	finite		77	Green [13]
E	shears	arb.	arb.	K commutative	80	Ellers [9]
N	axial affinities	arb.	arb.		80	Ellers [10]
G_m	hyperreflections and their inverses	arb.	arb.		80	Ellers [8]
PGL(V)	central collineations	arb.	arb.		80	Ellers [10]
O(V)	reflections	$\neq 2$	arb.	V regular	62	Becken [3]
O(V)	transvections	$= 2$	finite	V regular, $\|K\| \geq 2n$	68	Meyer [15]
O(V)	reflections	$\neq 2$	finite		69	Ahrens, Dress, Wolff [1]
O(V)	reflections	$\neq 2$	infinite		75	Nolte [16]
O(V)	reflections	$= 2$	arb.		80	Günther, Nolte [14]
O(V)	simple isometries	arb.	arb.	index $V \leq 1$	79	Ellers [7]
O(V)	simple isometries	arb.	arb.		80	Ellers, Nolte [11]
$O(\bar{V})$	reflections in $\bar{V}$	$\neq 2$	arb.	rad $B(\pi) \subset \bar{V} \subset V$	77	Nolte [17]
$O(\bar{V})$	reflections in $\bar{V}$	$= 2$	arb.	rad $B(\pi) \subset \bar{V} \subset V$	80	Günther, Nolte [14]
U(V)	simple isometries	arb.	finite	$\|K\| \geq 5$, V regular, antiaut. $\neq 1$	67	Böge-Becken [4]
U(V)	simple isometries	arb.	finite	antiaut. $\neq 1$	68	Götzky [12]

Group	Generators	Char K	dim V	Remarks	Date	Reference
U(V)	transvections	= 2	arb.	antiaut. = 1	78	Ellers [6]
Sp(V)	transvections	≠ 2	finite		75	Spengler [18]
Sp(V)	transvections	= 2	arb.	$\|K\| > 2$	78	Ellers [6]

§4. References

[1] J. AHRENS, A. DRESS, and H. WOLFF, Relationen zwischen Symmetrien in orthogonalen Gruppen, J. reine angew. Math. 234 (1969) 1-11.

[2] F. BACHMANN, Aufbau der Geometrie aus dem Spiegelungsbegriff, 2nd edition, Springer-Verlag, New York-Heidelberg-Berlin, 1973.

[3] S. BECKEN, Spiegelungsrelationen in orthogonalen Gruppen, J. reine angew. Math. 210 (1962) 205-215.

[4] S. BÖGE née BECKEN, Definierende Relationen zwischen Erzeugenden der klassischen Gruppen, Abh. Math. Sem. Univ. Hamburg 30 (1967) 165-178.

[5] J. DIEUDONNÉ, La géométrie des groupes classiques, Springer-Verlag, Berlin-Göttingen-Heidelberg, 1955.

[6] E.W. ELLERS, Relations in Classical Groups, J. Algebra 51 (1978) 19-24.

[7] ________, Radical Relations in Unitary, Symplectic, and Orthogonal Groups, J. reine angew. Math. 306 (1979) 1-6.

[8] ________, Relations in Hyperreflection Groups, Proc. Amer. Math. Soc. (to appear).

[9] ________, Defining Relations for the Equiaffine Group, Geometriae Dedicata (to appear).

[10] ________, Relations in the Projective General Linear Group and in the Affine Subgroup (to appear).

[11] ________ and W. NOLTE, Radical Relations in Orthogonal Groups (to appear).

[12] M. GÖTZKY, Unverkürzbare Produkte und Relationen in unitären Gruppen, Math. Z. 104 (1968) 1-15.

[13] S.M. GREEN, Generators and Relations for the Special Linear Group over a Division Ring, Proc. Amer. Math. Soc. 62 (1977) 229-232.

[14] G. GÜNTHER and W. NOLTE, Defining Relations in Orthogonal Groups of Characteristic Two, Canadian J. Math. (to appear).

[15] K. MEYER, Transvektionsrelationen in metrischen Vektorräumen der Charakteristik 2, J. reine angew. Math. 233 (1968) 189-199.

[16] W. NOLTE, Spiegelungsrelationen in den engeren orthogonalen Gruppen, J. reine angew. Math. 273 (1975) 150-152.

[17] ________, Das Relationenproblem für eine Klasse von Untergruppen orthogonaler Gruppen, J. reine angew. Math. 292 (1977) 211-220.

[18] U. SPENGLER, Relationen zwischen symplektischen Transvektionen, J. reine angew. Math. 274/275 (1975) 141-149.

[19] G. THOMSEN, Grundlagen der Elementargeometrie in gruppenalgebraischer Behandlung, Hamburger Mathematische Einzelschriften, 16. Heft, B.G. Teubner-Verlag, Leipzig-Berlin, 1933.

SOME COMBINATIONAL PROBLEMS IN GEOMETRY

P. Erdös

Mathematical Institute
Hungarian Academy of Sciences

In this short survey I mainly discuss some recent problems which occupied me and my colleagues and collaborators for the last few years. I will not give proofs but will either give references to the original papers or to the other survey papers. I hope I will be able to convince the reader that the subject is "alive and well" with many interesting, challenging and not hopeless problems. Development in geometry of the kind which Euclid surely would recognize as geometry was perhaps not very significant - one very striking theorem was proved about 75 years ago by Morley which states that if one trisects the angles α, β, γ of a triangle the pairs of trisectors of the angles meet in an equilateral triangle. Surely a striking and unexpected result of great beauty. In fact because the trisection can not be carried out by ruler and compass it is not quite sure if Euclid would recognize this as legitimate (if and when I meet him [soon?] I plan to ask him).

On the other hand, many new results have been found on geometric inequalities - I won't deal with them in this paper and state only one of them, the so called Erdös-Mordell inequality (which was one of my first conjectures - I conjectured it in 1932 - two years later Mordell found the first proof). Let A, B, C be any triangle, O a point in its interior, OX is perpendicular to AB, OY to BC and OZ to AC. Then,

$$\overline{OA} + \overline{OB} + \overline{OC} \geq 2(\overline{OX} + \overline{OY} + \overline{OZ})$$

equality only if ABC is equilateral and O is the center.

I first of all give references to some earlier papers and books which deal with similar or related questions:

P. Erdös, "On Some Problems of Elementary and Combinational Geometry," *Annali di Mat. Ser IV, V 103* (1975), p. 99-108. We will refer to this paper as I.

Very interesting problems and results are in the monograph of B. Grünbaum, "Arrangements and Spreads," *Conference Board on Math. Sciences*, Amer. Math. Soc., No.10. This monograph has a very extensive and useful bibliography.

The book of Hadwiger, Debrunner and Klee, *Combinatorial Geometry in the Plane*, Holt, Rinehart and Winston, New York, 1964, contains much interesting information about geometric and combinational results. It can be used as a textbook to learn the subject but contains few unsolved problems.

Assoc. Sympos. Pure Math., Vol.7 (Convexity), 1963, contains many papers on related problems - in particular the beautiful papers of Danzer, Grünbaum and Klee are relevant to our subject matter.

Very interesting geometric questions of a related but somewhat different kind are in the books of L. Fejes-Toth, *Lagerungen in der Ebene, auf der Kugel und im Raum*, Springer-Verlag, Berlin 1953, and *Regular Figures*, Pergamon Press Macmillan Co., New York, 1964.

G. Purdy and I plan to write a book on some of the questions and their extensions which we considered in our joint papers - if we live - the book should appear sometime in the next decade.

I apologize to the reader and to the authors for the many references which I omitted here, these omissions are partly due to limitations of time and space and partly to ignorance.

1. Let $f_k(n)$ be the largest integer so that there are n distinct points $x_1, \ldots, x_n$ in k-dimensional Euclidean space E_k so that there are $f_k(n)$ pairs x_i, x_j with $d(x_i, x_j) = 1$. ($d(x_i, x_j)$ denotes the distance between x_i and x_j). $g_k(n)$ is the largest integer so that for every such choice of n points in E_k there are at least $g_k(n)$ distinct numbers among the $d(x_i, x_j)$. These problems are extensively studied in I, here I just state the outstanding open problems (we restrict ourselves to $k = 2$).

$$n^{1+c/\log\log n} < f_2(n) = o(n^{3/2}) \tag{1}$$

Whereas $f_2(n) < C\,n^{3/2}$ is very simple the seemingly slight improvement (due to E. Szemerédi is very difficult). I conjecture that the lower bound in (1) gives the right order of magnitude and I offer 250 dollars for a proof or disproof. I give 100 dollars for a proof of $f_2(n) < n^{1+\epsilon}$.

Perhaps the following stronger conjecture holds: There always is an x_i so that there are at most $o(n^\epsilon)$ ($n^{c/\log\log n}$?) x_j's equidistant from it.

$$c_1 n^{2/3} < g_2(n) < c_2 \frac{n}{(\log n)^{1/2}} \tag{2}$$

The lower bound in (2) is due to L. Moser. I believe that the upper bound gives the correct order of magnitude and I offer 250 dollars for a proof or disproof and 100 dollars for $g_2(n) > n^{1-\epsilon}$. (This would of course be implied by $f_2(n) < n^{1+\epsilon}$).

I conjectured and Altman proved that if the x_i are the vertices of a convex n-gon then $g_2(n) = \left[\frac{n}{2}\right]$. I also conjectured that there is an x_i so that there are at least $\left[\frac{n}{2}\right]$ distinct distances amongst the $d(x_i, x_j)$, $j=1,\ldots n$, $j \neq i$. This problem is still open. I also conjectured that there is an x_i which does not have three other x's equidistant from it. This was disproved by Danzer (unpublished), but perhaps there is always an x_i which does not have four other vertices equidistant from it.

Szemerédi made the pretty conjecture that Altman's $g_2(n) = \left[\frac{n}{2}\right]$ remains true if we only assume that no three of the x_i are on a line, but he only proved $g_2(n) \geq \left[\frac{n}{3}\right]$ in this case.

L. Moser and I conjectured that if the x_i are the vertices of a convex polygon then $f_2(n) < Cn$. It is annoying that no progress was made with this elementary conjecture. We have a simple example which shows $f_2(5n+1) \geq 3n$ and as far as I know this is all that is known.

G. Purdy and I observed that if no three of the x_i are on a line then $f_2(n) > cn\log n$ is possible, this follows easily by a method of Kárteszi. We have no idea for the exact order of magnitude of $f_2(n)$ in this case.

Most of the results stated in this Chapter are referenced in I.

REFERENCES

S. Jozsa and E. Szemerédi, "The Number of Unit Distances in the Plane," Coll. Math. Soc. Janos Bolyai, *Finite and Infinite Sets*, Keszthely, Hungary, 1973, p. 939-950, North Holland.

P. Erdos and G. Purdy, "Some Extremal Problems in Geometry," *IV Assoc. Seventh Southeastern Conference on Combinations Graph Theory and Computing*, 1976, p.307-322. (Congress Numerantium XVII). (For related problems see our paper III and V, same conference, 1975 and 1977, p. 291-308 and p. 569-578).

2. An old problem of E. Klein (Mrs. Szekeres) states: Let $H(n)$ be the smallest integer so that every set of $H(n)$ points in the plane, no three on a line, contains the vertices of a convex n-gon (it is not at all obvious that $H(n)$ exists for every n). She proved $H(4) = 5$ and Szekeres conjectured $H(n) = 2^{n-2} + 1$. Makai and Turan proved $H(5) = 9$ and Szekeres and I proved

$$2^{n-2} + 1 \leq H(n) \leq \binom{2n-4}{n-2}.$$

All this is in I (See Introduction). Recently I found the following interesting modification of this problem: Let M_n be the smallest integer so that every set of M_n points no three on a line always contains the vertices of a convex n-gon which contains no x_i in its interior. Trivially $M_4 = 5$, and Ehrenfeucht proved that M_5 exists, Harborth proved $M_5 = 10$. It is not at all clear that M_n exists and at present it is possible that even M_6 does not exist (in other words, for every m there are m points in the plane no three on a line so that every convex hexagon determined by these points contains at least one other point in its interior).

REFERENCES

H. Harborth, "Konvexe Funfeck in Ebenen Punktmengen," *Elemente der Math.* *33* (1978), 116-118.

3. Let there be given n points in the plane not all on a line. Is it true that there always is a line which goes through precisely two of the points? Such a line is called an ordinary line. This beautiful result was conjectured by Sylvester in 1893. I rediscovered the conjecture in 1933 and a few days later T. Gallai found a proof. The first proof was published by Melchior who rediscovered it quite independently in 1940. Extensive references and the history of this problem can be found in II and I and in a paper of Motzkin. Here I only state a few recent results and problems. Denote by $t_k(n)$ the largest integer for which there is a set of n points in the plane for which there are $t_k(n)$ lines containing exactly k of the points. $t_3(n)$ Has been studied for more than 150 years. The sharpest results on $t_3(n)$ are due to Burr, Grünbaum and Sloane. They conjecture that

$$t_3(n) = 1 + [n(n-3)/6] \quad \text{for} \quad n \neq 7, 11, 16, 19. \tag{1}$$

They prove ( ] × [is the least integer ≥ x)

$$1 + \left[\frac{n(n-3)}{6}\right] \le t_2(n) \le \left[\left(\binom{n}{2} -]\frac{3n}{7}[\right)/3\right].$$

Croft and I prove that for every $k \geq 3$

$$t_k(n) > c_k n^2 \tag{2}$$

c_k is an absolute constant. The simple proof of (2) is given in II. The best possible value of c_k in (2) is not known. Denote by $t_k'(n)$ the largest integer for which there is a set of n points in the plane no k+1 of them on a line for which there are $t_k'(n)$ lines containing exactly k of the points. I conjectured that for $k > 3$, $t_k'(n) = o(n^2)$ and could not even prove $t_k'(n)/n \to \infty$. Karteszi proved $t_k'(n) > c_k\, n \log n$ and Grünbaum showed that $t_k'(n) > cn^{1+1/k-2}$. Further problem: Assume $k = [cn^{1/2}]$. Determine or estimate $t_k'(n)$. It is true that

$$t_k'(n) > \alpha n^{1/2}/c$$

where α is independent of n and c?

Let $x_1, \ldots, x_n$ be n points in E_2. Join every two of them. Prove (or disprove) that one gets at least ckn distinct lines where c is an absolute constant independent of n and k. This (and more) was proved by Kelly and Moser if $k < c_1 n^{1/2}$.

Let $x_1, \ldots, x_n$ be n points in the plane not all on a line and let $L_1, \ldots, L_m$ be the set of lines determined by these points. Graham conjectured that if S is a subset of $\{x_1, \ldots, x_n\}$ so that every line L_i intersects S, then for at least one i, $L_i \subset S$. This conjecture was recently proved by Rabin and Motzkin.

I then asked the following question: Does there exist for every k a finite set S of points in the plane so that if one colors the points of S by two colors in an arbitrary way, there always should be a line which contains at least k points and all whose points are of the same color. Graham and Selfridge gave an affirmative answer for $k = 3$, but the cases $k > 3$ seem to be open.

Finally, I want to call attention to a nearly forgotten problem of Serre: Let

A_n be the projective n space over the complex numbers. A finite subset is a Sylvester-Gallai configuration if every line through two of its points also goes through a third. Characterize all planar Gallai-Sylvester configurations. Is there a non-planar Gallai-Sylvester configuration?

For generalization of the Gallai-Sylvester theorem to matroids, see, e.g. the book of D.J.A. Welsh, *Matroid Theory*, p. 286-297, Academic Press, 1976.

For a generalization of different nature, see, e.g., M. Edelstein, "Generalizations of the Sylvester Problem," *Math. Magazine*, *43* (1970), p. 250-254, and M. Edelstein, F. Herzog, and L.M. Kelly, "A Further Theorem of the Sylvester Type," *Proc. Amer. Math. Soc.*, *14* (1963), p. 359-363.

REFERENCES

T.S. Motzkin, "The Lines and Planes Connecting the Points of a Finite Set, " *Trans. Amer. Math. Soc.*, *70* (1951), p. 451-464.

S.A. Burr, B. Grünbaum, and N.J.A. Sloane, "The Orchard Problem," *Geometriae Dedicata*, *2* (1974), p. 397-424. (This paper contains an extensive bibliography and many interesting historical remarks.)

J.T. Serre, "Problem 5359," *Amer. Math. Monthly*, *73* (1966), p. 89.

B. Grünbaum, "New Views on Old Questions of Combinatorial Geometry," *Teoriae Combinatorie*, *1*, ( ) p. 451-478.

4. In this last Chapter I state a few miscellaneous problems. Recently "we" (Graham, Montgomery, Rothschild, Spencer, Straus and I) published several papers on a subject which we called Euclidean Ramsey theorems. A subset S of E_m is called Ramsey if for every k there is an m_k so that if we decompose E_{m_k} into k subsets, $E_{m_k} = \bigcup_{i=1}^{k} S_i$ at least one S_i has a subset congruent to S. We prove that every brick (i.e., rectangular parallelepiped) is Ramsey and that every S which is Ramsey is inscribed in a sphere. The most striking open problems are: Is the regular pentagon Ramsey? Is there an obtuse angled triangle which is Ramsey? Are in fact all obtuse angled triangles Ramsey?

Let $S_1 \cup S_2$ be the plane. Is it true that if T is any triangle (with the possible exception of equilateral triangles of one fixed height) then either S_1 or S_2 contains the vertices of a triangle congruent to T? Many special cases of this startling conjecture have been proved by us and Schader but so far the general case eluded us. There surely will be interesting generalizations for higher dimensions but these have not yet been investigated.

Let S be a set of points in the plane no two points of S are at distance one. We conjectured that the complement of S contains the vertices of a unit square. This conjecture was proved by R. Juhász. She in fact showed that if X_1, X_2, X_3, X_4, are any set of four points then the complement of S contains a congruent copy. It is not known at present if this remains true for 5 points; she showed that there is a k so that it fails for k points.

Clearly many more problems can be stated here, and in fact many have been stated in our papers. I hope more people will work on this subject in the future and our results will soon become obsolete.

The following problem is due to Hadwiger and Nelson: Join two points of r-dimensional space if their distance is one. Denote by α_r the chromatic number of this graph. Is it true that $\alpha_2 = 4$? It is known that $4 \leq \alpha_2 \leq 7$. I am sure that $\alpha_2 > 4$ but cannot prove it. By a well known theorem of the Bruijn and myself if $\alpha_2 > 4$ then there is a finite set of points $x_1,\ldots,x_n$ in the plane so that the graph whose edges are (x_i, x_j), $d(x_i, x_j) = 1$ has chromatic number greater than four. The determination of such a graph may not be easy since perhaps n must be very large.

α_r for large r was first studied by Lavman and Rogers. The sharpest known result is due to P. Frankl, $\alpha_r > r^c$ for every c if $r > r_0(c)$. It seems certain that there is a fixed $\varepsilon > 0$ so that $\alpha_r > (1+\varepsilon)^r$. ($\alpha_r < 3^r$ is proved by Lavman and Rogers.) This conjecture would easily follow from the following purely combinatorial conjecture (which perhaps is very hard). Let $|S| = n$, $A_i \subset S$, $1 \leq i \leq u_n$ be a family of subsets of S satisfying for every $1 \leq i_1 < i_2 \leq u$, $|A_{i_1} \cap A_{i_2}| \neq [\frac{n}{4}]$. Then there is an $\varepsilon > 0$ independent of n for which

$$\max u_n < (2-\varepsilon)^n \tag{1}$$

(1) no doubt remains true if the assumption $|A_{i_1} \cap A_{i_2}| \neq [\frac{n}{4}]$ is replaced by: There is a t, $\eta n < t < (\frac{1}{2} - \eta)n$ so that $|A_{i_1} \cap A_{i_2}| \neq t$ for every $1 \leq i_1 < i_2 \leq u_n$, only here ε will depend on η. At present no proof seems to be in sight.

Let $x_1,\ldots,x_n$ be n distinct points in the plane. Denote by $C(x_1,\ldots,x_n)$ the number of distinct circles of radius one which go through at least three of the x_i. Put

$$F(n) = \max C(x_1,\ldots,x_n) \tag{2}$$

where the maximum in (2) is taken for all possible choices of distinct points $x_1,\ldots,x_n$. I conjectured more than two years ago that

$$F(n)/n^2 \to 0\ ,\ F(n)/n \to \infty \tag{3}$$

It seems that (3) is trivial but I could not prove it and I have no idea about the true order of magnitude of $F(n)$, probably $F(n) < n^{1+\varepsilon}$ for every $\varepsilon > 0$, if $n > n_0(\varepsilon)$.

Let $x_1,\ldots,x_n$ be n points in E_r satisfying $d(x_i, x_j) \geq 1$. Determine or estimate

$$D_r(n) = \min \max_{1 \leq i < j \leq n} d(x_i, x_j)$$

where the minimum is taken over all choices of $x_1,\ldots,x_n$ in E_r satisfying $d(x_i, x_j) \geq 1$. The exact value of $D_r(n)$ is known only for very few values of r and n. A classical result of Thue states

$$\lim_{n=\infty} D_2(n)/n^{1/2} = \left(\frac{2\,3^{1/2}}{\pi}\right)^{1/2}.$$

The value of $\lim_{n=\infty} D_3(n)/n^{1/3}$ is not known and is an outstanding open problem in the geometry of numbers.

Let $x_1, \ldots, x_n$ be n points in the plane. Denote by $L_1, \ldots, L_m$ the set of lines determined by these points. Denote by u_i the number of points on L_i. $u_1 \geq u_2 \geq \ldots \geq u_m$. Clearly

$$\sum_{i=1}^{m} \binom{u_i}{2} = \binom{n}{2} \tag{4}$$

Let $\{u_i\}$ be a set of integers satisfying (4). It would be of interest to obtain nontrivial conditions on the u_i which would assure that there is a set of points in the plane for which there are u_i points on L_i. Perhaps there is no simple necessary and sufficient condition. Denote by $f(n)$ the number of distinct sequences $u_1 \geq \ldots \geq u_m$ (m is also a variable) for which there is a set of points $x_1, \ldots, x_n$ with u_i points on L_i. It is easy to see that

$$\exp[c_1 n^{1/2}] < f(n) < \exp[c_2 n^{1/2}] \tag{5}$$

I expect that the lower bound gives the correct order of magnitude in (5), but I had not the slightest success in proving this.

One can formulate this problem in a more combinatorial way. Let $|S| = n$, $A_i \subset S$, $1 \leq i \leq m$ are subsets of S ($|A_i| \geq 2$). Assume that every pair x,y of elements of S are contained in exactly one A_i. Put $|A_i| = u_i$, $u_1 \geq u_2 \geq \ldots \geq u_m$. Clearly (4) holds here too. Denote by $F(n)$ the number of possible choices for the u's. It is not hard to prove that (5) holds for $F(n)$ too, but here I expect that the upper bound gives the correct order of magnitude, but again I had no success. ($F(n) > f(n)$ easily follows since by Gallai-Sylvester $u_m = 2$ in the geometric case.)

A well known theorem of de Bruijn and myself states that (unless $|A_1| = n$) we must have $m \geq n$. This easily implies that there are $c_1 n^{1/2}$ A_i's of the same size. I believe that this is best possible, in other words: There is a system of subsets $A_i \subset S$ $m > 1$, every pair of elements of S is contained in exactly one A_i and there are at most $c_2 n^{1/2}$ values of i for which the A_i are of the same size. Perhaps it is not hard to construct such a design and my lack of success was due to lack of experience with construction of block designs.

Assume $u_1 \leq Cn^{1/2}$. Purdy and I recently obtained fairly accurate asymptotic formulas in the general combinatorial case for

$$\max \sum \binom{u_i}{3}$$

in terms of u_1. On the other hand, we had no success in the geometric case (i.e., when the x_i are points in the plane and the L_i are lines). We conjectured that if $u_1 < c_1 n^{1/2}$ then

$$\sum \binom{u_i}{3} < c_2 n^{3/2}$$

where $c_2 = c_2(c_1)$.

REFERENCES

P. Erdös, R.L. Graham, Montgomery, B. Rothschild, J. Spencer and E. Straus, "Euclidean Ramsey Theorems I, II, III," *J. Com. Theory A 14* (1973), p. 341-363, *Proc. Conf. Finite and Infinite Sets*, June 1973, Keszthely, Hungary, p. 529-557, and p.558-584.

L.E. Shader, "All Right Triangles are Ramsey in E_2, *Proc. 7th Southeastern Conf. Combinatorics* (1974), p.476-480.

D.G. Lavman, and C.A. Rogers, "The Realization of Distances within Sets in Euclidean Space," *Mathematika 19* (1972), p. 1-24.

D.G. Lavman, "A Note on the Realization of Distances within Sets in Euclidean Space," *Comment. Math. Helvetici, 53* (1978), p. 529-539.

N.G. de Bruijn and P. Erdös, "A Colour Problem for Infinite Graphs and a Problem in the Theory of Relations," *Indag. Math. 13* (1951), p. 371-373, and *Nederl. Akad. Wetensch Proc. 57* (1948), p. 1277-79.

D.R. Woodall, "Distances realized by Sets Covering the Plane," *Journal Comb. Theory (A) 14* (1973), p. 187-200.

L.M. Kelly and W. Moser, "On the Number of Ordinary Lines Determined by n Points," *Canad. J. Math. 10* (1958), p. 210-219.

P. Bateman and P. Erdös, "Geometrical Extrema Suggested by a Lemma of Besicovitch," *Amer. Math. Monthly, 58* (1951), p. 306-314.

ÜBER DIE ALGORITHMISCHE LÖSUNG DES STEINITZPROBLEMS EINER INNEREN KENNZEICHNUNG POLYTOPALER SPHÄREN

Günter Ewald

1. Einleitung: Für zellzerlegte zweidimensionale Sphären gibt der bekannte Satz von Steinitz Bedingungen dafür an, daß die Sphären polytopal sind, d.h. so auf den Rand konvexer Polytope topologisch abgebildet werden können, daß die Zellzerlegung in den Randkomplex des Polytops übergeht. Für höherdimensionale Sphären sind entsprechende Bedingungen unbekannt (Steinitzproblem). B. Grünbaum ([4], S. 91) hat mit Hilfe eines Satzes von Tarski wenigstens gezeigt, daß ein Algorithmus existiert, der von einem (durch sein Eckenschema) gegebenen Zellkomplex zu entscheiden gestattet, ob er polytopal ist oder nicht (vgl. hierzu auch [3] , II.5 und [2]). Dieser Algorithmus, mit dem quadratische Ungleichungen und Gleichungen aufgelöst werden, ist indessen sehr unübersichtlich und wahrscheinlich kaum so zu vereinfachen, daß man ihn praktisch verwenden kann, etwa für eine Untersuchung der von Altshuler und Steinberg [1] angegebenen 3-Sphären mit 10 Ecken.
Wir entwickeln im folgenden die Grundgedanken eines geometrisch durchsichtigen Algorithmus für den simplizialen Fall. Von diesem ist zu hoffen, daß er wesentlich verbessert und für Berechnungen der genannten Art herangezogen werden kann.*

2. Projektionen: Ausgangspunkt für unsere Überlegungen ist der folgende Satz (vgl. Grünbaum [4], S. 72): Jedes k-dimensionale konvexe Polytop P, kurz k-Polytop genannt, läßt sich durch eine Parallelprojektion φ aus einem Simplex T^n gewinnen, wenn n+1 die Zahl der Ecken von P ist: $\varphi(T^n) = P$.
Wir spalten nun φ für simpliziale Polytope P (d.h. solche, deren Randkomplex $\mathcal{B}(P)$ simplizial ist) wie folgt auf:

Satz 1: Sei $P \subset \mathbb{R}^k$ ein simpliziales k-Polytop mit n+1 Ecken (k > 1). Dann gibt es in $\mathbb{R}^n \supset \mathbb{R}^k$ ein n-Simplex T^n und in $\mathcal{B}(T^n)$ einen zu $\mathcal{B}(P)$ isomorphen Komplex $\mathcal{C}$, sowie Parallelprojektionen $\varphi_1,\ldots,\varphi_{n-k}$, so daß folgendes gilt:
(a) $\varphi(T^n) := \varphi_{n-k}\varphi_{n-k-1}\cdots\varphi_1(T^n) = P$.
(b) Für $P_j := \varphi_j \cdots \varphi_1(T^n)$; $j=1,\ldots,n-k$, gilt: $n=\dim T^n = 1+\dim P_1 = 2+\dim P_2 = \ldots = n-k+\dim P_{n-}$
(c) Jedes P_j ist simplizial; $j=1,\ldots,n-k$.

---

* Anmerkung bei der Drucklegung: Inzwischen konnte der Algorithmus mit Hilfe der Methode der Gale-Diagramme weiterentwickelt und duch ein einfaches Rechenverfahren der praktischen Verwendbarkeit nähergebracht werden. Wir beschränken uns daher im folgenden auf einen Beweis des für den Algorithmus grundlegenden Satzes 1 und eine kurze Darlegung der geometrischen Gestalt des Algorithmus, die bei der Verwendung von Gale-Diagrammen in den Hintergrund tritt.

(d) Die k+1 in $\mathbb{R}^k$ gelegenen Ecken von P bleiben unter jedem φ_j fest; die übrigen Ecken liegen in keiner affinen Hülle einer echten Seite von P_j; j=1,...,n-k-1.

(e) In jedem $\mathfrak{B}(P_j)$ gibt es einen zu $\mathcal{C}$ isomorphen Teilkomplex $\mathcal{C}_j$, so daß für die durch φ_j induzierten Komplexabbildungen $\hat{\varphi}_j$ gilt:

$$\mathcal{C} =: \mathcal{C}_0 \xrightarrow[\hat{\varphi}_1]{} \mathcal{C}_1 \xrightarrow[\hat{\varphi}_2]{} \mathcal{C}_2 \xrightarrow[\hat{\varphi}_3]{} \cdots \xrightarrow[\hat{\varphi}_{n-k}]{} \mathcal{C}_{n-k} = \mathfrak{B}(P);$$

dabei sind die $\hat{\varphi}_j$ alle bijektiv; j=1,...,n-k. (Die $\mathcal{C}_j$ sind also jeweils in den "Schattengrenzen" von P_j bei der Projektion $\hat{\varphi}_j$, j=1,...,n-k-1, enthalten).

Beweis: Seien F,F' Facetten, d.h. (k-1)-Seiten von P, so daß $F \cap F'$ eine (k-2)-Seite von P darstellt, seien $0 = a_o, a_1, \ldots, a_{k-2}$ die Ecken von $F \cap F'$ und a_{k-1}, a_k die übrigen Ecken von $F \cup F'$. Als Vektoren aufgefaßt bilden $a_1, \ldots, a_k$ eine Basis von $\mathbb{R}^k$. Wir erweitern diese zu einer Basis $\{a_1, \ldots, a_k, a_{k+1}, \ldots, a_n\}$ von $\mathbb{R}^n \supset \mathbb{R}^k$. Die nicht in $F \cup F'$ gelegenen Ecken von P bezeichnen wir mit $e_{k+1}, \ldots, e_n$.

Wir setzen (mit $\mathbb{R}x := \{tx \mid t \in \mathbb{R}\}$):

$U_i = \emptyset$ für $i \le 0$

$U_1 := \{0\}$

$U_2 := \mathbb{R}(a_{k+1} - e_{k+1})$

$U_3 := U_1 + \mathbb{R}(a_{k+2} - e_{k+2})$

.

.

.

$U_{n-k} := U_{n-k-2} + \mathbb{R}(a_{n-1} - e_{n-1})$.

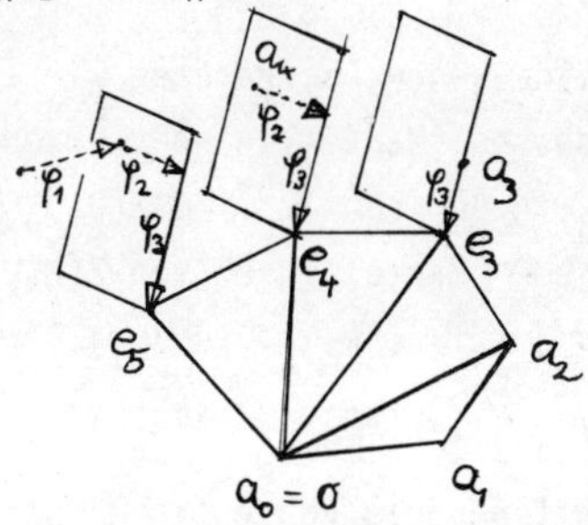

Fig. 1

Die Projektion φ_1 legen wir wie folgt fest: $\varphi_1(a_i) = a_i$ für i=0,1,...,n-1. $\varphi_1(a_n) \in e_n + U_{n-k}$, aber $\notin e_n + U_{n-k-1}$. Durch affine Fortsetzung auf $\mathbb{R}^n$ wird φ_1 zu einer Parallelprojektion von $\mathbb{R}^n$ auf $\mathbb{R}^{n-1}$.

Um φ_2 zu definieren, setzen wir $\varphi_2(a_i) = a_i$ für i=1,...,n-2, wählen $\varphi_2\varphi_1(a_n) \in e_n + U_{n-k-1}$, aber $\notin e_n + U_{n-k-2}$ sowie $\varphi_2(a_{n-1}) \in e_{n-1} + U_{n-k-1}$, aber $\notin e_{n-1} + U_{n-k-2}$, und zwar so, daß $\mathbb{R}(a_{n-1} - \varphi_2(a_{n-1})) = \mathbb{R}(\varphi_1(a_n) - \varphi_2\varphi_1(a_n))$. Durch affine Fortsetzung wird φ_2 wieder zu einer Parallelprojektion von $\mathbb{R}^{n-1}$ auf $\mathbb{R}^{n-2}$.

In dieser Weise fahren wir fort und setzen für jedes $j \in \{3, \ldots, n-k\}$:

$\varphi_j(a_i) = a_i$ für i=1,...,n-j-1 und

$\varphi_j \varphi_{j-1} \cdots \varphi_1(a_n) \in e_n + U_{n-k-j+1}$, aber $\notin e_n + U_{n-k-j}$

$\varphi_j \varphi_{j-1} \cdots \varphi_2(a_{n-1}) \in e_{n-1} + U_{n-k-j+1}$, aber $\notin e_{n-1} + U_{n-k-j}$

$\varphi_j(a_{n-j}) \in e_{n-j} + U_{n-k-j+1}$, aber $\notin e_{n-j} + U_{n-k-j}$,

und zwar so, daß jeweils

$$\mathbb{R}(\varphi_{j-1} \cdots \varphi_1(a_i) - \varphi_j \varphi_{j-1} \cdots \varphi_1(a_i)) = \mathbb{R}(\varphi_{j-1} \cdots \varphi_1(a_{i+1}) - \varphi_j \varphi_{j-1} \cdots \varphi_1(a_{i+1}))$$

$i=n-j,\dots,n-1$. Dann ergibt sich durch affine Fortsetzung eine Parallelprojektion $\varphi_j:\mathbb{R}^{j+1} \to \mathbb{R}^j$.

Für $j=n-k$ ergibt sich, wenn man $\varphi := \varphi_{n-k}\cdots\varphi_1$ setzt:

$$\varphi(a_i) = e_i,\ i=k+1,\dots,n.$$

Beschränkt man die affine Fortsetzung auf die konvexe Fortsetzung für die Eckenmenge von T^n, dann erhält man:

$$\varphi(T^n) = P$$

Dies beweist (a). Da bei jedem φ_j die Dimension um 1 erniedrigt wird, gilt auch (b).

Bei der Festlegung von φ_j kann man $\varphi_j(a_{n-j}) \in e_{n-j}+U_{n-k-j+1}$ in einer kleinen Umgebung relativ zu $e_{n-j}+U_{n-k-j+1}$ variieren und die übrigen Bildpunkte entsprechend festsetzen. Dabei kann man erreichen, daß für $j < n-k$ die Punkte $\varphi_j\cdots\varphi_1(a_n),\dots,\varphi_j(a_{n-j})$ zusammen mit den Ecken von p zu je $n-j+1$ affin unabhängig sind. Das ergibt (c).

Um (d) einzusehen, betrachten wir einen Zylinder $Z(P,U_{n-k})$ mit dem Rand von P als Basis, dessen "Mantellinien" (Erzeugenden) Parallen von U_{n-k} sind. Wegen (a) ist jedes P_j in dem von $\mathbb{R}^{n-j} \cap Z(P,U_{n-k})$ begrenzten Vollzylinder enthalten. Wegen (c) trifft jede "Mantellinie" von $Z(P,U_{n-k})$ das Polytop P_j in genau einem Punkt. Wir setzen $\mathcal{C}_{n-j} := (\hat{\varphi}_{n-k}\cdots\hat{\varphi}_{j-1})^{-1}(\mathcal{B}(P))$. Dann ist

$$\operatorname{set} \mathcal{C}_{n-j} := (\varphi_{n-k}\cdots\varphi_{j-1})^{-1}(\operatorname{set} \mathcal{B}(P)) = Z(P,U_{n-k}) \cap P_j$$

die Schattengrenze von P_j in Richtung $\varphi_{n-k}\cdots\varphi_{j-1}$. Ferner ist $\hat{\varphi}_{n-k}\cdots\hat{\varphi}_{j-1}$ eine Bijektion von $\mathcal{C}_{n-j}$ auf $\mathcal{B}(P)$, also ist

$$\begin{aligned}\hat{\varphi}_j\,\mathcal{C}_{n-j+1} = \hat{\varphi}_j(\hat{\varphi}_{n-k}\cdots\hat{\varphi}_j)^{-1}(\mathcal{B}(P)) &= \hat{\varphi}_j\,\hat{\varphi}_j^{-1}\,\hat{\varphi}_{j-1}^{-1}\cdots\hat{\varphi}_{n-k}^{-1}(\mathcal{B}(P))\\ &= \hat{\varphi}_{j-1}^{-1}\cdots\hat{\varphi}_{n-k}^{-1}(\mathcal{B}(P))\\ &=(\hat{\varphi}_{n-k}\cdots\hat{\varphi}_{j-1})^{-1}(\mathcal{B}(P))\\ &= \mathcal{C}_{n-j},\end{aligned}$$

ferner $\hat{\varphi}_j^{-1}\,\mathcal{C}_{n-j} = \hat{\varphi}_j^{-1}\,\hat{\varphi}_{j-1}^{-1}\cdots\hat{\varphi}_{n-k}^{-1}(\mathcal{B}(P))$

$$= (\hat{\varphi}_{n-k}\cdots\hat{\varphi}_j)^{-1}(\mathcal{B}(P)) = e_{n-j+1}.$$

Da $\hat{\varphi}_j$ eine Abbildung einer endlichen Menge ist, folgt somit die Bijektivität von $\hat{\varphi}_j : \mathcal{C}_{n-j+1} \to \mathcal{C}_{n-j}$. - Damit ist Satz 1 bewiesen.

3. Der Algorithmus. Sei ein simplizialer Komplex $\mathcal{C}$ mit $n+1$ Ecken (= 0-Zellen) durch sein Eckenschema gegeben. Wir ordnen die Ecken von $\mathcal{C}$ bijektiv den Ecken $a_0,a_1,\dots,a_n$ eines n-Simplex T^n zu. Dadurch wird eine Einbettung von $\mathcal{C}$ in $\mathcal{B}(T^n)$ festgelegt. Wir

können annehmen, daß (k-1)-Zellen F,F' von $\mathcal{C}$ existieren, so daß $F \cap F'$ eine (k-2)-Zelle von $\mathcal{C}$ darstellt. Wir setzen $T^k := \mathrm{conv}(F \cup F')$ und $\mathbb{R}^k := \mathrm{aff}\, T^k \subset \mathrm{aff}\, T^n =: \mathbb{R}^n$.

Wir suchen, in einer Art Umkehrung von Satz 1, Projektionen $\varphi_1,\ldots,\varphi_{n-k}$, die die in Satz 1 angegebenen Eigenschaften (a) - (d) besitzen. Um ein algorithmisches Verfahren zu erhalten, brauchen wir dabei eine repräsentative Auswahl von endlich vielen Projektionsrichtungen, mit denen entschieden werden kann, ob $\mathcal{C}$ polytopal ist oder nicht. Der geometrische Grundgedanke ist hierbei folgender:
Die erste Projektion φ_1 läßt $T^{n-1} = \mathrm{conv}\,\{a_o,a_1,\ldots,a_{n-1}\}$ punktweise fest. $\varphi_1(a_n)$ wird in einer der 2^n-1 Zusammenhangskomponenten gewählt, in die die Trägerhyperebenen der Facetten von T^{n-1} den $\mathbb{R}^{n-1} := \mathrm{aff}\, T^{n-1}$ zerteilen. Man kann zeigen, daß es genügt, in jeder Komponente einen repräsentativen Punkt für die Auswahl von $\varphi_1(a_n)$ festzulegen.

Im Falle von φ_2 lassen wir $T^{n-2} := \mathrm{conv}\,\{a_o,a_1,\ldots,a_{n-2}\}$ punktweise fest. Wir betrachten die Menge $X^{n-1} := \{x \mid x \in \mathrm{aff}\, Z,\ Z \in \mathrm{skel}_{n-3}\, T^{n-1}\}$ und verbinden $\varphi_1(a_n)$ mit jedem Punkt von X^{n-1} durch eine Gerade. Den so entstehenden Kegel verschieben wir in der Weise, daß das Bild von $\varphi_1(a_n)$ mit a_{n-1} zusammenfällt. Der neue Kegel werde mit $\mathfrak{X}$ bezeichnet. Setzt man noch
$Y^{n-1} := \{x \mid x \in \mathrm{aff}\, Z,\ Z \in \mathrm{skel}_{n-3}\, \mathrm{conv}\{a_o,a_1,\ldots,a_{n-1},\varphi_1(a_n)\}\}$ und bezeichnet den Kegel der Verbindungsgeraden von a_{n-1} mit den Punkten von Y^{n-1} mit $\mathfrak{X}'$, dann zerteilen die (n-3)-Ebenen von $(\mathfrak{X} \cup \mathfrak{X}') \cap \mathbb{R}^{n-2}$) (Anzahl $2\binom{n}{2} = n(n-1)$) den $\mathbb{R}^{n-2}$ in Zusammenhangskomponenten. Deren Anzahl beträgt, wie man aus [5], S.45 durch Umrechnung entnimmt (vgl. auch [6]) höchstens $\sum_{j=o}^{n-2} \binom{n(n-1)-1}{j}$.

(Vgl. Fig. 2 für n=4 und k=2.) In jeder Komponente wählt man wieder einen repräsentativen Punkt, im Falle beschränkter Komponenten etwa den Schwerpunkt, als möglichen Bildpunkt $\varphi_2(a_{n-1})$.-

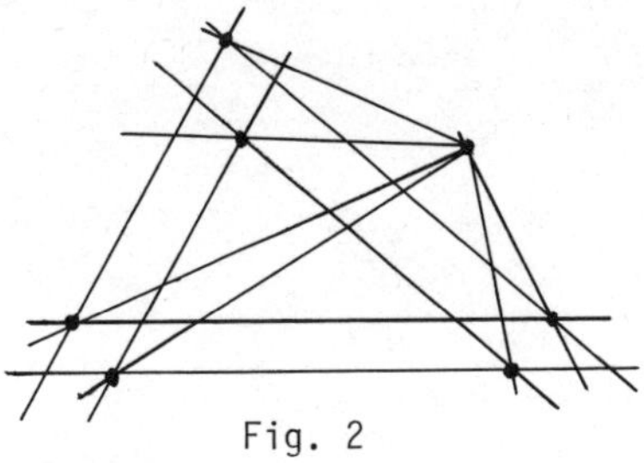

Fig. 2

In dieser Weise fährt man fort. Den Nachweis, daß mit der angegebenen repräsentativen Wahl von Projektionsrichtungen wirklich alle Fälle erfaßt werden, führen wir an anderer Stelle mit Hilfe der Methode der Gale-Diagramme.
Bei der i-ten Projektion wählt man diejenigen repräsentativen Richtungen für φ_{i+1} aus, bei denen $\varphi_{i+1}(\mathrm{set}\, \mathcal{C}_j)$ auf dem Rand seiner konvexen Hülle liegt, $i=o,\ldots,n-k-1$ ($\mathcal{C}_o=\mathcal{C}$). Gibt es für ein i_o keine derartigen Richtungen, dann ist

nach Satz 1 entschieden, daß $\mathcal{C}$ nicht polytopal ist. Andernfalls ist $\varphi_{n-k}\cdots\varphi_1(\mathrm{set}\,\mathcal{C})$ Rand eines k-Polytops P.

Literatur:

[1] Altshuler, A. u. Steinberg, L.; An enumeration of combinatorial 3-manifolds with nine vertices, Discrete Math. 16 (1976), 91-108

[2] Danaraj, G. u. Klee, V.; Which spheres are shellable?, Ann. Discrete Math. 2 (1978), 33-52

[3] Ewald, G., Kleinschmidt, P., Pachner, U. u. Schulz, C.; Neuere Entwicklungen in der kombinatorischen Konvexgeometrie. In: Beiträge zur Geometrie. Proceedings des Geometrie-Symposiums vom 28. Juni bis 1. Juli 1978 in Siegen, Birkhäuser Verlag 1979

[4] Grünbaum, B.; Convex Polytopes, Interscience Publishers, New York 1967

[5] Grünbaum, B.; Arrangements of Hyperplanes, Preprint 1970

[6] Kleinschmidt, P. u. Pachner, U.; Shadow-boundaries and cuts of Convex Polytopes, Manuskript 1979.

Regular Permutation Geometries

by

Heinz-Richard Halder (München)

Let $\tau \neq \emptyset$ be a m-ary relation on the set T, $J \neq \emptyset$ a set of indices, $P := \times_{j\in J} T$ and $\rho := \times_{j\in J} \tau$ the product of the relations τ in P. Let further $\mathcal{K}$ be a system of subsets of P fullfilling the following condition: For each $K \in \mathcal{K}$ there exist permutations γ_j in the symmetric group of T for $j \in J$ with $\gamma_j(\tau) = \tau$ and $K = \{(\gamma_j(x))_{j\in J};\ x \in T\}$. The pair $(P, \mathcal{K})$ is called a regular permutation geometry, if for all $(x_1, \ldots, x_m) \in \rho$ there exists exactly one $K \in \mathcal{K}$ with $x_1, \ldots, x_m \in K$.

Affine spaces, Minkowski planes, 2-structures, hyperbola-structures, weak affine spaces, (B_n^*, G_n, S_n)-geometries and the geometry of all Veronese-manifolds V_n^s on a Segre-manifold $S_{(n)s}$ can be interpreted as examples for regular permutation geometries.

Let $(P, \mathcal{K})$ be in the following a regular permutation geometry. For $i, j \in J$ we define $\Gamma^{i,j} := \{\gamma_j \gamma_i^{-1};\ \exists K \in \mathcal{K},\ K = \{(\gamma_j(x))_{j\in J};\ x \in T\}\}$.

Lemma 1. $\Gamma^{i,j}$ operates transitively on τ for $i \neq j$.

Proof. Let be $(x_{1,l}, \ldots, x_{m,l}) \in \tau$ for all $l \in J$. Then there exists an element $(x_1, \ldots, x_m) \in \rho$ with $x_n = (x_{n,l})_{l\in J}$ for $n = 1, \ldots, m$ and an element $K \in \mathcal{K}$ with $x_1, \ldots, x_m \in K$. For $K = \{(\gamma_j(x))_{j\in J};\ x \in T\}$ we get $\gamma_j \gamma_i^{-1}(x_{n,i}) = x_{n,j}$ $(n = 1, \ldots, m)$ and therefor $\gamma_j \gamma_i^{-1}((x_{1,i}, \ldots, x_{m,i})) = (x_{1,j}, \ldots, x_{m,j})$. Thereby follows the transitivity of $\Gamma^{i,j}$ on τ.

In the proof of Lemma 1 we get in the case $J = \{i, j\}$ that K is the unique element of $\mathcal{K}$ containing the points $(x_{1,i}, x_{1,j}), \ldots, (x_{m,i}, x_{m,j}) \in P$. So we get

Lemma 2. $\Gamma^{i,j}$ operates regularily on τ for $i \neq j$ and $J = \{i, j\}$.

These results lead to the following questions: Does $\Gamma^{i,j}$ operate regularily on τ for $i \neq j$? Are there at least two indices i and j such that $\Gamma^{i,j}$ operates regularily on τ? What happens if all $\Gamma^{i,j}$ $(i \neq j)$ operate regularily on τ?

The first two questions can be answered by counterexamples which can be easily constructed by the following theorem, which answers also the third question.

Theorem. All $\Gamma^{i,j}$ $(i \neq j)$ operate regularily on τ if and only if one of the following two conditions is fullfilled:

a) $|J| \leq 2$

b) All $\Gamma^{i,j}$ $(i \neq j)$ are isomorphic permutation groups.

Proof. By reason of Lemma 2 there is in the main only to show that the regularity of $\Gamma^{i,j}$ for all $i \neq j$ implies the validity of statement b).
In cause of $(\gamma_k \gamma_j^{-1})(\gamma_j \gamma_i^{-1}) = \gamma_k \gamma_i^{-1}$ holds $\Gamma^{j,k}\Gamma^{i,j} \supset \Gamma^{i,k}$. By the regularity of the permutation sets $\Gamma^{j,k}$, $\Gamma^{i,j}$ and $\Gamma^{i,k}$ on τ we get the equality $\Gamma^{j,k}\Gamma^{i,j} = \Gamma^{i,k}$. Without loss of generality we can assume that $\{(x)_{j \in J};\ x \in T\}$ is an element of K. Therefore we get $id_T \in \Gamma^{j,k}$, $id_T \in \Gamma^{i,j}$ and $id_T \in \Gamma^{i,k}$. Thereby follows (*) $\Gamma^{j,k} = \Gamma^{i,j} = \Gamma^{i,k} =: \Gamma$ and (**) $\Gamma = \Gamma^{-1}$. In cause of the equations (*) and (**) Γ is a group. So we get the result.

ON CHARACTERISATIONS OF KINEMATIC SPACES BY PARALLELISMS

Armin Herzer

For a K-vector space V let $\mathcal{P} = {}^{V}/_{K}$ be the projective geometry corresponding to V, the projective subspaces of $\mathcal{P}$ being considered as sets of points. For a set $\mathcal{R}$ of points of $\mathcal{P}$ define

$$\mathcal{P}\setminus\mathcal{R} := \{\ \mathcal{U}\setminus\mathcal{R} \mid \mathcal{U}\in\mathcal{P}\ \}.$$

A parallelism $\parallel$ on a set $\mathcal{S}$ of subspaces of $\mathcal{A}=\mathcal{P}\setminus\mathcal{R}$ is an equivalence relation on $\mathcal{S}$ satisfying the "euklidean parallel axiom": For every $\mathcal{U}\in\mathcal{S}$ and every point p of $\mathcal{A}$ there exists exactly one $\mathcal{U}'\in\mathcal{S}$ with $p\in\mathcal{U}'$ and $\mathcal{U}\parallel\mathcal{U}'$. - In what follows let us suppose that $|K| \geqq 4$.

Let A be a K-algebra (associative and with unit element 1), E the group of units of A and $Q = A\setminus E$ the set of the singular elements in A. Put $\mathcal{P} = {}^{A}/_{K}$ and $\overline{Q} = \{\langle a\rangle \mid a\in Q,\ a\neq o\}$. Then $\overline{E} = {}^{E}/_{K^*}$ operates on $\mathcal{A} = \mathcal{P}\setminus\mathcal{Q}$ as a regular group of collineations both by left- and by right-multiplication. The pair $(\mathcal{A},\overline{E})$ is called a <u>two-sided incidence group</u> ([8],[3]).

For every subalgebra B of A define $\overline{B} := \{\langle b\rangle \mid b\in E\cap B\}$. Then the subgeometry $\mathcal{C}$ of all translates $x\overline{B}$, $x\in\overline{E}$, can be furnished with a parallelism $\parallel$, such that

$$x\overline{B} \parallel y\overline{C} \quad\Leftrightarrow\quad B = C.$$

There arises the following question: Under which circumstances can the algebra A be reconstructed from the geometry with parallelism $(\mathcal{C},\parallel)$, considered as subgeometry of $\mathcal{A}$? A partial answer is given in [3] §5 for the case, where $\overline{Q}$ is a subspace of $\mathcal{P}$ ("$\mathcal{A}$ is a slit geometry").

In each case the reconstruction is made in three steps: 1.Construction of the regular group of collineations $\overline{E}$, 2.Extension of these collineations from $\mathcal{A}$ to $\mathcal{P}$, 3.Reconstruction of A.

Important for the second step is the following

PROPOSITION. <u>Let P be a projective geometry and let $A := P \setminus R$ for some pointset R. Let the rank of A be greater than n+2 for a natural number n. If for every subspace W of A of rank n the geometry $G^W = \{X \in A \mid W \leqq X\}$ is a projective geometry, then every collineation of A can be uniquely extended to a collineation of P.</u>

The proof is an immediate consequence of [2] Korollar p.470 and of the construction of the imbedding projective geometry from A. This proposition includes the analogous results concerning slit spaces ([7]) and biaffine spaces ([9]).

An interesting class of K-algebras in regard to geometry consists of the so-called kinematic algebras. A kinematic algebra is defined by the property, that for every element c of the algebra there exists a polynomial f of degree 2 in K[x] such that f(c)=o. Let A be a kinematic algebra; then $(A,\overline{E})$ constructed as above is called a <u>kinematic space</u>. Kinematic spaces and algebras were investigated in [1],[4],[5].
Here $\overline{Q}$ is a quadric ([5]), and every line through <1> is a subgroup of $\overline{E}$. One can define two parallelisms by using the left and the right cosets of the lines through <1>; the structure obtained this way is called a <u>double-space</u>. Conversely some classes of double-spaces arising from kinematic algebras have been characterized in [6] and [8].
Here I will show, that geometries $(C,\|)$ obtained from kinematic algebras also can be characterized along the lines of [3]. So we can characterize those geometries by only one parallelism.

In the sequel let $(C,\|)$ be defined by a kinematic algebra as above.

(1) <u>If h_i are different parallel lines intersecting a line g, then</u> $h_1 \leqq h_2 + h_3$.

<u>Proof</u>: We can choose $o \neq a_i \leqq g \cap h_i$ with $a_1 = a_2 + a_3$. There are $x,y \in E$ with $a_2 = xa_1$ and $a_3 = ya_1$ and x+y=1. Therefore $h_2 = xh_1$ and $h_3 = yh_1$ and thus

$$h_1 = (x+y)h_1 \leqq xh_1+yh_1 = h_2+h_3.$$

Define two lines g,g' to be adjacent, if there exist three different parallel lines meeting both g and g'.

(2) If g and g' are adjacent lines and t and t' are parallel lines such that t meets g and g', and t' meets g, then t' and g' are complanar. Moreover, if g and g' are skew, then t' meets g'.

Proof: Let g and g' be different lines which meet three parallel lines h_i, i=1,2,3. Without loss of generality let $g \cap h_1 = \langle 1 \rangle$ and $x,y \in g$ with $x+y=1$, $h_2=xh_1$ and $h_3=yh_1$. For $o \neq b \in g' \cap h_1$ define $g''=gb$. Then $g'' \cap h_2 = \langle xb \rangle$ and $g'' \cap h_3 = \langle yb \rangle$, and so g and g'' are adjacent. Let $g \cap t = \langle c \rangle$, $g'' \cap t = \langle c' \rangle$, and $g \cap t' = \langle d \rangle$. Then $\langle d \rangle = \langle zc \rangle$ for some $z \in g$, so that $t'=zt$ and $g'' \cap t' = \langle zc' \rangle$. Now if h_2 and h_3 are skew, then h_1,h_2,h_3 are mutually skew, hence $g'=g''$. If on the other hand h_2 and h_3 are contained in a plane U, then g,g',g'' and also t,t' are contained in U.

We shall call a parallelism on the lineset of a geometry $A = P \setminus R$ regular, if it satisfies (1) and (2).

Now, let P be a pappian projective geometry and R be a quadric of P (i.e. R can also be the pointset of either a pair of hyperplanes or else of a proper subspace of P). Let $\parallel$ be a regular parallelism on the lineset of $A = P \setminus R$. For every line h of A, on the set of all lines meeting h, we shall define "locally" a parallelism $\underset{h}{\parallel}$, which is consistent with the second parallelism given by right multiplication in case that $(A,\parallel)$ arises from a kinematic space. Without loss of generality let g be a line with $g \cap h = \langle 1 \rangle$; we define $S(g,h)$ to be the set of all lines parallel to h, which meet g. By (1) $S(g,h)$ spans a subspace U of rank 3 or 4. We consider U and its subspaces as subspaces of P, and the points of R we call singular points of P.

1st case: U has rank 4. Then by (2) $S(g,h)$ is contained in a regulus of a hyperbolic quadric H of U. For a line g' define $g \underset{h}{\parallel} g'$, if and only if g' belongs to the opposite regulus of H.

In fact, if $(A,\|)$ is constructed from a kinematic space in the above manner, then h (without singular points) is a subgroup of $\overline{E}$, fixing every line of $S(g,h)$. So $g \underset{h}{\|} g'$ if and only if $g'=gx$ for an appropriate element x of h.

2nd case: U has rank 3. Then by (2) for every two parallel lines t and t' with $t \leqq U$ and $t' \cap U \neq \emptyset$ holds that $t' \leqq U$.

(3) Every line has at most 2 singular points, for R is a quadric.

(4) Let t and t' be parallel lines of U. Then $t \cap t'$ is a singular point, by definition of parallelism.

(5) Any set of parallel lines in U belongs to some pencil of lines of U through a singular point.

For, suppose there are three parallel lines t_i in U, which form the sides of a triangle. So $A_{ij} = t_i \cap t_j$ for $1 \leqq i < j \leqq 3$ are three different points. There exists a point P of U (nonsingular), which is not contained in any of these three lines. Let t_0 be the parallel line through P. Then t_0 meets at least one of the lines t_i in some point different from the three points A_{ij} - a contradiction to (3).

(6) The centers of the pencils of all classes of parallel lines in U are exactly the points of some (improper) line of U.

For, if there were a line ℓ containing two different centers of pencils of parallel lines, then ℓ would belong to two different classes of parallels. This contradicts the definition of parallelism. Now apply (5).

The set of the singular points of U forms a conic, which by (6) contains all points of a line hence consists of the points of either one line or two lines.

(a) If the singular points of U are the points of a line, then $U \setminus R$ formes an affine plane. For two lines ℓ, ℓ' of U define

$$\ell \underset{h}{\|} \ell' \quad \Leftrightarrow \quad \ell \| \ell'.$$

(b) If the singular points of U are the points of two different lines

t_1, t_2 and, say, the centers of the pencils of parallel lines are the points of t_1, then for two different lines ℓ, ℓ' of U define

$$\ell \underset{h}{\parallel} \ell' \quad \Leftrightarrow \quad \ell \cap \ell' \leqq t_2 .$$

In fact, if $(A,\parallel)$ arises from a kinematic algebra, then $U \cap \overline{E}$ is a subgroup of $\overline{E}$. In case (b) this subgroup operates on U by right multiplication necessarily as the group $\Gamma(t_1,t_2)$ of collineations of U with axis t_2 and center on t_1; on the other hand it acts on U by left multiplication as the group $\Gamma(t_2,t_1)$ of all collineations of U with axis t_1 and center on t_2. This follows from the fact that for points X,Y of U not collinear with $O:=\langle 1\rangle$ the multiplication of X by Y from the right gives

$$XY = (((O+X)\cap t_1)+Y)\cap(((O+Y)\cap t_2)+X);$$

this is the same as the multiplication of Y by X from the left.
The subalgebra corresponding to this subgroup can be described as the algebra of the upper triangular two by two matrices with entries in K.
In case (a) the situation is simpler, for an affine plane defines a parallelism uniquely.
So in every case the local parallelism $\underset{h}{\parallel}$ can be uniquely defined.

We formulate the incidence propositions (D1) and (D2) of [3] for the special case of our $(A,\parallel)$:
Let A_i, B_i be pairwise different points and define

$$h_i = A_i+B_i, \qquad 1 \leqq i \leqq 3,$$

$$\left.\begin{array}{l} s_{ij} = A_i+A_j \\ t_{ij} = B_i+B_j \end{array}\right. \qquad 1 \leqq i < j \leqq 3.$$

(D1) If the lines h_i are parallel and different, then the conditions

$$s_{12} \underset{h_1}{\parallel} t_{12}$$

and $$s_{23} \underset{h_2}{\parallel} t_{23}$$

imply $$s_{13} \underset{h_1}{\parallel} t_{13}.$$

(D2) If $s_{ij} \parallel t_{ij}$ for $1 \leq i < j \leq 3$, then the conditions

$$\begin{array}{ll} & h_1 \underset{s_{12}}{\parallel} h_2 \\ \text{and} & h_2 \underset{s_{23}}{\parallel} h_3 \\ \hline \text{imply} & h_1 \underset{s_{13}}{\parallel} h_3 . \end{array}$$

The regular parallelism of $(A,\parallel)$ is called <u>arguesian</u>, if it satisfies (D1) and (D2).

We can state now the following

THEOREM. <u>Let P be a pappian projective geometry and R a quadric of P. Let $\parallel$ be a parallelism on the lineset of $A = P \setminus R$. Then there exists a kinematic space $(A,\bar{E})$, such that the parallelism is defined by the left cosets of $(A,\bar{E})$, if and only if this parallelism is regular and arguesian.</u>

This theorem can be proved in a manner completely analogous to the proof of [3], (5.7) and (5.11) (case of a slit geometry).

EXAMPLE. We consider the parallelism of the kinematic space A obtained from the algebra of all two-by-two matrices with entries from a quadraticly closed field K. Then A arises from a projective 3-space by omitting the points of a ruled quadric H; the two parallelism are constructed by using the two reguli of H: Every pair (g,h) of lines belonging to a fixed regulus of H represents a class of parallels of A. This is in case $g \neq h$ the hyperbolic congruence of all lines of A meeting g and h, - and in case $g=h$ the parabolic congruence of all lines of A tangent to H in points of g.

LITERATURE.

1. L.Bröcker: Kinematische Räume. Geom.Dedic. 1, 241-268(1972).

2. A.Herzer: Projektiv darstellbare stark planare Geometrien vom Rang 4. Geom.Dedic.5, 467-484(1976).

3. A.Herzer: Halbprojektive Translationsgeometrien. Mitt.Math.Sem.Giessen, Heft 127 (1977).

4. H.Karzel: Kinematic spaces. Symposia Mathematica 11, 413-439(1973).

5. H.Karzel: Kinematische Algebren und ihre geometrischen Ableitungen. Abh.Math.Sem.Hamburg 41, 158-171(1974).

6. H.Karzel, H.J.Kroll und K.Sörensen: Invariante Gruppenpartitionen und Doppelräume. J.reine angew.Math. 262/263, 153-157(1973).

7. H.Karzel und H.Meißner: Geschlitzte Inzidenzgruppen und normale Fastmoduln. Abh.Math.Sem.Hamburg 31, 69-88(1967).

8. H.Karzel und I.Pieper: Bericht über geschlitzte Inzidenzgruppen. Jber.Deutsch.Math.Verein.72, 70-114(1970).

9. E.M.Schröder: Zur Theorie subaffiner Inzidenzgruppen. Journal of Geometry 3, 31-69(1973).

ON REALIZATIONS OF VECTOR PRODUCTS BY POLYNOMIALS WHICH ARE IDENTITIES FOR MATRIX RINGS

Yehiel Ilamed
Soreq Nuclear Research Centre
Yavne, Israel

1. INTRODUCTION

Let R denote the real field, R^n the real n-space and $<,>:R^n \to R$ a positive definite inner product in R^n. A continuous vector product of r vectors in R^n, $0<r<n$, as defined by Eckmann [1], is a function $v(u_1,\ldots,u_r)$ of r variable vectors $u_1,\ldots,u_r \in R^n$ with range R^n, having the following properties:

(i) $v(u_1,\ldots,u_r)$ is continuous in $u_1,\ldots,u_r \in R^n$,

(ii) $<v(u_1,\ldots,u_r),u_j> = 0$, for $j=1,\ldots,r$ and

(iii) $<v(u_1,\ldots,u_r),v(u_1,\ldots,u_r)> = \det(<u_i,u_j>)$, where $\det(<u_i,u_j>)$ means the determinant of the $r \times r$ matrix whose ij-th entry is $<u_i,u_j>$.

Let F be a field of characteristic different from 2. The Brown and Gray [2] definition of a <u>multilinear vector product</u> of r vectors is obtained by substituting in the above definition of a continuous vector product: F for R, "multilinear" for "continuous" and "nondegenerate symmetric bilinear form" for "positive definite inner product".

The existence of continuous and multilinear vector products is expressed by the following two theorems.

THEOREM 1. Continuous vector products of r vectors in R^n exist only in the following cases: (i) n even, $r=1$, (ii) n arbitrary, $r=n-1$, (iii) $n=7$, $r=2$ and (iv) $n=8$, $r=3$, [1].

THEOREM 2. Multilinear vector products of r vectors in F^n exist in the same cases as in Theorem 1, [2].

The proof of Theorem 1 based on deep theorems in topology is given in Refs. [1] and [3] while an elementary exposition of the results is given in Ref. [4]. An algebraic proof of Theorem 2 is given in Ref. [2].

A generalized multilinear vector product was defined and studied by Vanhecke [5]. More about these definitions, and the existence and the importance of these vector products can be found in Refs. [1-9].

Let $x_1, y_1, x_2, y_2, \ldots$ denote noncommutative indeterminates and let R_m denote the vector space of $m \times m$ real matrices. Let V^n denote an n-dimensional subvector space of R_m. Let us define a positive definite inner product in V^n by

(1) $$\langle A,B\rangle = \mathrm{trace}(AB^T), \quad A, B \in V^n$$

where B^T means the transpose of B. We say that $v(x_1,\ldots,x_r;y_1,\ldots,y_s)$, an r-linear polynomial in $x_1,\ldots,x_r$, is a realization of a vector product of r vectors $A_1,\ldots,A_r \in V^n \subset R_m$ if there exists a pair n,m and matrices $B_1,\ldots,B_s \in R_m$ so that $(v(A_1,\ldots,A_r;B_1,\ldots,B_s))^T \in V^n$ is the vector product of arbitrary $A_1,\ldots,A_r \in V^n \subset R_m$; we call this an (r,n,m)-VP. For example, the commutator $x_1x_2-x_2x_1$ is a realization of a vector product of two vectors since for A_1, A_2 skew symmetric real 3×3 matrices $(A_1A_2-A_2A_1)^T$ is a $(2,3,3)$-VP, (the usual vector product).

Let $s_r = s_r(x_1,\ldots,x_r)$ and $v_r = v_r(x_1,\ldots,x_r;y_1,\ldots,y_{r+1})$ be defined by

$$s_r(x_1,\ldots,x_r) = \Sigma_\sigma (\mathrm{sg}\sigma) x_{\sigma(1)} x_{\sigma(2)} \cdots x_{\sigma(r)}$$ and

$$v_r = v_r(\underline{x};\underline{y}) = \Sigma_\sigma \Sigma_\gamma (\mathrm{sg}\sigma)(\mathrm{sg}\gamma) y_{\gamma(1)} x_{\sigma(1)} y_{\gamma(2)} x_{\sigma(2)} \cdots y_{\gamma(r)} x_{\sigma(r)} y_{\gamma(r+1)}$$

where σ ranges over the permutations on $\{1,\ldots,r\}$, γ ranges over the cyclic permutations on $\{1,\ldots,r+1\}$ and $\mathrm{sg}\sigma$ means the sign of σ. It is known that the standard polynomial s_{2n} is the minimal polynomial identity for R_n, [10], and that v_r is an identity for R_m if $r \geq m^2$ [11].

The main result of this research note is that the v_r polynomials, or real linear combinations of v_r polynomials ($y_1,\ldots,y_{r+1}$ are parameters) are realizations of vector products of r vectors for some of the pairs r,n specified in Theorem 1; examples of corresponding (r,n,m)-VP's ((r,n,m)-vector products) are given in section 3 and in the above example.

In the next section we define and construct a natural basis for the polynomial realizations of vector products.

2. VP-POLYNOMIALS

We say that $p(x_1,\ldots,x_r;y_1,\ldots,y_s)$ is a VP-polynomial if the following two conditions are satisfied:

(a) $p(x_1,\ldots,x_r;y_1,\ldots,y_s)$ is r-linear in $x_1,\ldots,x_r$

(b) $\langle (p(A_1,\ldots,A_r;B_1,\ldots,B_s))^T, A_j\rangle = 0$, $j=1,\ldots,r$ and $A_1,\ldots,A_r,B_1,\ldots,B_s$ any matrices in R_m; where $\langle , \rangle$ is defined by Eq. (1) and m is arbitrary.

Adding new conditions to (a) and (b) we may obtain, as in the following two theorems, uniquely determined VP-polynomials.

THEOREM 3. If $p(x_1,\ldots,x_r)$ satisfies the conditions:

(i) $p\equiv p(x_1,\ldots,x_r)$ is a VP-polynomial,

(ii) the monomial $x_1x_2\ldots x_r$ appears in p with coefficient $+1$ and

(iii) the number of monomials appearing in p is minimal,

then $p = s_{2k}(x_1,\ldots,x_{2k})$ for $r=2k$ and $p = 0$ for $r=2k+1$, $k=1,2,\ldots$.

THEOREM 4. If $q = q(x_1,\ldots,x_r;y_1,\ldots,y_{r+1})$ satisfies:

(i) q is a VP-polynomial,

(ii) $y_1x_1y_2x_2\ldots y_rx_ry_{r+1}$ appears in q with coefficient $+1$ and

(iii) the number of monomials appearing in q is minimal,

then $q(x_1,\ldots,x_r;y_1,\ldots,y_{r+1}) = v_r(x_1,\ldots,x_r;y_1,\ldots,y_{r+1})$.

PROOF OF THEOREMS 3 AND 4. $v_{2k}(x_1,\ldots,x_{2k};1,1,\ldots,1) = s_{2k}(x_1,\ldots,x_{2k})$ and $v_{2k+1}(x_1,\ldots,x_{2k+1};1,1,\ldots,1) = 0$, hence it remains to prove only Theorem 4.

Starting with $y_1x_1y_2x_2\ldots y_rx_ry_{r+1}$, we add a monomial to obtain a polynomial p_1 so that $tr(p_1x_1) = 0$. By definition a polynomial has trace (tr) zero if the polynomial can be expressed as a sum of commutators. Using the fact that the trace of a monomial is invariant with respect to a cyclic permutation of its factors we obtain

$$p_1 = y_1x_1y_2x_2\ldots y_rx_ry_{r+1} - y_2x_2\ldots y_rx_ry_{r+1}x_1y_1 .$$

Starting with p_1 we obtain, in the same way, a polynomial p_2 so that $tr(p_2x_2) = tr(p_2x_1) = 0$. Continuing to construct, step by step, the simplest p_{i+1} from p_i , we obtain at the last step $p_r=v_r(\underline{x};\underline{y})$.

Another proof of Theorem 4 is given in Ref. [12] and of Theorem 3 in Ref. [13].

COROLLARY. $v_r(x_1,\ldots,x_r;y_1,\ldots,y_{r+1})$ is a VP-polynomial of $x_1,\ldots,x_r$ for each $(r+1)$-list of parameters $y_1,\ldots,y_{r+1}$.

It follows that any finite sum $\Sigma_{i=0}a_iv_r(\underline{x};y_{1+i},y_{2+i},\ldots,y_{r+1+i})$ is a VP-polynomial, (real a_i's). In the next section we use this freedom to construct polynomials that are realizations of vector products.

3. VECTOR PRODUCTS

Let $a_1,a_2,\ldots$ denote real numbers and let (ij) denote the matrix with 1 in the ij-th entry and zeroes in all other entries. In the following we give examples of (r,n,m)-VP's and of the corresponding polynomials.

(a) Let $a_1(11) + a_2(22) = A \in R_2$, then $((12)A(21) - (21)A(12))^T$ is an (1,2,2)-VP corresponding to $v_1(x;y_1,y_2)$.

(b) Let $a_1(11) + a_2(22) + a_3(33) + a_4(44) = A$ be any diagonal matrix in R_4 and let $(12) + (34) = B \in R_4$, then $(BAB^T - B^TAB)^T$ is an (1,4,4)-VP corresponding to $v_1(x;y_1y_2)$.

(c) Let $a_1(11) + a_2(12) + a_3(21) + a_4(22) = A$ be any matrix in R_2, then $((11)A(22) - (22)A(11) + (12)A(21) - (21)A(12))^T$ is an (1,4,2)-VP corresponding to $v_1(x;y_1,y_2) + v_1(x;y_3,y_4)$.

(d) Let $\Sigma_{i=1}^{2n} a_i(ii) = A$ be any diagonal matrix in R_{2n} and let $(12)+(34)+ \ldots +(2n-1\ 2n) = B \in R_{2n}$, then $(BAB^T - B^TAB)^T$ is an (1,2n,2n)-VP corresponding to $v_1(x;y_1y_2)$.

(e) Let $a_{i1}(12)+a_{i2}(21)+(\surd 2)^{-1}a_{i3}((11)-(22)) = A_i$, i=1,2, be any two trace zero matrices in R_2 , then $(A_1A_2 - A_2A_1)/(\surd 2)$ is an (2,3,2)-VP corresponding to $(x_1x_2 - x_2x_1)/(\surd 2)$.

(f) Let $A_1,\ldots,A_r$ be any matrices in R_m and let $B_1,\ldots,B_{r+1}$ be r+1 linearly independent matrices in R_m , then for $r = m^2-1$ $v_r(A_1,\ldots,A_r;B_1,\ldots,B_{r+1})$ is, up to a real factor, an (m^2-1, m^2,m)-VP corresponding to $v_r(x_1,\ldots,x_r;y_1,\ldots,y_{r+1})$, $r=m^2-1$.

COMMENT. We expect that it is possible to construct (r,n,m)-VP's for all the cases mentioned in Theorem 1.

REFERENCES

[1] B. Eckmann, Stetige Lösungen linearer Gleichungssysteme, Comment. Math. Helv. 15 (1943), 318-339.

[2] R.B. Brown and A. Gray, Vector cross products, Comment. Math. Helv. 42 (1967) 222-236.

[3] G. Whitehead, Note on cross-sections on Stiefel manifolds, Comment. Math. Helv. 37 (1962), 239-240.

[4] B. Eckmann, Continuous Solutions of Linear Equations - Some Exceptional Dimensions in Topology, in Battelle Rencontres 1967, ed. by C.M. DeWitt and J.A. Wheeler, Benjamin, New York, 1968, p. 512-526.

[5] L. Vanhecke, Vectoriële π-producten en veralgemeende (4,±2)-structuren in Liber Amicorum Prof. Em. Dr. H. Florin, Leuven, 1975, 269-287.

[6] P. Zwengrowski, A 3-fold vector product in R^8, Comment. Math. Helv. 40 (1965/66), 149-152.

[7] A. Gray, Vector cross products on manifolds, Trans. Amer. Math. Soc., 141 (1969), 465-504.

[8] A. Gray, Vector cross products, Rend. Sem. Mat. Univ. e Politec. Torino, 35 (1976-77), 69-75.

[9] L. Vanhecke, On r-fold (4,2)- and f-products, Kōdai Math. Sem. Rep. 28 (1977), 162-181.

[10] S.A. Amitsur and J. Levitzki, Minimal identities for algebras, Proc. Amer. Math. Soc. 1 (1950), 449-463.

[11] S.A. Amitsur, On a central identity for matrix rings, J. London Math. Soc. (2), 14 (1976) 1-6.

[12] Y. Ilamed, On identities for matrix rings and polynomials orthogonal to their arguments, Ring Theory, Proc. 1978 Antwerp Conf., ed. by F. Van Ostaeyen, Marcel Dekker, New York, 1979, p. 81-86.

[13] Y. Ilamed, A characterization of the even degree standard polynomial, Proc. Fifth Int. Coll. Group Theoretical Methods in Physics, Academic Press, New York, 1977, p. 623-626.

ON THE CONSTRUCTION OF ARCHIMEDEAN ORDERS OF A FREE PLANE

J. Joussen
Universität Dortmund
Abteilung Mathematik

D-4600 Dortmund, Germany

I intend to report on the following theorem which is the result of a long series of investigations concerning the construction of orders of free planes.

> Theorem: Every finitely generated free plane admits at least one archimedean order.

S. Priess-Crampe [4] has characterized the projective planes with an archimedean order using the concept of " plane planes " due to H. Salzmann: she proved that the archimedean projective planes are exactly the projective subplanes of the plane planes. By " plane planes " we mean those topological projective planes which are homeomorphic with the real projective plane.

Using this result of Mrs. Priess-Crampe the above theorem yields the

> Corollary: Every finitely generated free plane can be embedded as a projective subplane in a plane plane.

The aim of my lecture is to give an impression of the construction of an archimedean order of a free plane.

First let me recall the definition of an "order" of a projective plane Π: Let $[a,b \mid c,d]$ be a function with values $+1$ or -1 which is defined on the quadruples of collinear points a,b,c,d such that $a,b \neq c,d$. Assume that the following conditions are satisfied:

(1) $[a,b \mid c,d] = [b,a \mid c,d] = [a,b \mid d,c]$

(2) $[a_1,a_2 \mid c,d] \cdot [a_2,a_3 \mid c,d] \cdot [a_3,a_1 \mid c,d] = +1$

(3) $[a,b \mid c,d] = [a',b' \mid c',d']$, whenever the quadruples a,b,c,d and a',b',c',d' are perspectively situated

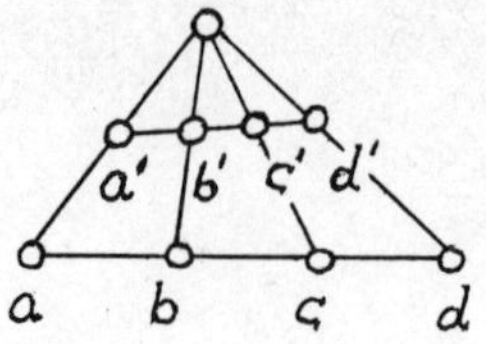

(4) $Z(a,b,c,d) = 1$, whenever a,b,c,d are four distinct collinear points. Here $Z(a,b,c,d)$ denotes the number of values $[a,b \mid c,d]$, $[a,c \mid b,d]$ and $[a,d \mid b,c]$ which are equal to -1 .

A function satisfying properties (1) to (4) is called an "order" of Π; and according as $[a,b \mid c,d]$ is equal to -1 or $+1$, we say that the pair a,b does or does not "separate" the pair c,d .

An order of Π is called an <u>"archimedean order"</u> if it has the following property: given any distinct points o,a,b,c,d,ω and distinct lines g,h such that

$$o,a,b,\omega \, I \, g; \qquad \omega,c,d \, I \, h ,$$

there always exists a number $k \in \mathbb{N}$ such that

$$[o,b \mid a_k, \omega] = +1 .$$

Here a_k is constructed as follows: draw lines from o to d and from a to c ; join the point of intersection with ω ; then construct $a_1, a_2, a_3, \ldots$ and so on according the accompanying figure .

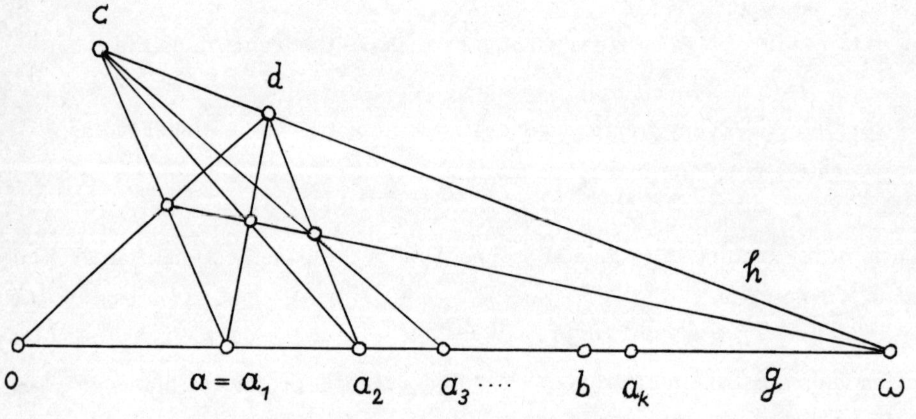

The meaning is that if h is interpreted as the line at infinity and we add the distance from o to a repeatedly to itself, then a_k is the resulting point of k times the distance from o to a . The condition above means, therefore, that we can exceed any point b by this construction. This is the original geometric meaning of the concept of archimedean order.

Next let me recall the definition of finitely generated <u>"free planes"</u>, that is, the free planes in the sense of M. Hall [1]. Given a natural number $m \geq 4$; the free plane $F_m = F(J^m_0)$ is the union of the incidence structures J^m_n which are defined as follows: J^m_0 consists of m points and one line which is incident with all the points except for two. Furthermore: J^m_{n+1}, $n \geq 0$, is connected with the preceding structure J^m_n by the formula $J^m_{n+1} = \overline{J^m_n}$. Define $\overline{J}$ as follows: take all pairs of points not joined in J as new "lines" and all pairs of lines which do not intersect in J as new "points". Let all new points and lines be distinct from each other. $\overline{J}$ shall then consist of the elements of J and these new points and lines. For example:

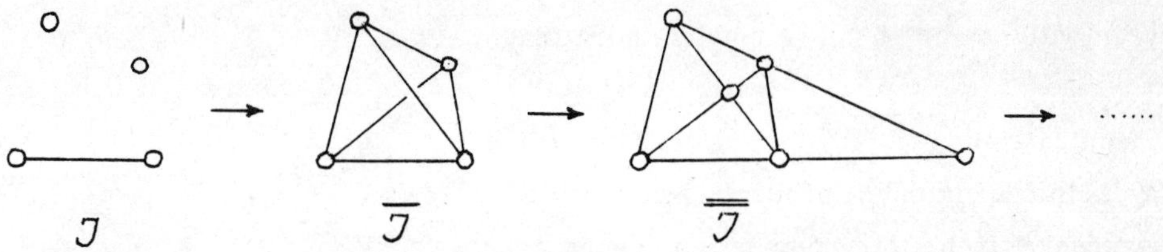

The union of the points and lines of $J, \overline{J}, \overline{\overline{J}}, \ldots$ is the free plane F_4.

The construction of orders of a free plane F_m can be done by a step-by-step-process of defining orders on its generating incidence structures J^m_n. For this purpose we need a suitable notion of "order of an incidence structure". Formally one could take the same definition given in the beginning of this lecture for the orders of a projective plane, namely, a function $[a, b \mid c, d]$ satisfying (1) to (4). But this is not sufficient for our purposes. For in an incidence structure there are not enough perspectivities in general, and requirement (3) then is too weak. Instead we shall make use of a so-called <u>"ordering function"</u> in the sense of E. Sperner [2], that is a function

$$f: \begin{cases} (L \times P)' \to \{+1, -1\} \\ (h, a) \to f(h, a) =: ha \end{cases}$$

defined on the set $(L \times P)'$ of non-incident pairs (h, a) ($h =$ line, $a =$ point) with values $ha = +1$ or -1. In the affine case one may think of an ordering function as a division of an ordered plane into parts by each line in the usual way, but with no difficulty ordering functions can be defined on projective planes also, or even, as is our intention, on incidence structures.

With the aid of an ordering function we are able to define an order in the original sense as follows: first, for any two lines g, h and any two points a, b not incident with the lines, define

$$[g, h \mid a, b] := (ga)(gb)(ha)(hb) .$$

Secondly, let a, b, c, d be four collinear points such that $a, b \neq c, d$. Select a line g incident with c, and a line h incident with d, neither line being incident with a or b. Put

$$[c, d \mid a, b] := [g, h \mid a, b] .$$

Here we have to make an assumption in order that the symbol $[c, d \mid a, b]$ be well-defined: if g' is another line, incident with c but not with a or b, we must require that

$$(ga)(gb) = (g'a)(g'b)$$

or, equivalently,

$$[g, g' \mid a, b] = +1 .$$

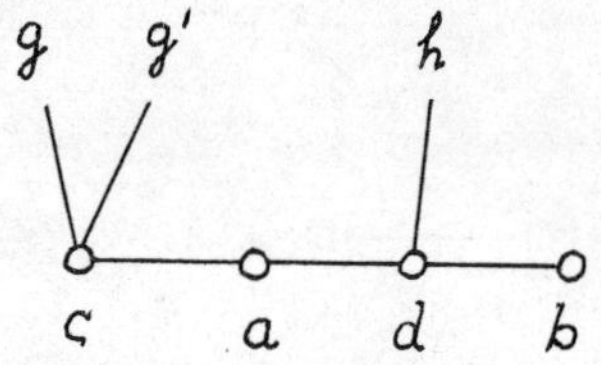

This is the so-called "Geradenrelation". Conversely, if the Geradenrelation generally holds, then the function $[c, d \mid a, b]$ is well-defined. Moreover, it fulfills the conditions (1), (2) and (3) set up in the beginning of the lecture. Condition (4), however, must be postulated separately.

By the preceding explanation the problem of extending orders from J to $\bar{J}$ has been reduced to the less difficult problem of extending ordering functions. To make this last problem solvable in general, we need one more assumption. Our ordering functions must be "definite", which means the following: there exists a constant $\varkappa$ (= +1 or -1) such that $\prod_{i \neq k} (h_i a_k) = \varkappa$ for all sextuples $h_1, h_2, h_3; a_1, a_2, a_3$, which satisfy one of the following configurations (the second configuration is the dual of the first one).

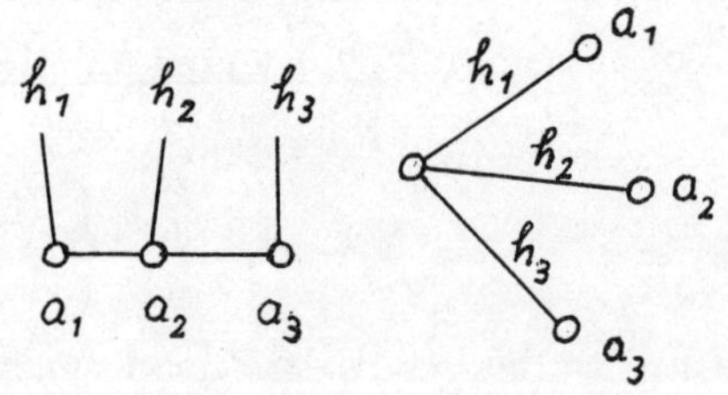

In a projective plane, the definiteness of an ordering function turns out to be equivalent to the Geradenrelation, but in general it is only true that the Geradenrelation is a consequence of definiteness ; definiteness is thus the stronger postulate .

According as the constant $\varkappa$ is equal to -1 or $+1$, a definite ordering function is also called a "harmonic" or a "non-harmonic" ordering function .

We are now able to define the notion of "<u>order of an incidence structure</u>" precisely.

Assume that the incidence structure J is regular and half-projective . That means that each element of J is incident with at least two other elements of J, and that either every two lines have a point of intersection or every two points are joined by a line. Then we define an "order" of J to be a class of harmonic ordering functions of J satisfying $Z \equiv 1$. Here "class" is understood with reference to the equivalence relation

$$f_1 \backsim f_2 \ :\Leftrightarrow\ [g,h \mid a,b\,]_{f_1} \equiv [g,h \mid a,b]_{f_2}$$

and Z is the function defined on quadruples of distinct collinear points as in (4) and dually on quadruples of distinct concurrent lines .

It should be remarked that the definition of order of an incidence structure, given thereby, is a real extension of the notion of order of a projective plane ; that is, if J is a projective plane Π, the orders of J in the sense of the foregoing definition are identical with the orders of Π in the usual sense .

The possibility of constructing archimedean orders of a free plane is a consequence of the following

<u>Theorem:</u> Suppose that J is finite, regular and half-projective. Let there be given distinct points o, a, b, ω, c, d and distinct lines g, h of J such that

$$o, a, b, \omega \ I\ g \ ; \qquad \omega, c, d \ I\ h \ .$$

If φ is an order of J , then there exist $k, n \in \mathbb{N}$ and an order φ_n of J_n (J_n being defined by $J_0 := J, \dots, J_n := \overline{J_{n-1}}$) satisfying

(i) $a_k \in J_n$ (a_k constructed in the plane $F(J) = \bigcup_{n=0}^{\infty} J_n$ as above) ;

(ii) φ_n is an extension of φ ;

(iii) $[\,o, b \mid a_k, \omega]_{\varphi_n} = +1$.

I shall dispense with the details of the proof of this theorem. It is very complicated and would need considerable time to even give a rough impression of it. The proof is to be published in Resultate der Mathematik, Birkhäuser, Basel-Stuttgart.

Let me close my lecture with the remark, that the problem of constructing archimedean orders of a finitely generated free plane can be obviously solved with the aid of this theorem. For in such a plane there is at most a countable number of configurations as given in the theorem ; so we are able to perform a step-by-step-construction of an order of the full plane, which satisfies the conditions of the theorem for all those configurations and is therefore an archimedean order.

References :

[1] Hall, M.: Projective Planes. Trans. Amer. Math. Soc. 54, 229-277 (1943).

[2] Sperner, E.: Die Ordnungsfunktionen einer Geometrie. Math. Ann. 121, 107 - 130 (1949).

[3] Joussen, J.: Die Anordnungsfähigkeit der freien Ebenen. Abh. Math. Sem. Univ. Hamburg 29, 137 - 184 (1966).

[4] Priess-Crampe, S.: Archimedisch angeordnete projektive Ebenen, Math. Z. 99, 305 - 348 (1967).

RECTANGULAR SPACES

Helmut Karzel
Technische Universität München

The concept of a parallelogram space with congruence $(P,\mathfrak{L},\equiv,\parallel)$ will be introduced and it will be shown, that such a space can be represented as a commutative kinematic space (theorem (2.12)). By forming the quotient any commutative kinematic space appears as a substructure of a vector space (theorem (3.6)). Parallelogram spaces with congruence fulfilling additional axioms are called rectangular spaces. This concept covers euclidean spaces and planes, Dehn spaces, and rectangular planes. § 4 is a survey of results that so far have been achieved for these structures.

§ 1 Introduction and fundamental concepts

Let $(P,\mathfrak{L},\equiv,\alpha)$ be an absolute geometry where P designates the set of points, $\mathfrak{L}$ the set of lines, $\equiv$ the congruence, and α the order relation. Then the pair $(P,\mathfrak{L})$ is an <u>incidence space</u>, that means:

(<u>I1</u>) To any two distinct points $a,b \in P$ there is exactly one line $L \in \mathfrak{L}$ with $a,b \in L$.

(<u>I2</u>) Every line $L \in \mathfrak{L}$ contains at least two points.

As usual we use the following notations (cp.[6]): For $a,b \in P$ let $\overline{a,b}$ be the line uniquely determined by (<u>I1</u>), if $a \neq b$, and $\overline{a,b} = \{a\}$ if $a = b$. A subset $T \subset P$ is called a <u>subspace</u> if for all $a,b \in T$ we have $\overline{a,b} \subset T$. The set $\mathfrak{T}$ of all subspaces is $\cap$-closed. Therefore $-: \mathfrak{p}(M) \longrightarrow \mathfrak{T};\ S \longrightarrow \overline{S} := \cap\{T: S \subset T \in \mathfrak{T}\}$ is a closure operation, called the <u>incidence-hull</u> and for $T \in \mathfrak{T}$ the number $\dim T := \inf\{|S| : \overline{S} = T\} - 1$ is called the dimension where $|S|$ designates the cardinal number of the set S. Subspaces of dimension 2 are <u>planes</u>. Two lines $A,B \in \mathfrak{L}$ are <u>perpendicular</u>, in symbols $A \perp B$, if $|A \cap B| = 1$ and if $a,a' \in A$, $b,b' \in B$ with $(a,b) \equiv (a',b)$ implies $(a,b') \equiv (a',b')$.

An absolute geometry has the following properties:

(L) For any $L \in \mathfrak{L}$ and $p \in P \setminus L$ there is exactly one line M, designated by $\{p \perp L\} := M$, with $p \in M$ and $M \perp L$.

(W) For any five distinct points $a,b,c,d,d' \in P$ with $\{a,b,c\}$ collinear and $(a,b,d) \equiv (a,b,d')$ it follows that $(c,d) \equiv (c,d')$.

(S) For any five distinct points $a,b,c,d,e \in P$ with $\{a,b,c\}$ non collinear $(a,b) \equiv (a,c)$, $d \in \overline{a,b}$, $e \in \overline{a,c}$ and $\{a \perp \overline{b,c}\} = \{a \perp \overline{e,d}\}$, then $(a,d) \equiv (a,e)$.

For the set $\mathfrak{L}$ of lines we can define two relations of parallelism $\|_1$ and $\|_2$:

1. $A \|_1 B \Leftrightarrow A = B$ or $A \cap B = \emptyset$ and $\dim \overline{A \cup B} = 2$
2. $A \|_2 B \Leftrightarrow \forall X \in \mathfrak{L}$ with $X \perp A$ and $X \cap B \neq \emptyset$ we have $X \perp B$.

By (P1), (P2) and (T) we mean the statements:

(P1) (<u>Parallelaxiom</u>) For any $p \in P$ and $L \in \mathfrak{L}$ there is exactly one $M \in \mathfrak{L}$, designated by $\{p \| L\} := M$, with $p \in M$ and $M \| L$.

(P2) For any $a,b,c \in P$ with $b \neq a,c$ it follows that $\{a \| \overline{b,c}\} \cap \{c \| \overline{a,b}\} \neq \emptyset$.

(T) For any four distinct points $a,b,c,d \in P$ with $\overline{a,b} \| \overline{c,d}$ and $\overline{a,b} \neq \overline{c,d}$ we have $(a,b) \equiv (c,d) \Leftrightarrow \overline{a,c} \| \overline{b,d}$ or $\overline{a,d} \| \overline{b,c}$.

An absolute geometry is an <u>euclidean geometry</u> if $\|_1 = \|$ fulfils the parallelaxiom (P1). In a euclidean geometry one can prove the theorem:

(R) Let $A,B,C,D \in \mathfrak{L}$ be four lines of a plane with $A \perp B \perp C \perp D$. Then $A \perp D$.

At the end of the last century the following question arose. Can one replace the parallelaxiom (P1) by the condition (R)? Assuming the order relation α fulfils the archimedean axiom, then (P1) is a consequence of (R) (cp.[6], p. 130). On the other hand M. Dehn [1] showed that there are examples of absolute planes with a nonarchimedean order relation fulfilling (R) but not (P1). An absolute geometry shall be called <u>Dehn geometry</u> if (R) holds. Then the following theorems can be proved:

(1.1) <u>In a euclidean geometry</u>, $\|_1 = \|_2$ <u>and</u> $\|_1$ <u>is an equivalence relation</u> (cp.[2]) <u>satisfying</u> (P2) <u>and</u> (T).

(1.2) <u>In a Dehn geometry</u>, $\|_2 \subset \|_1$, <u>and</u> $\|_2$ <u>is an equivalence relation satisfying</u> (P1), <u>and</u> (P2), <u>and</u> (T).

And in a Dehn geometry the property (S) can be expressed for $\| := \|_2$ in the form:

(S') $\forall\, a,b,c,d,e \in P$ with $\{a,b,c\}$ non collinear, $(a,b) \equiv (a,c)$, $d \in \overline{a,b}$, $e \in \overline{a,c}$ and $\overline{b,c} \parallel \overline{e,d}$ we have $(a,d) \equiv (a,e)$.

§ 2 Parallelogram spaces with congruence

A triple $(P,\mathfrak{L},\parallel)$ will be called <u>parallelogram space</u> or shortly p-space if $(P,\mathfrak{L})$ is an incidence space (defined by (I1), (I2)) with $|\mathfrak{L}| \geq 2$, and if $\parallel$ is an equivalence relation on $\mathfrak{L}$ such that the axioms (P1) and (P2) are valid.

<u>Remarks</u>. For any Dehn geometry, $(P,\mathfrak{L},\parallel_2)$ is a p-space, according to (1.2). Every p-space $(P,\mathfrak{L},\parallel)$ is a double space $(P,\mathfrak{L},\parallel_l,\parallel_r)$ with $\parallel_l := \parallel_r := \parallel$ in the sense of [4] and therefore also a parallelstructure in the sense of J. André, or a weak affine space, in the sense of E. Sperner, if all lines have the same cardinality.

An automorphism δ of a p-space $(P,\mathfrak{L},\parallel)$ is called <u>dilatation</u>, if for all $X \in \mathfrak{L}$ one has $\delta(X) \parallel X$; a dilatation τ is called <u>translation</u> if $\tau = id$ or if for all $x,y \in P$ one has $\tau(x) \neq x$ and $\overline{\tau(x),x} \parallel \overline{\tau(y),y}$. An involutory dilatation δ fixing a point a is called <u>reflection in the point</u> a.
As usual, one proves

(2.1) <u>The set Δ of all dilatations forms a group; a dilatation with two different fixed points is the identity; for two points</u> $a,b \in P$ <u>there is at most one translation τ with</u> $\tau(a) = b$.

<u>Problem</u>: Can a point have more then one reflection in the point a ?

In every p-space we can introduce the following operations:
Let $(P^3)'$ be the set of all triples of non collinear points, $o \in P$ a distinguished point and $(P^2)' := \{(a,b) \in P^2 : (a,b,o) \in (P^3)'\}$.
Then $\omega : (P^3)' \longrightarrow P; (a,b,c) \longrightarrow a\,b\,c := \{a \parallel \overline{b,c}\} \cap \{c \parallel \overline{b,a}\}$ is a ternary and $+ : (P^2)' \longrightarrow P; (a,b) \longrightarrow a+b := a\,o\,b$ a binary operation with $a\,b\,c = c\,b\,a$, resp. $a+b = b+a$ for $(a,b,c) \in (P^3)'$, resp. $(a,b) \in (P^2)'$.
A p-space is called <u>Fanoean</u>, if for all $(a,b,c) \in (P^3)'$ one has $a\,b\,c = b\,a\,c$.

A quadruple $(P,\mathfrak{L},\parallel,\equiv)$ is called <u>p-space with congruence</u>, if $(P,\mathfrak{L},\parallel)$ is a p-space with $(P^3)' \neq \emptyset$ and $\equiv$ is a congruence relation on P^2

such that (T) is valid (a congruence relation $\equiv$ is an equivalence relation such that for $a,b,c \in P$ one has $(a,b) \equiv (b,a)$, and $(a,a) \equiv (b,c) \Leftrightarrow b = c$).

From now on let $(P, \mathfrak{L}, \parallel, \equiv)$ be a p-space with congruence.
An automorphism α of $(P, \mathfrak{L}, \parallel)$ is called *congruence preserving* or *motion*, if for all $a,b \in P$ one has $(\alpha(a), \alpha(b)) \equiv (a,b)$.

(2.2) *For* $a,b \in P$ *with* $a \neq b$ *one has* $|\{x \in \overline{a,b} : (a,x) \equiv (a,b)\}| \leq 2$, *and* $\{x \in \overline{a,b} : (a,x) \equiv (a,b)\} = \{b\}$ *if* $(P, \mathfrak{L}, \parallel)$ *is Fanoean.*

Proof. Let $a' \in P \setminus \overline{a,b}$ and $b' := a'ab$. Then $(a',b') \equiv (a,b)$ by (T). For $x \in \overline{a,b}$ with $(a,x) \equiv (a,b) \equiv (a',b')$ one has $\overline{x,a} = \overline{a,b} \parallel \overline{a',b'}$, hence $x = a'b'a$ or $x = b'a'a = b$ by (T).

A consequence of (2.2) is

(2.3) *To a point* a *there is at most one congruence preserving reflection in the point* a.

(2.4) *If there is a triple* $(a,b,c) \in (P^3)'$ *with* $abc = bac =: d$ *then* $(P, \mathfrak{L}, \parallel, \equiv)$ *is Fanoean.*

Proof. It is enough to show: For all $x \in P \setminus \overline{a,b}$ one has $xab = xba$. Let $y := xab$, $z := yab$ and $a' := byz$. Then one has $\overline{d,c} \parallel \overline{a,b} \parallel \overline{x,y} = \overline{y,z} \parallel \overline{a',b}$, and by (T), $(d,c) \equiv (a,b) \equiv (y,z) \equiv (a',b)$, and again by (T), $\overline{d,a'} \parallel \overline{c,b}$ or $\overline{d,b} \parallel \overline{a',c}$, i.e. $d = a'bc$ or $d = ba'c$. This implies $a' = a$ because $d = abc = bac$. Now $a = a' = byz$ implies $z = yba = x$ because $y = xab$. Again $x = z = yab$ gives us $y = xba = xab$.

(2.5) *Two distinct points* $a,b \in P$ *have at most one mid-point* m (i.e. $m \in \overline{a,b}$ and $(m,a) \equiv (m,b)$) *and none, if* $(P, \mathfrak{L}, \parallel)$ *is Fanoean.*

Proof. (2.1) implies that there isn't any mid-point if $(P, \mathfrak{L}, \parallel)$ is Fanoean. Therefore we assume that $(P, \mathfrak{L}, \parallel)$ is non-Fanoean and by (2.4) we know $abc \neq bac$ for all $(a,b,c) \in (P^3)'$. Let m be a mid-point of a,b. It is enough to show: For $c \in P \setminus \overline{a,b}$ and $d := acb$ one has $m \in \overline{c,d}$. For $e := cab$ and $f := cam$ it follows that $\overline{c,f} = \overline{c,e} \not\ni d$ by (2.4), $\overline{c,f} \parallel \overline{a,m} = \overline{m,b}$ and $(c,f) \equiv (a,m) \equiv (m,b)$. By (T) this implies $\overline{c,m} \parallel \overline{f,b}$ or $\overline{c,b} \parallel \overline{f,m}$. Since $\overline{f,m} \parallel \overline{c,a} \neq \overline{c,b}$ we get $\overline{c,m} \parallel \overline{f,b}$. Because of $d = acb$ and $f = cam$ we have $\overline{m,f} \parallel \overline{c,a} = \overline{d,b}$ and $(m,f) \equiv (c,a) \equiv (d,b)$, implying $\overline{m,d} \parallel \overline{f,b}$ or $\overline{m,b} \parallel \overline{f,d}$ by (T). Since $\overline{m,b} \parallel \overline{c,f} \not\ni d$, one gets $\overline{m,b} \not\parallel \overline{f,d}$.

But $\overline{m,d} \parallel \overline{f,b} \parallel \overline{c,m}$ implies $c \in \overline{m,d}$.

(2.6) *Any Fanoean p-space with congruence* $(P,\mathfrak{L},\parallel,\equiv)$ *is a prismatic space*, i.e. *for all* $(a,b,c) \in (P^3)'$, $a' \in P\setminus(\overline{a,b} \cup \overline{a,c})$, $b' := a'ab$, $c' := a'ac$ *one gets* $\overline{b,c} \parallel \overline{b',c'}$ (cp.[4]).

Proof. By definition and (T) we have $\overline{b',b} \parallel \overline{a,a'} \parallel \overline{c,c'}$ and $(b',b) \equiv (a,a') \equiv (c',c)$. With (T) this implies $\overline{b,c} \parallel \overline{b',c'}$ or $\overline{b,c'} \parallel \overline{b',c}$, i.e. $c' = cbb'$ or $c' = cb'b$. Since $(P,\mathfrak{L},\parallel,\equiv)$ is Fanoean we get $c' = cbb' = cb'b$ and therefore $\overline{b,c} \parallel \overline{b',c'}$.

Now we consider the non-Fanoean case.

(2.7) *To any point* $a \in P$ *there is exactly one congruence preserving reflection* $\tilde{a}$ *in the point* a *and for every* $x \in P\setminus\{a\}$ *and* $y \in P\setminus\overline{x,a}$ *one gets* $\tilde{a}(x) = (axy)ya$ *and* a *is the mid-point of* x *and* $\tilde{a}(x)$.

Proof. For $z := axy$ and $x' := (axy)ya$ one gets $\overline{a,x} \parallel \overline{y,z} \parallel \overline{a,x'}$; i.e. $x' \in \overline{a,x}$ and $(a,x) \equiv (y,z) \equiv (a,x')$ by (T). Now $x = x' = (axy)ya$ would imply $axy = xay$. Since $(P,\mathfrak{L},\parallel,\equiv)$ is non-Fanoean we have $x' \neq x$, and because of (2.2) x' does not depend of the choice of $y \in P\setminus\overline{x,a}$. Hence we have: To any point $x \in P\setminus\{a\}$ there is exactly one point x' such that a is the mid-point of x and x' and if there is a congruence preserving reflection $\tilde{a}$ in a (cp.(2.3)) then $\tilde{a}(x) = x'$. The map $': P \longrightarrow P;\ x \longrightarrow x'$ with $a' = a$ is involutory and fixes the lines through a. For $x,y \in P$ with $(x,y,a) \in (P^3)'$ and $z := axy$ one gets $x' = (axy)ya = zya = ayz$, $y' = (ayz)za = x'za$ implying $\overline{x,y} \parallel \overline{a,z} \parallel \overline{x',y'}$ and $(x,y) \equiv (z,a) \equiv (x',y')$.
Thus we have proved:

1. For all $X \in \mathfrak{L}$ one has $X' \in \mathfrak{L}$ and $X' \parallel X$.
2. For all $x,y \in P$ with $(x,y,a) \in (P^3)'$ or $x = a$ or $y = a$ one has $(x,y) \equiv (x',y')$.

Now let $x,y \in P\setminus\{a\}$ with $y \in \overline{x,a}$. For $u \in P\setminus\overline{x,a}$ and $v = xyu$ we obtain by (T), 2. and 1.: $(x,y) \equiv (u,v) \equiv (u',v') \equiv (x',y')$. Hence $\tilde{a} = {}'$ is the congruence preserving reflection in the point a.

(2.8) *For any* $a,b \in P$ *with* $a \neq b$ *the map* $\tau := \tilde{a}\tilde{b}$ *is a congruence preserving translation with the direction* $\overline{a,b}$.

Proof. By (2.7) and (2.1) τ is a congruence preserving dilatation. For $x \in P\setminus\overline{a,b}$, $x' := \tilde{b}(x)$, $x'' := \tilde{a}(x') = \tau(x)$ and $y := ax'b = bx'a$ we get $x'' = (ax'b)ba = yba$ and $x = \tilde{b}(x') = (bx'a)ab = yab$, hence

$\overline{x,y} \parallel \overline{a,b} \parallel \overline{x'',y}$ and therefore $\overline{x,\tau(x)} = \overline{x,x''} \parallel \overline{a,b}$. By (2.4), $x'' = y\,b\,a \neq y\,a\,b = x$. Now $x \in \overline{a,b}$ and $x = \tilde{a}\tilde{b}(x)$ implies $x' := \tilde{a}(x) = \tilde{b}(x)$, hence $a = b$ by (2.5).

(2.9) *To any* $a,b,c \in P$ *there is exactly one* $d \in P$ *with* $\tilde{a}\tilde{b}\tilde{c} = \tilde{d}$; *furthermore for* $(a,b,c) \in (P^3)'$, $d = a\,b\,c$.

Proof. Because of (2.7) we may assume $b \neq a,c$.
Case 1): $c \notin \overline{a,b}$. For $d := a\,b\,c$ and $d' := \tilde{a}\tilde{b}\tilde{c}(d)$ we get:
a) $\tilde{c}(d) \in \overline{c,d} = \{d \parallel \overline{a,b}\}$ by (2.7), hence $d' \in \{d \parallel \overline{a,b}\}$ by (2.8).
b) $\tilde{b}\tilde{c}(d) \in \{d \parallel \overline{b,c}\} = \overline{a,d}$, hence $d' \in \overline{a,d}$. a) and b) gives us $d = \tilde{a}\tilde{b}\tilde{c}(d)$. Since $\tilde{a}\tilde{b}\tilde{c}$ is a congruence preserving dilatation fixing d we obtain by (2.1),(2.2), and (2.7), $\tilde{a}\tilde{b}\tilde{c} = \tilde{d}$ or $\tilde{a}\tilde{b}\tilde{c} = \mathrm{id}$. The last equation implies $\tilde{a}\tilde{b} = \tilde{c}$, a contradiction to (2.8).
Case 2): $c \in \overline{a,b}$. For $x \in P\setminus\overline{a,b}$, $y := a\,b\,x$ and $d := y\,x\,c$ we get $\tilde{y} = \tilde{a}\tilde{b}\tilde{x}$ and $\tilde{d} = \tilde{y}\tilde{x}\tilde{c}$, because of case 1), hence $\tilde{a}\tilde{b}\tilde{c} = \tilde{a}\tilde{b}\tilde{x}\tilde{x}\tilde{c} = \tilde{y}\tilde{x}\tilde{c} = \tilde{d}$.

Since for $(a,b,c) \in (P^3)'$, $a\,b\,c = c\,b\,a$, so it follows from (2.8) and (2.9) that:

(2.10) $T := \{\tilde{a}\tilde{b} : a,b \in P\}$ *is a commutative group consisting of congruence preserving translations*.

(2.11) $(P,\mathfrak{L},\parallel)$ *is a prismatic space*.

Proof. For $(a,b,c) \in (P^3)'$, $a' \in P\setminus(\overline{a,b} \cup \overline{a,c})$, $b' := a'a\,b$ and $c' := a'a\,c$ it is true that (cp.(2.9)) $\tilde{b'} = \tilde{a'}\tilde{a}\tilde{b} = \tilde{b}\tilde{a}\tilde{a'}$ and $\tilde{c'} = \tilde{a'}\tilde{a}\tilde{c} = \tilde{c}\tilde{a}\tilde{a'}$, implying $\tilde{b}\tilde{c}\tilde{c'} = \tilde{b}\tilde{c}\tilde{c}\tilde{a}\tilde{a'} = \tilde{b'}$ i.e. $b' = b\,c\,c'$, thus $\overline{b',c'} \parallel \overline{b,c}$.

From (2.6),(2.11) and [4] we obtain the parts a),b),c) of the theorem:

(2.12) 1. *For any p-space* $(P,\mathfrak{L},\parallel,\equiv)$ *with congruence we have*:

a) *The operation* $+$ *can be extended on the whole set* P^2 *such that* $(P,\mathfrak{L},+)$ *becomes a commutative kinematic space* (i.e. $(P,+)$ is a commutative group, for every $a \in P$ the map $a^+ : P \to P;\ x \to a + x$ is an automorphism of $(P,\mathfrak{L})$ and every line $L \in \mathfrak{L}$ containing o is a subgroup of $(P,+)$).

b) *For* $A,B \in \mathfrak{L}$ *one has* $A \parallel B \Leftrightarrow A - A = B - B$.

c) *The set of translations of* $(P,\mathfrak{L},\parallel)$ *coincides with the set of maps* $\{a^+ : a \in P\}$.

d) *For any* $(a,b,c) \in (P^3)'$ *we have* $a\,b\,c = a - b + c$.

e) For any $a,b,c,d \in P$ with $a-b=c-d$ we get $(a,b)\equiv(c,d)$.

f) Every translation is a motion.

2. Let $(P,\mathfrak{L},+)$ be a commutative kinematic space, $\parallel$, resp., $\equiv$ defined by b), resp., e), then $(P,\mathfrak{L},\parallel,\equiv)$ is a p-space with congruence.

Proof. d): The translation $(c-b)^+$ maps b onto c and a onto $d := a-b+c$. Therefore $\overline{a,b}\parallel\overline{d,c}$ and $\overline{a,d}\parallel\overline{b,c}$, i.e. $d = a-b+c = a\,b\,c$.
e) For $c \notin \overline{a,b}$, we have $\overline{a,b}\parallel\overline{d,c}$ and $\overline{b,d}\parallel\overline{a,c}$, since $(d-b)^+(b)=d$, $(d-b)^+(a)=c$, hence $(a,b)\equiv(c,d)$ by (T). For $c\in\overline{a,b}$ and $x\notin\overline{a,b}$ we have $d=c-a+b\in\overline{a,b}$, $y:=x-a+b=x-c+d$, and therefor $(a,b)\equiv(x,y)\equiv(c,d)$.
f) is a consequence of e).

Supplements:

(2.13) Let $(P,\mathfrak{L},\parallel,\equiv)$ be a p-space with congruence and $(P,\mathfrak{L},+)$ the associated commutative kinematic space. The following statements are equivalent:

a) $\exists(a,b,c)\in(P^3)'$ with $a\,b\,c = b\,a\,c$.

b) $(P,\mathfrak{L},\parallel)$ is Fanoean.

c) $(P,+)$ is a group of exponent 2.

d) There aren't any congruence preserving reflections in the points.

Also, the following are equivalent:

e) $(P,\mathfrak{L},\parallel)$ is non-Fanoean.

f) $(P,+)$ hasn't any involutory elements.

g) For any point $a\in P$ there is a congruence preserving reflection $\tilde{a}$ in a, and for $(a,b,c)\in(P^3)'$, if $d=a\,b\,c$, then $\tilde{d}=\tilde{a}\,\tilde{b}\,\tilde{c}$.

§ 3 Quotients

At first we consider a non empty set M and a set Φ of maps of M in M such that any two maps of Φ commute. Let Δ be the semigroup generated by Φ and the identity 1, $\Delta_* := \{\delta\in\Delta : \delta \text{ injective}\}$ and Γ a semigroup of Δ_* with $1\in\Gamma$. For $a\in M$ let $\Delta(a) := \{x\in M : \exists\,\alpha,\xi\in\Delta : \alpha(a)=\xi(x)\}$. Then we get:

(3.1) $\{\Delta(x) : x\in M\}$ is a decomposition of M in pair-wise disjoint classes and for $a\in M$, $\alpha\in\Delta$ we have $\alpha(\Delta(a))\subset\Delta(a)=\Delta(\alpha(a))$ and

$\alpha(\Gamma(a)) \subset \Gamma(\alpha(a))$.

(3.2) Let $a \in M$ and $\alpha, \beta \in \Delta$ with $\alpha(a) = \beta(a)$. Then the restrictions $\alpha|\Gamma(a)$ and $\beta|\Gamma(a)$ are equal.

Proof. For $x \in \Gamma(a)$ there are $\gamma, \delta \in \Gamma$ with $\gamma(x) = \delta(a)$. This gives us $\gamma\alpha(x) = \alpha\gamma(x) = \alpha\gamma(a) = \delta\alpha(a) = \delta\beta(a) = \gamma\beta(x)$ hence $\alpha(x) = \beta(x)$ because γ is injective.

We can enlarge the set M and the semigroup Δ by forming quotients: $(a,\alpha) \sim (b,\beta) \Leftrightarrow \alpha(b) = \beta(a)$, resp., $(\gamma,\alpha) \sim (\delta,\beta) \Leftrightarrow \alpha\delta = \beta\gamma$, defines an equivalence relation on $M \times \Gamma$, resp., $\Delta \times \Gamma$. The equivalence class determined by (a,α), resp., (γ,α) shall be designated by $\frac{a}{\alpha}$, resp., $\frac{\gamma}{\alpha}$ and the set of equivalence classes by $M_\Gamma := \{\frac{a}{\alpha} : a \in M, \alpha \in \Gamma\}$, resp., $\Delta_\Gamma := \{\frac{\gamma}{\alpha} : \gamma \in \Delta, \alpha \in \Gamma\}$. Since the maps $M \longrightarrow M_\Gamma;\ a \longrightarrow \frac{a}{1}$, and $\Delta \longrightarrow \Delta_\Gamma;\ \alpha \longrightarrow \frac{\alpha}{1}$ are injective, we consider M as subset of M_Γ and Δ as subset of Δ_Γ. With regard to the multiplication $\Delta_\Gamma \times \Delta_\Gamma \longrightarrow \Delta_\Gamma$; $(\frac{\alpha}{\beta}, \frac{\gamma}{\delta}) \longrightarrow \frac{\alpha\gamma}{\beta\delta}$, the quotient set Δ_Γ becomes a commutative semigroup. By defining the external operation $\Delta_\Gamma \times M_\Gamma \longrightarrow M_\Gamma$; $(\frac{\alpha}{\beta}, \frac{a}{\gamma}) \longrightarrow \frac{\alpha(a)}{\beta\cdot\gamma}$, the semigroup Δ_Γ can be considered as a set of maps of M_Γ in M_Γ. The set $\Gamma_\Gamma := \{\frac{\gamma}{\alpha} : \gamma, \alpha \in \Gamma\}$ forms a subgroup of Δ_Γ, and Γ_Γ consists of permutations of M_Γ.

(3.3) a) $\forall \frac{a}{\gamma} \in M_\Gamma \quad \exists b \in M: \ \Delta_\Gamma(\frac{a}{\gamma}) = \Delta_\Gamma(b) \supset \Delta(b)$.

b) For $a, b \in M$ with $\Delta(a) \cap \Delta(b) = \emptyset$ we have $\Delta_\Gamma(a) \cap \Delta_\Gamma(b) = \emptyset$.

c) If $\Delta = \Delta_*$ and $\Delta(a) = M$ for an $a \in M$ then Δ_Δ operates regularly on M_Δ.

Now we consider the special case that $M = (P,+)$ is a commutative group, with $|P| \geqq 2$, and Φ a set of monomorphisms of $(P,+)$, such that any two monomorphisms of Φ commute. Let Δ be the semigroup generated by Φ, the identity 1, and the monomorphism $-1 : P \longrightarrow P$; $x \longrightarrow -x$, and let $<\Delta>_+$ be the subring of the endomorphism ring of $(P,+)$ generated by Δ. For $A \subset P$ let $A^* := A \setminus \{0\}$ and let $<\Delta>_+^* := <\Delta>_+ \setminus \{0\}$.

(3.4) Suppose that for every $a \in P^*$ the set $[a] := \Delta(a) \cup \{0\}$ is a subgroup of $(P,+)$. Then we have:

a) $\{[a] : a \in P^*\}$ is a partition of $(P,+)$, and hence, $(P,+,\circledcirc)$, with

$\mathfrak{G} := \{a + [b] : a \in P, b \in P^*\}$, is a commutative kinematic space.

b) For $a + [b] \in \mathfrak{G}$ and $\alpha \in <\Delta>_+$ we have $\alpha(a + [b]) = \alpha(a) + \alpha([b]) \subset \alpha(a) + [b]$.

c) For $\alpha, \beta \in <\Delta>_+$ and $a \in P^*$, with $\alpha(a) = \beta(a)$, we have $\alpha = \beta$.

d) Every $\alpha \in <\Delta>_+^*$ is a monomorphism.

e) Every $\alpha \in <\Delta>\setminus\{1\}$ hasn't any fixed point in P^*.

f) For every $a \in P$ we have $\Delta(a) = <\Delta>_+^*(a)$.

g) $<\Delta>_+$ is a commutative integral domain.

h) $\hat{P} := P_{<\Delta>_+^*}$ is a commutative group with respect to the addition

$\frac{a}{\alpha} + \frac{b}{\beta} := \frac{\beta(a) + \alpha(b)}{\alpha \cdot \beta}$, and $(\hat{P}, K)$ with $K := {}^{<\Delta>_+}{}_{<\Delta>_+^*}$, is a vector space over the commutative field K. To every $\frac{a}{\alpha} \in P^*$ there is an element $b \in P$ such that $K \cdot \frac{a}{\alpha} = K\,b$.

Proof. a) and b) follow from our assumption together with (3.1) and the fact that $<\Delta>_+$ consists of endomorphisms of $(P,+)$.
c) From (3.2) we get $\alpha|\Delta(a) = \beta|\Delta(a)$. If $\Delta(a) = P^*$ then $\alpha = \beta$. Suppose $\Delta(a) \neq P^*$. For $x \in P^*\setminus\Delta(a)$ we have $\{x\} = [x] \cap (a + [x - a])$ by a), hence $\alpha(x), \beta(x) \in [x] \cap (\alpha(a) + [x - a]) = [x] \cap (\beta(a) + [x - a])$ by b). This gives us $\alpha(x) = \beta(x)$ because $|[x] \cap (\alpha(a) + [x - a])| \leq 1$ by a). Hence $\alpha = \beta$. d) and e) are consequences of c) because $0,1 \in <\Delta>_+$, and f) and g) follow from d). We get h) from g) and (3.3).

The above results shall be applied on commutative kinematic spaces $(P, \mathfrak{L}, +)$ with $|\mathfrak{L}| \geq 2$. Let $\mathfrak{R} := \{X \in \mathfrak{L} : 0 \in X\}$ be the associated partition of subgroups of $(P,+)$, $\Delta_o := \{\sigma \in \mathrm{Aut}(P, \mathfrak{L}, +) : \forall X \in \mathfrak{R}, \sigma(X) = X\}$, $R := <\Delta_o>_+$ the subring of the endomorphism ring of $(P,+)$ generated by Δ_o and $\Delta := Z(R)\setminus\{0\}$, where $Z(R)$ designates the center of R. Then one has:

(3.5) R is an integral domain and (P,R) is a unitary R-Module with the properties:

a) $\forall \sigma \in R$, $\forall X \in \mathfrak{R}$, $\forall a \in P$, we have $\sigma(a + X) \subset \sigma(a) + X$.

b) Δ_o is equal to the set of units of R, and is also equal to the set of dilatations fixing 0.

c) The semigroup Δ consists of monomorphisms of $(P,+)$ and $<\Delta>_+ = \Delta \cup \{0\} = Z(R)$.

d) For every $a \in P^*$ the set $\Delta(a) \cup \{0\}$ is a subgroup of $\overline{0,a}$.

Proof. Since Δ_o is a group and $(-1) \in \Delta_o$ we have $R = \{\sigma_1 + \dots + \sigma_n : \sigma_i \in \Delta_o, n \in \mathbb{N}\}$. For $X \in \mathfrak{K}$ and $\sigma = \sigma_1 + \dots + \sigma_n \in R$, with $\sigma_i \in \Delta_o$, we obtain

$\sigma(X) \subset \sigma_1(X) + \sigma_2(X) + \dots + \sigma_n(X) = X + X + \dots + X = X$, and that is why $\sigma(a + X) = \sigma(a) + \sigma(X) \subset \sigma(a) + X$. Let $\sigma \in R$ and $a \in P^*$ with $\sigma(a) = 0$. For $A := \overline{0,a}$ $x \in P \setminus A$, $X := \overline{0,x}$ and $b := \overline{0,(x-a)}$ we get $x = X \cap (a + B)$, $0 = X \cap B$, and therefore, $\sigma(x) \in \sigma(X) \cap \sigma(a + B) \subset X \cap (\sigma(a) + B) = X \cap B = \{0\}$. Hence R is an integral domain, because $|\mathfrak{L}| \geqq 2$. a) tells us that Δ_o consists of dilatations fixing 0. On the other hand every dilatation preserves parallelograms. Therefore any dilatation fixing 0 is an automorphism of $(P,+)$, hence an element of Δ_o. Any unit σ of R is a permutation of P and thus a dilatation by a). With this we have now proved b). c) is valid because $\Delta \subset R^*$.

d) For $x_1, x_2 \in \Delta(a)$ there are $\alpha_i, \xi_i \in \Delta$ with $\alpha_i(a) = \xi_i(x_i)$. Since $(-1) \in \Delta$ we have $\xi_i(-1) \in \Delta$ and $\alpha_i(a) = \xi_i(-1)(-x_i)$, hence $-x_i \in \Delta(a)$. Further $\xi_1\xi_2(x_1 + x_2) = \xi_2\alpha_1(a) + \xi_1\alpha_2(a) = (\xi_2\alpha_1 + \xi_1\alpha_2)(a)$, hence $x_1 + x_2 \in \Delta(a) \cup \{0\}$, because $\xi_1\xi_2, \xi_2\alpha_1 + \xi_1\alpha_2 \in Z(R)$. From a) we get $\Delta(a) \cup \{0\} \subset \overline{0,a}$.

Because of (3.4)h) and (3.5) we get the theorem:

(3.6) Every commutative kinematic space $(P,+,\mathfrak{L})$, with $|\mathfrak{L}| \geqq 2$, can be embedded in a vector space $(\hat{P},K) := (P_\Delta, Z(R)_\Delta)$ over the commutative field K such that we have:

a) For $A \in \mathfrak{K}$ the set $\hat{A} := \{\frac{a}{\varkappa} : a \in A, \varkappa \in \Delta\}$ is a vector subspace of $(\hat{P},K)$, with $A \leqq \hat{A}$, and is an $\hat{R}$-submodule for $\hat{R} := R_\Delta$.

b) For $A, B \in \mathfrak{K}$, with $A \neq B$, one has $\hat{A} \cap \hat{B} = \{0\}$.

c) $\hat{\mathfrak{K}} := \{\hat{A} : A \in \mathfrak{K}\}$ is a partition consisting of vector subspace of $(\hat{P},K)$, and $(\hat{P},+,\hat{\mathfrak{L}})$, with $\hat{\mathfrak{L}} := \{a + \hat{X} : a \in \hat{P}, X \in \mathfrak{K}\}$ is a commutative kinematic space.

d) The map $\mathfrak{L} \longrightarrow \hat{\mathfrak{L}};\ a + X \longrightarrow \frac{a}{1} + \hat{X}$ is an injection.

e) $\dim_K(\hat{A}) = 1 \Leftrightarrow \forall x,y \in A^* \ \exists \sigma, \tau \in \Delta : \sigma(x) = \tau(y)$.

§ 4 Rectangular spaces

A p-space with congruence $(P,\mathfrak{L},\equiv,\|)$ shall be called a <u>rectangular space</u>, or shortly <u>r-space</u>, if also the axioms (W),(S'), and the following axiom (E) are valid:

(E) In any plane E there are at least two rhombi $(a_i,b_i,c_i,d_i)\in E^4$ with $\overline{a_1,c_1}\neq\overline{a_2,c_2},\overline{b_2,d_2}$ (a rhombus (a,b,c,d) is a quadruple consisting of four distinct non-collinear points with $(a,b)\equiv(b,c)\equiv(c,d)\equiv(d,a)$).

The foundation of r-spaces has been accomplished only to some extend:

(4.1) <u>Suppose</u> $(P,\mathfrak{L},\equiv,\|)$ <u>is an r-space with</u> $\|=\|_1$, i.e. $(P,\mathfrak{L})$ is <u>an affine space</u> (cp.[6]). <u>Then we have</u>:

a) 1. <u>If</u> $(P,\mathfrak{L})$ <u>is an affine plane</u> ($(P,\mathfrak{L},\equiv)$ <u>is then called a euclidean plane</u>, <u>and</u> $A\parallel B$ <u>means here</u> $|A\cap B|\neq 1$), <u>then there is a separable quadratic field extension</u> (L,K) <u>with</u>

$$(*)\quad L=P,\quad \mathfrak{L}=\{a+Kb : a,b\in L,\ b\neq 0\},\ \text{and}$$
$$(a,b)\equiv(c,d) \Leftrightarrow (a-b)(\overline{a}-\overline{b})=(c-d)(\overline{c}-\overline{d}),$$

<u>whereby</u> $x\longrightarrow\overline{x}$ <u>designates the involutory field automorphism of</u> L <u>fixing exactly every element of</u> K.

2. <u>If</u> (L,K) <u>is a separable quadratic field extension then</u> $(P,\mathfrak{L},\equiv)$, <u>defined by</u> $(*)$, <u>is a euclidean plane</u> ([3],[7]).

b) 1. <u>If</u> $\dim(P,\mathfrak{L})\geqq 3$ ($(P,\mathfrak{L}\equiv)$ is then called euclidean space), <u>then there is a vector space</u> (V,K) <u>over a commutative field</u> K <u>of</u> $\mathrm{Char}(K)\neq 2$ <u>and a definite quadratic form</u> $Q:V\longrightarrow K$ (i.e. $Q(x)=0$ implies $x=0$) <u>such that</u>

$$(**)\quad P=V,\quad \mathfrak{L}=\{a+Kb : a,b\in V,\ b\neq 0\},\ \text{and}$$
$$(a,b)\equiv(c,d)\Leftrightarrow Q(a-b)=Q(c-d).$$

2. <u>If</u> (V,K) <u>is a vector space over a commutative field</u> K <u>of</u> $\mathrm{Char}(K)\neq 2$ <u>with</u> $\dim(V,K)\geq 3$, <u>and if</u> $Q:V\longrightarrow K$ <u>is a definite quadratic form, then</u> $(P,\mathfrak{L},\equiv)$, <u>defined by</u> $(**)$, <u>is a euclidean space</u> ([9],[2],[10]).

<u>Remarks</u>: For the definition of a euclidean space $(P,\mathfrak{L},\equiv)$ it is enough to claim:

1. $(P,\mathfrak{L})$ is a pseudo-affine space; i.e. every plane of $(P,\mathfrak{L})$ is an affine plane (cp.[10]).

2. For $\|:=\|_1$ and $\equiv$ the axioms (T),(W),(S') and (E) are valid ([2],[10]).

(4.2) 1. Suppose $(P,\mathfrak{L},+)$ is a commutative kinematic space and suppose to each $A \in \mathfrak{R} := \{X \in \mathfrak{L} : 0 \in X\}$ there is exactly one involutory automorphism $\widetilde{A}$ of $(P,\mathfrak{L},+)$ such that the two axioms (A1),(A2) are valid:

(A1) For all $A \in \mathfrak{R}$, $x \in P$, we get $\widetilde{A}(x) = x \Leftrightarrow x \in A$;

(A2) For all $A,B,C \in \mathfrak{R}$ there is a $D \in \mathfrak{R}$ with $\widetilde{A}\widetilde{B}\widetilde{C} = \widetilde{D}$.

(Structures $(P,\mathfrak{L},+,\widetilde{\mathfrak{R}})$, with these properties, are called rectangular planes). Then there is a separable quadratic field extension (L,K), such that the two following conditions are valid:

a) $1 \in P$ and $(P,+) \leqq (L,+)$.

b) For every $x \in L$ with $x\overline{x} = 1$ we have $x \cdot P \subset P$ (here $\overline{x}$ designates the images of x by applying the K-automorphism of $(L,+,\cdot)$ distinct from the identity).

2. Suppose (L,K) is a separable quadratic field extension, and P is a subset of L, such that the conditions a) and b) are valid. Let $\mathfrak{L} := \{(a+Kb) \cap P : a,b \in P, b \neq 0\}$ and for $A := Ka \cap P$ with $a \in P^*$, let $\widetilde{A} : P \longrightarrow P;\ x \longrightarrow \frac{a}{\overline{a}} \cdot \overline{x}$. Then $(P,\mathfrak{L},+,\widetilde{\mathfrak{R}})$ is a rectangular plane ([8]).

Remarks. 1. Part 1. of this theorem can be proved by applying (3.4). As it was done in [3] one has to show that the set $\Phi_1 = \{\widetilde{A}\widetilde{B} : A,B \in \mathfrak{R}\}$ forms a commutative group consisting of automorphisms of $(P,+)$, and that every $\alpha \in \Phi_1$, with $\alpha \neq 1$, fixes only 0. Therefore Φ_1 and $\Phi_2 := \{\alpha - 1 : \alpha \in \Phi_1 \setminus \{1\}\}$ generate a commutative semigroup Δ consisting of monomorphisms of $(P,+)$. Then one has to prove that $\Delta(a) = P^*$ for $a \in P^*$. That means that the assuptions of (3.4) are valid, and by (3.4)f) and h) one sees that $(\hat{P},L) := (P_{<\Delta>_+^*}, <\Delta>_+ {<\Delta>_+^*})$ is a one dimensional vector space. Now let $e \in P^*$ and $\alpha = \widetilde{\overline{0,e}}$ the involutory automorphism of $(P,\mathfrak{L},+,\widetilde{\mathfrak{R}})$ at the line $\overline{0,e}$. Then $\hat{P}$ and L can be identified with the bijection $\iota : L \longrightarrow \hat{P};\ x \longrightarrow x \cdot e$, and α can be extended to an involutory automorphism of $(L,+,\cdot)$.

2. To get examples of rectangular planes one has to start from a separable quadratic field extension (L,K). If the subring $<L_1>_+$ generated by $L_1 := \{x \in L : x\overline{x} = 1\}$ is equal L, one gets by theorem (4.2) a euclidean plane; otherwise, one gets a proper rectangular plane.

Recently R. Stanik [11] proved the theorem:

(4.3) Suppose $(P,\mathfrak{L},+,\widetilde{\mathfrak{R}})$ is a rectangular plane and (L,K) is the associated separable quadratic field extension according to (4.2).

a) Let α be an order relation such that $(P,\mathfrak{L},\alpha)$ is an ordered plane in the sense of [6] p.82,83. Then $\{z \in L : z \cdot P \subset P\}$ is a valuation ring and there is an ordering $\leqslant$ of K such that $(K,+,\cdot,\leqslant)$ is an ordered field, and for each $\lambda \in K$, with $0 \leqslant \lambda \leqslant 1$, one has $\lambda \cdot P \subset P$.

b) If $(K,+,\cdot,\leqslant)$ is an ordered field and if for each $\lambda \in K$, with $0 \leqslant \lambda \leqslant 1$, one has $\lambda \cdot P \subset P$ then there is an order relation α such that $(P,\mathfrak{L},\alpha)$ is an ordered plane.

References

[1] DEHN,M.: Die LEGENDRE'schen Sätze über die Winkelsumme im Dreieck. Math. Ann. 53 (1900) 404-439

[2] KARZEL,H.: Zur Begründung euklidischer Räume.To appear in Mitt. der Math. Gesellsch. in Hamburg

[3] —, and G. KIST: Zur Begründung metrisch-affiner Ebenen. Abh. Math. Sem. Univ. Hamburg, to appear

[4] —, KROLL,H.-J.; SÖRENSEN,K.: Invariante Gruppenpartitionen und Doppelräume. Journal für reine und angew. Mathematik, Bd. 262/263 (1973) 153-157

[5] —, and K. SÖRENSEN: Rectangular and Pseudorectangular Planes and Their Representation by v-local Systems. To appear in Proceedings of a Conference on Geometry in Silivri (Turkey) 1978

[6] —, SÖRENSEN,K; WINDELBERG,D.: Einführung in die Geometrie. Göttingen 1973

[7] —, and R. STANIK: Metrische affine Ebenen. Abh. Math. Sem. Univ. Hamburg, to appear

[8] —, and R. STANIK: Rechtseitebenen und ihre Darstellung durch Integritätssysteme. To appear in Mitt. der Math. Gesellsch. in Hamburg

[9] KROLL,H.-J. and K. SÖRENSEN: Pseudo-euklidische Ebenen und euklidische Räume. J. of Geometry, Vol.8, 1/2 (1976) 95-115

[10] SÖRENSEN,K.: Euklidische Räume der Ordnung 3. To appear in Mitt. der Math. Gesellsch. in Hamburg

[11] STANIK,R.: Anordnung in Rechtseitebenen und Integritätssystemen. To appear

ON SHARPLY 2-TRANSITIVE PERMUTATION SETS

Harold Kühlbrandt

(M,Γ) is called a *permutation set* if M is a set and Γ a set of permutations of M.
A permutation set (M,Γ) is called *sharply n-transitive* if $n \in \mathbb{N}$ and for all $x_1,\dots,x_n, y_1,\dots,y_n \in M$ with $|\{x_1,\dots,x_n\}| = |\{y_1,\dots,y_n\}| = n$ there is exactly one $\gamma \in \Gamma$ with $\gamma(x_i) = y_i$ $\forall i \in \{1,\dots,n\}$.
A permutation set (M,Γ) is called a *permutation group* if Γ is a subgroup of the symmetric group of M.

Let us begin with some results on sharply n-transitive permutation sets and groups in the case $n=2$ or $n=3$.

There is a one-to-one correspondence between

R 1) sharply 2-transitive permutation sets and 2-structures ([5]),

R 2) sharply 2-transitive permutation groups and
(i) rectangular 2-structures ([5]),
(ii) near-domains ([4]),

R 3) sharply 3-transitive permutation sets and hyperbola-structures ([2],[11]),

R 4) sharply 3-transitive permutation groups and
(i) rectangular hyperbola-structures ([11]),
(ii) KT-fields ([7]),

R 5) symmetric sharply 3-transitive permutation sets (M,Γ) and ovoidal Minkowsky planes. In this case the set Γ is isomorphic to the group PGL(2,K) where K is a commutative field ([1], [6]).

Here a permutation set (M,Γ) is called *symmetric* if

(ΓS) For any two $\alpha,\beta \in \Gamma$ the existence of an $x \in M$ with $\alpha(x) \neq \beta(x)$ and $\alpha^{-1}\beta(x) = \beta^{-1}\alpha(x)$ implies $a^{-1}\beta = \beta^{-1}\alpha$.

The symmetry axiom (ΓS) for sharply 3-transitive permutation sets (M,Γ) has been introduced by W. BENZ [2].

Result R5) shows that it may be interesting to study permutation sets which are not necessarily groups. In this connection we are interested in the following axioms on a permutation set (M,Γ).

(ΓR) $\Gamma \neq \emptyset$ and $(\alpha, \beta \in \Gamma \Rightarrow \alpha\beta^{-1} \in \Gamma)$

(ΓR^*) $id \in \Gamma$ and $(\alpha, \beta \in \Gamma \Rightarrow \alpha\beta^{-1}\alpha \in \Gamma)$

(ΓS^*) Every $\alpha \in \Gamma$ which interchanges two distinct elements of M is an involution (i.e. $\alpha^2 = id \neq \alpha$).

Obviously (ΓR) implies (ΓR^*) for every permutation set (M,Γ) and from R5) we get $(\Gamma S) \Rightarrow (\Gamma R)$ for every sharply 3-transitive permutation set (M,Γ). Therefore the question arises whether $(\Gamma S) \Rightarrow (\Gamma R)$ is true for sharply 2-transitive permutation sets or not and in [5] H. KARZEL asked whether $(\Gamma R^*) \Leftrightarrow (\Gamma R)$ holds for sharply 2-transitive sets. These questions are answered (among others) by the following

Theorem 1. Let (M,Γ) be a sharply 2-transitive set. Then

a)
$$\begin{array}{ccccc} (\Gamma R) & \Rightarrow & (\Gamma R^*) & \Rightarrow & (\Gamma S) \\ & & \Downarrow & & \Downarrow \\ & & \Gamma = \Gamma^{-1} & \Rightarrow & (\Gamma S^*) \end{array}$$

b) None of these implications is reversible and neither $(\Gamma S) \Rightarrow \Gamma = \Gamma^{-1}$ nor $\Gamma = \Gamma^{-1} \Rightarrow (\Gamma S)$ is true.

Remarks: α) It is not known whether $(\Gamma R^*) \Rightarrow (\Gamma R)$ is true for sharply 3-transitive permutation sets or not.

β) From result R1) follows that each of the axioms (ΓR), (ΓR^*), (ΓS), $\Gamma = \Gamma^{-1}$, and (ΓS^*) corresponds to a configuration theorem in the associated 2-structure. Theorem 1 shows the logical interdependence of these configuration theorems (see [10]).

Theorem 1b can be proved by establishing suitable algebraic models of sharply 2-transitive permutation sets.

The following theorem is helpful to construct such models. In this theorem the characterization of sharply 2-transitive permutation groups by near-domains (see result R2) (ii)) is generalized to an algebraic representation of sharply 2-transitive permutation sets which satisfy (ΓS^*) and $\mathfrak{J}\Gamma \cup \Gamma\mathfrak{J} \subset \Gamma$ (where $\mathfrak{J} := \{\omega \in \Gamma;\ \omega^2 = id \neq \omega\}$) by quasi-domains.

$(Q,+,\cdot)$ is called a <u>quasi-domain</u>, if

$(Q,+)$ is a loop with neutral element 0,

$a+b=0 \Rightarrow b+a=0 \quad \forall a,b \in Q$,

$(Q^*,\cdot)$ (where $Q^* := Q\setminus\{0\}$) is a loop with neutral element 1,

$a(b+c) = ab+ac \quad \forall a,b,c \in Q$,

$0\cdot a = 0 \quad \forall a \in Q$,

for any two $a,b \in Q$ there is a $d_{a,b} \in Q$ with

a) $a+(b+x) = (a+b)+d_{a,b}x \quad \forall x \in Q$

b) $c(d_{a,b}x) = (cd_{a,b})x \quad \forall c,x \in Q$.

We will be interested in the following additional properties of a quasi-domain $(Q,+,\cdot)$:

(Q10) $(-a)b = -ab \quad \forall a,b \in Q$

(Q12) $ab = 1 \Rightarrow a(bx) = x \quad \forall a,b,x \in Q$

(Q13) $a(b(ax)) = (a(ba))x \quad \forall a,b,x \in Q$ ("BOL-identity").

A quasi-domain $(Q,+,\cdot)$ is a
- <u>quasi-field</u> iff $(Q,+)$ is a group
- <u>near-domain</u> iff $(Q^*,\cdot)$ is a group
- <u>near-field</u> iff $(Q,+)$ and $(Q^*,\cdot)$ are groups.

<u>Theorem 2</u>. a) Let (M,Γ) be a sharply 2-transitive permutation set with

$(*)$ (ΓS^*) and $\mathfrak{J}\Gamma \cup \Gamma\mathfrak{J} \subset \Gamma$.

Then an addition $+$ and a multiplication $\cdot$ can be defined in M such that $(M,+,\cdot)$ is a quasi-domain with (Q10).
Furthermore Γ is the set of affine mappings
$\langle a,m\rangle : M \longrightarrow M,\ x \longrightarrow a+mx \quad (a,m\in M,\ m\neq 0)$ of $(M,+,\cdot)$.

b) Let $(Q,+,\cdot)$ be a quasi-domain and Γ the set of affine mappings of $(Q,+,\cdot)$. Then (Q,Γ) is a sharply 2-transitive permutation set and

b1) $(*) \Leftrightarrow$ (Q10)

b2) $\Gamma = \Gamma^{-1} \Leftrightarrow$ (Q12)

b3) $(\Gamma R^{*}) \Leftrightarrow$ (Q13)

b4) $(\Gamma R) \Leftrightarrow (Q^{*},\cdot)$ is a group $\Leftrightarrow (Q,+,\cdot)$ is a near-domain.

Remarks: α) b4) is a part of result R2)(ii).

β) The first statement of b) and b2) were proved by G. KIST [8].

γ) It can be shown that $(*)$ implies the symmetry axiom (ΓS) (see [10]) but it is not known whether $(*)$ and (ΓS) are equivalent for sharply 2-transitive permutation sets.

δ) It is not known whether there are quasi-domains which are not quasi-fields.

For the proofs of theorem 1 and 2 see [10].

References

[1] ARTZY,R.: A pascal theorem applied to Minkowsky Geometry. J. Geometry 3 (1973) 93-105

[2] BENZ,W.: Permutations and plane sections of a ruled quadric. In: Symposia Mathematica, Istituto Nazionale di Alta Matematica 5 (1970) 325-339

[3] HEISE,W. and H. KARZEL: Symmetrische Minkowski-Ebenen. J. Geometry 3 (1973) 5-20

[4] KARZEL,H.: Inzidenzgruppen. Lecture notes Universität Hamburg, 1965

[5] — Zusammenhänge zwischen Fastbereichen, scharf 2-fach transitiven Permutationsgruppen und 2-Strukturen mit Rechtecksaxiom. Abh. Math. Sem. Univ. Hamburg 32 (1968) 191-206

[6] KARZEL,H.: Symmetrische Permutationsmengen. Aequationes Mathematicae 17 (1978) 83-90

[7] KERBY,W. and H. WEFELSCHEID: Über eine scharf 3-fach transitiven Gruppen zugeordnete algebraische Struktur. Abh. Math. Sem. Univ. Hamburg 37 (1972) 225-235

[8] KIST,G.: Quasibereiche. To appear in : Beiträge zur Geometrie und Algebra, TUM-Berichte, TU München, Inst. f. Math.

[9] KÜHLBRANDT,H.: Automorphismen von 2-Strukturen. To appear in: Beiträge zur Geometrie und Algebra Nr.5, TUM-Berichte, TU München, Inst. f. Math.

[10] — Algebraisierung scharf 2-fach transitiver Permutationsmengen durch Quasibereiche. To appear in Aequationes Mathematicae

[11] SOPPA,R.: Scharf dreifach transitive Permutationsgruppen. Staatsexamensarbeit Hamburg 1969. (For a survey of the results see [3])

Harold Kühlbrandt
Institut für Mathematik
Technische Universität München

Postfach 20 24 20

D-8000 München 2

DIMENSION OF NEARAFFINE SPACES

Jürgen Misfeld and Helga Tecklenburg

Universität Hannover (Germany)

Nearaffine spaces were introduced by J. ANDRÉ (1975) in order to describe nearfield spaces. The operation of joining points is not commutative; in so far these structures differ from affine spaces. We give some illustrative examples, in which the Desargues-theorem is valid only for special configurations. Most of the known results are related to finite nearaffine spaces. ANDRÉ introduced the concepts of subspace and dimension only for finite nearaffine spaces. In our paper we define subspaces and the notion of dimension for the arbitrary case. A subspace is a subset U of the point set P with the property, that with two points the whole joining line is in U, and that two points are joinable by straight lines with respect to U. It is shown, that an arbitrary intersection of subspaces is a subspace, so there is given a closure operator on the subsets of P, which leads to the notion of dimension. Some further properties of the lattice of subspaces are studied.

§ 1. NEARAFFINE SPACES.

Let $F = (P, \longrightarrow, \|)$ be a structure consisting in a non-void point set P, a mapping

$$\longrightarrow : \begin{cases} P\times P \setminus \Delta_P \longrightarrow \mathcal{P}(P) \\ (x,y) \longrightarrow \overrightarrow{x,y} \end{cases}$$

, of which the elements $\overrightarrow{x,y} \in \mathcal{P}(P)$ are called *lines*, and an equivalence relation $\|$ on the set $L := \longrightarrow (P\times P \setminus \Delta_P)$ of all lines, called *parallelism*. F is called a *nearaffine space*, if the following axioms hold:

(L1) $x,y \in \overrightarrow{x,y}$ for every $(x,y) \in P\times P \setminus \Delta_P$.

(L2) From $z \in \overrightarrow{x,y} \setminus \{x\}$ follows $\overrightarrow{x,y} = \overrightarrow{x,z}$.

(P1) (Euclid's axiom of parallelism). To every line L and every $x \in P$ exists exactly one line $\overrightarrow{x,y}$ with $\overrightarrow{x,y} \| L$, which we denote by $(x \| L)$.

(P2) $\overrightarrow{x,y} \parallel \overrightarrow{y,x}$ for every $(x,y) \in P{\times}P \setminus \Delta_P$.

(T) (Tamaschke-condition). Let $x,y,z \in P$ be pairwise different and $x',y' \in P$ with $\overrightarrow{x,y} \parallel \overrightarrow{x',y'}$. Then $(x' \parallel \overrightarrow{x,z}) \cap (y' \parallel \overrightarrow{y,z}) \neq \emptyset$ holds.

(G1) From $\overrightarrow{x,y} = \overrightarrow{y,x} \neq \overrightarrow{x',y'}$ follows $|\overrightarrow{x,y} \cap \overrightarrow{x',y'}| \leq 1$.

(G2) (Chain condition). To every two points $x,y \in P$, $x \neq y$, there exist finitely many points $x_o, x_1, \ldots, x_n$ with $x_o := x$, $x_n := y$ and $\overrightarrow{x_{i-1},x_i} = \overrightarrow{x_i,x_{i-1}}$ for $i = 1,2,\ldots,n$.

(R) To every line there exist two points not lying on this line.

From now on we assume $F = (P, \rightarrow, \parallel)$ to be a nearaffine space. The line $\overrightarrow{x,y}$ is called *joining line* or *join from* x *to* y, and the point x is called *base point* of L; we denote this by $x \prec L$.

J. ANDRÉ proved the following theorems ([1; p. 69,73], [5; p. 3]):

(1.1) (Veblen condition). Let x,y,z be pairwise different points and $y' \in \overrightarrow{x,y}$. Then $\overrightarrow{x,z} \cap (y' \parallel \overrightarrow{y,z}) \neq \emptyset$ holds. □

(1.2) (Condition for closed parallelograms)
For every three points x,y,z we have $(z \parallel \overrightarrow{x,y}) \cap (y \parallel \overrightarrow{x,z}) \neq \emptyset$. □

(1.3) From $\overrightarrow{x,y} = \overrightarrow{y,x} = \overrightarrow{x,z}$ follows $\overrightarrow{x,z} = \overrightarrow{z,x}$. □

(1.4) From $\overrightarrow{x,y} = \overrightarrow{y,x}$ and $\overrightarrow{x,y} \parallel \overrightarrow{x',y'}$ follows $\overrightarrow{x',y'} = \overrightarrow{y',x'}$. □

(1.5) For every two points x,y the following statements are equivalent:
(1) $\overrightarrow{x,y}$ possesses at least two base points.
(2) Every point of $\overrightarrow{x,y}$ is base point.
(3) $\overrightarrow{x,y} = \overrightarrow{y,x}$. □

The lines from F, which have one of the properties given in (1.5), are called *straight lines*. In this case we write $\overline{x,y}$ instead of $\overrightarrow{x,y}$. By G we denote the set of all straight lines. The elements of $L \setminus G$ are called *proper lines*. The set G_v of all straight lines, which go through the point v, is called the *pencil of straight lines generated by* v or the *pencil of straight lines with center* v. For $U \subset P$ we define $L(U) := \{L \in L \mid L \subset U\}$, $G(U) := L(U) \cap G$ and $G(U)_v := G_v \cap G(U)$. We assume, that there exists a line L with $|L| \geq 2$; otherwise F is the affine space over GF(2).

§ 2. EXAMPLES OF NEARAFFINE SPACES.

Nearaffine spaces can be constructed in the same way as Desarguesian affine spaces over fields if one takes near vector spaces over nearfields (see [2; p. 59]). Further examples ANDRÉ has given in [1] and [3]. In the following we introduce two classes of examples, which are instructive and which have the property, that the Desarguesian theorem is valid only for special configurations. They show that a classification of nearaffine planes similar to the LENZ-BARLOTTI classification of projective planes leads to non-void classes. These problems will be discussed in a further note.

The points of lines of an affine space are exchangeable, while lines of nearaffine spaces contain special points, the base points. Therefore for the construction of lines in nearaffine spaces such curves of the usual real plane are suitable, which contain a distinguished point; such curves are par example parabolas. This motivates the following examples.

Take $P := \mathbb{R}^2$ as point set, and let be $n \in \mathbb{N}$. Let be $a,b \in P$, $a=(a_1,a_2)$, $b=(b_1,b_2)$. We define in the case

$$a_1 \neq b_1 : \quad d(a,b,n) := (b_2-a_2)\cdot|b_1-a_1|^{-n}\cdot \operatorname{sign}(b_1-a_1) \text{ and}$$

$$L(a,b,n) := \{(x_1,x_2) \in \mathbb{R}^2 \mid x_2-a_2 = d(a,b,n)\cdot|x_1-a_1|^n\cdot \operatorname{sign}(x_1-a_1)\}$$

and for

$$a_1 = b_1 : \quad L(a,b) := \{(x_1,x_2) \in \mathbb{R}^2 \mid x_1-a_1 = 0\}.$$

The sets $L(a,b,n)$ are called n-*lines*. We define a parallelism between lines as follows:

$L(a,b,n) \parallel L(c,d,m)$ iff $d(a,b,n) = d(c,d,m)$ and $n=m$ or $d(a,b,n) = d(c,d,m) = 0$ holds.

$L(a,b) \parallel L(c,d)$ for every $a,b,c,d \in P$.

We now define one class of nearaffine spaces. For points $a,b \in P$, $a \neq b$, we define joining lines $\overrightarrow{a,b}$ with base point a. On the set L of all lines we define a parallelism as given above. Let be $n,m \in \mathbb{N}$. Then define

$$\overrightarrow{a,b} := \begin{cases} L(a,b,n) & \text{iff } d(a,b,n) \geq 0 \\ L(a,b,m) & \text{iff } d(a,b,n) < 0 \\ L(a,b) & \text{iff } a_1=b_1. \end{cases}$$

In the case $n=m=1$ we have the usual euclidean affine plane, but in the case $n \neq 1$ or $m \neq 1$ we get proper nearaffine spaces. Straight lines are the 1-lines, the n- or m-lines with $d=0$, and the lines $L(a,b)$. The n-or m-lines with $n,m \neq 1$ and $d \neq 0$ are proper lines with base point a, namely "parabolas". In order to have examples, for which the Desarguesian theorem doesn't hold, we consider - similar to the classical Moulton-planes - broken parabolas.

For this purpose we define for $a,b \in P$, $k_1,k_2 \in \mathbb{R}_+$ with $0 < k_1,k_2 \leq 1$, $n \in \mathbb{N}$, in the case

$$a_1 \neq b_1:\ d(a,b,k_1,k_2,n) := (\beta_2 b_2 - \alpha_2 a_2)\cdot|\beta_1 b_1 - \alpha_1 a_1|^{-n}\cdot \mathrm{sign}(b_1-a_1) \text{ and}$$

$$L(a,b,k_1,k_2,n) := \{(x_1,x_2) \in \mathbb{R}^2 \mid x_2 - \delta_2\alpha_2 a_2 = \delta_2\cdot d(a,b,k_1,k_2,n)\cdot |\delta_1 x_1 - \alpha_1 a_1|^n \cdot \mathrm{sign}(x_1-a_1)\}$$

and in the case

$$a_1 = b_1:\ L(a,b) := \{(x_1,x_2) \in \mathbb{R}^2 \mid x_1 - a_1 = 0\},$$

with

$$\alpha_1 := \begin{cases} 1 & \text{iff } a_1 \leq 0 \\ k_1^{-1} & \text{iff } a_1 > 0 \end{cases} \qquad \alpha_2 := \begin{cases} 1 & \text{iff } a_2 \leq 0 \\ k_2 & \text{iff } a_2 > 0 \end{cases}$$

$$\beta_1 := \begin{cases} 1 & \text{iff } b_1 \leq 0 \\ k_1^{-1} & \text{iff } b_1 > 0 \end{cases} \qquad \beta_2 := \begin{cases} 1 & \text{iff } b_2 \leq 0 \\ k_2 & \text{iff } b_2 > 0 \end{cases}$$

$$\delta_1 := \begin{cases} 1 & \text{iff } x_1 \leq 0 \\ k_1^{-1} & \text{iff } x_1 > 0 \end{cases} \qquad \delta_2 := \begin{cases} 1 & \text{iff } x_2 \leq 0 \\ k_2^{-1} & \text{iff } x_2 > 0 \,. \end{cases}$$

We assume $k_i = 1$ for at least one $i = 1,2$. Let be $n,m \in \mathbb{N}$.
As above we define

$$\overrightarrow{a,b} := \begin{cases} L(a,b,k_1,k_2,n) & \text{iff } d(a,b,k_1,k_2,n) \geq 0 \\ L(a,b,k_1,k_2,m) & \text{iff } d(a,b,k_1,k_2,n) < 0 \\ L(a,b) & \text{iff } a_1 = b_1 \,. \end{cases}$$

In the case $n=m=k_2=1$ we get the classical Moulton-plane.

§ 3. Dimension of nearaffine spaces.

In this paragraph we define a notion of dimension of nearaffine spaces analogous to affine spaces. We define an operator on the set of pencils of straight lines, which has similar properties as a closure operator.

For this purpose we consider subspaces of nearaffine spaces, as they were introduced by J. ANDRÉ in [1]. We show that any subspace of a nearaffine space, which contains at least two points, possesses a base. Further all bases of a subspace have the same cardinality.

Definition. Let be $U \subset P$, $v \in P$ and $S_v \subset G_v$. Two points x,y are called *joinable in* U, denoted by $x \underset{U}{\sim} y$, resp. *joinable with respect to* S_v, denoted by $x \underset{S_v}{\sim} y$, if $x = y$ or there exist finitely many points $x =: x_o, x_1, \ldots, x_n := y$ such that $\overline{x_{i-1}, x_i} \in G(U)$ resp. $(v \| \overline{x_{i-1}, x_i}) \in S_v$ for $i=1,\ldots,n$.

Remarks.

1. $\underset{U}{\sim}$ resp. $\underset{S_v}{\sim}$ are equivalence relations on U resp. on P.
2. Condition (G2) means that $x \underset{P}{\sim} y$ is valid for all $x,y \in P$.

Definition. A subset $U \subset P$ of a nearaffine space is called a *subspace of* F, denoted by $U < P$, if the following conditions hold:

(U1) $x,y \in U$ and $x \neq y$ imply $\overrightarrow{x,y} \subset U$.

(U2) Any two points of U are joinable in U.

The set of all subspaces of F is denoted by $\mathcal{U}$. In § 4 we show that any subset $U \subset P$ satisfying (U1) is a subspace of F.

Without using the finiteness of the point set ANDRÉ [1; p.79] has proved for finite nearaffine spaces:

(3.1) Let be $U \in \mathcal{U}$, $L \in L(U)$ and $x \in U$. Then $(x \| L) \in L(U)$. □

(3.2) *Lemma.* Let U be a subspace of F satisfying (R). Then

$$F(U) := (U, \rightarrow\big|_{U \times U \setminus \Delta_U}, \|\big|_{L(U) \times L(U)})$$

is a nearaffine space. □

(3.3) *Lemma.* Let be $v \in P$ and $U \in \mathcal{U}$. The mapping

$$< \ > : \begin{cases} P(G_v) \longrightarrow \mathcal{U} \\ S_v \longrightarrow \langle S_v \rangle \end{cases} \quad \text{with}$$

$$\langle S_v \rangle := \begin{cases} \{x \in P \mid v \underset{S_v}{\sim} x\} & \text{if } S_v \neq \emptyset, \\ \emptyset & \text{if } S_v = \emptyset \end{cases}$$

has the following properties:

(1) $S_v \subset G(U)$ implies $<S_v> < U$.

(2) $S_v \subset G(<S_v>)$.

(3) $R_v \subset S_v$ implies $<R_v> < <S_v>$.

Proof. At first we have to show that $<S_v> \in U$ holds for any $S_v \in P(G_v)$. Assume $x,y \in <S_v>$ and $x \neq y$. Because of the transitivity of $\underset{S_v}{\sim}$ there are points $x =: x_o, x_1, \ldots, x_n := y$ such that $(v \| \overline{x_{i-1}, x_i}) \in S_v$ and $x_i \neq x$ for $i = 1,\ldots,n$. We prove by induction:

$$(*) \qquad \overrightarrow{x,x_i} \subset <S_v> .$$

$(v \| \overline{x,x_1}) \in S_v$ and $v \underset{S_v}{\sim} x$ imply $\overline{x,x_1} \subset <S_v>$. Now let be $\overrightarrow{x,x_{i-1}} \subset <S_v>$. For any $p \in \overrightarrow{x,x_i}$ there exists a point $q \in \overrightarrow{x,x_{i-1}} \cap (p \| \overline{x_{i-1},x_i})$ by (1.1). We get $v \underset{S_v}{\sim} q$ and $v \underset{S_v}{\sim} p$, hence $p \in <S_v>$.

Now (*) yields $\overrightarrow{x,y} = \overrightarrow{x,x_n} \subset <S_v>$ and $x_i \in <S_v>$ for $i = 1,\ldots,n$. From this follows $\overline{x_{i-1},x_i} \subset <S_v>$ for $i = 1,\ldots,n$, thus $x \underset{<S_v>}{\sim} y$.

For the proof of property (1) assume $S_v \subset G(U)$ and $x \in <S_v>$. There are points $v =: x_o, x_1, \ldots, x_n := x$ such that $(v \| \overline{x_{i-1},x_i}) \in S_v \subset G(U)$ for $i = 1,\ldots,n$. By induction we get $x_n \in U$:
$x_1 \in \overline{v,x_1} \in G(U)$ is true, and $x_{n-1} \in U$ implies $x_n \in \overline{x_{n-1},x_n} = (x_{n-1} \| (v \| \overline{x_{n-1},x_n})) \subset U$ by (3.1).
The properties (2) and (3) are obvious. □

Definition. Let U be a subspace of F, $v \in U$ and $S_v \subset G_v$.

(1) S_v *generates* U if $<S_v> = U$.

(2) S_v is called *independent* if $G \in S_v$ implies $G \not< <S_v \setminus \{G\}>$.

(3) S_v is called a *base of* U if S_v generates U and is independent.

(3.4) Let be $v,w \in P$, $S_v \subset G_v$, $S_v \neq \emptyset$ and $S_w := \{(w \| G) \in G_w \mid G \in S_v\}$. Then either $<S_v> \cap <S_w> = \emptyset$ or $<S_v> = <S_w>$.

Proof. If there is a point $x \in <S_v> \cap <S_w>$, then for any $y \in <S_v>$ there exist points $x =: x_o, x_1, \ldots, x_n := y$ with $(v \| \overline{x_{i-1},x_i}) \in S_v$ for $i = 1,\ldots,n$. We have $(w \| \overline{x_{i-1},x_i}) = (w \| (v \| \overline{x_{i-1},x_i})) \in S_w$ $(i=1,\ldots,n)$, hence $x \underset{S_w}{\sim} y$, and $y \in <S_w>$.
By symmetric arguments we get $<S_w> \subset <S_v>$. □

Lemma (3.3) implies:

(3.5) (Finiteness conditions). The following statements are valid for any $v \in P$ and $S_v \subset G_v$:

(1) For any $x \in \langle S_v \rangle$ there is a finite subset $R_v \subset S_v$ such that $x \in \langle R_v \rangle$.

(2) For any $G \in G(\langle S_v \rangle)_v$ there exists a finite subset $R_v \subset S_v$ with $G < \langle R_v \rangle$. □

(3.6) Let be $U \in U$ and $G \in G$ with $G \cap U \neq \emptyset$. Then $\bigcup_{y \in U} (y \| G)$ is also a subspace.

ANDRÉ [1; p. 80] has shown the validity of this statement for finite nearaffine spaces, but he doesn't use the finiteness of the point set. □

(3.7) Let be $v \in P$, $G \in G_v$, $S_v \subset G_v$, $S_v \neq \emptyset$.
Then $\langle S_v \cup \{G\} \rangle = \bigcup_{y \in \langle S_v \rangle} (y \| G)$.

Proof. Due to $v \in \langle S_v \rangle \cap G$ and (3.3) we have $S_v \cup \{G\} \subset G(\langle S_v \rangle) \cup \{G\} \subset G(\bigcup_{y \in \langle S_v \rangle} (y \| G))$, hence $\langle S_v \cup \{G\} \rangle \subset \bigcup_{y \in \langle S_v \rangle} (y \| G)$ by (3.6) and (3.3).

On the other hand, (3.1) and (3.3) give $\bigcup_{y \in \langle S_v \rangle} (y \| G) \subset \langle S_v \cup \{G\} \rangle$. □

(3.8) *Lemma*. (Exchange condition). Let be $v \in P$, $S_v \subset G_v$ and $G, H \in G_v$. Then $H < \langle S_v \cup \{G\} \rangle$ and $H \not< \langle S_v \rangle$ imply $G < \langle S_v \cup \{H\} \rangle$.

Proof. We may assume $H \neq G$, thus $S_v \neq \emptyset$. Let be $x \in H \setminus \langle S_v \rangle$. By (3.7) there is a $y \in \langle S_v \rangle$ such that $x \in (y \| G)$. Due to (3.1) $\overline{x,y} < \langle S_v \cup \{H\} \rangle$ implies $G = (v \| \overline{x,y}) < \langle S_v \cup \{H\} \rangle$. □

(3.9) Let be $v \in P$, $S_v \subset G_v$ independent and $G \in G_v \setminus G(\langle S_v \rangle)$. Then $S_v \cup \{G\}$ is also independent.

Proof. If $S_v \cup \{G\}$ is dependent, then there exists a straight line $H \in S_v \cup \{G\}$ with $H < \langle (S_v \cup \{G\}) \setminus \{H\} \rangle$. $G \not< \langle S_v \rangle$ implies $H \in S_v$, hence $H < \langle (S_v \setminus \{H\}) \cup \{G\} \rangle$. Because S_v is independent, we obtain the contradiction $G < \langle S_v \rangle$ by (3.8). □

(3.10) *Theorem*. (Existence of a base). Let U be a subspace containing at least two points and $v \in U$. Then for any independent set $S_v \subset G(U)_v$ there exists a base $B_v \subset G(U)_v$ of U with $S_v \subset B_v$.

The proof is analogous to the corresponding theorem in affine spaces (cf. [7; Thm. (8.7)]) by application of (3.3), (3.5), (3.9) and Zorn's Lemma. □

Thus any subspace of a nearaffine space containing at least two points possesses a base.

Exactly as in the theorem of equal cardinality of bases in affine spaces (cf. [7; Thm. (8.8)]) it may be shown by (3.3), (3.5), (3.8), (3.9) and (3.10):

(3.11) Any two bases of a subspace of F belonging to the same pencil of straight lines have the same cardinality. □

(3.12) *Theorem.* (Equal cardinality of bases). Any two bases of a subspace of a nearaffine space have the same cardinality.

Proof. Assume $U \in \mathcal{U}$ and $v,w \in U$. Let $\mathcal{B}_v \subset \mathcal{G}(U)_v$ and $\mathcal{B}_w \subset \mathcal{G}(U)_w$ be two bases of U. For $\mathcal{B}'_w := \{(w\|G) \in \mathcal{G}_w \mid G \in \mathcal{B}_v\}$ we have obviously $|\mathcal{B}'_w| = |\mathcal{B}_v|$. It suffices to show that $\mathcal{B}'_w$ is a base of U, because this yields $|\mathcal{B}_w| = |\mathcal{B}'_w| = |\mathcal{B}_v|$ by (3.11).

$w \in \langle \mathcal{B}_v \rangle \cap \langle \mathcal{B}'_w \rangle$ implies $\langle \mathcal{B}'_w \rangle = \langle \mathcal{B}_v \rangle = U$ by (3.4). Suppose $\mathcal{B}'_w$ is dependent. Then there exists a straight line $G \in \mathcal{B}'_w$ such that $G < \langle \mathcal{B}'_w \setminus \{G\} \rangle$. We obtain $v \in U = \langle \mathcal{B}'_w \rangle < \langle \mathcal{B}'_w \setminus \{G\} \rangle$ by (3.3). (3.1) and (3.4) yield $(v\| G) < \langle \mathcal{B}'_w \setminus \{G\} \rangle = \langle \mathcal{B}_v \setminus \{(v\| G)\} \rangle$, contradicting the independence of $\mathcal{B}_v$. □

(3.10) and (3.12) lead to

Definition. Let U be a subspace of a nearaffine space. Then the cardinal number

$$\dim U := \begin{cases} -1 & \text{if } U = \emptyset, \\ 0 & \text{if } |U| = 1, \\ |\mathcal{B}| & \text{if } \mathcal{B} \text{ is a base of } U \end{cases}$$

is called the *dimension of* U.

In finite nearaffine spaces, affine spaces and Desarguesian semiregular nearaffine spaces this concept of dimension is equivalent to the usual notion (cf. [1; II Def.1.2], [7; p.49], [2; I § 3, III § 8]).

§ 4. Properties of subspaces.

J. ANDRÉ [1; II Thm. 5.1] has shown that any subset U of a finite nearaffine space F satisfying (U1) already satisfies (U2), i.e. U is a subspace of F. This statement is valid for arbitrary nearaffine spaces. It follows: The intersection of any number of subspaces of a nearaffine space is a subspace, and the set of all subspaces forms a lattice.

A consequence of (3.3), (3.7), (3.8), (3.10) is:

(4.1) The following conditions are equivalent for any $H \in \mathcal{U}$:

(1) H is a proper maximal subspace of F.

(2) There exist $v \in H$, $G \in \mathcal{G}_v \setminus \mathcal{G}(H)$ and $\mathcal{B}_v \subset \mathcal{G}_v$ with $<\mathcal{B}_v> = H$ and $<\mathcal{B}_v \cup \{G\}> = P$.

(3) There exists a $G \in \mathcal{G}$ such that $|G \cap H| = 1$ and $\bigcup_{y \in H} (y \| G) = P$. □

Definition. A subspace H satisfying one of the conditions in (4.1) is called a *hyperplane*. The set of all hyperplanes of a nearaffine space F is denoted by $\mathcal{H}$.

In finite nearaffine spaces and affine spaces this concept of hyperplanes is identical with the theory of hyperplanes, as it is established in [1; II Def. 1.3], resp. [7; p. 52].

Definition. A subset $U \subset P$ of a nearaffine space is called a *flat* if any line incident with two points of U lies in U.

(4.2) Any hyperplane of a nearaffine space is a flat.

Proof. Let be $H \in \mathcal{H}$, $L \in \mathcal{L} \setminus \mathcal{L}(H)$, $v \lessdot L$ and $w \in L \cap H$. We have to prove: $L \cap H = \{w\}$. We may assume $v \notin H$. (4.1) implies the existence of $G \in \mathcal{G}$, $x \in H$ with $|G \cap H| = 1$ and $v \in (x \| G)$. Due to $x \underset{H}{\sim} w$ there exist points $x =: x_o, x_1, \dots, x_n := w$ all belonging to H such that $\overline{x_{i-1}, x_i} \in \mathcal{G}(H)$ $(i = 1, \dots, n)$. It suffices to show: $\overrightarrow{v, x_i} \cap H = \{x_i\}$ $(i = 0, \dots, n)$. $\overline{v, x_o} \cap H = \{x_o\}$ is obvious. Now assume $i > 0$ and $\overrightarrow{v, x_{i-1}} \cap H = \{x_{i-1}\}$. For any point $y \in \overrightarrow{v, x_i} \cap H$, there exists a $z \in \overrightarrow{v, x_{i-1}} \cap (y \| \overline{x_{i-1}, x_i})$ by (1.1). $(y \| \overline{x_{i-1}, x_i}) \subset H$ implies $z \in \overrightarrow{v, x_{i-1}} \cap H$. The induction hypothesis yields $z = x_{i-1}$, hence $y = x_i$. □

(4.3) *Lemma.* Any subspace of a nearaffine space F is a flat.

Proof. Let be $U \in \mathcal{U}$, $L \in \mathcal{L} \setminus \mathcal{L}(U)$, $v \lessdot L$ and $w \in L \cap U$. We may assume $|U| \geq 2$ and $v \notin U$. Let $\mathcal{B}_w \subset \mathcal{G}_w$ be a base of U. Using (G2) we obtain the existence of points $w =: x_o, x_1, \dots, x_n := v$ with $G_i := (w \| \overline{x_{i-1}, x_i}) \in \mathcal{G}$ $(i = 1, \dots, n)$. For $i = 1, \dots, n$ we define $V_i := <\mathcal{B}_w \cup \{G_1, \dots, G_i\}>$ and $V_o := <\mathcal{B}_w>$. According to $v \in V_n \setminus V_o$ there exists a natural number

$k \in \{1,\dots,n\}$ with $v \in V_k \setminus V_{k-1}$. Because of $G_k \not\subset V_{k-1}$ it is easy to see that V_{k-1} is a hyperplane of the nearaffine space $F(V_k)$. The line $L = \overrightarrow{v,w}$ is contained in $F(V_k)$ but not in V_{k-1}, thus $|L \cap V_{k-1}| \le 1$ by (4.2), and especially $|L \cap U| \le 1$. □

(4.4) *Lemma*. If a hyperplane H and a line L have exactly one point in common, then every line L' beeing parallel to L has exactly one point in common with the hyperplane H.

Proof. Let be $H \in \mathcal{H}$, $L \in \mathcal{L}$ with $L \cap H = \{w\}$ and $x \in P$. There exists an $u \in L$ with $L \| \overrightarrow{u,w}$.

We have

(1) $(x \| L) \not\subset H$,

for otherwise $u \in \overrightarrow{w,u} = (w \| (x \| L)) \subset H$, contradicting $u \notin H$. Further

(2) $(x \| L) \cap H \neq \emptyset$,

for the assumption $(x \| L) \cap H = \emptyset$ leads to a contradiction: According to $H \in \mathcal{H}$ there exist $G \in \mathcal{G}$ and $v,y \in H$ with $|G \cap H| = 1$, $u \in (v \| G)$ and $x \in (y \| G)$. Applying (T) to the points u,v,w and x,y we obtain the existence of a point $z \in (x \| \overrightarrow{u,w}) \cap (y \| \overrightarrow{v,w})$, hence $z \in (x \| L) \cap H$. Using (1) and (2) we obtain $|(x \| L) \cap H| = 1$ by (4.2). □

(4.5) Let be $U \subset P$ satisfying (U1), $v \in U$, $G \in \mathcal{G}_v$, $S_v \subset \mathcal{G}_v$ and $x \in U \cap \langle S_v \cup \{G\} \rangle$ with $\overrightarrow{v,x} \in \mathcal{L} \setminus \mathcal{G}$. Then there exist points $y \in U \setminus \{x\}$ and $z \in \langle S_v \rangle \setminus \{v,y\}$ such that $z \in \overrightarrow{y,x}$.

Proof. We may assume $x \notin \langle S_v \rangle$. $\overrightarrow{v,x} \in \mathcal{L} \setminus \mathcal{G}$ is equivalent to $\overrightarrow{x,v} \not\subset \overrightarrow{v,x}$. (We remark this without proof.) Hence there exists a point $y \in \overrightarrow{x,v} \setminus \overrightarrow{v,x}$. Then we have $\overrightarrow{v,x} \cap \langle S_v \rangle = \{v\}$ due to (4.3). By (4.4) the hyperplane $\langle S_v \rangle$ of the nearaffine space $F(\langle S_v \cup \{G\} \rangle)$ has exactly one point z in common with the line $\overrightarrow{y,x}$, which is parallel to $\overrightarrow{v,x}$. $y \notin \overrightarrow{v,x}$ yields to $z \neq v$. Moreover, $\overrightarrow{v,x} \not\subset \langle S_v \rangle$ implies $z \neq y$ by (4.3). □

(4.6) Let be $U \subset P$ satisfying (U1), $L \in \mathcal{L}(U)$, $v \in L$ and $y \in U \setminus \{v\}$. Then $x \in \overrightarrow{y,v} \setminus \{y\}$ implies $(x \| L) \subset U$.

Proof. $x \in U$ is obvious. In the case $y \in L$ we have $(x \| L) \| L = \overrightarrow{v,y} \| \overrightarrow{y,v} = \overrightarrow{y,x} \| \overrightarrow{x,y}$, hence $(x \| L) = \overrightarrow{x,y} \subset U$. Assume now $y \notin L$ and $z \in (x \| L) \setminus \{x,y\}$. We can find a $w \in \overrightarrow{y,z} \cap (v \| \overrightarrow{x,z})$ by (1.1). $L \| (x \| L) = \overrightarrow{x,z} \| (v \| \overrightarrow{x,z})$ implies $L = (v \| \overrightarrow{x,z})$, hence $w \in L \subset U$. Thus we have $z \in \overrightarrow{y,w} \subset U$, whence $(x \| L) = \overrightarrow{x,z} \subset U$. □

(4.7) *Lemma.* Let be $U \subset P$ satisfying (U1). Then $u \in U$ and $L \in \mathcal{L}(U)$ imply $(u \| L) \subset U$.

Proof. Let be $u \in U$, $L \in \mathcal{L}(U)$, $v \lessdot L$ and $u \neq v$. According to (G2) there exist points $v =: x_o, x_1, \ldots, x_n := u$ with $G_i := (v \| \overline{x_{i-1}, x_i}) \in \mathcal{G}$ for $i = 1, \ldots, n$. By $u \in U \cap \langle\{G_1, \ldots, G_n\}\rangle$ it suffices to show:
(*) $x \in U \cap \langle\{G_1, \ldots, G_i\}\rangle$ implies $(x \| L) \subset U$ $(i = 1, \ldots, n)$.
Due to (4.6) (*) is valid for $i = 1$. Now let (*) be valid for $i-1$ with $1 < i \leq n$, and $x \in U \cap \langle\{G_1, \ldots, G_i\}\rangle$. By (4.6) we may assume $\overrightarrow{v,x} \in \mathcal{L} \setminus \mathcal{G}$. (4.5) implies the existence of $y \in U \setminus \{x\}$ and $z \in \langle\{G_1, \ldots, G_{i-1}\}\rangle \setminus \{v,y\}$ with $z \in \overrightarrow{y,x}$. The induction - hypothesis leads to $(z \| L) \subset U$, hence $(x \| L) = (x \| (z \| L)) \subset U$ by (4.6). □

(4.8) Let be $U \subset P$ satisfying (U1) with $|U| \geq 2$ and $v \in U$. Then $\mathcal{G}(U)_v \neq \emptyset$.

Proof. Assume $\mathcal{G}(U)_v = \emptyset$. For any $w \in U \setminus \{v\}$ there exist $v =: x_o, x_1, \ldots, x_n := w$ such that $G_i := (v \| \overline{x_{i-1}, x_i}) \in \mathcal{G}$ $(i=1,\ldots,n)$. We show:
(*) $|U \cap \langle\{G_1, \ldots, G_i\}\rangle| \geq 2$ implies
$|U \cap \langle\{G_1, \ldots, G_{i-1}\}\rangle| \geq 2$ $(i=2,\ldots,n)$.
Let be $i \in \{2,\ldots,n\}$ and $x \in U \cap \langle\{G_1, \ldots, G_i\}\rangle$ with $x \neq v$. (4.5) gives the existence of $y \in U \setminus \{x\}$ and $z \in \langle\{G_1, \ldots, G_{i-1}\}\rangle \setminus \{v,y\}$ with $z \in \overrightarrow{y,x}$, hence $z \in U \cap \langle\{G_1, \ldots, G_{i-1}\}\rangle$, i.e. $|U \cap \langle\{G_1, \ldots, G_{i-1}\}\rangle| \geq 2$.
Applying (*) to $v, w \in U \cap \langle\{G_1, \ldots, G_n\}\rangle$ several times we obtain $|U \cap G_1| \geq 2$, thus $G_1 \subset U$ contradicting our hypothesis. □

A consequence of [6; Thm. 5.4] is

(4.9) Let be $v \in P$, $\mathcal{S}_v \subset \mathcal{G}_v$ and $x, y \in P$ with $x \neq y$, $x \underset{\mathcal{S}_v}{\sim} y$. Then there exist points $x =: x_o, x_1, \ldots, x_n := y$ with $(v \| \overline{x_{i-1}, x_i}) \in \mathcal{S}_v$ for $i=1,\ldots,n$ satisfying the following conditions:

(E1) $x_i \neq x_j$ for $i,j \in \{0,\ldots,n\}$, $i \neq j$.

(E2) $\overline{x_{i-1}, x_i} \cap \overline{x_{j-1}, x_j} = \emptyset$ for $i,j \in \{1,\ldots,n\}$, $i \neq j-1, j, j+1$.

(E3) $\overline{x_{i-1}, x_i} \nparallel \overline{x_{j-1}, x_j}$ for $i,j \in \{1,\ldots,n\}$, $i \neq j$. □

(4.10) Let be $v \in P$, $F \in \mathcal{G}_v$ and $\mathcal{S}_v \subset \mathcal{G}_v$. If the set $\mathcal{S}_v$ is non-void and independent, then there exists a straight line $G \in \mathcal{S}_v$ with $F \nless \langle \mathcal{S}_v \setminus \{G\} \rangle$.

Proof. In the case $F \nless \langle \mathcal{S}_v \rangle$ choose an arbitrary straight line $G \in \mathcal{S}_v$. Now assume $F < \langle \mathcal{S}_v \rangle$ and $x \in F \setminus \{v\}$. Using (4.9) we can find points

$v =: x_o, x_1, \dots, x_n := x$ with $(v \| \overline{x_{i-1}, x_i}) \in S_v$ $(i=1,\dots,n)$ satisfying the condition (E3). Define $G := \overline{v, x_1}$. By (E3) we have $G \not\parallel \overline{x_{i-1}, x_i}$, hence $(v \| \overline{x_{i-1}, x_i}) \in S_v \setminus \{G\}$ for $i=2,\dots,n$. Thus we get $x_1 \underset{S_v \setminus \{G\}}{\sim} x$ and $F \nless < S_v \setminus \{G\} >$, because S_v is independent. □

(4.11) *Theorem.* Any subset of P satisfying (U1) is a subspace of F.

Proof. Assume there is a subset $U \subset P$ satisfying (U1) but not (U2). Then there exists a $v \in U$ so that $U' := \{x \in U | \neg (v \underset{U}{\sim} x)\}$ is non-void. Let $B_v \subset G_v$ be a base of P. For any $x \in U'$ we define
$m_x := \min \{|S_v^x| \mid S_v^x \subset B_v, |S_v^x| \in \mathbb{N}, x \in < S_v^x > \}$, and
$m := \min \{m_x \mid x \in U'\}$. There is a $w \in U'$ and a $S_v \subset B_v$ such that $|S_v| = m$ and $w \in < S_v >$. Obviously S_v is independent. From $w \in U' \cap < S_v >$ follows $|S_v| \geq 2$. Using (4.8) and (4.10) we get the existence of straight lines $F \in G(U)_v$, $G \in S_v$ with $F \nless < S_v \setminus \{G\} >$. Because of (4.4) there is exactly one point $z \in (w \| F) \cap < S_v \setminus \{G\} >$. $F \subset U$ implies $z \in (w \| F) \subset U$ by (4.7). Due to the minimality of m, the conditions $z \in U$ and $v \underset{S_v \setminus \{G\}}{\sim} z$ lead to $v \underset{U}{\sim} z$, hence $v \underset{U}{\sim} w$, contradicting the hypothesis $w \in U'$. □

(4.12) *Corollary.* The intersection of a family of subspaces of a nearaffine space F is a subspace of F. □

This leads to

(4.13) For any nearaffine space the mapping

$$- : \begin{cases} P(\mathbb{P}) \longrightarrow \mathcal{U} \\ S \longrightarrow \bigcap\limits_{S \subset U \in \mathcal{U}} U \end{cases}$$

is a *closure operation* on P. □

It is easy to see:

(4.14) *Theorem.* $(\mathcal{U}, <)$ is a complete lattice. □

The lattice of all subspaces of an arbitrary nearaffine space F is relatively atomic, relatively complemented and upper continuous. $(\mathcal{U}, <)$ is "semi-modular in the sense of Wilcox" iff F is an affine space. The same is true for the "exchange property".

Finaly we remark that the following dimension-theorem may be proved:

(4.15) If U_1 und U_2 are two finite-dimensional subspaces of a near-affine space such that $U_1 \cap U_2 \neq \emptyset$, then
$\dim \overline{U_1 \cup U_2} + \dim U_1 \cap U_2 = \dim U_1 + \dim U_2$.

The assumption $U_1 \cap U_2 \neq \emptyset$ is necessary, because in any proper near-affine space there are subspaces U_1, U_2, V_1, V_2 such that
$\dim \overline{U_1 \cup U_2} + \dim U_1 \cap U_2 \lneqq \dim U_1 + \dim U_2$ and
$\dim \overline{V_1 \cup V_2} + \dim V_1 \cap V_2 \gneqq \dim V_1 + \dim V_2$. □

REFERENCES.

[1] ANDRÉ, J.: On finite non-commutative affine spaces. In: Combinatorics, ed. by M.HALL jun. and J.H. VAN LINT, 2nd ed., Mathematical Centre, 65-113, Amsterdam 1975.

[2] ANDRÉ, J.: Affine Geometrien über Fastkörpern. Mitt. Math. Sem. Giessen 114, 99p. (1975).

[3] ANDRÉ, J.: Some new results on incidence structures. Atti dei Convegni Lincei 17, Colloquio internazionale sulle teorie combinatorie II, 201-222 (1976).

[4] ANDRÉ, J.: Über verschiedene Klassen von Unterräumen in Räumen mit nichtkommutativer Verbindung. In: Beitr.geom./Algebra, Proc. Symp. Duisburg 1976, 11-23 (1977).

[5] ANDRÉ, J.: Introduction to non-commutative affine Geometry. Lectures held at Kuwait University, March 1979.

[6] HISCHER, D.: Schließungsaussagen in fastaffinen Räumen. Mitt. Math. Sem. Giessen 131, 95 p. (1978).

[7] KARZEL, H., K. SÖRENSEN, D. WINDELBERG: Einführung in die Geometrie. Göttingen 1973.

GENERATING CRYPTOMORPHIC AXIOMATIZATIONS OF MATROIDS

G. Nicoletti
Istituto di Geometria
"L.Cremona"
Università di Bologna
(ITALIA)

There exist many ways to define equivalently (cryptomorphically) the concept of matroid (in his "Matroid Theory", D.J.A. Welsh writes: "Deciding which set of axioms would be the most natural to start with was difficult"). In this note I show a deep symmetry between these axiomatizations: in this symmetry the family of bases has a central role; it is possible to define recursively new families of sets which axiomatize cryptomorphically the concept of matroid.

A *matroid* is an ordered pair $(S,\mathcal{I})$, where $S\neq\emptyset$ is a finite set, and $\mathcal{I}$ is a collection of subsets of S such that:

i1) $I_1 \varepsilon \mathcal{I},\ I_2 \subseteq I_1 \rightarrow I_2 \varepsilon \mathcal{I}$; ($\mathcal{I}$ is a descending family of subsets of S);

i2) $\mathcal{I} \neq \emptyset$;

i3) $\forall\, I_1, I_2 \varepsilon \mathcal{I}, |I_1| < |I_2|,\ \exists x \varepsilon I_2 - I_1 : I_1 \cup x \ \varepsilon\ \mathcal{I}$.

The subsets in $\mathcal{I}$ are called *independent sets*. If $A\subseteq S$, the maximal independent sets contained in A have the same cardinality. Hence, we can define the *rank* of a subset A as follows:

$$r(A) := \max\{|I|; I\subseteq A,\ I \varepsilon \mathcal{I}\};$$

the rank satisfies the following conditions:

r1) $r(\emptyset) = 0$;

r2) $r(A) \leq r(A\cup x) \leq r(A)+1$;

r3) $r(A\cup x) = r(A\cup y) = r(A) \rightarrow r(A\cup x\cup y) = r(A)$.

We have $I \varepsilon \mathcal{I} \longleftrightarrow r(I) = |I|$.

The family of *bases* is the collection

$$\mathcal{B} := \{B \varepsilon \mathcal{I} \mid B \text{ maximal}\}.$$

The family $\mathcal{B}$ has the following properties:

b1) $B_1, B_2 \varepsilon \mathcal{B}, B_1 \subseteq B_2 \rightarrow B_1 = B_2$; ($\mathcal{B}$ is an antichain of subsets of S);

b2) $\mathcal{B} \neq \emptyset$;

b3) $\forall B_1, B_2 \varepsilon \mathcal{B}, \forall x \varepsilon B_1\ \exists y \varepsilon B_2 : (B_1 - x) \cup y \varepsilon \mathcal{B}$ (exchange axiom).

The bases have obviously the same cardinality, which is defined to be the *rank* of the matroid.

The independent sets are precisely all subsets of bases:

$$\mathcal{I} = \{I\subseteq S \mid \exists B \varepsilon \mathcal{B} : I\subseteq B\}.$$

The family of *spanning sets* is the collection of all supersets of bases:

$$\mathcal{G}:=\{G\subseteq S \mid \exists B\in\mathcal{B}: I\subseteq B\};$$

the family $\mathcal{G}$ has the following properties:

g1) $G_1\in\mathcal{G}, G_2\supseteq G_1 \rightarrow G_2\in\mathcal{G}$; ($\mathcal{G}$ is an ascending family of subsets of S);

g2) $\mathcal{G}\neq\emptyset$;

g3) $\forall G_1,G_2\in\mathcal{G}, |G_1|>|G_2|, \exists x\in G_1-G_2: G_1-x\in\mathcal{G}$.

The bases are precisely the minimal spanning sets:

$$\mathcal{B}=\{G\in\mathcal{G} \mid G \text{ minimal}\}.$$

Dependent sets are the subsets of S which are not independent sets; the family $\mathcal{D}$ of all dependent sets satisfies the following properties:

d1) $D_1\in\mathcal{D}, D_2\supseteq D_1 \rightarrow D_2\in\mathcal{D}$; ($\mathcal{D}$ is an ascending family of subsets of S);

d2) $\emptyset\notin\mathcal{D}$;

d3) $\forall D_1,D_2\in\mathcal{D}: D_1\cap D_2\notin\mathcal{D}\rightarrow\forall x\in S: D_1\cup D_2-x\in\mathcal{D}$.

The minimal dependent sets are called *circuits*; the family $\mathcal{C}$ of all circuits satisfies the following properties:

c1) $C_1,C_2\in\mathcal{C}, C_1\subseteq C_2 \rightarrow C_1=C_2$; ($\mathcal{C}$ is an antichain of subsets of S);

c2) $\emptyset\notin\mathcal{C}$;

c3) $\forall C_1,C_2\in\mathcal{C}, C_1\neq C_2, \forall x\in S\ \exists C_3\in\mathcal{C}: C_3\subseteq C_1\cup C_2-x$.

Dependent sets are precisely all supersets of circuits:

$$\mathcal{D}=\{D\subseteq S \mid \exists C\in\mathcal{C}: C\subseteq D\}.$$

A subset A is called a *closed set* if $\forall x\notin A: r(A\cup x)=r(A)+1$; a maximal closed set different from S is called a *hyperplane*. The family $\mathcal{H}$ of all hyperplanes satisfies the following properties:

h1) $H_1,H_2\in\mathcal{H}, H_1\subseteq H_2 \rightarrow H_1=H_2$; ($\mathcal{H}$ is an antichain of subsets of S);

h2) $S\notin\mathcal{H}$;

h3) $\forall H_1,H_2\in\mathcal{H}, H_1\neq H_2, \forall x\in S\ \exists H_3\in\mathcal{H}: H_3\supseteq(H_1\cap H_2)\cup x$.

Circuits, dependent sets, independent sets, bases, spanning sets and hyperplanes can be used equivalently to axiomatize matroids: their respective axiom systems are given by the properties listed above. Now, we have three antichain of subsets of S, namely $\mathcal{C}$, $\mathcal{B}$, $\mathcal{H}$, two acsending families of subsets of S, namely $\mathcal{D}$, $\mathcal{G}$, and only one descending family of subsets of S, namely $\mathcal{I}$.

Now, the situation is the following:

$\mathcal{C}$: circuits, or minimal dependent sets;

$\mathcal{D}$: dependent sets, or supersets of circuits, or non-independent sets;

$\mathcal{I}$: independent sets, or non-dependent sets, or subsets of bases;

$\mathcal{B}$: bases, or maximal independent sets, or minimal spanning sets;

$\mathcal{G}$: spanning sets, or supersets of bases.

What about hyperplanes and subsets of hyperplanes? It has been shown (M.Barnabei, G.Nicoletti: Axiomatizing Matroids by Means of the Set of Non-Generators, to appear in Boll. U.M.I.) that if we denote by $\mathcal{N}$ the family of non-spanning sets, we can axio̲matize matroids by means of $\mathcal{N}$ also. Axioms for non-spanning sets are:

n1) $N_1 \in \mathcal{N},\ N_2 \subseteq N_1 \rightarrow N_2 \in \mathcal{N}$; ($\mathcal{N}$ is a descending family of subsets of S);

n2) $S \notin \mathcal{N}$;

n3) $\forall N_1, N_2 \in \mathcal{N}: N_1 \cup N_2 \notin \mathcal{N} \rightarrow \forall x \in S: (N_1 \cap N_2) \cup x \in \mathcal{N}$.

The maximal non-spanning sets are precisely the hyperplanes:

$$\mathcal{H} = \{N \in \mathcal{N} \mid N \text{ maximal}\}.$$

Hence, we can complete the diagram as follows:

$$\mathcal{C} \underset{\text{minimal}}{\overset{\text{supersets}}{\rightleftarrows}} \mathcal{D} \underset{\text{non-}}{\overset{\text{non-}}{\rightleftarrows}} \mathcal{I} \underset{\text{subsets}}{\overset{\text{maximal}}{\rightleftarrows}} \mathcal{B} \underset{\text{minimal}}{\overset{\text{supersets}}{\rightleftarrows}} \mathcal{G} \underset{\text{non-}}{\overset{\text{non-}}{\rightleftarrows}} \mathcal{N} \underset{\text{subsets}}{\overset{\text{maximal}}{\rightleftarrows}} \mathcal{H}$$

This diagram suggests us the following rules to generate new axiomatizing families for matroids:

i) if $\mathcal{A}_0$ is an antichain of subsets of S which axiomatize matroids, then the family $\mathcal{A}_1$ of all supersets of sets in $\mathcal{A}_0$, and the family $\mathcal{A}_{-1}$ of all subsets of sets in $\mathcal{A}_0$ axiomatize also cryptomorphically matroids;

ii) if $\mathcal{A}_1$ is an ascending family of subsets of S which axiomatize matroids, then the family $\mathcal{A}_0$ of all minimal sets in $\mathcal{A}_1$, and the family $\mathcal{A}_2$ of all subsets of S not in $\mathcal{A}_1$ axiomatize also cryptomorphically matroids;

iii) if $\mathcal{A}_{-1}$ is a descending family of subsets of S which axiomatize matroids, then the family $\mathcal{A}_{-2}$ of all subsets of S not in $\mathcal{A}_{-1}$, and the family $\mathcal{A}_0$ of all maximal sets in $\mathcal{A}_{-1}$ axiomatize also cryptomorphically matroids.

These rules immediately led to study axiom systems for the subsets of circuits and for the supersets of hyperplanes.

We observe now that in the axiom system for bases, the third axiom can be replaced by the following:

b3,1) $\forall B_1, B_2 \in \mathcal{B},\ \forall X \subseteq B_1,\ \forall Y \supseteq B_2,\ X \subseteq Y,\ \exists B_3 \in \mathcal{B}: X \subseteq B_3 \subseteq Y$; (middle basis-axiom);

to obtain an equivalent system of axioms.

We note that this new axiom system is symmetric with respect to the boolean operations that is, if we "reverse" the inclusion relation, the axiom system remains unchanged.

This fact leds us to state the following "duality principle":

> if in the statement of a theorem on matroids we replace to the words in the first column the correspondent words of the second column, we obtain a new theorem on matroids:

circuit	hyperplane
dependent set	non-spanning set
independent set	spanning set
basis	basis
spanning set	independent set
non-spanning set	dependent set
hyperplane	circuit
empty set	S
S	empty set

contains	is contained in
is contained in	contains
element of S	complement of an element of S
complement of an element of S	element of S

and so on.

As a consequence, if $\mathcal{B}$ is the family of bases of a matroid, the family $\mathcal{B}$ of all complements of bases in $\mathcal{B}$ is also a family of bases for a new matroid, which is called the *dual matroid.* This fact, whose classical proof is very tedious, is now an obvious consequence of the duality principle, because the complement is an anti-automorphism of the boolean algebra of the subsets of S in itself.

REFERENCES

M. Barnabei, G. Nicoletti, *Axiomatizing Matroids by Means of the Set of Non-Generators*, to appear in Boll. U.M.I.

T.H. Brylawski, *An Outline for the Study of Combinatorial Pregeometries*, Lecture Note Series, University of North Carolina, Chapel Hill, 1972.

H.H. Crapo, G.C. Rota, *On the Foundations of Combinatorial Theory: Combinatorial Geometries*, M.I.T. Press, Cambridge, Mass., 1970.

D.J.A. Welsh, *Matroid Theory*, Academic Press, London, 1976.

Partial planes with exactly two complete parallel classes

Günter Pickert

Mathematisches Institut der Justus-Liebig Universität
Arndtstr. 2, D 6300 Giessen, F.R. Germany

1. A <u>partial plane</u> is a triple $(P,\mathcal{L},I)$ of sets with $I \subseteq P \times \mathcal{L}$ (i.e. I is a relation from P to $\mathcal{L}$) and the property ($\vee$ meaning "or")

(PP) $\forall p_1,p_2 \in P,\ \forall L_1,L_2 \in \mathcal{L} : (\forall i,k \in \{1,2\}:\ p_i\, I\, L_k) \Rightarrow p_1 = p_2 \ \vee\ L_1 = L_2$.

With the usual terminology of points and lines (for the elements of P resp. $\mathcal{L}$) (PP) can be formulated as "Two lines have at most one point in common" or (equivalently and dually) "Through two points passes at most one line" (here and in the sequel "two" always includes inequality). By

$$L \parallel L' \Leftrightarrow L = L' \ \vee\ \forall p \in P:\ p \not I L \ \vee\ p \not I L'$$

we introduce the relation $\parallel$ (<u>parallel</u>) in $\mathcal{L}$, which obviously is reflexive and symmetric. Restricting the investigation to partial planes with the "weak parallel axiom"

(WP) $\quad \parallel$ <u>is transitive</u>

(i.e. through a given point passes at most one line parallel to a given line). $\parallel$ is an equivalence relation in $\mathcal{L}$. For the <u>parallel class</u> $\parallel L = \{L' \mid L \parallel L'\}$ of a line L there are two extreme cases, defined by

$$\parallel L \ \text{trivial} \Leftrightarrow \ \parallel L = \{L\},$$

$$\parallel L \ \text{complete} \Leftrightarrow \ \parallel L \neq \{L\} \ \wedge\ \forall p \in P,\ \exists L' \in \parallel L:\ p I L';$$

by (WP) L gives a partition of P in the second case, and therefore the full parallel axiom means, that all parallel classes are complete (if not all points are on one line). Deleting a point in a projective plane gives a partial plane with (WP) and exactly one complete parallel class, all others being trivial. But deleting a second point destroys (WP). Thus one wonders what structures will fulfil (PP), (WP) and

(CP) Two parallel classes are complete, all others trivial.

Deleting in such a partial plane the lines of the two complete parallel classes $\mathcal{L}_1$, $\mathcal{L}_2$ and dualizing leads to a linear space $(P^*, \mathcal{L}^*, I^*)$ with $P = \mathcal{L} \setminus (\mathcal{L}_1 \cup \mathcal{L}_2)$, $\mathcal{L}^* = P$, $I^* = I^{-1}$ and with two crossed parallelisms π_1, π_2, where "crossed" means

$$\forall L, M \in \mathcal{L}^* : |\pi_1 L \cap \pi_2 M| = 1$$

(see [3], Satz 6); here "linear space" is used in the sense, that every two points are joined by exactly one line (so we do not add the condition, that on every line there are at least two points). To avoid lines without points in this linear space we add the condition

(TP) $\forall p \in P,\ \exists L \in \mathcal{L} : pIL \quad \wedge \quad \| L = \{L\}$,

i.e. through every point passes at least one line belonging to a trivial parallel class. The partial planes with (WP), (CP), (TP), being the object of the following investigation, will be called DP-planes ("D" for "double", "P" for "partial" as well as for "parallel"). The order [p] of a point p and the order [L] of a line L are defined as the cardinal numbers of the set of lines $\notin \mathcal{L}_1 \cup \mathcal{L}_2$ resp. points incident with p resp. L.

2. Let $(P, \mathcal{L}, I)$ be a DP-plane with the two complete parallel classes $\mathcal{L}_1$, $\mathcal{L}_2$. Since $L' \in \mathcal{L}_1$, $L'' \in \mathcal{L}_2$ as non-parallel lines meet, we have

$$[L'] = |\mathcal{L}_2| > 1, \quad [L''] = |\mathcal{L}_1| > 1$$

and thus $P \neq \emptyset$. From (TP) then follows the existence of $L \in \mathcal{L}$ with $L \nparallel L', L''$ for all $L' \in \mathcal{L}_1$, $L'' \in \mathcal{L}_2$ and therefore

$$|\mathcal{L}_1| = [L] = |\mathcal{L}_2|.$$

Thus every line has the same order k (≥ 2), which will be called the

<u>order of the DP-plane</u>. Now we "coordinatize" the DP-plane by introducing a set C with $|C| = k$, numbering the lines of $\mathcal{L}_1$, $\mathcal{L}_2$ as follows

$$\mathcal{L}_1 = \{L'_x \mid x \in C\}, \quad \mathcal{L}_2 = \{L''_y \mid y \in C\}$$

and using (x,y) $(\in C^2 = C\times C)$ as the <u>pair of coordinates</u> for the point $p \, I \, L'_x, L''_y$. In the sequel every point will be replaced by its pair of coordinates, every line by the set of coordinate pairs of the points incident with the line and thus I by $\in$. So the point set is now C^2, the <u>axial lines</u> L'_x, L''_y are the subsets $\{x\}\times C$ resp. $C\times\{y\}$ of C^2, every other line is a permutation of C, and a DP-plane of order k can be described as a pair (C,G), where G is a set of permutations of the set C with $|C| = k \geqq 2$. Conditions (PP) for p_1, p_2 on an axial line, (WP) and the first part of (CP) are trivially fulfilled, (TP) is translated into

(Tr) G <u>is transitive on</u> C,

(i.e. for every $(x,y) \in C^2$ there is $g \in G$ with $g(x) = y$ and the second part of (CP) together with (PP) for p_1, p_2 not on an axial line into

(FP) <u>For every two</u> $g, g' \in G$ <u>the permutation</u> $g^{-1} \circ g'$ <u>has exactly one fixed point.</u>

Here the "product" $g \circ h$ of mappings $(:C \to C)$ is defined by

$$\forall x \in C: \; (g \circ h)(x) = g(h(x)).$$

Since permutation groups are much better known than permutation sets in general, we consider at first the special case, where $(G, \circ)$ is a group; here (FP) can be simplified to

(FP') <u>Every element of</u> G, <u>different from the identity has exactly one fixed point</u>.

A well-known procedure describes a transitive permutation group G on C as operating by left multiplication on the sets of left cosets of the stabilizer $G_o = \{g \mid g(o) = o\}$, o arbitrarily chosen in C; here we have for the unit element 1 of G

(1) $$\{1\} = \bigcap_{h \in G} hG_oh^{-1} ,$$

and vice versa every pair of groups (G,G_o) with $G_o < G$ (i.e. G_o subgroup $\neq G$ of G) and (1) describes in this way a transitive permutation group. The condition $G_o \neq G$ is of course only needed, because we have supposed $|C| > 1$. Translating (FP') in this language gives

(2) $\forall h,h' \in G,\ \forall g \in G \setminus \{1\}: (g \in hG_oh^{-1} \cap h'G_oh'^{-1} \Rightarrow h'^{-1}h \in G_o)$

together with

(3) $$G = \bigcup_{h \in G} hG_oh^{-1} .$$

Now from (2) we get with $h' = 1$

(4) $$\forall h \in G: (hG_oh^{-1} = G_o \Leftrightarrow h \in G_o),$$

i.e. G_o is its own normalizer. With (4) the conclusion in (2) can be reformulated:

$$h'^{-1}h \in G \Leftrightarrow h'^{-1}hG_oh^{-1}h' = G_o \Leftrightarrow hG_oh^{-1} = h'G_oh'^{-1} ,$$

and thus, assuming (4), the condition (2) is equivalent to

(5) $\forall h,h' \in G: (hG_oh^{-1} \cap h'G_oh'^{-1} \neq \{1\} \Rightarrow hG_oh^{-1} = h'G_oh'^{-1}),$

which obviously includes (1), since $G_o \neq G$ and (4) imply the existence of different conjugates of G_o. So the DP-planes with (FP') can be described by group pairs (G,G_o) with $G_o < G$ and (3-5), that is

(FP") G_o is its own normalizer in G, and the conjugates of G_o in G form a partition of the group G.

As I have been told by Prof. Kegel (Freiburg), one can construct pairs of infinte groups with (FP"), using a procedure of H. and B.H.Neumann. On the other hand there are no group pairs (G,G_o) with $G_o < G$, (FP") and finite index $(G:G_o)$. This can be seen geometrically, since every point (x,y) of the DP-plane described by (G,G_o) has the order

$$|\{g \mid g \in G,\ g(x) = y\}| = |G_o| \in \mathbb{N},$$

and this is impossible according to

Proposition 1. In a finite DP-plane of order k ($\in \mathbb{N} \setminus \{1\}$) the order of every point is $\leq k-3$, and not all points have the same order; with b as the number of non-axial lines the inequality $5 \leq k < b$ holds.

Proof. Suppose [p] = r for all $p \in P$. Then by removing the axial lines and dualizing we get a (b,r,1)-block-design with k as the order of each point (= non-axial line of the DP-plane) and k^2 as the number of blocks. The Fisher equations for block-designs therefore give

$$bk = k^2 r, \quad b - 1 = k(r - 1)$$

and from these k = 1 in contradiction to $k \geq 2$. Let $\{p_1, \ldots, p_k\}$ be an axial line L. Then, since every non-axial line intersects L in one point, we have

$$b = \sum_{j=1}^{k} [p_j] \geq k;$$

but b = k is impossible, because then we would have [p] = 1 for the points p of every axial line and therefore for all $p \in P$. Now we consider a point p and a non-axial line L, not going through p. All the [p] + 2 lines through p intersect L in different points, which gives $[p] + 2 \leq k$, therefore $[p] < k - 1$. For the number v_p of points on lines through p we have therefore

$$v_p = [p](k-1) + 2k - 1 < (k-1)^2 + 2k - 1 = k^2.$$

Thus there exists a point q, not joined with p by a line. Counting the points on a non-axial line L through q we get, since $p \notin L$, $[p] + 2 < k$, therefore $[p] \leqq k-3$ and thus $k \geqq 4$, since $[p] \geqq 1$. For $k = 4$ this would give $[p] = 1$ for all $p \in P$, which is impossible.

3. Whether finite DP-planes exist, is not known. A partial result in the search for such structures is (a slight progress compared with Satz 8 and its proof in [3])

Proposition 2. In a finite DP-plane of order k with b as the number of non-axial lines there are points with order at least 4 and the following inequalities hold:

$$k \geqq 7\ , \quad k+4 \leqq b \leqq k(k-4).$$

Proof. We use the description, given in 2, by a coordinate set C with $|C| = k$, and define for $r \in \mathbb{N}$, $x \in C$ and $L \in G$ (=set of non-axial lines)

$v_r = |\{p|\ [p] = r\}|$, $v_{r,L} = |\{p|\ p \in L,\ [p] = r\}|$, $v_{rx} = |\{y|[(x,y)] = r\}|$.

Taking into account, that r can be restricted to numbers $\leqq$ k-3 (see Prop. 1) we get by the usual counting arguments

$$(1) \qquad \sum_{r=1}^{k-3} v_{r,L} = k$$

$$(2) \qquad \sum_{r=1}^{k-3} v_{rx} = k$$

$$(3) \qquad \sum_{L \in G} v_{r,L} = rv_r$$

$$(4) \qquad \sum_{r=2}^{k-3} (r-1)v_{r,L} = b-1$$

$$(5) \qquad \sum_{x \in C} v_{rx} = v_r\ ;$$

e.g. to prove (3), one has to count the incident pairs (p,L) with $p \in P$, $L \in G$, $[p] = r$. Putting $d = b-k(\geqq 1$ according to Prop. 1) we get from (2), $\sum_{y \in C} [(x,y)] = b$ and the definition of v_{rx}

(6) $$\sum_{r=2}^{k-3} (r-1)v_{rx} = d.$$

Comparing (1), (4) leads to determine the maximum value of $\sum_{r=2}^{k-3} (r-1)x_r$ under the conditions $\sum_{r=1}^{k-3} x_r = k$, $x_r \geqq 0$: The maximum $k(k-4)$ is attained only for $x_1 = \ldots = x_{r-4} = 0$, $x_{r-3} = k$. But since $v_{r,L} = 0$ for all $r \leqq k-4$ and all $L \in G$ means $[p] = k-3$ for all $p \in P$, which is impossible according to Prop. 1, we have $b - 1 < k(k-4)$ and thus $b \leqq k(k-4)$. This together with $k < b$ (see Prop. 1) gives $k - 4 > 1$ and therefore already $k \geqq 6$. (3,4) and (5,6) lead to

(7) $$\sum_{r=2}^{k-3} r(r-1)v_r = b(b-1),$$

(8) $$\sum_{r=2}^{k-3} (r-1)v_r = kd.$$

If j ($\in \mathbb{N}$) is chosen so that

(9) $$\forall r \in \mathbb{N} \ (r > j \Rightarrow v_r = 0)$$

(according to Prop.1 (9) holds for $j \geqq k-3$), then multiplying (8) with j and subtracting from (7) gives

(10) $$0 \geqq \sum_{r=2}^{k-3} (r-j)(r-1)v_r = k(k-1) + d(d-1) - (j-2)kd.$$

Multiplying with $(dk)^{-1}$ leads to

(11) $$\left(\frac{k}{d} + \frac{d}{k}\right) - \left(\frac{1}{d} + \frac{1}{k}\right) \leqq j - 2,$$

and here the first sum is $\geqq 2$, the second sum $\leqq \frac{1}{2} + \frac{1}{6} < 1$, if $d > 1$, so that in this case (11) gives $j \geqq 4$, which also results from (10) and $k > 2$ in the case $d = 1$. Since (9) holds for $j = k-3$, this gives $k \geqq 7$. From (6,5) follows

$$r-1 > d \Rightarrow \forall x \in C: (v_{rx} = 0) \Rightarrow v_r = 0,$$

so that (9) holds for $j = d+1$, and with this value for j we derive

$$k \leqq d(d-1)$$

from (10), which together with $7 \leqq k$ gives $d \geqq 4$.

In this proof no explicit use was made of the second complete parallel class; but the essentially used facts, that every axial line has order k and that $[p] \leqq k-3$ for all $p \in P$, depend on the existence of a second complete parallel class. By explicitly using both complete parallel classes one can also prove the impossibility of $k = 7$: Here (10) with $j = k-3 = 4$ gives

(12) $\quad 2(v_2 + v_3) = d(15-d) - 42 \geqq 0$

and therefore

$$4 \leqq d \leqq 11.$$

Combining the equation in (12) with (8) gives

(13) $\quad 3v_4 - v_2 = (d-8)d + 42$

and therefore

$$3v_4 \geqq (d-8)d + 42.$$

By combining this result with the inequality $v_4 \leqq 7\left[\frac{d}{3}\right]$, following from (5,6), the values 4,5,11 for d are ruled out. But every one of the remaining values 6,7,8,9,10 has to be treated differently.

<u>4.</u> For a DP-plane $(P, \mathcal{L}, I)$ of order k, respresented by (C,G) as in <u>2</u>, we consider besides the finiteness condition

(F) $\quad k \in \mathbb{N}$

the following three other conditions:

(Gr) $\quad$ (G,o) <u>is a group</u>.

(D) $\quad$ G <u>operates doubly transitive on</u> C.

(A) $\quad$ <u>There exists an affine plane with point set</u> P, <u>line set</u> $\supset \mathcal{L}$ <u>and incidence relation</u> $\supset I$; we say for short: there exists a

<u>point conserving embedding into an affine plane.</u>

It is easy to see, that (Gr) can be geometrically interpreted by the "Rechtecksaxiom" of Karzel [2]. We stated already in <u>2</u>, that (Gr) can be fulfilled, but not (Gr) $\wedge$ (F). Considering all the four conditions and their conjunctions a nearly complete result is given in

<u>Proposition 3.</u> <u>DP-planes with resp.</u> (Gr),(D),(A) <u>exists, but none with resp.</u> (Gr) $\wedge$ (F), (D) $\wedge$ (F), (A) $\wedge$ (F), (A) $\wedge$ (D), (Gr) $\wedge$ (D).

<u>Proof.</u> a) (D) means geometrically, that the DP-plane is a linear space: every two points are joined by a line. Beginning with a set of four points, we apply the well-known procedure of constructing the free plane generated by this set with a slight modification: From the set of four points we make at first an affine plane $\mathcal{A}_2$ of order 2, calling the lines ot two parallel classes of $\mathcal{A}_2$ "lines of first resp. second class". This is the first stage. Generally at the n-th stage we have a linear space with the properties:

(i) Through every point there is exactly one line of first class and one line of second class:

(ii) two lines of the same class do not meet;

(iii) through every point goes a line, which belongs neither to the first nor to the second class.

The (n+1)-th stage now is reached in three steps:

1) For every two lines L,M,not meeting and not both belonging to the same class, the set {L,M} is added as new point, incident only with L and M;

2) For every two points (new or old) p,q, which are not yet joined, we add

$\{p,q\}$ as new line, incident only with p,q;

3) For every (new) point p, which is not incident with a line of first (resp.second) class, we add a new line, incident only with p, calling it a line of first (resp.second) class.

This is once more a linear space with (i),(ii),(iii). Only the proof of (iii) is not trivial in the case of a new point $\{L,M\}$, where L is of first and M of second class: There is a point p of $\mathcal{A}_2$, not on L and not on M, and thus the line $\{p,\{L,M\}\}$ is constructed in step 2. It is clear, that the union of all stages gives a DP-plane, which is also a linear space.

b) To construct a DP-plane with (A) we start from an infinite affine plane, coordinatized by a set C, so that we can take C^2 as the point set and - distinguishing two elements $o,e \in C$ - represent the parallel classes of lines by the two coordinate axes and the lines joining the "origin" (o,o) with (e,u) for $u \in C \setminus \{o\}$, u called the slope of the lines. C being infinite, we have $|C \setminus \{o\}| = |C^2|$ and therefore a bijection ρ from $C \setminus \{o\}$ onto C^2. We construct L_u (for $u \in C \setminus \{o\}$) as the line of slope u through the point $\rho(u)$. Then, because of

$$L_u \neq L_{u'} \Rightarrow u \neq u' \Rightarrow L_u \nparallel L_{u'}$$

the lines $\{a\} \times C$, $C \times \{b\}$ $(a,b \in C)$, L_u $(u \in C \setminus \{o\})$ form a DP-plane with point set C^2, and due to its construction there is a point conserving embedding into the affine plane, we have started from. Of course ρ need not be a bijection, but only a surjection, and one could restrict u to an equipotent subset of $C \setminus \{o\}$. Another construction starts with the affine plane over a field $(C,+,\cdot)$, in which every quadratic or cubic equation has a root: With $L_u = \{(x,y) \mid y = u(x+u+u^2)\}$ for $u \in C \setminus \{o\}$

we have once more a DP-plane with (A).

c) Deleting the axial lines in a DP-plane of order k with (D) $\wedge$ (F) gives a $(k^2,k,1)$-block design and thus $[p] = k+1$ for all $p \in P$ in contradiction to Prop. 1. Since DP-planes with (D) are 2-structures in the sense of Karzel [2], one could also use prop.(1.2) in [2]: From this follows, that a finite DP-plane with (D) would be an affine plane, which is impossible.

d) (A) $\wedge$ (F) means, that we have a point conserving embedding into an affine plane $\mathcal{A}$ of finite oder k. The axial lines of the DP-plane constitute two of the k+1 parallel classes of $\mathcal{A}$. The b(>k) non-axial lines belong to the remaining k-1 parallel classes; but they should be pairwise non-parallel, which is a contradiction.

e) Let there be a point conserving embedding of a DP-plane into an affine plane $\mathcal{A}$. We take a non-axial line L of the DP-plane and a different line $M \parallel L$ of $\mathcal{A}$. Then M does not belong to the DP-plane and thus two of its points are not joined in the DP-plane. So (A) $\wedge$ (D) can never hold.

f) Let G be a doubly transitive permutation group on C with (FP') and $|C| \geqq 2$. We choose $a,b \in C$ with $a \neq b$. Then $g \in G$ exists with $g(a) = b$, $g(b) = a$ and therefore $g^2(a) = a$, $g^2(b) = b$. From (FP') follows $g^2 = 1$ and, since $g \neq 1$, the existence of only one $c \in C$ with $g(c) = c$. Since $c \neq a$, there is $h \in G$ with $h(a) = c$, $h(c) = a$ and therefore $h^2 = 1$. From $gh(c) = g(a) = b \neq c$ follows $gh \neq 1$, so that gh has exactly one fixed point d. Since $gh(a) = g(c) = c$, we have $c \neq d$. But from $gh(h(d)) = g(d) = g(gh(d)) = h(d)$ follows, that g(d) is fixed point of gh and therefore $= d$, which is impossible, since g has only the fixed point c.

<u>5.</u> The question remains open, if a DP-plane with (Gr) $\wedge$ (A) exists.

Hoping to find such a DP-plane, we strengthen (A) by supposing the embedding plane to be a <u>translation plane</u> (see e.g.[4]): $(C,+,\cdot)$ is a quasifield with 0,1 as neutral elements of addition resp. multiplication; the lines are the point sets $[c] = \{c\}\times C$, $[u,v] = \{(x,y) \mid y = ux + v,\ x \in C\}$. All lines $[c]$, $[o,v]$ belong as axial lines to the DP-plane, and the non-axial lines form a group $(G, \circ)$. Since

$$[u,v] \parallel [u',v'] \Leftrightarrow u = u'$$

there is a mapping f of a subset $C' \in C \setminus \{o\}$ onto C such that the

$$L_u = [u,f(u)] \quad (u \in C')$$

are the non-axial lines of the DP-plane. (TP) can be formulated as

(TP') $\qquad \forall x,y \in C \quad \exists u \in C': \quad y = ux + f(u).$

Vice versa every surjection $f: C' \to C$ with $C' \subseteq C \setminus \{o\}$ and (TP') defines in this way a DP-plane with (A), embedded in the translation plane over $(C,+,\cdot)$, Condition (Gr), stating that $(G, \circ)$ with $G = \{L_u \mid u \in C'\}$ is a group, now implies, that for $u,v \in C'$ there is $w \in C'$ with

$$L_u \circ L_v = L_w,$$

that is:

(1) $\qquad \forall x \in C: u(vx + f(v)) + f(u) = wx + f(w).$

With x = o this gives

(2) $\qquad uf(v) + f(u) = f(w)$

and for x = 1 then, using (2) and the one distributive law valid in quasifields,

$$uv = w.$$

We have therefore

(3) $\qquad uf(v) + f(u) = f(uv),$

(4) $$L_u \circ L_v = L_{uv}$$

for all $u,v \in C'$. Because of (4) and since $u \to L_u$ is injective, the inverse of this mapping is an isomorphism of the group $(G, \circ)$ into the loop $(C \setminus \{o\}, \cdot)$ and therefore $(C', \cdot)$ a subgroup of this loop and isomorphic to $(G, \circ)$. To simplify the notations, we identify C' in the sequel with G. So the construction of a DP-plane with (Gr) $\wedge$ (A), embedded in the translation plane over the quasifield $(C,+,\cdot)$ needs a subgroup of the multiplicative loop of the quasifield and a surjection $f:G \to C$ with (3) for all $u,v \in G$; this functional equation means, that f is a onedimensional C-cocycle of the group G, operating by left multiplication on the abelian group $(C,+)$. So we are led to

Proposition 4. *The DP-planes with* (Gr) $\wedge$ (A), *embedded in the translation plane over the quasifield* $(C,+,\cdot)$ *are given by the pairs* (G,f), *where* G *is subgroup of the loop* $(C \setminus \{o\},\cdot)$ *and* f *a onedimensional C-cocycle of* G *with* $fG = C$, *such that the* $[u,f(u)]$ $(u \in G)$ *are the non-axial lines.*

Proof. We have only to check, that the $[u,f(u)]$, which are pairwise non-parallel, fulfill indeed (TP) and (Gr). Since $fG = C$ and G is a group, elements $x,y \in C$ can be represented as $x = f(v)$, $y = f(uv)$ with $u,v \in G$, and (3) gives $y = ux + f(u)$. Thus (TP') and therefore (TP) holds. To prove (Gr), one has only to show (4), since then $(\{L_u \mid u \in G\}, \circ)$ is isomorphic to $(G,\cdot)$. But (4) means (1) with $w = uv$ and this equation follows from (3) by the distributive law and

(5) $$\forall u,v \in G \quad \forall x \in C: \quad u(vx) = (uv)x.$$

To prove (5), we write $x = f(w)$, $w \in G$. using (3), the associativity of multiplication in G and the distributive law, we get

$f((uv)w) = (uv)x + uf(v) + f(u)$

$= f(u(vw)) = u(vx) + uf(v) + f(u)$

and therefore (5).

Since $\qquad L_u(o) = o \Leftrightarrow f(u) = o,$

we have $G_o = f^{-1}\{o\}$ for the subgroup G_o, introduced in 2, and from (3) follows, if we put $u' = uv$:

(6) $\qquad f(u) = f(u') \Leftrightarrow uG_o = u'G_o.$

Thus f induces a bijection of the set of the left cosets uG_o onto C. Since 2(3) together with $G_o \neq G$ excludes $|G_o| = 1$, f itself is not injective. A direct proof of $|G_o| \neq 1$, using the cocycle equation (3) runs as follows: (3) with $u = 1$ gives $f(1) = o$; since $fG = C$, there is $u \in G \setminus \{1\}$, and from the quasifield-definition (see [4],p.92) follows the existence of $x \in C$ with $ux + f(u) = x$; with $f(v) = x$, $v \in G$ and $u' = v^{-1}uv \in G \setminus \{1\}$ this gives

$f(v) = uf(v) + f(u) = f(uv) = f(vu') = vf(u') + f(v)$

and therefore $f(u') = o$, $u' \in G_o$. (For another proof in the special case, that C is a skew-field and $G = C \setminus \{o\}$, see [1].)

References

[1] T.M.K.DAVISON, J. ACZEL, 5. Remark (P171R1), aequationes mathematicae 17, 374/5 (1978)

[2] H.KARZEL, Zusammenhänge zwischen Fastbereichen, scharf zweifach transitiven Permutationsgruppen und 2-Strukturen mit Rechtecksaxiom. Abh.Math.Sem.Hamburg 32, 191-206 (1968).

[3] G.PICKERT, Fastprojektive Ebenen. Mitt.Math.Sem. Giessen, Heft 123, 7-23, (1977).

[4] G.PICKERT, Projektive Ebenen. Springer Verlag, 2.Aufl. 1975.

A Problem of Free Mobility

Günter Pickert

Mathematisches Institut der Justus-Liebig-Universität
Arndtstr. 2, D 6300 Giessen, F.R.Germany

Let V be a 2-dimensional vector space over the field K and Γ a group of automorphisms of V. We want to define "congruence" ($\cong$) of vectors in V by Γ: $\forall \vec{x},\vec{y} \in V: (\vec{x} \cong \vec{y} \;\Leftrightarrow\; \exists \gamma \in \Gamma: \gamma(\vec{x}) = \pm\vec{y})$.

Obviously $\cong$ is an equivalence relation in V. In order to ensure also the "natural" property

$$\forall \vec{x} \in V \setminus \{0\},\ \forall r \in K: (\vec{x} \cong \vec{x}r \;\Leftrightarrow\; r^2 = 1),$$

one has to demand

(EV) No element of $K \setminus \{1,-1\}$ is eigenvalue of an element of Γ.

But to be able to "compare" linear independent vectors with regard to congruence, Γ must allow "free mobility" in the following sense:

(FM) Γ is transitive on the set of onedimensional subspaces of V.

Now the question is, under what conditions for K there exists Γ with (EV),(FM). Of course for a pythagorean field K (i.e. $1 + x^2$ is a square $\neq 0$ for every $x \in K$) such a group Γ exists. This remains true also for a weakly pythagorean field K, that is a field K with

(WP) $\forall x \in K: (1 + x^2 \text{ square} \;\Leftrightarrow\; -(1 + x^2) \text{ non-square})$:

Relative to a given basis of V we describe Γ by the group of matrices

(1) $\begin{pmatrix} a & -b \\ b & a \end{pmatrix}$ with $a,b \in K$, $a^2 + b^2 = \pm 1$;

according to (WP) $-b^2$ is a square in K only for $b = 0$, which gives (EV), and for every $c \in K$ (WP) gives $c^2 \neq -1$, therefore according to (WP) an element $b \in K$ exists with $b^2(1 + c^2) = \pm 1$, that is $a^2 + b^2 = \pm 1$ for $a = bc$, thus proving (FM). To connect this with the results of P.KUSTAANHEIMO,

On the relation of congruence in finite geometries, Rend.d.Math. 16, 286-291(1957), one has to put there $k = -1$, getting $x^2 + y^2 = 1$ and $x^2 + y^2 = -1$ as equations of the two "measure curves". For a field, which is formally real, the concepts "pythagorean" and "weakly pythagorean" coincide: From (WP) follows, that $1 + x^2$ must be a square (and then $\neq 0$), since otherwise we would have $y \in K$ with $-(1 + x^2) = y^2$ and thus $1 + x^2 + y^2 = 0$, contrary to the assumption "formally real". On the other hand every finite field K, in which -1 is not a square (i.e. $|K| \equiv 3 \bmod. 4$), is weakly pythagorean, but of course not pythagorean: In such a field the squares $\neq 0$ form a subgroup of index 2 in the multiplicative group, whence we get not only (WP) but also the stronger property "weakly euclidean":

(WE) $\quad \forall x \in K \setminus \{0\}$: (x square $\Leftrightarrow$ -x non-square).

If -1 is not a square in K, the group of matrices (1) operating on K^2 is isomorphic to a subgroup G of the multiplicative group L^* of a quadratic extension field L of K, operating on L by multiplication. Specializing our problem to such groups, the conditions (EV),(MF) are translated into

(EV') $\quad G \cap K^* \subseteq \{1,-1\}$,

(FM') $\quad GK^* = L^*$

with K^* as the multiplicative group of K.

For a finite field K a subgroup G of L^* (the multiplicative group of a quadratic extension field L of K) with (EV'),(MF') exists if and only if $|K| \not\equiv 1 \bmod. 4$.

To prove this, we suppose $L^* = \langle\omega\rangle$, $G = \langle\omega^k\rangle$. Then with $|K| = q$ we have $|L| = q^2$, $|L^*| = |K^*| \cdot (q+1)$, thus $K^* = \langle\omega^{q+1}\rangle$ and with

$d = \text{g.c.d.}(k,q+1)$, $m = \text{l.c.m.}(k,q+1) = d^{-1}k(q+1)$ therefore

$$G \cap K^* = \langle\omega^m\rangle, \tag{2}$$

$$GK^* = \langle\omega^d\rangle. \tag{3}$$

Now because of (3) the condition (FM') means, that no common divisor $\neq 1$ of $k, q+1$ divides $q^2 - 1$, and this is equivalent with $d = 1$. If q is odd, (2) und (EV') give $\frac{1}{2}(q^2 - 1) \mid m$ and therefore $\frac{1}{2}(q - 1) \mid k$. Thus $\frac{1}{2}(q-1)$ cannot be even, since otherwise 2 would be a common divisor of $k, q+1$ contrary to $d = 1$. But for odd $\frac{1}{2}(q - 1)$, that is $q \equiv 3$ mod.4 we know already the existence of a subgroup G ($\leqq L^*$) with (EV'),(MF'); besides, from (2), (3) one can easily see, that $G = \langle\omega^k\rangle$ with $k = \frac{1}{2}(q-1)$ fulfils (EV'),(MF'). If q is even, (EV') means $(q-1) \mid k$, according to (2) and $d = 1$. Therefore, since $q - 1$, $q + 1$ are relatively prime, conditions (EV'), (MF') are fulfilled for $G = \langle\omega^{q-1}\rangle$. To construct also in this case an isomorphic group of matrices, we use $(1,\rho)$ with $\rho = \omega^{q-1}$ as basic of $L|K$, and with $r = \rho + \rho^{-1} \in K$ we get $\rho^2 = r\rho + 1$. Thus the matrix $R = \begin{pmatrix} 0 & 1 \\ 1 & r \end{pmatrix}$ describes the multiplication with ρ, and the automorphism of V, determined by R relative to a given basis, generates therefore a group fulfilling (EV, (MF). This automorphism, given by R, can be written as a product of two shears, since the matrix

$$\begin{pmatrix} r+1 & 1 \\ r & 1 \end{pmatrix} = \begin{pmatrix} 1 & 1 \\ 0 & 1 \end{pmatrix} \begin{pmatrix} 1 & 0 \\ r & 1 \end{pmatrix}$$

is similar to R, having same determinant and trace as R.

If K has characteristic $\neq 2$, the existence of a group Γ with (EV), (FM) implies, that not all elements of K are squares. To prove this, we assume to the contrary, that every element of K is a square, so that according to (EV) the elements of Γ must be affine reflections, shears, $\vec{x} \to -\vec{x}$ or the products of this "point reflection" with shears. That with these transormations alone (FM)cannot be fulfilled, follows from the impossibility of the following three cases:

a) Γ contains two shears with different axes.

A basis of V can be chosen so, that the matrices of the shears are $\begin{pmatrix}1 & c\\ 0 & 1\end{pmatrix}$ and $\begin{pmatrix}1 & 0\\ 1 & 1\end{pmatrix}$ with $c \neq 0$. Their product $\begin{pmatrix}1+c & c\\ 1 & 1\end{pmatrix}$ has the characteristic equation

$$\lambda^2 - (2+c)\lambda + 1 = 0.$$

This must be equivalent to $\lambda = -1$, so that $c = -4$ results. But Γ contains also the shear with matrix $\begin{pmatrix}1 & -c\\ 0 & 1\end{pmatrix}$ implying $c = 4$, which is a contradiction.

b) Γ contains an affine reflection and a shear, whose axis differs from the axes of the reflection.

A basis of V can be chosen so, that the matrices of the two mappings are $\begin{pmatrix}0 & 1\\ 1 & 0\end{pmatrix}$ and $\begin{pmatrix}1 & c\\ 0 & 1\end{pmatrix}$ with $c \neq 0$.

Their product $\begin{pmatrix}0 & 1\\ 1 & c\end{pmatrix}$ has the characteristic equation

$$\lambda^2 - c\lambda - 1 = 0,$$

and this must have the roots 1, -1, which is impossible.

c) Γ contains two affine reflections with different axes.

A basis of V can be chosen so, that one reflection has the matrix $\begin{pmatrix}1 & 0\\ 0 & -1\end{pmatrix}$. Then the matrix of the other is of the form $\begin{pmatrix}a & (1-a^2)b^{-1}\\ b & -a\end{pmatrix}$ with $a^2 \neq 1$.

The product of the matrices has the characteristic equation

$$\lambda^2 - 2a\lambda + 1 = 0,$$

which gives the contradiction $a = \pm 1$, since $\lambda = \pm 1$.

The question remains open, if in the case of characteristic 0 the existence of quadratic irrationalities really is necessary for the fulfilment of (EV), (FM) or if a group Γ with (EV), (FM) exists even for the field of rational numbers as K.

A UNIFIED APPROACH TO MIQUEL'S THEOREM AND ITS DEGENERATIONS

By

H.-J. Samaga

Universität Hamburg

"Die allgemeine Miquel'sche Kreisfigur ... kann bekanntlich kurz so beschrieben werden: Legt man für vier Punkte eines Kreises eine zyklische Folge fest und durch je zwei so aufeinanderfolgende Punkte jedesmal irgendeinen Kreis, so gehören die vier weiteren Schnittpunkte benachbarter dieser Kreise selbst einem Kreis an." With these words L. PECZAR began his paper [5] , published in 1950, in which he gave a short algebraic proof of the following theorem:

Let L be an associative, commutative algebra over the field K with multiplicative identity 1_L, such that $1_K = 1_L$ [1)] and $K \neq L$. If for pairwise different $a,b,c,d,e,f,g,h \in L$ a-d, b-c, e-h, f-g, a-h, f-c, e-d, b-g, e-g, f-h are regular and some other differences are not zero divisors, and if a,b,c,d; a,d,e,h; b,c,f, g; a,c,f,h; b,d,e,g are concircular, then e,f,g,h are concircular too.

By definition the four elements r,s,t,u are called concircular, if and only if

$$\begin{bmatrix} r & s \\ t & u \end{bmatrix} := \frac{r-u}{r-t} \cdot \frac{s-t}{s-u} \in K^* := K \setminus \{0\} .$$

In the meantime this theorem and the way to prove it by cross ratio has been enlarged in different ways.

At first L was embedded in the projective line over L

$\mathbb{P}(L) := \{U(x_1,x_2) \mid \langle x_1,x_2\rangle = L\}$. By U we denote the group of all units of L. The elements of $\mathbb{P}(L)$ are called points. Two points $A = U(a_1,a_2)$, $B = U(b_1,b_2)$ are called parallel $(A \parallel B) :\Leftrightarrow a_1b_2 - a_2b_1 \notin U$. If L is a field, the parallel-relation is equalized to the relation of identity. If L is the ring of the dual numbers, parallel is an equivalence-relation. In case of $L = K \times \ldots \times K$ parallel is divided into n equivalence classes $\parallel_1, \ldots, \parallel_n$, and $P = Q \Leftrightarrow P \parallel_i Q$ for all $i=1,\ldots,n$.

Definition. Let A,B,C,D be points such that $A \nparallel C$, $B \nparallel D$. Then

$$\begin{bmatrix} A & B \\ C & D \end{bmatrix} := \begin{vmatrix} a_1 & a_2 \\ d_1 & d_2 \end{vmatrix} \cdot \begin{vmatrix} a_1 & a_2 \\ c_1 & c_2 \end{vmatrix}^{-1} \cdot \begin{vmatrix} b_1 & b_2 \\ c_1 & c_2 \end{vmatrix} \cdot \begin{vmatrix} b_1 & b_2 \\ d_1 & d_2 \end{vmatrix}^{-1} \qquad \text{(cross-ratio)}$$

Let $\mathbb{P}(K)$ be the projective line over K . Because of

1) This - necessary - assumption L. PECZAR missed in [5] .

$$K^*(k_1,k_2) \in \mathbb{P}(K) \quad \to \quad U(k_1,k_2) \in \mathbb{P}(L)$$

$\mathbb{P}(K)$ can be embedded in $\mathbb{P}(L)$. We define

$$\mathbb{K}^o := \{\mathbb{P}(K)^\gamma \mid \gamma \in \Gamma(L)\} \subset \mathbb{P}(L) \quad {}^{2)} .$$

The elements of $\mathbb{K}^o$ are called ordinary circles or regular circles or chains. Abbreviating in this paper we want to write o-circle. Altogether we have constructed a chain-geometry $\Sigma(K,L)$ [3]. If $K = \mathbb{R}$ and $L = \mathbb{C}$ or $\mathbb{D}$ (ring of the dual numbers over $\mathbb{R}$) or $\mathbb{R} \times \mathbb{R}$ we get the classical geometries of Möbius or Laguerre or Minkowski.

Next we give some remarks valid in chain-geometries.

a) For every three pairwise nonparallel points there is a unique o-circle containing them.

b) Different points on an o-circle are nonparallel.

c) A,B,C,D are pairwise different points on an o-circle $\Rightarrow \begin{bmatrix} A & B \\ C & D \end{bmatrix} \in K^*$.

From c) we get that in chain-geometries we can use the PECZAR-definition of concircular points in an analogous way. Another important idea is that one dealing with tangency of o-circles. In this paper we want to define tangency only in case of $\dim_K L = 2$ in an exact way.

Definition Let $\Sigma(K,L)$ be a chain-geometry, $\dim_K L = 2$, $k,\ell \in \mathbb{K}^o$. k and ℓ tangent eachother in the point

$$A \quad (k \; A \; \ell) \; :\Leftrightarrow \; k \cap \ell = \{A\} \quad \text{or} \quad k = \ell .$$

From now on we assume a two dimensional chain-geometry $\Sigma(K,L)$.
The proofs of the following remarks can be found in W. BENZ [4] .

d) (tangency theorem) For each o-circle k, each point $A \in k$, and each point B nonparallel to A there is a unique o-circle ℓ containing A and B such that $k \; A \; \ell$.

e) In each point the tangency relation is an equivalence relation.

f) Let $A,B,C \in k \in \mathbb{K}^o$ be paiwise different points, and $A,D,E \in \ell \in \mathbb{K}^o$ pairwise different, too. Then are equivalent

(i) $k \; A \; \ell$

(ii) $\begin{bmatrix} A & B \\ D & C \end{bmatrix} - \begin{bmatrix} A & B \\ E & C \end{bmatrix} \in K^*$

(iii) $\begin{bmatrix} A & D \\ B & E \end{bmatrix} - \begin{bmatrix} A & D \\ C & E \end{bmatrix} \in K^*$

2) It is $\Gamma(L) \cong GL(2,L)/_{Z}\, GL(2,L)$, Z the center of the group $GL(2,L)$.

3) For a detailed description of the theory of chain-geometries we refer to the book [4] by W. BENZ.

Using the remarks c) and f) W. BENZ proved the so called full theorem of Miquel by cross ratio ([2], [4]). We want to write down this theorem in the following way:

<u>Theorem 1</u> Let $k_o, k_1, k_2, k_3, k_4 \in \mathbb{K}^o$ and let $A,B,C,D,E,F,G,H \in \mathbb{P}(L)$ such that $|\{A,B,\dots,G\}| = 7$, $H \in \mathbb{P}(L) \setminus \{B,C,\dots,G\}$, and $E \nparallel F$ or $G \nparallel H$.

$$\text{If } k_o \cap k_i = \left\{\begin{array}{l} A,\ C \\ B,\ C \\ B,\ D \\ A,\ D \end{array}\right\},\ k_i \cap k_{i+1} = \left\{\begin{array}{l} C,\ F \\ B,\ G \\ D,\ E \\ A,\ H \end{array}\right\} \text{ for } i = \left\{\begin{array}{l} 1 \\ 2 \\ 3 \\ 4\ (5=1) \end{array}\right. \qquad (*)$$

then there is a unique o-circle k such that

$$k \cap k_i = \left\{\begin{array}{l} F,\ H \\ F,\ G \\ E,\ G \\ E,\ H \end{array}\right\} \text{ for } i = \left\{\begin{array}{l} 1 \\ 2 \\ 3 \\ 4 \end{array}\right. \qquad (**)$$

Because the assumptions (*) and the statements (**) are not very intuitive we like to explain this kind of configuration by drawing a cube. The six faces of a cube represent the o-circles $k_o,\dots,k_4,k$ of the configuration, and the eight vertices represent the points. We like to draw the cube in such a way that the bottom face of the cube represents the o-circle k closing the configuration (see fig. 1).

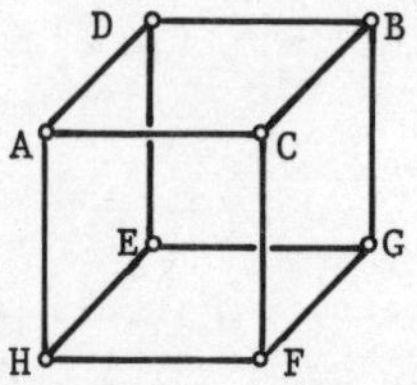

fig. 1

Using the remarks c) and f) consequently W. SCHAEFFER proved in 1974 more general theorems than theorem 1 (see [7]). By some considerations of incidence we summarize these theorems to one theorem which runs as follows:

<u>Theorem 2</u> Let $k_o, k_1, k_2, k_3, k_4 \in \mathbb{K}$ and let $A,B,C,D,E,F,G,H \in \mathbb{P}(L)$ such that $|\{A,B,E,F\}| = 4$, and $E \parallel F$ or $G \parallel H$. If the assumption (*) (theorem 1) holds true, then there is a unique o-circle k such that (**) (theorem 1) holds true (see fig.1).

The assumption about nonparallelism of points makes some trouble, but it cannot be dropped, because there exist counterexamples. So the idea of the o-circles was enlarged by E, SCHRÖDER in some papers ([8], [9], [10]), if $\dim_K L = 2$.

As is well known there are three types of twodimensional commutative algebras over K, namely L field or $L = \mathbb{D}$ or $L = K \times K$. Only if L is not a field there are different nonparallel points. In these algebras E. SCHRÖDER defined so called irregular or extraordinary circles. In this paper we want to give a different, but equivalent definition.

Definition Let $\Sigma(K,L)$ be a chain-geometry, $\dim_K L = 2$.

(i) $L = K \times K$: For each $P \in \mathbb{P}(L)$
$k = k(P) := \{X \in \mathbb{P}(L) \mid X \parallel P\}$ is called irregular circle. P is called the supporter of k .

(ii) $L = \mathbb{D}$: For every pair $P,Q \in \mathbb{P}(L)$, $P \nparallel Q$,
$k = k(P,Q) := \{X \in \mathbb{P}(L) \mid X \parallel P \text{ or } X \parallel Q\}$
is called irregular circle.

We want to abbreviate irregular circle by i-circle. We denote the set of all i-circles by $\mathbb{K}^i$ and $\mathbb{K} := \mathbb{K}^o \cup \mathbb{K}^i$, the elements of $\mathbb{K}$ are called circles.
It is $\mathbb{K}^o \cap \mathbb{K}^i = \emptyset$.

We want to give an example. In the parabola-model of the Laguerre-geometry $\Sigma(\mathbb{R}, \mathbb{D})$ the o-circles are straight lines not in the direction of the y-axis, or parabolas with an axis parallel to the y-axis. The i-circles are pairs of parallel straight lines parallel to the y-axis (see fig.2). For more examples we refer to the end of this paper.

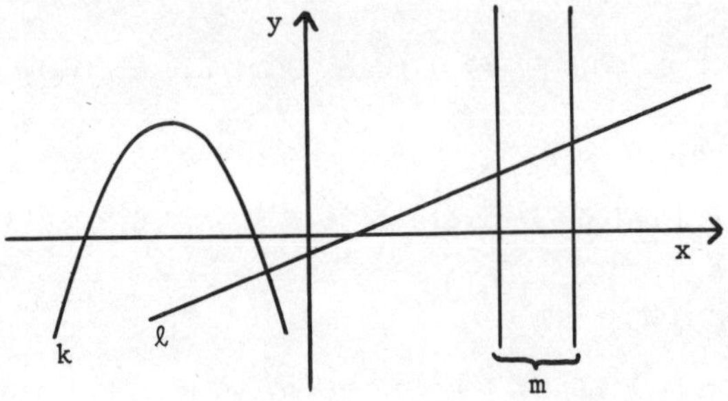

fig. 2: $k, \ell \in \mathbb{K}^o$, $m \in \mathbb{K}^i$

We summarize further remarks dealing with chain-geometries of dimension two.

g) Let $A,B,C \in \mathbb{P}(L)$, $A \nparallel B$. Then there exists a unique circle k containing A,B,C .

h) Let $k_o \in \mathbb{K}^o$, $k_1 \in \mathbb{K}^i \Rightarrow 1 \leq |k_o \cap k_1| \leq 2$. In case of $k_o \cap k_1 = \{P\}$, P is the supporter of k_1 .

i) Let $k_1,k_2 \in \mathbb{K}^i \Rightarrow |k_1 \cap k_2| \neq 1$. Except the minimal Laguerre-case [4] there exist different i-circles k_1,k_2 , such that $|k_1 \cap k_2| > 2$.

j) Let $k \in \mathbb{K} \Rightarrow |k| \geq 3$.

We pass over the simple proofs and come to the main part of our paper. Using i-circles E. SCHRÖDER enlarged the full theorem of Miquel once more.

---

4) This is $\Sigma(GF(2), \mathbb{D})$, $|\mathbb{D}| = 4$. There are six points, eight o-circles and three i-circles.

<u>Theorem 3</u> ($\dim_K L = 2$) Let $k_0,k_1,k_2,k_3,k_4 \in \mathbb{K}$ and let $A,B,C,D,E,F,G,H \in \mathbb{P}(L)$ such that $|\{A,B,E,F\}| = 4$. If the assumption (*) (theorem 1) holds true and if $\Sigma(K,L)$ is not the minimal Laguerre-case, then there is a unique circle k such that (**) (theorem 1) holds true (see fig.1).

With some restricting assumptions theorem 3 holds true even in the minimal Laguerre-case (see the proof of theorem 3 in this paper later on). Proving theorem 3 E. SCHRÖDER uses other methods than cross ratio.

Now we want to show:

A. Theorem 3 can be proved by cross ratio.

B. The assumption $|\{A,B,E,F\}| = 4$ can be replaced by the assumption $E \nparallel F$.
This cannot be dropped, we give a counterexample afterwards.

We need a connexion between cross ratio and points being incident with an i-circle. So our next goal is

<u>Theorem 4</u> Given a two dimensional chain geometry $\Sigma(K,L)$.
Let $A,B,C,D \in \mathbb{P}(L)$ at least three different points, $A \nparallel C$, $B \nparallel D$.
Then

(i) $\begin{bmatrix} A & B \\ C & D \end{bmatrix} \in \{0,1\} \implies$ there is a unique circle $k \in \mathbb{K}$ such that $A,B,C,D \in k$.

(ii) There is a unique i-circle $k \in \mathbb{K}^i$, $A,B,C,D \in k$
$\implies \begin{bmatrix} A & B \\ C & D \end{bmatrix} \in \{0,1\}$.

<u>Proof</u> Since $A = U(a_1,a_2)$, $a_i \in L$, and so on, we get by a simple calculation

$$\begin{bmatrix} A & B \\ C & D \end{bmatrix} = 0 \iff (a_1 d_2 - a_2 d_1)(b_1 c_2 - b_2 c_1) = 0 \qquad (\text{o})$$

$$\begin{bmatrix} A & B \\ C & D \end{bmatrix} = 1 \iff (a_1 b_2 - a_2 b_1)(c_1 d_2 - c_2 d_1) = 0 \qquad (\text{oo})$$

(i) If $|\{A,B,C,D\}| = 3$, because of $A \nparallel C$ according to remark g), there is a unique circle k being incident with A,B,C,D . If $|\{A,B,C,D\}| = 4$, there are $(a_1 d_2 - a_2 d_1)$, $(b_1 c_2 - b_2 c_1)$, respectively $(a_1 b_2 - a_2 b_1)$, $(c_1 d_2 - c_2 d_1)$ zero divisors and it follows $A \parallel D$, $B \parallel C$ in case of (o), respectively $A \parallel B$, $C \parallel D$ in case of (oo). If L is a field, the assumptions cannot be realized. If L is the ring of the dual numbers, we have $A,B,C,D \in k(A,C) = k(B,D) \in \mathbb{K}^i \subseteq \mathbb{K}$ (o) or $A,B,C,D \in k(A,D) = k(B,C)$ (oo) . If L is the ring of the double numbers, the zero divisors by (o) respectively (oo) are belonging to different maximal ideals. (Remember that the relation of parallelism is divided into two equivalence relations $\parallel_1$, $\parallel_2$.) So we can assume without loss of generality $A \parallel_1 D$, $B \parallel_2 C$ respectively $A \parallel_1 B$, $C \parallel_2 D$. Let $P := U((a_{11},b_{12}) , (a_{21},b_{22}))$,

$Q := U((a_{11}, c_{12}), (a_{21}, c_{22}))$.[5)] Thus we have $P, Q \in \mathbb{P}(L)$ and $A,B,C,D \in k(P) \in \mathbb{K}^i \subseteq \mathbb{K}$ respectively $A,B,C,D \in k(Q)$. The uniqueness of the circle follows directly by remark g).

(ii) Let $L = K \times K$. Using the assumption $A \not\parallel C$, $B \not\parallel D$, from the definition of the i-circles we obtain $A \parallel_1 D$, $B \parallel_2 C$ (†) or $A \parallel_1 B$, $C \parallel_2 D$ (††). If L is the ring of the dual numbers we also obtain (†) or (††) with the additional quality $\parallel_1 = \parallel_2$. In both geometries (†) respectively (††) implies $(a_1 d_2 - a_2 d_1) \cdot (b_1 c_2 - b_2 c_1) = 0$ respectively $(a_1 b_2 - a_2 b_1)(c_1 d_2 - c_2 d_1) = 0$. Thus the assertion follows from (o), (oo).

We shall need the following corollary of theorem 4:

k) Let $A,B,C,D \in \mathbb{P}(L)$ pairwise different and $C \not\parallel A \not\parallel D \not\parallel B \not\parallel C$. Then are equivalent

(i) $\begin{bmatrix} A & B \\ C & D \end{bmatrix} = 1$

(ii) There is a unique i-circle $k \in \mathbb{K}^i$ such that $A,B,C,D \in k$.

Now up to the proof of theorem 3 (with the assumption $E \neq F$ instead of $|\{A,B,E,F\}| = 4$). For a better understanding we use the cube-model.

1. The assumption $|k_i \cap k_j| \leq 2$ for some $i \neq j$ and the exclusion of the minimal Laguerre-case imply that different points, which lie on a common edge (of the cube) have to be nonparallel. Only in the minimal Laguerre-case

$k_i \cap k_j = \{A,B\}$, $A \parallel B$, $A \neq B$ (□)

can happen (see remark i). If (□) is excluded, theorem 3 is true in the minimal Laguerre-case, too.

2. We prove that the assumptions (*) (see theorem 1) and $E \neq F$ imply $|\{A,B,E,F\}| = 4$. At first we show $A \neq B$. Assume $A = B$. This yields $A \in \bigcap_{i=0}^{4} k_i$.

So we have α) $C = A = D$ or β) $C = A = E$ or γ) $F = A = D$ or δ) $F = A = E$. From these possibilities δ) is in contradiction to the assumption $E \neq F$. Further β) and γ) don't differ substantially, and they can be identified by changing the denotations of the points.

To α). $C = A = D$ implies for all $i \in \{1,2,3,4\}$ $k_0\ A\ k_i$. This yields $k_i\ A\ k_{i+1}$ (put 5=1). So we get $F = G = E = H$, contradiction ($E \neq F$).

To β). It is $k_0\ A\ k_i$ for i=1,2. It follows $k_1\ A\ k_2$ and $E = A = F$, contradiction. So $A \neq B$ is proved.

By arguments of symmetry now we only have to show $A \neq E$. Assume $A = E$. Then $A \in k_0 \cap k_3 = \{B,D\}$. As shown before $A \neq B$. So $A = D$ yields $k_0\ A\ k_4$,

5) Because of $L = K \times K$ it is $A = U(a_1, a_2)$, $a_i = (a_{i1}, a_{i2})$, $a_{ij} \in K$, and so on.

k_3 A k_4 . Thus k_0 A k_3 (see remark e)). This implies A = B , contradiction. So we have A $\neq$ E .

3. We prove that considerations of incidence and the remarks d) and e) imply theorem 3, if points coincide, which have not one and only one common edge in the cube model.

α) Let $k_2 \ni G = A \in k_1$. Because of $k_1 \cap k_2 = \{C,F\}$ and A $\neq$ F (see 2.) we have A = C = G . Thus k_0 A k_1 . Similarly (A $\neq$ E, A $\in k_3 \cap k_4$) we get A = D and thus k_0 A k_4 . Now remark e) yields k_1 A k_4 . So the assumption $k_1 \cap k_4 = \{A,H\}$ implies A = H . Finally, because of E $\not\parallel$ A $\not\parallel$ F there is a unique circle k containing the pairwise different points E,F,A (see remark g)). This circle (ordinary or irregular) fulfils the demanded assumptions (**) of theorem 1.

β) Let C = D . By A $\neq$ B we can assume without loss of generality B $\neq$ C . Repeating our arguments we have C = G = H . If $|\{C,E,F\}| = 3$ 3. is proved as in α) . If C = E ($\neq$ F) there exists a unique circle k obtaining F $\notin$ k_4 with k C k_4 . Because of $k_3 \cap k_4 = \{C\}$ this implies k C k_3 . So the assertion is proved. The remaining case C = F can be proved in the same way.

γ) The assumptions C = G , G = H can be referred to α) or β) and so 3. is proved up to changing of denotation.

4. Using the remarks c), f), and k) now we are able to complete our proof of theorem 3 up to changing of denotation by verifying the following identities of cross ratio. From now on different denoted points shall be different. The identities are

(i) Let $|\{A,B,\ldots,H\}| = 8$:

$$\begin{bmatrix} A & B \\ C & D \end{bmatrix}\begin{bmatrix} A & F \\ H & C \end{bmatrix}\begin{bmatrix} F & B \\ G & C \end{bmatrix}\begin{bmatrix} E & B \\ D & G \end{bmatrix}\begin{bmatrix} A & E \\ D & H \end{bmatrix}\begin{bmatrix} E & F \\ G & H \end{bmatrix} = 1$$

(ii) Let A = H :

$$\begin{bmatrix} A & B \\ C & D \end{bmatrix}\left(\begin{bmatrix} A & F \\ D & C \end{bmatrix} - \begin{bmatrix} A & F \\ E & C \end{bmatrix}\right)\begin{bmatrix} F & B \\ G & C \end{bmatrix}\begin{bmatrix} E & B \\ D & G \end{bmatrix}\begin{bmatrix} E & F \\ G & A \end{bmatrix} = 1$$

(iii) Let A = H, D = E, and let X be a further point lying on k . X exists because of remark j) :

$$\begin{bmatrix} A & B \\ C & D \end{bmatrix}\left(\begin{bmatrix} A & F \\ X & C \end{bmatrix} - \begin{bmatrix} A & F \\ D & C \end{bmatrix}\right)\left(\begin{bmatrix} D & B \\ X & G \end{bmatrix} - \begin{bmatrix} D & B \\ A & G \end{bmatrix}\right)\begin{bmatrix} F & B \\ G & C \end{bmatrix}\begin{bmatrix} D & F \\ G & A \end{bmatrix} = 1$$

(iv) Let A = H , B = G :

$$\begin{bmatrix} A & B \\ C & D \end{bmatrix}\left(\begin{bmatrix} A & F \\ D & C \end{bmatrix} - \begin{bmatrix} A & F \\ E & C \end{bmatrix}\right)\left(\begin{bmatrix} B & E \\ C & D \end{bmatrix} - \begin{bmatrix} B & E \\ F & D \end{bmatrix}\right)\begin{bmatrix} E & F \\ B & A \end{bmatrix} = 1$$

(v) Let A = H , B = D :

$$\left(\begin{bmatrix} B & A \\ G & C \end{bmatrix} - \begin{bmatrix} B & A \\ E & C \end{bmatrix}\right)\left(\begin{bmatrix} A & F \\ B & C \end{bmatrix} - \begin{bmatrix} A & F \\ E & C \end{bmatrix}\right)\begin{bmatrix} F & B \\ G & C \end{bmatrix}\begin{bmatrix} E & F \\ G & A \end{bmatrix} = 1$$

(vi) Let $A = H$, $B = D$, $E = G$, and let X be a further point lying on k_3 :

$$\begin{bmatrix} F & B \\ E & C \end{bmatrix} \left(\begin{bmatrix} B & A \\ X & C \end{bmatrix} - \begin{bmatrix} B & A \\ E & C \end{bmatrix} \right) \left(\begin{bmatrix} A & F \\ B & C \end{bmatrix} - \begin{bmatrix} A & F \\ E & C \end{bmatrix} \right) \left(\begin{bmatrix} E & A \\ B & F \end{bmatrix} - \begin{bmatrix} E & A \\ X & F \end{bmatrix} \right) = 1$$

(vii) Let $A = H$, $B = D$, $F = G$:

$$\left(\begin{bmatrix} B & F \\ A & E \end{bmatrix} - \begin{bmatrix} B & F \\ C & E \end{bmatrix} \right) \left(\begin{bmatrix} F & A \\ B & E \end{bmatrix} - \begin{bmatrix} F & A \\ C & E \end{bmatrix} \right) = \begin{bmatrix} A & F \\ B & C \end{bmatrix} - \begin{bmatrix} A & F \\ E & C \end{bmatrix}$$

(viii) Let $A = H$, $B = D$, $E = G$, and let $X \in k_1$, $Y \in k_3$ be further points:

$$\left(\begin{bmatrix} C & E \\ A & B \end{bmatrix} - \begin{bmatrix} C & E \\ X & B \end{bmatrix} \right) \left(\begin{bmatrix} B & Y \\ A & E \end{bmatrix} - \begin{bmatrix} B & Y \\ C & E \end{bmatrix} \right) \left(\begin{bmatrix} E & Y \\ A & B \end{bmatrix} - \begin{bmatrix} E & Y \\ C & B \end{bmatrix} \right) = \begin{bmatrix} A & X \\ B & C \end{bmatrix} - \begin{bmatrix} A & X \\ E & C \end{bmatrix}$$

(ix) Let $A = H$, $B = G$, $C = F$, $D = E$, and let $W \in k_1$, $X \in k_2$, $Y \in k_3$, $Z \in k_4$ be further points:

$$\begin{bmatrix} A & B \\ D & C \end{bmatrix}^2 \left(\begin{bmatrix} D & Z \\ Y & A \end{bmatrix} \cdot \begin{bmatrix} D & Z \\ B & A \end{bmatrix} \right) \left(\begin{bmatrix} C & X \\ A & B \end{bmatrix} - \begin{bmatrix} C & X \\ W & B \end{bmatrix} \right) \left(\begin{bmatrix} B & C \\ D & X \end{bmatrix} - \begin{bmatrix} B & C \\ Y & X \end{bmatrix} \right) = \begin{bmatrix} A & W \\ D & C \end{bmatrix} - \begin{bmatrix} A & W \\ Z & C \end{bmatrix}$$

These identities only prove the existence of the wanted circle k . The uniqueness follows immediately by the fact, that through four nonparallel points there is at most one circle, or by the uniqueness of touching circles. So theorem 3 is proved completely.

It was already mentioned that in theorem 3 the assumption $E \neq F$ cannot be dropped. Now we want to present a counterexample in the case of the classical Möbiusgeometry $\Sigma(\mathbb{R},\mathbb{C})$. Let $A,B,\dots,H \in \mathbb{P}(\mathbb{C})$ such that $E = F = G = H$ and let $k_0,k_1,\dots,k_4 \in \mathbb{K}$. Let the assumptions (*) of theorem 3 be fulfilled. Then there exists no circle k such that $k \cap k_i = \{E\}$ for all $i=1,2,3,4$: Otherwise because of remark e) we should get $k_1 \mathrel{E} k_4$ for instance, in opposition to the assumption $E \neq A \in k_1 \cap k_4$ (see fig.3).

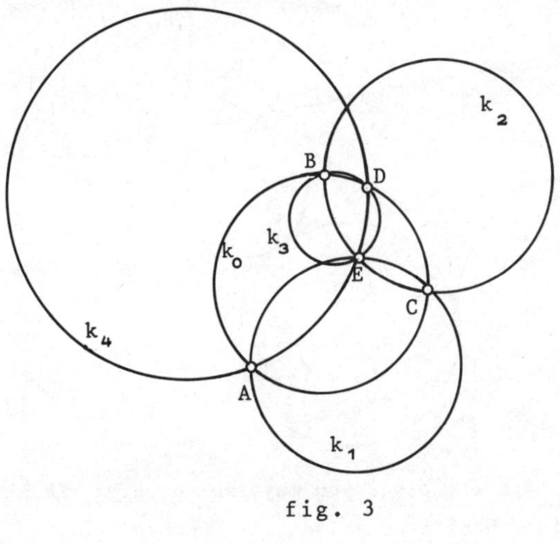

fig. 3

Finally we want to give two examples instead of the big number of statements summarized in theorem 3: If L is the ring of the double numbers, and if we distinguish between o-circles and i-circles, the generalized full theorem of Miquel comprehends more than 80 different statements!

Example A: R. ARTZY introduced in his paper [1] an incidence assumption π, and he proved that a Laguerre plane [6)] in which π holds for a fixed point X is isomorphic to a chain geometry $\Sigma(K, \mathbb{D}_K)$. The theorem π is the following specialization of theorem 3: Let A = C, F = H, and let the other points be pairwise different, let $k_i \in \mathbb{K}^o$ for i=0,1,2,4, let $k_3 \in \mathbb{K}^i$. Figure 4 shows the cube model belonging to π, figure 5 shows the same facts in the so called Speer/Zykel model of the classical Laguerre plane. At this model the i-circle k_3 exists out of two different classes of parallel spears, the spears B and D are represents of these classes. The main idea of π is the touching of the cycles k_1 and k in the spear F.

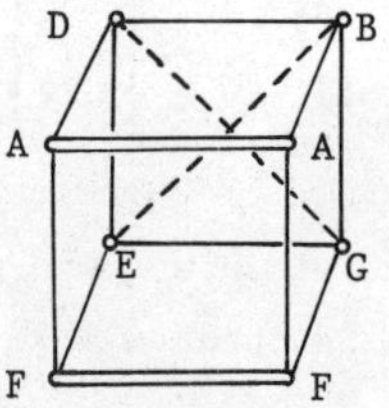

fig. 4

Example B: The at first in the paper [3] by W. BENZ defined B^*-geometries are isomorphic to the chain geometries $\Sigma(K, K \times K)$, if and only if a symmetry condition (S) is fulfilled. (S) is the specialization $|\{A,B,\ldots,H\}| = 8$, $k_o \in \mathbb{K}^o$, $k_j \in \mathbb{K}^i$ for j=1,2,3,4 of theorem 3. We want to

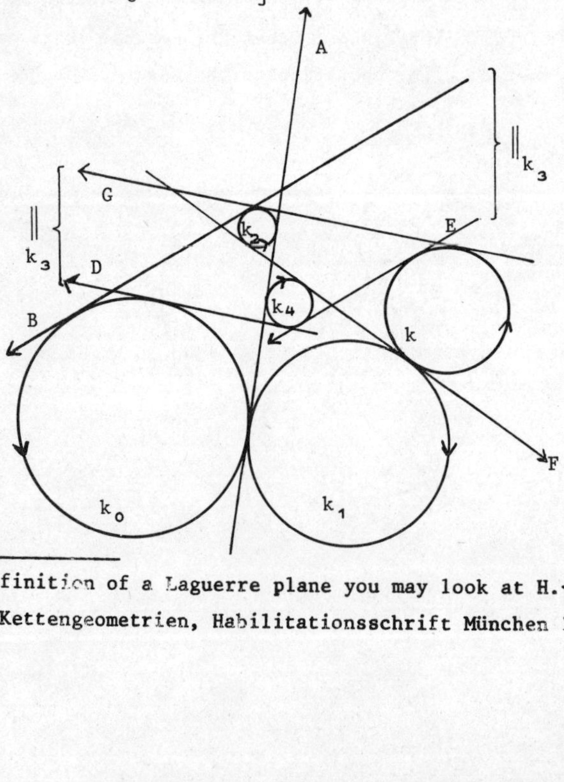

fig. 5

6) For the definition of a Laguerre plane you may look at H.-J. KROLL: Anordnungsfragen in Kettengeometrien, Habilitationsschrift München 1974.

give another specialization of theorem 3 which is isomorphic to (S) and therefore leads to the chain geometries $\Sigma(K, K \times K)$: Let $|\{A,B,\ldots,H\}| = 8$, $k_0, k_3 \in \mathbb{K}^o$, $k_1, k_2, k_4 \in \mathbb{K}^i$. The essential statement is the proof of the existence of a (necessary o-) circle k such that $E,F,G \in k$. Figure 6 shall represent these facts in the hyperbola model of the classical Minkowski geometry $\Sigma(\mathbb{R}, \mathbb{R} \times \mathbb{R})$.

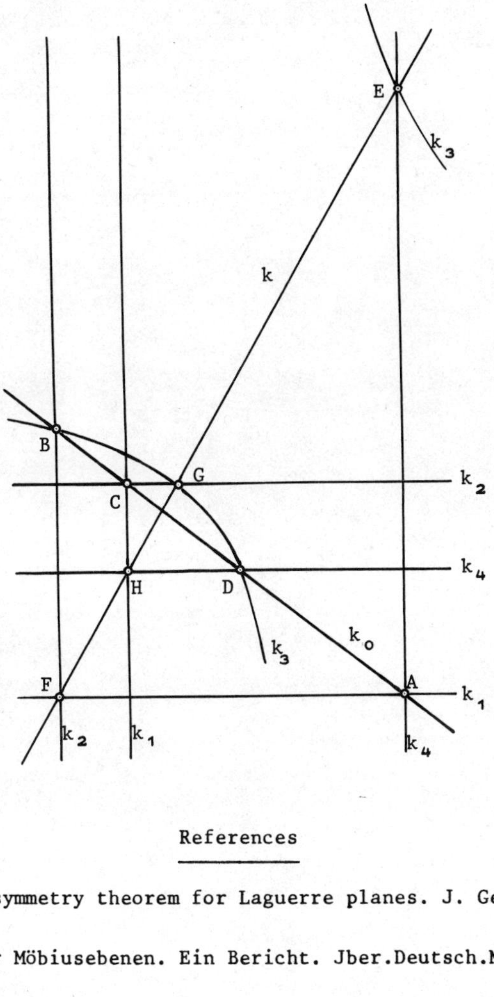

fig. 6

References

[1] ARTZY, R., A symmetry theorem for Laguerre planes. J. Geometry 5, 109 - 116 (1974)

[2] BENZ, W., Über Möbiusebenen. Ein Bericht. Jber.Deutsch.Math.Verein. 63, 1 - 27 (1960)

[3] BENZ, W., Permutations and plane sections of a ruled quadric. Symposia Mathematica 5, 325 - 339 (1970)

[4] BENZ, W., Vorlesungen über Geometrie der Algebren. Berlin-Heidelberg-New York: Springer 1973

[5] PECZAR, L., Über eine einheitliche Methode zum Beweis gewisser Schließungssätze. Monatsh. Math. 54, 210 - 220 (1950)

[6] SAMAGA, H.-J., Miquelsche Sätze in der Minkowski-Geometrie (nicht veröffentlicht).(1978)

[7] SCHAEFFER, H., Benz - Peczar - Doppelverhältnisidentitäten zum allgemeinen Satz von Miquel. Abh. Math. Sem. Univ. Hamburg 42, 228 - 235 (1974)

[8] SCHRÖDER, E., Geometrie. Vorlesungsausarbeitung. Hamburg 1975/76

[9] SCHRÖDER, E., Kreisgeometrische Darstellung metrischer Ebenen und verallgemeinerte Winkel- und Distanzfunktionen. Abh.Math.Sem.Univ.Hamburg 42, 154 - 186 (1974)

[10] SCHRÖDER, E., Eine allgemeine Fassung des Satzes von Miquel (nicht veröffentlicht). (1978)

[11] v. SCHWANENFLÜGEL, M., Schließungssätze in der Minkowski'schen Geometrie. Dissertation Hamburg 1978

Dr. H.-J. Samaga
Mathematisches Seminar der
Universität Hamburg
Bundesstr. 55
D-2000 Hamburg 13
Bundesrepublik Deutschland

AUTOMORPHISMS OF LAGUERRE-GEOMETRY AND CONE-PRESERVING MAPPINGS OF METRIC VECTOR SPACES

by

Helmut Schaeffer
Mathematisches Seminar
der Universität
Bundesstraße 55
2000 Hamburg 13
West-Germany

A pair (V,q), where $q:V \to K$ is a quadratic form on the finite-dimensional vector-space V over K is what we call a metric vector-space. We denote by f the bilinear form associated to q, so

$$f(\mathfrak{x},\mathfrak{y}) = q(\mathfrak{x}+\mathfrak{y}) - q(\mathfrak{x}) - q(\mathfrak{y}) .$$

The radical of (V,q) is

$$\mathrm{rad}(V,q) := \{ \mathfrak{x} \in V \mid q(\mathfrak{x}) = 0 = f(\mathfrak{x},V) \} .$$

(V,q) is non-singular, iff its radical is zero. A mapping $\varphi : V \to V$ is called a semi-similarity of (V,q), iff it is bijective and semilinear (with respect to an automorphism α of K) and there exists $r \in K$ with:

$$\forall \mathfrak{x} \in V : q(\varphi(\mathfrak{x})) = r\, q(\mathfrak{x})^{\alpha} .$$

In a recent paper by J. Lester the following is proved:

THEOREM ([2]). Let (V,q) be a non-singular metric vector-space of dimension at least 3 over K, char $K \neq 2$. Assume that there exists $\mathfrak{x} \in V$, $\mathfrak{x} \neq 0$, with $q(\mathfrak{x}) = 0$. Then any bijection $\varphi : V \to V$, such that

(a) $\varphi(0) = 0$

(b) $\forall \mathfrak{x},\mathfrak{y} \in V : q(\mathfrak{x}-\mathfrak{y}) = 0 \Longleftrightarrow q(\varphi(\mathfrak{x}) - \varphi(\mathfrak{y})) = 0$

holds, is a semi-similarity of (V,q).

In this note we first define what we understand by "Laguerre-geometry" and then prove, that automorphisms of this geometry can be identified with the mappings in J. Lester's theorem. This generalizes known relations between classical Laguerre-geometry and Lorentz-transformations (see [1] for references).

1. Quadrics in projective spaces. Cones.

Let $\mathbb{P}$ be a projective space associated to a vector-space W of dimension at least 3 over K. Points of $\mathbb{P}$ (these are 1-dimensional subspaces of W) will be denoted by capital letters $A,B,\ldots\ldots$

$(A = \langle \mathfrak{a} \rangle$, $B = \langle \mathfrak{b} \rangle$, ..., $\mathfrak{a}, \mathfrak{b}$, ... $\in W)$.

If in addition (W,q) is a metric vector-space then

$$\mathfrak{q} := \{ A \in \mathbb{P} \mid A = \langle \mathfrak{a} \rangle ,\ 0 \neq \mathfrak{a} \in W,\ q(\mathfrak{a}) = 0 \}$$

is a <u>quadric</u> in $\mathbb{P}$. We like to call $\mathfrak{q}$ a <u>cone</u>, iff

(a) $|\mathfrak{q}| > 1$

(b) $\dim \mathrm{Rad}(W,q) = 1$.

The point $N = \mathrm{Rad}(W,q)$ is then the <u>vertex</u> of $\mathfrak{q}$.

A quadric $\mathfrak{q}$ is called <u>non-singular</u> (resp. <u>non-defective</u>) iff (W,q) is non-singular (resp. non-defective). "(W,q) non-defective" means that $f(\mathfrak{x},W) = 0$ implies $q(\mathfrak{x}) = 0$.

The following is well-known:

If P is a point on $\mathfrak{q}$, then any line through P is either a <u>tangent-line</u> (it meets $\mathfrak{q}$ in exactly one point or is contained in $\mathfrak{q}$) or is a <u>secant-line</u> (it meets $\mathfrak{q}$ in exactly two different points). If $P \in \mathfrak{q}$ is a non-singular point (that means $P \notin \mathrm{rad}(W,q)$) the set of points on all tangent-lines through P is a hyperplane of $\mathbb{P}$.

The next lemma has a corollary which we will use later.

1.1. <u>Lemma</u>. Let $q_i : W \to K$ $(i=1,2)$ be arbitrary quadratic forms on the n-dimensional vector-space W over K $(n \geq 2)$. Assume that

(a) $\exists\ \mathfrak{x} \in W : q_1(\mathfrak{x}) = 0 \neq f_1(\mathfrak{x}, W)$

(b) $\forall\ \mathfrak{x} \in W : q_1(\mathfrak{x}) = 0 \Leftrightarrow q_2(\mathfrak{x}) = 0$.

Then there exists $d \in K$, $d \neq 0$, such that

$$\forall\ \mathfrak{x} \in W : q_1(\mathfrak{x}) = d \cdot q_2(\mathfrak{x})$$

Proof. It follows from (a) that V has a basis $\mathfrak{a}_1, \ldots, \mathfrak{a}_n$ with $q_1(\mathfrak{a}_i) = 0$ $(i=1,\ldots,n)$ and $f_1(\mathfrak{a}_1, \mathfrak{a}_j) = 1$ $(j=2,\ldots,n)$.

We now use the (well-known) fact that a line joining $P = \langle \mathfrak{x} \rangle$ and $Q_j = \langle \mathfrak{a}_j \rangle$ is a tangent-line to the quadric $\mathfrak{q}_i$ defined by (W,q_i), iff $f_i(\mathfrak{x}, \mathfrak{a}_j) = 0$ $(i=1,2)$. Because of $\mathfrak{q}_1 = \mathfrak{q}_2$ we have:

$$\forall \mathfrak{x} \in W,\ \forall j : f_1(\mathfrak{x}, \mathfrak{a}_j) = 0 \Leftrightarrow f_2(\mathfrak{x}, \mathfrak{a}_j) = 0$$

and, therefore, the two linear mappings

$$\mathfrak{x} \to f_i(\mathfrak{x}, \mathfrak{a}_j) \qquad (i=1,2)$$

have (for any fixed j) the same kernel. This implies the existence of $d_j \in K$ such that

$$\forall \mathfrak{x} \in V : f_1(\mathfrak{x}, \mathfrak{a}_j) = d_j f_2(\mathfrak{x}, \mathfrak{a}_j) .$$

Because of

$$d_1 f_2(\mathfrak{a}_1, \mathfrak{a}_j) = d_1 f_2(\mathfrak{a}_j, \mathfrak{a}_1) = f_1(\mathfrak{a}_j, \mathfrak{a}_1) = f_1(\mathfrak{a}_1, \mathfrak{a}_j) = d_j f_2(\mathfrak{a}_1, \mathfrak{a}_j)$$

we have $d_1 = \ldots = d_n =: d$ and, finally,

$$q_1\left(\sum_{j=1}^{n} k_j \mathfrak{a}_j\right) = d\, q_2\left(\sum_{j=1}^{n} k_j \mathfrak{a}_j\right),$$

for all $k_j \in K$, follows.

1.2. Corollary. Let $\mathfrak{q}$ be a quadric defined by (W,q) in the projective space $\mathbb{P}$ over W, $\dim W \geq 3$. Assume that $\mathfrak{q}$ contains at least one non-singular point, then any collineation of $\mathbb{P}$, which maps $\mathfrak{q}$ onto itself, is induced by a semi-similarity of (W,q).

Proof. Let the collineation of $\mathbb{P}$ be induced by $\lambda : W \to W$, where λ is semilinear with respect to the automorphism α of K.
Define $q'(\mathfrak{r}) := q(\lambda(\mathfrak{r}))^{\alpha^{-1}}$, then $q' : V \to K$ is a quadratic form, and an application of 1.1 gives that λ is a semi-similarity of (W,q).

2. Laguerre-geometry $L(\mathbb{P},\mathfrak{q})$.

To define $L(\mathbb{P},\mathfrak{q}) = (\mathcal{P},\mathcal{K})$ we start with a projective space $\mathbb{P}$ of dimension $n \geq 3$ and a cone $\mathfrak{q}$ in $\mathbb{P}$ with vertex N defined by (W,q).

(a) Points of $L(\mathbb{P},\mathfrak{q})$ are points of $\mathbb{P}$ belonging to $\mathfrak{q}$ but different from N

(b) Circles of $L(\mathbb{P},\mathfrak{q})$ are hyperplanes of $\mathbb{P}$ not passing through N.

Incidence between points and circles is given by ordinary inclusion. We denote by $\mathcal{P}$ (resp. $\mathcal{K}$) the set of points (resp. circles).
To obtain another model $(\mathcal{P}',\mathcal{K}')$ of $L(\mathbb{P},\mathfrak{q})$ *) we assume for the rest of this section that $\mathfrak{q}$ is non-defective. We choose a hyperplane $\mathfrak{M}$ in $\mathbb{P}$, $\mathfrak{M} \not\ni N$, and consider the affine space $\mathbb{A}$ induced by $\mathbb{P}$ and $\mathfrak{M}$. The set $\mathfrak{q}' := \mathfrak{q} \setminus \mathfrak{M}$ is then an (affine) cone in $\mathbb{A}$ and we define $\mathcal{P}'$ and $\mathcal{K}'$ as follows:

(a) elements of $\mathcal{P}'$ are the hyperplanes of $\mathbb{A}$, which are parallel to tangent-hyperplanes of $\mathfrak{q}'$,

(b) elements of $\mathcal{K}'$ are the points of $\mathbb{A}$.

It is easy to see that $(\mathcal{P},\mathcal{K})$ and $(\mathcal{P}',\mathcal{K}')$ are isomorphic:
We can assume $W = V \oplus K$ $(= \{(\mathfrak{r}, x) \mid \mathfrak{r} \in V,\ x \in K\})$, where (V,q) is n-dimensional non-singular, non-defective, and N has $(0,1)$ as a basis: $N = \langle (0,1) \rangle$.
Let the affine space $\mathbb{A}$ be induced by the hyperplane

$$\mathfrak{M} := \{\langle(\mathfrak{r},0)\rangle \mid 0 \neq \mathfrak{r} \in V\},$$

*) In an unpublished note J. Lester suggests to consider $(\mathcal{P}',\mathcal{K}')$ as a generalization of classical Laguerre-geometry.

and define mappings

$$\mathscr{f} : \mathscr{P} \to \mathscr{P}' , \quad \mathscr{t} : \mathscr{K} \to \mathscr{K}' :$$

(a) If $A \in \mathscr{P}$ $(= \mathscr{Q} \setminus \{N\})$, $A = \langle(\mathscr{X}_1, x_1)\rangle \subset W$, $q(\mathscr{X}_1) = 0$, $\mathscr{X}_1 \neq 0$,

then $\mathscr{f}(A)$ is that hyperplane of $\mathbb{A}$, which has equation

$$f(\mathscr{X}, \mathscr{X}_1) + x_1 = 0 .$$

(b) If $\ell \in \mathscr{K}$, where ℓ has equation $f(\mathscr{X}, \mathscr{X}_0) + x = 0$, then $\mathscr{t}(\ell) = \mathscr{X}_0$.

Obviously $\mathscr{f}: \mathscr{P} \to \mathscr{P}'$ and $\mathscr{t}: \mathscr{K} \to \mathscr{K}'$ are bijections, and

$$A \in \ell \iff \mathscr{t}(\ell) \in \mathscr{f}(A)$$

holds, for all $A \in \mathscr{P}$, $\ell \in \mathscr{K}$. This shows, that $(\mathscr{f},\mathscr{t})$ is an isomorphism of $(\mathscr{P},\mathscr{K})$ onto $(\mathscr{P}', \mathscr{K}')$.

3. Automorphisms of L ($\mathbb{P}$, $\mathscr{Q}$).

Let $\mathbb{P}$, $\mathscr{Q}$, N and (W,q) be as in section 2. We assume that $\mathscr{Q}$ is non-defective. Of course any collineation σ of $\mathbb{P}$ such that $\sigma(\mathscr{Q}) = \mathscr{Q}$ fixes the point N and induces an automorphism of L($\mathbb{P}$, $\mathscr{Q}$).
The following converse holds also true:

3.1. Theorem. If $|K| \geq 3$ and (σ_1,σ_2) is an automorphism of L($\mathbb{P}$, $\mathscr{Q}$) = $= (\mathscr{P},\mathscr{K})$, $\sigma_1 : \mathscr{P} \to \mathscr{P}$, $\sigma_2 : \mathscr{K} \to \mathscr{K}$, then there exists a collineation σ of $\mathbb{P}$ such that σ_1 is the restriction of σ to $\mathscr{P}$.

The proof can be based completely on the method used in [4] and runs as follows:
Start with the automorphism (σ_1,σ_2) of L ($\mathscr{P}$, $\mathscr{Q}$). Then σ_1 (and σ_1^{-1} as well) maps sets which are sections of $\mathscr{Q}$ with hyperplanes not passing through N onto sets of the same kind. This generalizes to:

3.1.1. Independent points $A_1,\dots,A_m \in \mathscr{P}$ generating a subspace of $\mathbb{P}$ not containing N $(3 \leq m \leq n)$ are mapped under σ_1 (and σ_1^{-1}) onto points of the same kind.

Next one has to prove 3.1.2, which is a consequence of 3.1.1 (σ_1 is extended to a mapping $\sigma_1 : \mathscr{Q} \to \mathscr{Q}$ by defining $\sigma_1(N) := N$).

3.1.2. Points $A,B,C,D \in \mathscr{Q}$ belonging to a plane are mapped by σ_1 (and σ_1^{-1}) onto points belonging to a plane.

Now we are in the position to follow the proof in section 4 of [4] (p. 342), which finally leads to the desired result.

4. Cone preserving mappings as Laguerre-automorphisms.

Let (V,q) be a non-singular, non-defective metric vector space of dimension at least 3 such that there exists $0 \neq \mathfrak{x} \in V$ with $q(\mathfrak{x}) = 0$. Which are the bijections $\sigma : V \longrightarrow V$ with

(a) $\sigma(0) = 0$

(b) $\forall\, \mathfrak{a} \in V : \sigma(C_{\mathfrak{a}}) = C_{\sigma(\mathfrak{a})}$,

where $C_{\mathfrak{a}} := \{\mathfrak{x} \in V \mid q(\mathfrak{x} - \mathfrak{a}) = 0\}$?

In [2] , where this question is answered, a lemma is proved (lemma 3.2., p. 1251) which says, that σ maps a hyperplane with equation

$$f(\mathfrak{x}-\mathfrak{a}, \mathfrak{x}_0-\mathfrak{a}) = 0, \quad \mathfrak{a}, \mathfrak{x}_0 \in V , \quad q(\mathfrak{x}_0-\mathfrak{a}) = 0$$

onto a hyperplane of the same kind (with equation $f(\mathfrak{x} - \sigma(\mathfrak{a}), \sigma(\mathfrak{x}_0) - \sigma(\mathfrak{a})) = 0$).

Now these hyperplanes are exactly the points of Laguerre-geometry $(\mathfrak{P}', \mathfrak{K}')$ (see section 2).

As a corollary to 3.1, 1.2 and the isomorphism established in section 2 it follows that σ is a semisimilarity of (V,q) if $|K| \geq 3$.

Literature

[1] Benz,W., Zur Charakterisierung der Lorentz-Transformationen, J. of Geometry, 9, (1977), 29-37.

[2] Lester,J., Cone preserving mappings for quadratic cones over arbitrary fields, Can. J. Math., 24 (6), (1977), 1247 - 1253.

[3] Mäurer,H., Möbius- und Laguerre-Geometrien über schwach konvexen Semiflächen, M. Z., 98, (1967), 355 - 386.

[4] Wagner,R., Projektivitäten auf Quadriken, M. Z., 83, (1964), 336 - 344.

BOUNDS FOR THE NUMBER OF SOLUTIONS OF CERTAIN PIECEWISE LINEAR EQUATIONS

Ruben Schramm

Department of Mathematical Sciences
Tel-Aviv University, Tel-Aviv, Israel.

1. Introduction

EAVES [2], CHARNES, GARCIA and LEMKE [1], and SCHRAMM [6] have obtained theorems about the number of solutions of piecewise linear equations. In this paper we present further results which belong to the same sphere of interest. These results provide specific upper bounds for the number of solutions of certain piecewise linear equations. Throughout we shall consider only piecewise linear functions and piecewise linear equations which are associated with determinants of the same sign, a restriction which will be implied also when it is not explicitly mentioned.

The number of solutions of a piecewise linear equation equals another number, which occurs in a geometrical context. With each p.l. function we shall associate a certain p.l. manifold, the *Riemann manifold* of that function. The latter in general consists of a number of "sheets", glued to each other at those points of the manifold which correspond to the branch points of the function represented by the manifold. The number of sheets in the Riemann manifold of a p.l. function F turns out to equal the number of solutions of each of the equations $F(x) = \gamma$ when γ does not belong to a certain exceptional set. In the article we shall utilize this equality in both directions.

After having introduced the necessary notation in section 2, we shall in section 3 cite those theorems of [6] on which the results of this paper rest. In section 4 we shall define the Riemann manifold of a given piecewise linear function. The main results of the paper are derived in section 5, in which upper bounds are set up for the number of solutions of piecewise linear equations associated with determinants of the same sign.

2. Notation and Definitions

The closure of $A \subset R^n$ is $\bar{A}$ and its interior is A^o. The Euclidean norm of $x \in R^n$ is $|x|$.

Definition 1. A continuous function $F: R^n \to R^n$ is piecewise linear, p.ℓ. for short, when a finite set $H = \{H_1, \ldots, H_p\}$ of hyperplanes exists, such that $R^n \setminus \cup H$ is the disjoint union of finitely many open polyhedral sets $C_1, \ldots, C_q$, and for every $i = 1, \ldots, q$ the restriction F_i of F to $\bar{C}_i$ equals the restriction of an affine function to $\bar{C}_i$. Let $C = \{C_1, \ldots, C_q\}$. The values of F in $\bar{C}_i$ are computed from the expression $F(x) = F_i(x) = A_i x + b_i$, where A_i is a constant $n \times n$ matrix, and b_i a constant $n \times 1$ vector. The linear continuation of

F_i to R^n, unique for non singular A_i, is $F_i^L = F|_{C_i}^L$.

The sets $C, H, \{A_1, \ldots, A_q\}$ and $\{\det A_1, \ldots, \det A_q\}$, as well as corresponding elements of these sets, are said to be associated with F and with each other.

When $\gamma \in R^n$ is a constant, $F(x) = \gamma$ is a piecewise linear equation.

Definition 2. Let E be a set of hyperplanes in R^n. Then E^* is the set of points in R^n each of which belongs to at least two hyperplanes in E.

The set of hyperplanes $F_j^L(H_i) = F\Big|_{C_j}^L(H_i)$, with $H_i \in H$, $C_j \in C$, partitions image space into a collection of polyhedral sets. It is the set of these polyhedral sets which plays a prominent role in the theory of piecewise linear functions.

Definition 3. Let F_j^L be non singular for $j = 1, \ldots, q$, let $G = \{G_1, \ldots, G_s\}$ be the arbitrarily ordered set of hyperplanes $\{F_j^L(H_i) | j = 1, \ldots, q, \ i = 1, \ldots, p\}$, and $\Gamma = \{\Gamma_1, \ldots, \Gamma_t\}$ the set of open polyhedral sets which have $R^n \setminus \cup G$ as their disjoint union. Thus $s \leq pq$. We shall also refer to Γ as "the set of polyhedral sets generated by H through F."

Notation is required for certain subsets of H, C, Γ and R^n. It is expedient to define generally:

Definition 4. Let $E = \{E_1,\ldots,E_\pi\}$ be the finite set of hyperplanes in R^n, P the set $\{P_1,\ldots,P_\rho\}$ of open polyhedral sets which have $R^n\setminus\cup E$ as their disjoint union, and $a\in R^n$. Then $\bar{P} = \{\bar{P}_1,\ldots,\bar{P}_\rho\}$. Also, $E(a)$ is the set of hyperplanes $E_i\in E$ with $a\in E_i$, considered empty when $a\notin\cup E$, $P(a)$ the set of polyhedral sets $P_j\in P$ such that $a\in\bar{P}_j$, and $\bar{P}(a) = \{\bar{P}_j\,|\,P_j\in P(a)\}$. Finally, for $P_i\in P$ let $C(a,P_i)$ be the cone determined by P_i at a, and $P^c(a)= \{C(a,P_i)\,|\,P_i\in P(a)\}$.

The mathematical symbols introduced in this section will retain their assigned definitions throughout the whole paper. Therefore generally these definitions will not be repeated.

3. Earlier Results Connected with Solutions of Piecewise Linear Equations Used in Sections 4 and 5.

The polyhedral sets $\Gamma_k\in\Gamma$ and $F(C_j)$, for $C_j\in C$, stand in a simple topological relationship to each other:

Proposition A (Prop. 1 of [6]). *Let* $F:R^n\to R^n$ *be p.ℓ. and assume that none of the determinants associated with* F *vanishes. Let* $C_j\in C$ *and let* $\Gamma_k\in\Gamma$ *meet* $F(C_j)$. *Then* $\Gamma_k\subset F(C_j)$.

With every p.ℓ. function F and every point $a \in R^n$ we associate the p.ℓ. function F^a which in every cone $C(a,C_i)\in C^c(a)$ continues $F|_{C_i}$ linearly:

Proposition B (Prop. 2 of [6]). *Let* $a \in R^n$ *and let* F_j^L *be non singular for all* $C_j \in C(a)$. *For each* $C(a,C_i) \in C^c(a)$ *let* $F_i^a : \overline{C(a,C_i)} \to R^n$ *be the linear continuation of* F_i *to* $\overline{C(a,C_i)}$. *Then a unique function* $F^a : R^n \to R^n$ *exists, such that* $F^a|_{\overline{C(a,C_i)}} = F_i^a$ *for all* $C(a,C_i) \in C^c(a)$.

Theorem A (Th.1. of [6]). *Let* $F : R^n \to R^n$ *be p.l. Let* a *be a point in* R^n *such that the determinants associated with* F *in* $C(a)$ *have the same sign. Define* F^a *as in Proposition B,* H^a *as the set of hyperplanes in* H *which contain facets of polyhedral sets in* $C(a)$, *and* Γ^a *as the set of polyhedral sets generated by* H^a *through* F^a *(c.f. Def. 3). Then* $\cup\Gamma^a[F(a)] \subset F[\cup\bar{C}(a)]$.

Proposition C (Prop. 3 of [6]). *Let the different polyhedral sets* $C_i \in C$ *and* $C_j \in C$ *be associated with the p.l. function* F, *and let* $\overline{C_i}, \overline{C_j}$ *have a common facet. When* $\det A_i \cdot \det A_j > 0$ *the restriction of* F *to* $\overline{C_i} \cup \overline{C_j}$ *is a homeomorphism.*

Theorem B (Th. 2 of [6]). *Let the p.l. function* $F : R^n \to R^n$ *be associated with the set of hyperplanes* H *and the set of polyhedral sets* C. *Also, let the determinants associated with* F *have the same sign. Then the following is true: (i) For* $\gamma \in R^n \setminus F(H^*)$ *the number* m *of solutions of the equation*

$F(x) = \gamma$ in R^n is independent of γ . (ii) When Γ is the set of polyhedral sets generated by H through F, the number $\mu(\Gamma_i)$ of polyhedral sets $C_j \in C$ with $\Gamma_i \subset F(C_j)$ is m for all $\Gamma_i \in \Gamma$. (iii) For $\gamma \in F(H^*)$ the equation $F(x) = \gamma$ has at most m solutions in R^n.

Theorem C (Th. 3 of [6]). *Define F^a as in Proposition B. Then the following is true: (i) Under the assumptions of Theorem A(ii), for $\delta \in A = \cup\Gamma^a[F(a)] \setminus F^a([H(a)]^*)$ the number $\mu(a)$ of solutions of the equation $F(x) = \delta$ in $\cup\bar{C}(a)$ is independent of δ. (ii) For each $\Gamma^a_\ell \in \Gamma^a[F(a)]$ exactly $\mu(a)$ polyhedral sets $C_j \in C(a)$ satisfy $\Gamma^a_\ell \subset F(C_j)$. (iii) For $\delta \in (\cup\bar{\Gamma}^a[F(a)]) \cap F^a([H(a)]^*)$ the equation $F(x) = \delta$ has at most $\mu(a)$ solutions in $\cup\bar{C}(a)$.*

Definition 5. Let $C^{(b)}$, $C(\infty)$ be the sets of bounded and unbounded polyhedral sets $C_j \in C$, respectively. Arrange C such that $C(\infty) = \{C_1,\dots,C_u\}$. Thus $u \leq q$. Also, let E be the set of hyperplanes $\{F^L_j(H_i) | j = 1,\dots,u,\ i = 1,\dots,p\}$, Δ the set of open polyhedral sets which have $R^n \setminus \cup E$ as their disjoint union, and $\Delta(\infty)$ the set of unbounded polyhedral sets in Δ.

Theorem D (Th. 4 of [6]). *Let the determinants associated with the p.l. function $F: R^n \to R^n$ in $C(\infty)$ have the same sign, and define*

$r = \max\{|F(x)| \,|\, x \in \cup\bar{C}^{(b)}\}$, $B = \{x \in R^n \,|\, |x| \leq r\}$. Then the following holds: (i) For $\delta \in T = R^n \setminus B \setminus F(H^*)$ the number σ of solutions of the equation $F(x) = \delta$ in R^n is independent of δ. (ii) For every $\Delta_i \in \Delta(\infty)$ exactly $\nu(\Delta_i) = \sigma$ polyhedral sets $C_j \in C(\infty)$ satisfy $\Delta_i \subset F(C_j)$. (iii) For each $\delta \in (R^n \setminus B) \cap F(H^*)$ the equation $F(x) = \delta$ has at most σ solutions in R^n.

Theorem E (Th. 6 of [6]). Let the p.l. function $F: R^n \to R^n$ be associated with the sets of hyperplanes H and polyhedral sets C. (i) Let $b \in \cup H$ be a point such that the hyperplanes $H_i \in H$ which contain b have linearly independent normals. Then for $F|_{\cup\bar{C}(b)}$ to be a homeomorphism it is necessary and sufficient that the determinants associated with F in $C(b)$ have the same sign. (ii) Let Φ be the set of those closed facets of polyhedral sets in C which have no proper subfaces and $X \subset \cup\Phi$ a set of points representative of the faces in Φ. Let (i) be satisfied by every $b \in X$. Then for F to be a homeomorphism it is necessary and sufficient that the determinants associated with F have the same sign.

4. The Riemann Manifold of a Piecewise Linear Function

When all the determinants associated with a piecewise linear function F have the same sign, F can be represented by a piecewise linear manifold M, best thought of as an analogy to the Riemann surface of a complex function. The condition that the determinants associated with F all have the same sign is not necessary for the construction of M. However, for the purposes of this paper it will be helpful to consider only functions of the type indicated. For these, as in the complex case, the inverse F^{-1} of F is uniquely determined on the manifold that represents F. We shall now describe the construction of M.

The polyhedral sets $F(\bar{C}_1),\ldots, F(\bar{C}_q)$ serve as the building blocks for the construction of M. As in the complex case, however, these polyhedral sets are not considered just as point sets in R^n. For each $x \in R^n$ we shall discriminate between the point $F(x) \in R^n$ and the point $F^*(x) \in M$ which represents $F(x)$ on M. For each $\bar{C}_k \in \bar{C}$ we may visualize $F^*(\bar{C}_k) = F_k^*(\bar{C}_k)$ by $F(\bar{C}_k)$, but $F^*(C_k)$ and $F^*(C_\ell)$ are always considered disjoint when $1 \leq k < \ell \leq q$, even when $F(C_k)$ and $F(C_\ell)$ *do* meet. However, when φ, the intersection of $\bar{C}_k$ and $\bar{C}_\ell$, is not empty then $F^*(\bar{C}_k)$ is to be "glued" to $F^*(\bar{C}_\ell)$ at $F^*(\varphi)$. Stated differently,

$F_k^*(\varphi)$, $F_\ell^*(\varphi)$ and $F^*(\varphi)$ are identified with each other. The mapping $F^*: R^n \to M$ is a homeomorphism. In order to see this, add a point at infinity to each of R^n and M. The continuous function F^* can be extended to a continuous function on the compact set $R^n \cup \{\infty\}$, since $\lim_{|x|\to\infty} |F(x)| = \infty$. Since M is Hausdorff and since F^* is a bijection, F^* is a homeomorphism.

The following example, due to FUJISAWA and KUH [3] is, in a certain sense, the simplest example of a piecewise linear function F, such that the manifold which represents F has more than one sheet: we learn from Theorem E(ii) that when $n = 2$ and no point in H^* belongs to more than two lines $H_i \in H$ then F is a homeomorphism. Thus the manifold that represents F consists of one sheet. Hence, for the surface that represents F to have *more* than one sheet, some point in H^* must be common to at least three lines in H.

Example 1 [3, p. 324]. Let $R_1,\dots,R_6$ denote the clockwise numbered closed polygons generated by the lines $y=0$, $y=\sqrt{3}x$, $y=-\sqrt{3}x$, letting $R_1=\{(x,y)\,|\,y\geq 0, y\leq\sqrt{3}x\}$. Also, let $F_i=F|_{R_i}$,

$$A_1 = \begin{pmatrix}1, -2/\sqrt{3}\\ 0, \quad 1\end{pmatrix} \quad A_2 = \begin{pmatrix}0, -1/\sqrt{3}\\ \sqrt{3}, \quad 0\end{pmatrix} \quad A_3 = \begin{pmatrix}-1, -2/\sqrt{3}\\ 0, \quad -1\end{pmatrix}$$

$A_4 = -A_1$, $A_5=-A_2$, $A_6=-A_3$, $z = \begin{pmatrix}x\\y\end{pmatrix}$, and $F_i(z) = A_i z$. Then $F(R_1) = R_1 \cup R_2$, $F(R_2) = R_3 \cup R_4$, $F(R_3) = R_5 \cup R_6$, $F(R_4) = R_1 \cup R_2$,

$F(R_5) = R_3 \cup R_4$, $F(R_6) = R_5 \cup R_6$.

We immediately see that the surface which represents F is topologically equivalent to the Riemann surface of the complex function $w = \sqrt{z}$, the origin being the only branch point of both functions.

5. Upper Bounds for the Number of Solutions.

The following theorem is easily proved:

Theorem 1. *Let all the determinants associated with the piecewise linear function* $F:R^n \to R^n$ *have the same sign, let* H *contain a pair of non parallel hyperplanes and let* $\gamma \in R^n$. *Then the number of solutions of the equation* $F(x)=\gamma$, *which equals the number of sheets of the Riemann manifold that represents* F, *is* $[\frac{q}{2}]$ *at most. For* q *even this bound can be reduced to* $\frac{q}{2} - 1$.

Proof: Let the equation $F(x)=\gamma$ have exactly m solutions $x \in R^n$ for $\gamma \notin F(H^*)$ (Th. B(i)). Then the Riemann manifold of F has m sheets, each of which extends over the whole of R^n. Since each $\bar{C}_i \in \bar{C}$ is a proper subset of some half space, and $F^L_{\bar{C}_i}$ is an affine function, $F(\bar{C}_i)$ is also a proper subset of some half space. Therefore between them the q sets $F(\bar{C}_i)$ fill up less than q halves of a sheet, i.e. $[\frac{q}{2}]$ full sheets at most for all q, and $\frac{q}{2} - 1$ for q even.

We can also easily obtain an upper bound for the number of those sheets in the manifold of F which contain a given point $F^*(a)$, or equivalently, for the number of solutions of the equation $F(x)=\gamma$ in $\cup\bar{C}(a)$ when γ belongs to a sufficiently small neighbourhood of $F(a)$:

Theorem 2. *Let* a *be a point in* H^*, *let the cardinality of* $C(a)$ *be* q_a *and let the determinants associated with the p.l. function* $F:R^n \to R^n$ *in* $\cup\bar{C}(a)$ *all have the same sign. Define*

F^a, H^a and Γ^a as in Theorem A. Then q_a is even and for $\gamma \in \cup\Gamma^a[F(a)]$ the equation $F(x)=\gamma$ has at most $\frac{1}{2}q_a-1$ solutions in $\cup\bar{C}(a)$. Therefore the number of sheets of the Riemann manifold of F on which $F^*(a)$ lies also does not exceed $\frac{1}{2}q_a-1$.

Proof: The evenness of q_a is trivial to prove. The rest of the proof is obtained by a straightforward application of Theorem 1 to the piecewise linear function F^a.

We shall now develop two lemmas required for our next result, an improvement of Theorem 1.

Definition 7. A *slab* is the closure of the set of points in R^n contained between two fixed hyperplanes.

Definition 8. A set in R^n is *n-dimensionally unbounded* when it is not a subset of the union of a finite number of slabs.

In the sequel we shall assume that the equation of the hyperplane $H_i \in H$ is $\langle a_i, x\rangle = d_i$.

Lemma 1. *A polyhedral set $\bar{C}_\ell \in \bar{C}$ is n-dimensionally unbounded iff the interior of its characteristic cone $cc\bar{C}_\ell$ [4, p. 24] is not empty.*

Proof: Let $\bar{C}_\ell = \bigcap_1^p \{x \mid \langle a_i, x\rangle \geq d_i\}$. It is sufficient to show that

1) when $(cc\bar{C}_\ell)^o \neq \emptyset$ then $\bar{C}_\ell$ is n-dimensionally unbounded,

2) when $(cc\bar{C}_\ell)^o = \emptyset$ then $\bar{C}_\ell$ can be covered by a finite number of slabs.

Proof of 1) Let $\varphi_0 \in cc C_\ell$. Then φ_0 satisfies the inequalities

(1) $$\langle a_i, x\rangle > 0 \qquad i = 1,\dots,p,$$

and $\varepsilon > 0$ exists such that when $\varphi \in R^n$ satisfies $|\varphi-\varphi_o|<\varepsilon$ then φ satisfies (1) as well. The set $\{t\varphi | t\geq 0, |\varphi-\varphi_0|<\varepsilon\} \subset cc\bar{C}_\ell$, however, is obviously n-dimensionally unbounded. Also, $C_\ell \neq \emptyset$: when φ_o satisfies (1) then for $t_o>0$ sufficiently large $t_o\varphi_o \in C_\ell$. Since $t_o\varphi_o + ccC_\ell \subset C_\ell$, C_ℓ is n-dimensionally unbounded.

Proof of 2) We may assume that $C_\ell \neq \emptyset$, $cc\bar{C}_\ell \neq \emptyset$ or C_ℓ would be bounded [4, p. 23]. By assumption, for every solution $\varphi \in R^n$ of the inequalities

(2) $$\langle a_i, x\rangle \geq 0 \qquad i = 1,\dots,p$$

an index $1 \leq i(\varphi) \leq p$ exists, such that $a_{i(\varphi)} = 0$. We shall show that an index $1 \leq I \leq p$ exists such that *every* solution φ of (2) satisfies $\langle a_I, \varphi\rangle = 0$. If this were untrue, then for every $1 \leq k \leq p$ some solution ψ^k of (2) would satisfy $\langle a_k, \psi_k\rangle > 0$ and then $\psi = \sum_1^p \psi^k$ would satisfy (1), contrary to assumption. The relationship $\langle a_I, \varphi\rangle = 0$ means, however, that $cc\bar{C}_\ell$ is a subset of the hyperplane $\langle a_I, x\rangle = 0$. Also, whenever $a \in \bar{C}_\ell$, $\bar{C}_\ell$ is contained in the union of $a+cc\bar{C}_\ell$ and a finite number of slabs. Thus 2) is proved.

Lemma 2. *The number of n-dimensionally unbounded polyhedral sets $\bar{C}_\ell \in \bar{C}$ is even.*

Proof. For every non empty polyhedral set $C_\ell \in C$ exactly one point $\beta=(\beta_1,\ldots,\beta_p)\in\{-1,+1\}^p$ satisfies $C_\ell = C[\beta] = \bigcap_1^p \{x \mid \langle\beta_i a_i, x\rangle > d_i\}$. By Lemma 1 $\bar{C}[\beta]$ is n-dimensionally unbounded iff $(ccC[\beta])^\circ \neq \emptyset$. However, the inclusions

$$\varphi \in \bigcap_1^p \{x \mid \langle\beta_i a_i, x\rangle > 0\} = (cc\bar{C}_\ell)^\circ, \quad -\varphi \in \bigcap_1^p \{x \mid \langle(-\beta_i) a_i, x\rangle > 0\} = (ccC_\ell[-\beta])^\circ$$ are

equivalent. Therefore the same is true of the relationships $(cc\bar{C}[\beta])^\circ \neq \emptyset$, $(cc\bar{C}[-\beta])^\circ \neq \emptyset$. Also, when $(cc\bar{C}[-\beta])^\circ \neq \emptyset$ then $\bar{C}[-\beta] \neq \phi$ [4, p. 23]. The set of n-dimensionally unbounded polyhedral sets is thus the disjoint union of pairs of such sets.

Theorem 3. *Let the conditions of Theorem D be satisfied, let the number of n-dimensionally unbounded sets in C be q^*. Then for $\gamma \in T$ the equation $F(x)=\gamma$ has at most $\frac{1}{2}q^*-1$ solutions in R^n.*

Proof. For any polytope A with a finite number of facets let $|A|$ denote the n-dimensional Euclidean measure (volume) of A. Also, for $s>0$ let $Q(s)$ be the cube $\{x\in R^n \mid |x_i|\leq\frac{s}{2}, i=1,\ldots,n\}$. Define a measure μ on polyhedral sets $A\subset R^n$ by $\mu(A)=\lim_{s\to\infty}[|A\cap Q(s)|/s^n]$. We shall presently see that this limit always exists. Every subset of a slab has obviously zero μ-measure, while $\mu(R^n)=1$. Also, the μ-measure of a polyhedral set $\bar{C}_i\in\bar{C}$ is defined if and only if the μ-measure of the characteristic cone $cc\bar{C}_i$ is defined, since the set $(\bar{C}_i\setminus cc\bar{C}_i)\cup(cc\bar{C}_i\setminus\bar{C}_i)$ is the union of a finite number of slabs. By similarity $|cc\bar{C}_i\cap Q(s)|/s^n$ is independent of s, whence $\mu(cc\bar{C}_i)$ exists. Therefore $\mu(\bar{C}_i)$ exists too and $\mu(\bar{C}_i)=\mu(cc\bar{C}_i)$. Also, since H contains two non parallel hyperplanes, $\mu(\bar{C}_i)<\frac{1}{2}$, since the μ-measure of a half space is $\frac{1}{2}$. In order to conclude the proof of Th. 3 let m be the integer, such that each $\Gamma_i\in\Gamma$ is contained in exactly m different polyhedral sets $F(C_j)$ (Th. B(ii)). Since every $F(\bar{C}_j)$ is the union of sets $\bar{\Gamma}_k$ with disjoint interiors, we have

$$m=m\mu(R^n)=m\mu(\bar{\Gamma}_1+\ldots+\bar{\Gamma}_t)=\mu[F(\bar{C}_1)+\ldots F(\bar{C}_q)]=\mu[F(\bar{C}_1)]+\ldots+\mu[F(\bar{C}_q)].$$

Since $\mu(\bar{C}_i)=0$ whenever $\bar{C}_i$ is in a slab and q^* is the number of those polyhedral sets in $\bar{C}$ which are n-dimensionally unbounded, we have $m < \frac{1}{2}q^*$. By Lemma 2 q^* is even. Therefore $m \leq \frac{1}{2}q^*-1$.

Example 2. Let $n=2$ and let H consist of the lines $y=0$, $y=1$, $y=1-x$ and $y=1+x$. Here $q=9$. However, q^* is 6 only. Therefore, when F satisfies the relevant assumptions, the equation $F(x)=\gamma$ has $\frac{6}{2} - 1=2$ solutions at most.

Example 3. The configuration of Ex. 1 is easily generalized. We first take up the case $n=2$. Let p straight lines pass through the origin, thus generating $q^*=2p$ rays and q^* angular domains. Assume that consecutive rays are at an angle of $\frac{2\pi}{q^*}$ to each other. A p.ℓ. function F with positive determinants, analogous to the one of Ex. 1, can now be defined such that each of the angular domains is itself mapped on an angular domain of $\pi - \frac{2\pi}{q^*}$ radians. The image angles thus add up to $2\pi(\frac{1}{2}q^*-1)$ radians or $\frac{1}{2}q^*-1$ complete angles. Thus, for $\gamma \neq 0$ the equation $F(x)=\gamma$ has *exactly* $\frac{1}{2}q^*-1$ solutions.

Example 4. For a given even number q^* and for every $n>2$ a p.ℓ. function $F:R^n \to R^n$ with a Riemann manifold of $\frac{1}{2}q^*-1$ sheets can now easily be constructed. We shall obtain F from the function of Ex. 3, as follows:

Let $x=(x_1,\ldots,x_n)$, $F(x)=(u_1,\ldots,u_n)$. For u_1,u_2 employ the expressions which correspond to the function of Ex. 3, and for $i>2$ let $u_i=x_i$. Obviously the number of sheets of the Riemann manifold of F is $\frac{1}{2}q^*-1$.

The result of Theorem 3 is thus optimal. It would be of interest to investigate whether relevant multidimensional examples of p.ℓ. functions F exist, such that the Riemann manifold of F has $\frac{1}{2}q^*-1$ sheets and such that F is *not* reducible to a transformation of the plane.

The expressions given in Ths. 1-3 for the number of solutions of p.ℓ. equations involve properties of the set C only. A further improvement can be achieved when additional features of the p.ℓ. equation considered are taken into account. In the formulation of the relevant theorem we shall have recourse to the representation of a convex polyhedral set by an *irredundant family of half spaces* [4, p. 26]: let $\eta_1,\ldots,\eta_\ell$ be open half spaces and let $K=\bigcap_1^\ell \eta_i \neq \emptyset$ be a polyhedral set. We then say that the family $\{\eta_1,\ldots,\eta_\ell\}$ is irredundant if

$$\bigcap_{\substack{j=1\\ j\neq i}}^{\ell} \eta_j \neq K \quad \text{for all } i=1,\ldots,\ell.$$

Theorem 4. *Let the families of open half spaces* $\{h_{i,j} | j=1,\ldots,p_i\}$ *be irredundant for all* $i=1,\ldots,q$ *and let* $C_i = \bigcap_{j=1}^{p_i} h_{i,j}$. *Thus* $\partial h_{i,j} \in \cup H$ *and* $p_i \leq p$. *Also, for all* $i=1,\ldots,q$, $j=1,\ldots,p_i$, *let* $h_{ij} = \{x | \langle \nu_{i,j}, x \rangle > d_{i,j}\}$, *with suitable numbers* $d_{i,j}$ *and suitable column vectors* $\nu_{i,j}$, *normal to* $\partial h_{i,j}$. *Arrange* C *such that* $C(\infty) = \{C_1,\ldots,C_u\}$ *and let the conditions of Th. D be satisfied. For all* $i=1,\ldots,u$, $j,k=1,\ldots,p_i$, *define*

$$\varphi_{i,j,k} = \arccos \left| - \frac{\nu_{i,j}^t A_i^t A_i \nu_{i,k}}{|A_i \nu_{i,j}| \, |A_i \nu_{i,k}|} \right|$$

$$\psi_i = \frac{1}{2\pi} \min \{\varphi_{i,j,k} | j,k=1,\ldots,p_i, j \neq k\}.$$

Assume that H *contains a pair of non parallel hyperplanes. Then for* $\gamma \in T$ *the number of solutions of the equation* $F(x)=\gamma$ *in* R^n *is* $S_F = [\sum_1^u \psi_i]$ *at most, and* $S_F \leq \frac{q^*}{2} - 1$.

Proof: We have $C_i \subset h_{i,j} \cap h_{i,k}$ for all $i=1,\ldots,u$, $j,k=1,\ldots,p_i$. Also, $\varphi_{i,j,k}$ is the magnitude of the 2-angle subtended by the hyperplanes $H_{i,j} = \partial h_{i,j}$ and $H_{i,k} = \partial h_{i,k}$. Since H contains a pair of non parallel hyperplanes, we have $\varphi_{i,j,k} < \pi$, $\psi_i < \frac{1}{2}$. The proof is concluded as in the case of Th. 3.

References

1. A. Charnes, C.B. Garcia, C.E. Lemke, *Constructive Proofs of Theorems Relating to* $F(x)=y$, *With Applications*, Math. Programming 12 (1977), 328-343.

2. B.C. Eaves, *An Odd Theorem*, Proc. Amer. Math. Soc. 26 (1970), 509-513.

3. T. Fujisawa, E.S. Kuh, *Piecewise Linear Theory of Nonlinear Networks*, SIAM J. Appl. Math. 22 (1972), 307-328.

4. B. Gruenbaum, *Convex Polytopes*, Interscience Publishers, London, New York, Sydney, 1967.

5. R.S. Palais, *Natural Operations on Differential Forms*, Trans. Amer. Math. Soc. 92, (1959), 125-141.

6. R. Schramm, *On Piecewise Linear Functions and Piecewise Linear Equations*, (to appear in Mathematics of Operations Research).

ZUR TRANSLATIONSTRANSITIVITÄT IN AFFINEN HJELMSLEVEBENEN

Werner Seier

Eine affine Ebene ist translationstransitiv, wenn die Translationen in zwei verschiedenen Richtungen transitiv sind. Sind in einer affinen Hjelmslevebene die Translationen in zwei nicht benachbarten Richtungen transitiv, so erzeugen sie eine transitive Gruppe von Kollineationen. Wir untersuchen hier notwendige und hinreichende Bedingungen dafür, daß diese Kollineationen Translationen sind.

Wir beginnen mit den grundlegenden Definitionen. In einer Inzidenzstruktur $(\mathfrak{P}, \mathfrak{G}, \mathrm{I})$ heißen zwei Punkte benachbart (in Zeichen: $P \sim Q$), wenn sie mindestens zwei verschiedene Verbindungsgeraden haben. Zwei Geraden g und l heißen benachbart (in Zeichen: $g \sim l$), wenn es zu jedem Punkt P auf g einen Punkt Q auf l mit $P \sim Q$ und zu jedem Punkt R auf l einen Punkt S auf g mit $R \sim S$ gibt.

<u>1.Definition</u>: Es sei $(\mathfrak{P}, \mathfrak{G}, \mathrm{I})$ eine Inzidenzstruktur und $\parallel$ eine Parallelismus genannte Äquivalenzrelation auf der Geradenmenge. Dann heißt $\mathfrak{H} = (\mathfrak{P}, \mathfrak{G}, \mathrm{I}, \parallel)$ eine affine Hjelmslevebene (kurz: H-Ebene), wenn die folgenden Axiome erfüllt sind:

a) Zu je zwei Punkten P und Q gibt es eine Gerade g mit $P, Q \,\mathrm{I}\, g$.

b) Zu jedem Punkt P und jeder Geraden g gibt es genau eine Gerade l mit $P \,\mathrm{I}\, l$ und $l \parallel g$.

c) Zwei sich in einem Punkt P schneidende Geraden sind genau nicht benachbart, wenn P ihr einziger Schnittpunkt ist.

d) Es gibt einen Epimorphismus φ von $\mathfrak{H}$ auf eine affine Ebene $\overline{\mathfrak{H}}$ mit folgenden Eigenschaften:

(1) $\varphi(P) = \varphi(Q) \iff P \sim Q$.

(2) $\varphi(g) = \varphi(l) \iff g \sim l$.

(3) Haben g und l keinen Schnittpunkt, so sind $\varphi(g)$ und $\varphi(l)$ parallel.

Bekanntlich (siehe [4]) können wir dann jede Gerade mit der

Menge der auf ihr liegenden Punkte identifizieren und die Inzidenzrelation I durch die Elementbeziehung ersetzen, was hier stets geschehen soll.

Eine Kollineation τ einer H-Ebene $\mathcal{H}$ heißt <u>Translation</u> von $\mathcal{H}$, wenn $\tau(g) \parallel g$ für alle Geraden von $\mathcal{H}$ ist, und wenn aus $\tau(h)=h$ und $h \parallel l$ stets $\tau(l)=l$ folgt. Eine Parallelklasse $[g] = \{l \mid l \parallel g\}$ von Geraden heißt Richtung einer Translation τ, wenn $\tau(l)=l$ für ein l aus $[g]$ ist. Nach Definition ist dann $\tau(h)=h$ für alle h aus $[g]$. Jede Translation τ hat offenbar eine Richtung. Bildet τ einen Punkt P auf einen benachbarten Punkt Q ab (dann bildet τ jeden Punkt auf einen benachbarten Punkt ab), so hat τ mehrere Richtungen. τ heißt dann Nachbartranslation. Ist dagegen $\tau(P) \not\sim P$ für einen Punkt P, so ist die Richtung von τ eindeutig bestimmt.

Für Richtungen definieren wir eine Nachbarrelation wie folgt:

$$[g] \sim [l] \iff \exists g_1 \in [g] \text{ und } l_1 \in [l] \text{ mit } g_1 \sim l_1 .$$

In $\bar{\mathcal{H}}$ gilt dann $[\varphi(g)] = [\varphi(l)]$.

<u>2.Satz</u>: Es sei $\mathcal{H}$ eine H-Ebene, in der auf jeder Geraden mindestens drei paarweise nicht benachbarte Punkte liegen. Dann ist $\mathcal{H}$ genau dann translationstransitiv, wenn es drei paarweise nicht benachbarte Richtungen $[g]$, $[h]$ und $[l]$ gibt, so daß für je zwei nicht benachbarte Punkte P und Q,deren eindeutg bestimmte Verbindungsgerade PQ in $[g]$, $[h]$ oder $[l]$ liegt eine Translation τ mit $\tau(P)=Q$ existiert.

Beweis: Die Bedingung des Satzes besagt, daß $\mathcal{H}$ für die Richtungen $[g]$, $[h]$ und $[l]$ "im Großen" translationstransitiv ist. Wir zeigen, daß dies auch für jede Richtung, die weder zu $[g]$ noch zu $[h]$ noch zu $[l]$ benachbart ist, der Fall ist. Sei also $[k]$ eine solche Richtung und P und Q zwei nicht benachbarte Punkte mit $PQ=k_1$ aus $[k]$.Dann geht durch P eine Gerade g_1 aus $[g]$ und durch Q eine Gerade h_1 aus $[h]$. g_1 und h_1 schneiden sich in einem Punkt R und nach Voraussetzung über die Richtungen ist $P \not\sim R \not\sim Q$. Daher gibt es Translationen τ_1, τ_2 mit $\tau_1(P)=R$ und $\tau_2(R)=Q$, d.h. $\tau_2 \tau_1(P)=Q$. Wir haben zu zeigen, daß das Produkt $\tau_2 \tau_1$ eine Translation ist. Da $P \not\sim Q$ und PQ aus $[k]$ ist, müssen wir also zeigen, daß jede Gerade aus $[k]$ auf sich abgebildet wird. Es genügt offenbar, dies für solche Geraden aus $[k]$ zu zeigen, die nicht zu k_1 benachbart sind. Sei also k_2 aus $[k]$ und $k_2 \not\sim k_1$. Durch P geht eine Gerade l_1 aus $[l]$. Sie schneidet k_2 in einem Punkt A. Es ist $P \not\sim A$. Daher gibt es eine Translation σ mit $\sigma(P)=A$. Da τ_1 Translation ist, muß die Verbindungsgerade von A und $\tau_1(A)$ aus $[g]$ sein. Ferner ist $PA=l_1$ parallel zu $\tau_1(l_1)=R\tau_1(A)$, d.h. $(P,R,\tau_1(A),A)$ ist ein Parallelogramm. Da $\tau_1(l_1)$ aus $[l]$ ist, muß dann aber $\sigma(R)=\tau_1(A)$ sein. Analog erhält man $\sigma(Q)=\tau_2\tau_1(A)$. Da $\sigma(k_1)=k_2$ ist, erhält man $\tau_2\tau_1(A)\in k_2$, d.h. $\tau_2\tau_1(k_2)=k_2$.

Sei nun $[g']$ z.B. eine zu $[g]$ benachbarte Richtung. Wegen der Reichhaltigkeitsvoraussetzung gibt es dann eine Richtung $[k]$, die zu keiner der Richtungen $[g]$, $[h]$, $[l]$ benachbart ist. Dann erfüllen nach dem oben gezeigten auch $[k]$, $[h]$ und $[l]$ die Bedingungen des Satzes und wie oben können wir zeigen, daß $\mathcal{H}$ auch für die Richtung $[g']$ "im Großen" translationstransitiv ist. Also ist $\mathcal{H}$ für alle Richtungen "im Großen" translationstransitiv. Die Behauptung des Satzes ergibt sich nun ganz leicht, wenn man zeigt, daß in $\mathcal{H}$ die Translationen in einer festen Richtung stets eine Gruppe bilden. Dazu zitieren wir aus [6] folgendes

<u>3. Lemma</u>: Es sei $\mathcal{H}$ eine H-Ebene und $[a]$ eine Richtung von $\mathcal{H}$. Dann bilden die Translationen in Richtung $[a]$ eine Gruppe, wenn die folgende Bedingung erfüllt ist:

> Es seien a_1, a_2 aus $[a]$ mit $a_1 \not\sim a_2$ und A_1, A_2, B_1, B_2 vier Punkte mit $A_i, B_i \in a_i$ für $i=1,2$. Die eindeutig bestimmten Verbindungsgeraden von A_1 und A_2 bzw. B_1 und B_2 seien parallel. Ist dann l eine weitere Verbindungsgerade von A_1 und B_1 und h die Parallele zu l durch A_2, so liegt B_2 auf h.

Wir zeigen, daß $\mathcal{H}$ die Bedingungen des Lemma erfüllt. Sei dazu $[a]$ eine beliebige Richtung von $\mathcal{H}$ und a_i, A_i, B_i wie in Lemma 3 gegeben.

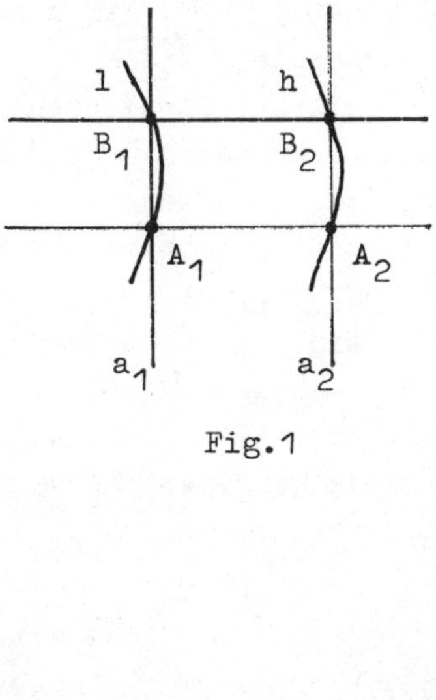

Fig. 1

Da $A_1 \not\sim A_2$ ist, gibt es eine Translation τ mit $\tau(A_1)=A_2$. Dann folgt $\tau(B_1)=B_2$, da $\tau(a_1)=a_2$ ist und B_1B_2 sowie A_1A_2 Richtungen von τ sind. Ist daher l eine weitere Verbindungsgerade von A_1 und B_1 und h die Parallele zu l durch A_2, so ist $\tau(l)=h$ und $\tau(B_1)=B_2$ liegt auf h. In $\mathcal{H}$ bilden die Translationen in einer festen Richtung also stets eine Gruppe. Sind dann P und Q benachbarte Punkte, so wählen wir ei-

nen Punkt R aus, der nicht zu P und Q benachbart ist und mit P und Q kollinear ist. Dann gibt es Translationen τ und σ mit $\tau(P)=R$ und $\sigma(R)=Q$. Da τ und σ die gleiche Richtung haben, ist dann $\sigma\cdot\tau$ eine Translation, die P auf Q abbildet.
Umgekehrt ist die Bedingung des Satzes natürlich auch notwendig.

Eine andere Kennzeichnung der Translationstransitivität ergibt sich mit Hilfe Des kleinen Satzes von Pappus. Dabei verstehen wir in einer H-Ebene unter dem kleinen affinen Satz von Pappus für die Richtung $[g]$ die folgende Aussage:

(p_g) Es seien g_1, g_2 aus $[g]$, $g_1 \not\sim g_2$ und $P_1, P_3, P_5 \in g_1$, $P_2, P_4, P_6 \in g_2$ derart, daß $P_1 \not\sim P_3$ ist. Ist dann P_1P_2 parallel zu P_4P_5 und P_2P_3 parallel zu P_5P_6 so ist auch P_3P_4 parallel zu P_6P_1.

Es gilt

4.Satz: Es sei $\mathcal{H}$ eine H-Ebene in der auf jeder Geraden mindestens drei paarweise nicht benachbarte Punkte liegen. $\mathcal{H}$ ist genau dann translationstransitiv, wenn es zwei nicht benachbarte Richtungen $[g]$ und $[h]$ gibt, sodaß (p_g) oder (p_h) gilt und zu je zwei nicht benachbarten Punkten P und Q mit $PQ \in [g] \cup [h]$ eine Translation τ mit $\tau(P)=Q$ existiert.

Beweis: Die Bedingungen des Satzes seien o.B.d.A. mit (p_g) erfüllt. Ferner sei $[l]$ eine Richtung mit $[g] \not\sim [l] \not\sim [h]$ und A und B zwei nicht benachbarte Punkte mit $AB \in [l]$. Wir zeigen, daß es dann eine Translation τ mit $\tau(A)=B$ gibt. Dazu legen wir durch A eine Gerade g_1 aus $[g]$ und durch B eine Gerade h_1 aus $[h]$. h_1 und g_1 schneiden sich in einem Punkt C. Es ist $A \not\sim C$ und $C \not\sim B$. Daher gibt es Translationen τ_1, τ_2 mit $\tau_1(A)=C$ und $\tau_2(C)=B$. Es folgt $\tau_2\tau_1(A)=B$. Es ist zu zeigen, daß $\tau_2\tau_1$ ei-

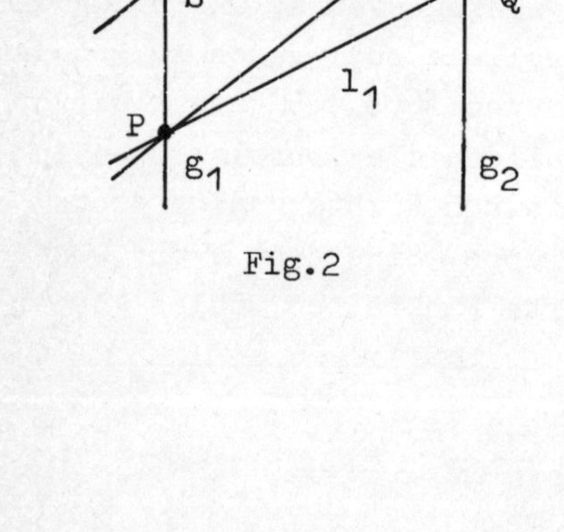

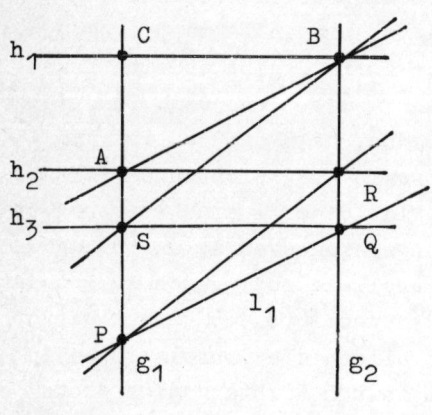

Fig.2

ne Translation ist. Sei dazu l_1 aus [1] mit $l_1 \nparallel AB$. Dann schneidet l_1 die Gerade g_1 in einem Punkt P. Durch B legen wir die Parallele g_2 zu g_1. h_2 sei die Parallele zu h_1 durch A. Sie schneidet g_2 in einem Punkt R. Es ist $\tau_1(R)=B$. Legen wir daher durch B die Parallele zu PR, so ist ihr Schnittpunkt S mit g_1 das Bild von P unter τ_1. Die Parallele h_3 zu h_1 durch S schneidet g_2 in einem Punkt Q und es ist $\tau_2(S)=Q$, also $\tau_2\tau_1(P)=Q$ und daher $\tau_2\tau_1(PQ)=PQ$. Weiter erfüllen die Geraden g_1,g_2 sowie die Punkte A,R,P,Q,S,B die Voraussetzungen von (p_g). Es folgt $PQ \parallel AB$, d.h. $l_1 \parallel PQ$ und daher $l_1=PQ$. Es folgt $\tau_2\tau_1(l_1)=l_1$. Daher ist $\tau_2\tau_1$ eine Translation.
Die Bedingung von Satz 4 ist also hinreichend. Daß sie auch notwendig ist zeigt

<u>5.Lemma</u>: Es sei $\mathcal{H}$ eine H-Ebene, in der es zu je zwei nicht benachbarten Punkten P und Q eine Translation τ mit $\tau(P)=Q$ gibt. Dann gilt in $\mathcal{H}$ die Bedingung (p_g) für alle Richtungen.

Beweis: $g_1,g_2,P_1,\dots\dots,P_6$ seien wie in den Voraussetzungen von (p_g) gegeben. Der Beweis verläuft dann ganz ähnlich wie in [2] für affine Ebenen. Da $P_1 \nsim P_3$ ist, sind auch P_1P_2 und P_2P_3 nicht benachbart. Daher schneiden sich die Parallele zu P_2P_3 durch P_1 und die Parallele zu P_1P_2 durch P_3 in einem Punkt Q. Nach Voraussetzung gibt es Translationen τ_1, τ_2 mit $\tau_1(P_6)=P_5$ und $\tau_2(P_5)=P_4$. Dann ist $\tau_1(P_2)=P_3$ und daher $\tau_1(P_1P_2)=P_3Q$, d.h. $\tau_1(P_1) \in P_3Q$. Ferner ist P_1Q nach Konstruktion und Voraussetzung parallel zu P_5P_6, d.h. $\tau_1(P_1Q)=P_1Q$. Es folgt $\tau_1(P_1) \in P_3Q$. Insgesamt ergibt sich $\tau_1(P_1)=Q$. Für τ_2 gilt $\tau_2(P_1)=P_2$, also $\tau_2(Q)=P_3$, da in dem Parallelogramm (P_1,Q,P_3,P_2) alle Schnittpunkte eindeutig sind. Es folgt $\tau_2\tau_1(P_1)=P_3$. Also gilt $\tau_2\tau_1(P_1P_6)=P_3P_4$. Da $\tau_2\tau_1$ jede Gerade auf eine Parallele abbildet, erhält man $P_1P_6 \parallel P_3P_4$.

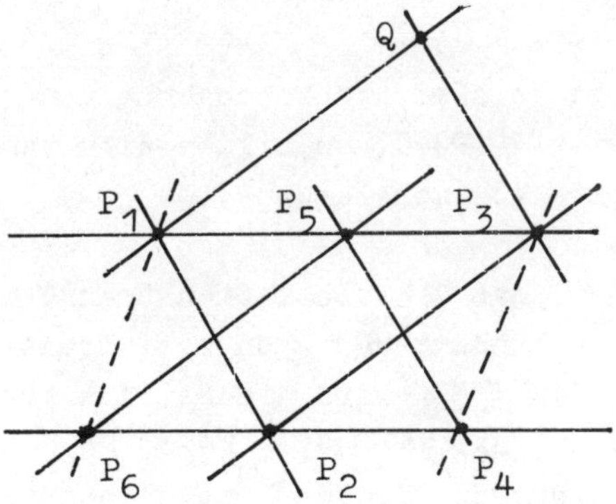

Fig.3

Unter stärkeren Schließungsvoraussetzungen kann man auf die Reichhaltigkeitsforderung von Satz 4 verzichten. Es gilt

6.Satz: Eine H-Ebene $\mathcal{H}$ ist genau dann translationstransitiv, wenn es zwei nicht benachbarte Richtungen $[g]$ und $[h]$ gibt, sodaß (p_g) und (p_h) erfüllt sind und zu je zwei nicht benachbarten Punkten P und Q mit $PQ \in [g] \cup [h]$ eine Translation τ mit $\tau(P)=Q$ existiert.

Beweis: Für jede Richtung $[k]$ mit $[g] \not\sim [k] \not\sim [h]$ zeigt man wie im beweis von Satz 4, daß es zu je zwei Punkten R und S mit $RS \in [k]$ eine Translation τ mit $\tau(R)=S$ gibt. Mit Hilfe der Richtungen $[k]$ und $[h]$ zeigt man dies auch für alle zu $[g]$ benachbarten Richtungen und mit Hilfe von $[g]$ und $[k]$ für alle zu $[h]$ benachbarten Richtungen. Der übrige Beweis ergibt sich dann wie in Satz 2.

In [5] wird gezeigt, daß die Translationen einer H-Ebene $\mathcal{H}$ jedenfalls dann eine Gruppe bilden, wenn $\mathcal{H}$ translationstransitiv ist und auf jeder Geraden von $\mathcal{H}$ mindestens drei paarweise nicht benachbarte Punkte liegen.Aus unseren Ergebnissen folgt nun, daß man die Voraussetzungen wesentlich abschwächen kann.

7.Satz: Gibt es in einer H-Ebene $\mathcal{H}$ zu je zwei nicht benachbarten Punkten P und Q eine Translation τ mit $\tau(P)=Q$, so bilden die Translationen eine scharf einfach-transitive abelsche Gruppe.

Beweis: Aus (5) und (6) folgt zunächst, daß $\mathcal{H}$ translationstransitiv ist. Wir haben dann zu zeigen, daß das Produkt $\sigma\tau$ von zwei Translationen wieder eine Translation ist. In [5] wird die Reichhaltigkeitsforderung nur für den Fall benutzt, daß σ und τ keine Nachbartranslationen sind und die Richtungen von σ und τ nicht benachbart sind. Im Beweis von (4) bzw. (6) wird aber gerade gezeigt, daß $\sigma\tau$ in diesem Fall eine Translation ist, wenn für eine der beiden Richtungen der kleine affine Pappus gilt. Nach (5) gilt aber in $\mathcal{H}$ (p_g) für alle Richtungen. Also ist $\sigma\tau$ eine Translation. Alle übrigen Fälle ergeben sich wie in [5]. Da die Gruppe der Translationen abelsch ist und eine Translation durch das Bild eines Punktes eindeutig bestimmt ist,ist der Satz vollständig bewiesen.

Mit Hilfe von (7) können wir nun zeigen, daß man in (5) bei den Voraussetzungen für die Pappusbedingung auf die Forderung $P_1 \quad P_3$ verzichten kann. Es gilt

<u>8.Lemma</u>:Es sei $\mathcal{H}$ eine H-Ebene in der es zu je zwei nicht benachbarten Punkten P und Q eine Translation τ mit $\tau(P)=Q$ gibt. Dann gilt in $\mathcal{H}$ der kleine affine Satz von Pappus:

(p) Sind g_1 und g_2 parallele Geraden mit $g_1 \not\sim g_2$ und $P_1,P_3,P_5 \in g_1$, $P_2,P_4,P_6 \in g_2$ derart,daß $P_1P_2 \parallel P_4P_5$ und $P_2P_3 \parallel P_5P_6$ ist, so ist auch $P_3P_4 \parallel P_6P_1$.

Beweis: Nach Satz 7 bilden die Translationen eine transitive abelsche Gruppe,woraus (siehe auch [3]) sofort die Behauptung folgt.

<u>Literatur</u>:

1) Klingenberg,W., Desarguessche Ebenen mit Nachbarelementen.Abh. Math. Sem. Univ. Hamburg 20, 97-111 (1955).
2) Lingenberg,R., Grundlagen der Geometrie I. Mannheim 1969.
3) Lorimer,J.W., Coordinate theorems for affine Hjelmslevplanes. Mathematical Report No.63, McMaster Univ.,Hamilton (Ontario) Canada (1973).
4) Lüneburg,H., Affine Hjelmslev-Ebenen mit transitiver Translationsgruppe. Math.Z. 79, 260-288 (1962).
5) Seier,W., Der kleine Satz von Desargues in affinen Hjelmslev-Ebenen.Geometriae Dedicata 3, 215-219 (1974).
6) Seier,W., Über Translationen in affinen Hjelmslev-Ebenen. Abh. Math. Sem. Univ. Hamburg 43 (1975).

Werner Seier
Mathematisches Seminar der Universität
Bundesstraße 55
2000 Hamburg 13

NEAR-RINGS WITH RIGHT INVERSE PROPERTY

Momme Johs Thomsen

1. Definitions and introduction

We start with the following definition:

A <u>near-ring</u> $F(+,\cdot)$ is a set F with two operations, addition and multiplication, such that:

1. $F(+)$ is a group.
2. $a(b+c) = ab + ac$, for all $a, b, c \in F$.

Certain special near-rings are used in geometry in several places. The two most important classes of near-rings in geometry are the class of near-fields and the class of quasi-fields (also often called Veblen-Wedderburn-systems). Let us recall the definitions:

A <u>near-field</u> is a near-ring $F(+,\cdot)$ such that:

1. $F(+,\cdot)$ is zero-symmetric, i. e.
   $0 \cdot b = 0$, for all $b \in F$
   (0 being the additive identity).
2. $F^*(\cdot)$ is a group ($F^* := F \setminus \{0\}$).

A <u>quasi-field</u> is a near-ring $F(+,\cdot)$ such that:

1. $F(+,\cdot)$ is zero-symmetric.
2. $F^*(\cdot)$ is a loop.
3. $F(+,\cdot)$ is planar, i. e.
   $-ax + bx = c$ has a unique solution for x ,
   given $a, b, c \in F$, $a \neq b$.

For both classes there are several essentially different proofs in the literature that such near-rings have abelian addition.

In [2] E. SPERNER constructed examples of weak-affine spaces by means of near-fields. These weak-affine spaces were thoroughly investigated by H. J. ARNOLD [1]. Later on in [3], [4] E. SPERNER constructed in a similiar way more general weak-affine spaces by using instead of near-fields more general systems, which are called in [7] right-S-systems and in [6] simply S-systems.

A right-S-system is a near-ring $F(+,\cdot)$ such that:

1. $F(+,\cdot)$ is zero-symmetric.
2. $F(+,\cdot)$ has a multiplicative identity $1 \neq 0$.
3. $F(+,\cdot)$ has the right inverse property, i. e. for each $b \in F^*$ there is at least one $b' \in F$ such that $(ab)b' = a$ for every $a \in F$.

Proper right-S-systems, that is right-S-systems which are neither near-fields nor quasi-fields, have been constructed by J. TIMM [6] by using a construction technique, which is a generalization of the method of Dickson-Zassenhaus-Karzel for the construction of near-fields. Since TIMM constructed right-S-systems by changing the multiplication in non-commutative fields without changing the addition, his systems have of course abelian addition. The question whether there exist right-S-systems without commutative addition has been open until now.

The purpose of this lecture is to prove that every right-S-system has abelian addition and to deduce some consequences of this property. The proof consisting of several steps is divided into two parts. In the first part the addition of any near-ring satisfying certain conditions is shown to be abelian. In the second we show that these conditions are satisfied by right-S-systems.

2. Commutativity theorems for near-rings

1. PROPOSITION. Let $F(+,\cdot)$ be a near-ring with an element e such that $(-e)x = -x$ for all $x \in F$. Then $F(+,\cdot)$ is abelian.

PROOF. The condition means, that the mapping $F \longrightarrow F : x \longmapsto -x$ is a left translation of the near-ring $F(+,\cdot)$, hence an endomorphism of the group $F(+)$. This is only the case in abelian groups.

2. PROPOSITION. Let $F(+,\cdot)$ be a near-ring with an element e such that

i) $\qquad (-e)x \neq x \quad$ for all $x \in F^*$

ii) $\qquad (-e)((-e)x) = x \quad$ for all $x \in F$.

Then the mapping

$$\delta : F \longrightarrow F : x \longmapsto x - (-e)x$$

has the following properties:

(1) $\qquad \delta$ is injective.

(2) $\qquad (-e)y = -y$ for all y in the image $F\delta$ of δ.

Moreover, $F(+,\cdot)$ is abelian if and only if

$$(-e)x = -x \quad \text{for all } x \in F .$$

PROOF. (1) is correct, since for all $x, w \in F$ we have: $x\delta = x - (-e)x = w - (-e)w = w\delta$ implies $-(-e)x + (-e)w = -x + w$, hence $(-e)(-x+w) = -x + w$. i) yields $-x + w = 0$, i. e. $x = w$.
(2) follows from ii): $(-e)(x\delta) = (-e)(x - (-e)x) =$
$= (-e)x - (-e)((-e)x) =$
$= (-e)x - x = -(x - (-e)x) = -x\delta$.
If $F(+,\cdot)$ is abelian, we have because of ii)
$(-e)(x + (-e)x) = (-e)x + (-e)((-e)x) = (-e)x + x = x + (-e)x$.
i) yields $x + (-e)x = 0$, i. e. $(-e)x = -x$.
Conversely, if the last equation holds $F(+,\cdot)$ is abelian by 1..

For finite near-rings the last proposition yields already a criterium for the commutativity of addition:

3. COROLLARY. A finite near-ring $F(+,\cdot)$ with an element e such that i) and ii) hold is abelian.

PROOF. The corollary follows from 2., since the mapping δ is by (1) and by finiteness surjective, hence by (2) $(-e)x = -x$ for all $x \in F$.

Before proceeding to the next proposition we note that in the last three propositions we postulated the existence of an element $-e$ with certain properties resembling those of the negative of the multiplicative identity in a skewfield. From now on we require that e is in fact a right multiplicative identity.

4. PROPOSITION. Let $F(+,\cdot,e)$ be a near-ring with a right identity e such that

iii) $\exists \; h \in F$ with $h(e+e) = e$

iv) $e+e$ is left regular

(i. e. $x_1(e+e) = x_2(e+e) \Longrightarrow x_1 = x_2$ for $x_1, x_2 \in F$).

Then for every $x \in F$

(3) $xh + xh = x$

(4) $(-e)x = -x \Longrightarrow (-e)(xh) = -xh$.

PROOF. (3) follows from iii):

$$xh + xh = x(h+h) = x(h(e+e)) = xe = x .$$

(4) is correct, for from $(-e)x = -x$ we have using (3)

$$\begin{aligned}((-e)(xh))(e+e) &= (-e)(xh) + (-e)(xh) = \\ &= (-e)(xh + xh) = (-e)x = \\ &= -x = -(xh + xh) = \\ &= (-xh) + (-xh) = \\ &= (-xh)(e+e) .\end{aligned}$$

Since $e+e$ is left regular, this yields $(-e)(xh) = -xh$.

5. THEOREM. Let $F(+,\cdot,e)$ be a near-ring with a right identity e such that the four conditions i) to iv) hold. Then $F(+,\cdot,e)$ is abelian and $(-e)x = -x$ for all $x \in F$.

PROOF. By 2. the two statements are equivalent. Let us show the second. δ being again the mapping of 2..
For every $x \in F$ we have by (2) $(-e)\cdot x\delta = -x\delta$, whence by (4)

$$(5) \qquad (-e)(x\delta\cdot h) = -(x\delta\cdot h) \quad .$$

This implies using (3)
$$\begin{aligned}(x\delta\cdot h)\delta &= x\delta\cdot h - (-e)(x\delta\cdot h) = \\ &= x\delta\cdot h - (-(x\delta\cdot h)) = \\ &= x\delta\cdot h + x\delta\cdot h = x\delta \quad ,\end{aligned}$$
whence using (1) $x\delta\cdot h = x$. Substituting this in (5) yields finally $(-e)x = -x$.

3. Near-rings with right inverse property

To be able to apply the last theorem to right-S-systems we have to derive some properties of these systems. For what follows we don't need the zero-symmetry of the definition of these systems.

6. PROPOSITION. Let $F(+,\cdot,1)$ be a near-ring with a left identity $1 \neq 0$ satisfying the right inverse property. For given $b \in F^*$ let b', b'' be elements of F such that $(ab)b' = a = (ab')b''$ for all $a \in F$. Then the following holds:

1) $(ab')b = a = (ab)b'$ for all $a \in F$, thus the right translation $R_b : F \longrightarrow F : x \longrightarrow xb$ is bijective with $R_b^{-1} = R_{b'}$.

2) $b\cdot b' = b'\cdot b = 1$.

3) $b_1 \in F$ and $b_1 \cdot b = 1$ $\Longrightarrow$ $b_1 = b'$.

4) b' is uniquely determined by $b \in F^*$, that is we are able to define the mapping
$F^* \longrightarrow F^* : b \longrightarrow b^{-1} := b'$.

5) $(b^{-1})^{-1} = b'' = b$

6) $b_r \in F$ and $b \cdot b_r = 1$ $\Longrightarrow$ $b_r = b^{-1}$.

7) $((-1)x)^{-1} = -x^{-1}$ for all $x \in F^*$

8) $(-1)((-1)x) = x$ for all $x \in F$

PROOF. 1) follows by three applications of the right inverse property: $(ab')b = (((ab')b)b')b'' = (ab')b'' = a$.
2) follows by setting $a = 1$ in 1).
3) follows from 2) and 1): $b_1 \cdot b = 1 = b' \cdot b$ $\Longrightarrow$ $b_1 = b'$.
4) If b^* is also an element of F satisfying $(ab)b^* = a$ for all $a \in F$, then by 2) $b^* \cdot b = 1 = b' \cdot b$, hence by 3) $b^* = b'$.
5) follows from 2) to 4).
6) From $b \cdot b_r = 1$ we have by 3) and 4) $b = b_r^{-1}$, hence by 5) $b_r = b^{-1}$.
7) We have $((-1)x) \cdot (-x^{-1}) = -((-1)x)x^{-1} = -(-1) = 1$, hence by 6) $((-1)x)^{-1} = -x^{-1}$.
8) We can assume $x \neq 0$, thus by 7) $(-1)x \neq 0$, so that by 1) there exists $y \in F$ such that $y((-1)x) = x$. This yields using the right inverse property and 7)
$y = (y((-1)x)) \cdot ((-1)x)^{-1} = x \cdot (-x^{-1}) = -x \cdot x^{-1} = -1$.

We are now ready to prove the main theorem.

7. THEOREM. Let $F(+,\cdot,1)$ be a near-ring with an identity $1 \neq 0$ satisfying the right inverse property (in particular, let $F(+,\cdot,1)$ be a right-S-system). Then the addition of $F(+,\cdot)$ is abelian. Moreover, $(-x)^{-1} = -x^{-1}$ for all $x \in F^*$ and $(-w) \cdot x = -w \cdot x$ for all $w, x \in F$.

PROOF. Since $1 = -1$ implies $x = x(-1) = -x$ for all $x \in F$ so that the theorem holds trivially, let us assume $1 \neq -1$. Then the conditions i) to iv) (with $e = 1$) are satisfied: ii) is 6.8) whereas the others follow from 6.1). Therefore by 5. the addition of $F(+,\cdot)$ is abelian and $(-1)x = -x$ for all $x \in F$. The last equation and 6.7) yield $(-x)^{-1} = -x^{-1}$ for every $x \in F^*$. Finally we deduce from the last equation using the right inverse property twice for given $w \in F$ and $x \in F^*$

$$\begin{aligned}(-w)\cdot x &= (((-w)\cdot x)\cdot(-x^{-1}))\cdot(-x) = \\ &= (-((-w)\cdot x)\cdot x^{-1})\cdot(-x) = (-(-w))\cdot(-x) = \\ &= w\cdot(-x) = -w\cdot x \quad .\end{aligned}$$

We give three corollaries of the last theorem.

8. COROLLARY. Let $F(+,\cdot,1)$ be a near-ring with an identity 1 such that $1+1 \neq 0$ and with the right inverse property. Then $F(+,\cdot,1)$ is a right-S-system.

PROOF. We have only to show that $F(+,\cdot)$ is zero-symmetric. With $h := (1+1)^{-1}$ we have for every $x \in F$ by (3) and 7.

$$\begin{aligned}0\cdot x &= 0\cdot(xh + xh) = 0\cdot(xh) + 0\cdot(xh) = \\ &= 0\cdot(xh) + (-0)\cdot(xh) = 0\cdot(xh) - 0\cdot(xh) = 0 \ .\end{aligned}$$

9. COROLLARY. Let $F(+,\cdot,1)$ be a near-ring with an identity $1 \neq 0$ and with the right inverse property. Let K be the set of all $k \in F$ such that $(xy)k = x(yk)$ and $(x+y)k = xk + yk$ hold for all $x, y \in F$. Then $K(+,\cdot)$ is a subskew-field of $F(+,\cdot)$ such that $F(+,\cdot)$ is a near-algebra over $K(+,\cdot)$ whose underlying near-module is a right vector space over $K(+,\cdot)$.

PROOF. Using the fact that the addition is abelian it is

straightforward to verify that $K(+,\cdot)$ is a skewfield. The rest of the statement follows readily from the definitions for which we refer to M. J. THOMSEN [5].

<u>10. COROLLARY</u>. Let $F(+,\cdot,1)$ be a right-S-system. If the dimension of the right vector space $F(+)$ over the subskewfield $K(+,\cdot)$ is finite, then $F(+,\cdot)$ is a quasifield.

<u>PROOF</u>. We first show that $F(+,\cdot)$ is planar, i. e. that the mapping $p_{a,b} : F \dashrightarrow F : x \dashrightarrow -ax + bx$ is bijective for given $a, b \in F$ with $a \neq b$. Now, $p_{a,b}$ is an endomorphism of the vector space, since we have for all $x, y \in F$ and $k \in K$

$$(x+y)p_{a,b} = -a(x+y) + b(x+y) = -ax - ay + bx + by =$$
$$= -ax + bx - ay + by = xp_{a,b} + yp_{a,b}$$

and

$$(xk)p_{a,b} = -a(xk) + b(xk) = -(ax)k + (bx)k =$$
$$= (-ax + bx)k = (xp_{a,b})k \quad .$$

Moreover, $p_{a,b}$ is injective, since from $-ax + bx = -ay + by$ we get $a(y-x) = ay - ax = by - bx = b(y-x)$, hence, by 6.1), $y-x = 0$. Consequently, by the finiteness of the dimension, $p_{a,b}$ is an automorphism of the vector space. In particular, every left translation $L_b : F \dashrightarrow F : x \dashrightarrow bx$ with $b \in F^*$ is bijective, since $L_b = p_{0,b}$, by zero-symmetry. From this and from 6.1) follows that $F^*(\cdot)$ is a loop.

References

[1] ARNOLD, H. J.: Algebraische und geometrische Kennzeichnung der schwach affinen Vektorräume über Fastkörpern. Abh. Math. Sem. Univ. Hamburg 32 (1968) 73 - 88.

[2] SPERNER, E.: On non-Desarguesian Geometries. Seminari dell' Istituto Nazionale di Alta Matematica (1962/63) 574 - 594.

[3] SPERNER, E.: Weak affine spaces and their algebraic representation. Wiskunde - Seminare, Universiteit van Stellenbosch (1966) 86 - 124.

[4] SPERNER, E.: Zur Geometrie der Quasimoduln. Istituto Nazionale di Alta Matematica, Symposia Matematica 5 (1971) 421 - 438.

[5] THOMSEN, M. J.: Zur Theorie der Fastalgebren. Mathematik-Arbeitspapiere Nr. 16. Universität Bremen, 1977.

[6] TIMM, J.: Eine Klasse schwacher binärer Doppelstrukturen. Abh. Math. Sem. Univ. Hamburg 33 (1969) 102 - 118.

[7] TIMM, J.: Zur Theorie der nicht notwendig assoziativen Fastringe. Abh. Math. Sem. Univ. Hamburg 35 (1970) 14 - 31.

Momme Johs Thomsen

Fachbereich Mathematik

Universität Bremen

Bibliothekstr.

D-2800 Bremen 33

ON REFLECTIONS IN MINKOWSKI-PLANES

H. Zeitler, Bayreuth

1. Introduction

1.1 Different ways to MINKOWSKI-planes

Investigations of MINKOWSKI-planes can be made in very different manners.

1.1.1 MINKOWSKI-plane axiomatically [1], [4]

In the beginning there is a system of axioms. Then the consequences of this system are investigated.
Here we present such a system of axioms.
Let $\mathfrak{P}$ be a set of points and $\mathfrak{Z}$, $\mathfrak{E}_1$, $\mathfrak{E}_2$ three subsets of the power set of $\mathfrak{P}$. The elements of $\mathfrak{Z}$ are called cycles, those of $\mathfrak{E}_1 \cup \mathfrak{E}_2$ generators. Two points are said to be parallel (unlinkable) if they are equal or elements of a generator.
The incidence structure $(\mathfrak{P}, \mathfrak{Z}, \mathfrak{E}_1, \mathfrak{E}_2, \in)$ is called a MINKOWSKI-plane if the following axioms are satisfied.

(M1) If $i \in \{1, 2\}$ and $P \in \mathfrak{P}$ then there exists exactly one generator $e_i \in \mathfrak{E}_i$ through P.

(M2) Every generator of $\mathfrak{E}_1$ has exactly one point in common with every generator of $\mathfrak{E}_2$ and vice versa.

(M3) There are 3 mutually non parallel points.

(M4) Any 3 mutually non parallel points lie on exactly one cycle.

(M5) Every generator has exactly one point in common with every cycle.

(M6) Touch axiom:
If a point P is on a cycle m and Q is not parallel to P and not on m, then there exists exactly one cycle n through Q with $m \cap n = \{P\}$.

A MINKOWSKI-plane is called miquelian, if the axiom of MIQUEL is satisfied.

(M7) Axiom of MIQUEL

Let A_1, A_2, A_3, A_4, B_1, B_2, B_3, B_4 be mutually non parallel points. If the quadrupels (A_1, A_2, B_1, B_2), (A_2, A_3, B_2, B_3), (A_3, A_4, B_3, B_4), (A_4, A_1, B_4, B_1), (A_1, A_2, A_3, A_4) are in each case elements of a cycle, then the points B_1, B_2, B_3, B_4 also lie on a cycle.

1.1.2 MINKOWSKI-plane as intersection-geometry [3]

Let a one sheet hyperboloid be embedded in the projective space. The points of this hyperboloid are called "points", the intersections of planes with the hyperboloid "cycles", respectively "generators". Then we obtain a MINKOWSKI-plane. This plane is studied by using theorems of the projective geometry.
This method of constructing the MINKOWSKI-plane consists in "labeling" within the projective space.

1.1.3 MINKOWSKI-plane algebraically [2], [5]

Starting with a field K , a ring R is constructed with convenient definitions and then a (K,R)-plane is developed, using only algebraic methods.

1.2 The aim of this paper

For this paper we have choosen the algebraic method, the way noted in 1.1.3. At first the main results of W. BENZ and E.M. SCHROEDER are sketched. The definition of reflections in cycles necessarily leads to the orthogonality of cycles and finally to pencils of cycles. These pencils will be classified. The three-reflection-theorem and the theorem about the minimal-decomposition-length of the product of reflections will conclude the work. The working with "unproper" elements is extremly important, because it enables us to define reflections in an exact manner. We note that all investigations are independent of the characteristic of the field K . Some theorems about finite MINKOWSKI-planes will be stated.

With respect to the contents of this paper, one can say that the works of W. BENZ are complemented by the investigations of reflections. The new theorems partially can be found in the works [4], [3(c)] using there the ways sketched in 1.1.1 and 1.1.2.

The real value of the paper however is a methodical one. In all parts of the work we "count", we "calculate" with "equations" of transformations and cycles. The analogy to the GAUSS-plane - as already stated in [7] - is very important. Theorems, which in the manner of 1.1.1 were directly deduced from an axiom system, now are proved by ordinary "counting". To enable us to perform such a work it is necessary to construct convenient tools. The main tools are the "equations of transformations", the "decomposition of cycle-transformations for simplification", the "conditions of orthogonality" and the "transformation of pencils for simplification". We will talk about these methods later.

Proofs of the stated theorems cannot be given here. Only the three-reflection-theorem will be proved to show in one example the methods used in this paper.

Altogether this work is only a report on various results.

2. Some algebra

Let K be a commutative field and R the set of all pairs (x_1, x_2) with $x_1, x_2 \in K$. Then in the set R we define equality, addition and multiplication as follows:

$$(x_1, x_2) = (y_1, y_2) :\Longleftrightarrow x_1 = y_1 \text{ and } x_2 = y_2$$

$$(x_1, x_2) + (y_1, y_2) := (x_1 + y_1, x_2 + y_2)$$

$$(x_1, x_2) \cdot (y_1, y_2) := (x_1 y_1, y_1 x_2 + x_1 y_2 - x_2 y_2)$$

We further define $(x_1, 0) = x_1$. Then R is an extension of K. Finally we introduce a special notation for the elements of R, namely $(x_1, x_2) = x_1 + \varepsilon x_2$. With this notation it follows $\varepsilon^2 + \varepsilon = 0$. The elements of the set R are written with capital letters and the elements of K with small letters.

It can be demonstrated that $(R, +, \cdot)$ is a commutative ring with the neutral element $(1, 0)$.

Aut R means the set of all automorphisms of R, $\overline{\mathrm{Aut}_K R}$ the set of all elements of Aut R, which fixes K in the whole, and $\mathrm{Aut}_K R$ finally the set of elements of $\overline{\mathrm{Aut}_K R}$ which fixes K element by element.

The element $\overline{X} = x_1 - x_2 - \varepsilon x_2 \in R$ is called conjugate to the element $X = x_1 + \varepsilon x_2 \in R$. The mapping $X \mapsto \overline{X}$ for all $X \in R$ is the only non identical automorphism of $\mathrm{Aut}_K R$. The mapping is involutorial.

The element $X\overline{X} = x_1^2 - x_1 x_2 \in K$ is called the norm of the element $X = x_1 + \varepsilon x_2 \in R$. We write $N(X)$. It follows $N(XY) = N(X) \cdot N(Y)$ for all $X, Y \in R$, further $N(0) = 0$ and $N(1) = 1$.

We now define $R^* = \{X \in R \,/\, N(X) = 0\}$. $(R^*, \circ)$ is an abelian group. All elements of K are norms of elements of R.

Let K be a finite field with $|K| = q = p^e$ (p prime, $e \in \mathbb{N}$), then with $K^* := K \setminus \{0\}$ we obtain the following statements

$|\overline{\mathrm{Aut}_K R}| = 2$, $|\mathrm{Aut}\, R| = 2e^2$, $|R^*| = (q-1)^2$,

$|\{X \in R \,/\, N(X) = 0\}| = 2q - 1$, $|\{X \in R \,/\, N(X) = a$ and $a \in K^*\}| = q - 1$.

3. The (K,R)-plane and its elements

3.1 The points

We consider the pairs of elements of the set R. Two such pairs (Z_1, Z_2), (W_1, W_2) with $Z_1, Z_2, W_1, W_2 \in R$ are equivalent, iff there exists $S \in R^*$ such that $W_1 = S Z_1$ and $W_2 = S Z_2$.

3.1.1 Proper points

The elements of $\mathcal{P}_1 = \{(Z_1, Z_2) \,/\, Z_1 \in R$ and $Z_2 \in R^*\}$ are called proper points.
It means no restriction to write $\mathcal{P}_1 = \{(Z_1, 1) \,/\, Z_1 \in R\}$.

3.1.2 Improper points

The elements of $\mathcal{Y}^* = \{(d + \varepsilon(1 + d), 1 + \varepsilon) / d \in K\}$ and of $\mathcal{J}^* = \{(1 + \varepsilon(1 + d), \varepsilon) / d \in K\}$ together with the pair $(1, 0) = P^*$ are called improper points.
For all these improper points (Z_1, Z_2) holds $N(Z_2) = 0$.
The set $\mathcal{P} = \{P^*\} \cup \mathcal{P}_1 \cup \mathcal{Y}^* \cup \mathcal{J}^*$ is called the set of all points.
This definition eliminates some special pairs of elements of R as forbidden "foolish points".

3.2 The hyperbolas

Each set of points

$\{(Z_1, Z_2) \in \mathcal{P} \,/\, Z_1\overline{Z}_1 - M\, Z_1\overline{Z}_2 - \overline{M}\, \overline{Z}_1 Z_2 + (M\overline{M} - c) Z_2\overline{Z}_2 = 0\}$

with $M \in R$, $c \in K^*$ is called hyperbola, (M,1) its centre, c its norm.

Each hyperbola contains exactly two improper points. With $M = m_1 + \varepsilon m_2$ we obtain for these points

$S_M^* = (m_1 + \varepsilon(1 + m_1) , 1 + \varepsilon)$ and $T_M^* = (1 + \varepsilon(1 + m_1 - m_2), \varepsilon)$.

3.3 The lines

Each set of points

$\{(Z_1, Z_2) \in \mathfrak{P} / \overline{M} Z_1\overline{Z}_2 + M \overline{Z}_1 Z_2 + d Z_2\overline{Z}_2 = 0\}$ with $M \in R^*$, $d \in K$

is called line.

Each line contains exactly one improper point, namely P^*.

The set $\mathfrak{Z}$ of all hyperbolas and lines we designate as the set of all cycles.

3.4 The generators

Each set of points

$s_M = \{((x_1, x_2), 1) \in \mathfrak{P} / x_1 - m_1 = 0\} \cup \{S_M^*\}$ and

$t_M = \{((x_1, x_1), 1) \in \mathfrak{P} / x_1 - x_2 - m_1 + m_2 = 0\} \cup \{T_M^*\}$

with $M = m_1 + \varepsilon m_2 \in R$, $S_M^* = (m_1 + \varepsilon(1 + m_1) , 1 + \varepsilon)$,

$T_M^* = (1 + \varepsilon(1 + m_1 - m_2), \varepsilon)$ is called proper generator, the two sets of points $s^* = \mathfrak{r}^* \cup \{P^*\}$ and $t^* = \mathfrak{g}^* \cup \{P^*\}$ improper generators.

The set of all generators s_M together with t^* forms the class $\mathfrak{t}_1$ of generators, analogously all generators t_M together with s^* form the class $\mathfrak{t}_2$.

3.5 The (K,R)-plane

The incidence structure ($\mathfrak{P}$, $\mathfrak{Z}$, $\mathfrak{t}_1$, $\mathfrak{t}_2$, $\in$) is called (K,R)-plane.

4. The cycle-preserving transformations

4.1 Cycle-preserving transformation

Each transformation, which maps points one to one onto points and cycles one to one onto cycles and preserves incidence, is called a cycle-preserving transformation, in short a cycle-transformation.

4.2 (K,R)-transformations

Maps which are represented by the following equations are called (K,R)-transformations.

$$\mu : \begin{cases} \mathfrak{P} \longrightarrow \mathfrak{P} \\ (Z_1, Z_2) \mapsto (S\,\rho(Z_1) + T\,\rho(Z_2),\; U\,\rho(Z_1) + V\,\rho(Z_2)) \end{cases}$$

with $S, T, U, V \in R$ and $\rho \in \overline{\mathrm{Aut}_K R}$.

For the determinant $\det = S V - T U$ we require $\det \in R^*$. It is determinated up to R^{*2}. Therefore we must write more exactly $\det \in R^* \cdot R^{*2}$.

To facilitate the working with transformations we introduce another notation.

$$\mu \begin{cases} Z_1' = S\,\rho(Z_1) + T\,\rho(Z_2) \\ Z_2' = U\,\rho(Z_1) + V\,\rho(Z_2) \end{cases}$$

We denote the set of all these transformations with $\mathfrak{M}$. In the special case $\rho(Z) = Z$ we write $\mathfrak{H}$ and speak on homographies.

4.3 Theorems about (K,R)-transformations

Theorem 1

Each (K,R)-transformation is a cycle-transformation.

To prove this theorem we make a "decomposition for simplification". We show that each (K,R)-transformation is a product of transformations of the following type:

$$\begin{cases} Z_1' = Z_2 \\ Z_2' = Z_1 \end{cases} \qquad \begin{cases} Z_1' = Z_1 + A\, Z_2 \\ Z_2' = Z_2 \end{cases} \quad \text{with } A \in R$$

$$\begin{cases} Z_1' = \rho(Z_1) \\ Z_2' = \rho(Z_2) \end{cases} \qquad \begin{cases} Z_1' = B\, Z_1 \\ Z_2' = Z_2 \end{cases} \quad \text{with } B \in R^*$$

Now these special transformations are applied to cycles.

Theorem 2

The sets $\mathfrak{H}$ and $\mathfrak{M} = \mathfrak{H} \cdot \overline{\mathrm{Aut}_K R}$ form groups, when the elements of these sets are composed.

Theorem 3

The group $\mathfrak{H}$ works sharpely triply transitive on $\mathfrak{P}$ in the following sense. Let (A, B, C), (A', B', C') be triples of non parallel points, then there exists one and only one transformation $\eta \in \mathfrak{H}$ such that $A' = \eta(A)$, $B' = \eta(B)$, $C' = \eta(C)$.

Theorem 4

The (K,R)-transformations $\mathfrak{M}$ are exactly the cycle-transformations.

5. Theorems about (K,R)-planes

Theorem 5

Two points $P = (P_1, P_2)$, $Q = (Q_1, Q_2)$ are parallel iff $N(P_1Q_2 - P_2Q_1) = 0$. We write $P \| Q$.

Theorem 6

The (K,R)-planes are exactly the miquelian MINKOWSKI-planes.

Theorem 7

All cycles have the same cardinality.

Theorem 8

In the finite case with $|K| = q = p^e$ there exist exactly $(q + 1)^2$ points, $q(q - 1)$ lines, $q^2(q - 1)$ hyperbolas and $2(q + 1)$ generators. Every cycle consists of $q + 1$ points. Each point is incident with two generators and exactly $q(q - 1)$ cycles, two non parallel points with exactly $q - 1$ cycles. The order of the groups $\mathcal{G}$ and $\mathcal{M}$ are $q^2(q^2 - 1)^2$ and $2e\, q^2(q^2 - 1)^2$.

The proof of theorem 4 uses the affine plane, the proofs of the theorems 1, 2, 3, 5, 7, 8 only consist of "counting", of "calculating" with our formulas for cycles and transformations. The more complicated proof of the important theorem 6 can be found in W. BENZ [2] and H. SCHAEFFER [6].

6. Reflections

Each cycle-transformation which leaves invariant each point of a cycle z and no other points is called reflection σ_z in z. The reflections are represented by the following equations.

Line reflection

$$z = \{(Z_1, Z_2) \in \mathcal{P} \; / \; \overline{M}\, Z_1\overline{Z}_2 + M\, \overline{Z}_1 Z_2 + d\, Z_2\overline{Z}_2 = 0\}$$

with $M \in R^*$, $d \in K$

$$\sigma_z \begin{cases} Z_1' = -\dfrac{M}{\overline{M}}\,\overline{Z}_1 - \dfrac{d}{\overline{M}}\,\overline{Z}_2 \\ Z_2' = \overline{Z}_2 \end{cases}$$

Hyperbola reflection

$$z = \{ (Z_1, Z_2) \in \mathcal{R} \,/\, Z_1\overline{Z}_1 - M\,\overline{Z}_1 Z_2 - \overline{M}\,Z_1\overline{Z}_2 + (M\,\overline{M} - c) Z_2\overline{Z}_2 = 0 \}$$

with $M \in R$, $c \in K^*$

$$\sigma_z \begin{cases} Z_1' = M\,\overline{Z}_1 + (c - M\,\overline{M})\,\overline{Z}_2 \\ Z_2' = \overline{Z}_1 - \overline{M}\,\overline{Z}_2 \end{cases}$$

The determinant of all these reflections is element of $K^* \cdot R^{*2}$.

For any cycle there exists one and only one reflection (to be proven!). Each homography whose determinant is element of $K^* \cdot R^{*2}$ is called normhomography.

Theorem 9

The product of an even number of reflections yields a normhomography, and, vice versa, each normhomography is a product of an even number of reflections.

To prove the last part of this theorem we make again a "decomposition for simplification". We show that each normhomography is a product of transformations of the following type:

translation:

$$\begin{cases} Z_1' = Z_1 + A\,Z_2 \\ Z_2' = Z_2 \end{cases} \quad \text{with } A \in R$$

reciprocation:

$$\begin{cases} Z_1' = Z_2 \\ Z_2' = Z_1 \end{cases}$$

spiral similarity:

$$\begin{cases} Z_1' = B\,Z_1 \\ Z_2' = Z_2 \end{cases} \quad \text{with } B \in K^* \cdot R^{*2} .$$

The spiral similarity is further decomposed in rotations (with $N(B) = 1$) and point dilatations (with $B \in K^*$).

$A = 0$ and $B = 1$ yields the identical mapping.

Now these transformations are represented as products of 2 or 4 reflections.

Theorem 10

All products of reflections form a group $\mathcal{L}$, the products of an even number of reflections form the subgroup $\mathcal{L}^+$.

Theorem 11

Let σ_z be a reflection in a cycle z and $\mu \in \mathfrak{M}$, then it holds

$$\sigma_{\mu(z)} = \mu \sigma_z \mu^{-1} .$$

Theorem 12

In the finite case with $|K| = q = p^e$ there is
$|\mathcal{L}^+| = q^2(q+1)^2(q-1)$ and $|\mathcal{L}| = 2e\,|\mathcal{L}^+|$.

It is very interesting to analyse the operating of a hyperbola reflection in all details. The results are analogous to those of circle-transformations in MÖBIUS-planes.
The definition of motions and similarities provides many possibilities for additional work.

7. Orthogonality

The cycle z_1 is called orthogonal to the cycle z_2 iff $z_1 \neq z_2$ and $\sigma_{z_1}(z_2) = z_2$. We write $z_1 \perp z_2$.

In the case Char K = 2 - and only in this case - we define that each cycle is orthogonal to itself. We write $z \perp z$.

Conditions of orthogonality

(a) Two different lines

$z_i = \{(Z_1, Z_2) \in \mathcal{P} \,/\, M_i \overline{Z}_1 Z_2 + \overline{M}_i Z_1 \overline{Z}_2 + d_i Z_2 \overline{Z}_2 = 0\}$

with $i \in \{1,2\}$, $M_i \in R^*$, $d_i \in K$ are orthogonal iff

$$\boxed{M_1 \overline{M}_2 + \overline{M}_1 M_2 = 0}$$

(b) The line

$$z_1 = \{ (Z_1, Z_2) \in \mathcal{R} \ / \ M_1\overline{Z}_1 Z_2 + \overline{M}_1 Z_1 \overline{Z}_2 + d_1 Z_2 \overline{Z}_2 = 0 \}$$

with $M_1 \in R^*$, $d_1 \in K$ and the hyperbola

$$z_2 = \{ (Z_1, Z_2) \in \mathcal{R} \ / \ Z_1\overline{Z}_1 - \overline{M}_2 Z_1 \overline{Z}_2 - M_2 \overline{Z}_1 Z_2 (M_2\overline{M}_2 - c_2) Z_2\overline{Z}_2 = 0 \}$$

with $M_2 \in R$, $c_2 \in K^*$ are orthogonal iff

$$\boxed{M_1\overline{M}_2 + \overline{M}_1 M_2 + d_1 = 0}$$

(c) Two different hyperbolas

$$z_i = \{ (Z_1, Z_2) \in \mathcal{R} \ / \ Z_1\overline{Z}_1 - \overline{M}_i Z_1 \overline{Z}_2 - M_i \overline{Z}_1 Z_2 + (M_i\overline{M}_i - c_i) Z_2\overline{Z}_2 = 0 \}$$

with $i \in \{1,2\}$, $M_i \in R$, $c_i \in K^*$ are orthogonal iff

$$\boxed{N(M_1 - M_2) = c_1 + c_2}$$

These conditions of orthogonality are fundamental for all the following proofs. They form an indispensable and useful tool. It is very surprising that in the case Char $K \neq 2$ there always exist concentric hyperbolas h_1, h_2 with $h_1 \neq h_2$ and $h_1 \perp h_2$. For example:

$h_1 = \{ (Z_1, Z_2) \in \mathcal{R} \ / \ Z_1\overline{Z}_1 = Z_2\overline{Z}_2 \}$,

$h_2 = \{ (Z_1, Z_2) \in \mathcal{R} \ / \ Z_1\overline{Z}_1 = -Z_2\overline{Z}_2 \}$.

<u>Theorem 13</u>

<u>Let</u> z_1, z_2 <u>be two cycles and</u> $\mu \in \mathcal{M}$, <u>then we have the equivalences</u>

(a) $z_1 \perp z_2 \iff z_2 \perp z_1$

(b) $z_1 \perp z_2 \iff \mu(z_1) \perp \mu(z_2)$

(c) $z_1 \perp z_2 \iff (\sigma_{z_1} \ \sigma_{z_2})^2$

Theorem 14

Let z_1, z_2 be two different cycles and $z_1 \perp z_2$, then
$|z_1 \cap z_2| \in \{0, 1\}$ in the case Char K = 2
$|z_1 \cap z_2| \in \{0, 2\}$ in the case Char K $\neq$ 2

Let z be a cycle. Then we define the bundle (orthogonalbundle) $z^\perp$ of z. $z^\perp := \{z_i \in \mathfrak{Z} \,/\, z_i \perp z\}$.

Theorem 15

In the finite case with $|K| = q$ each point of a cycle z is incident with exactly q different cycles of $z^\perp$ and the cardinality of $z^\perp$ is q^2.

(a) In the case Char K = 2 every cycle of $z^\perp$, non equal to z, has exactly one common point with z.

(b) In the case Char K $\neq$ 2, however, $\binom{q+1}{2}$ cycles of $z^\perp$ have exactly 2 points, the other $\binom{q}{2}$ cycles of $z^\perp$ don't have any point in common with z.

8. Pencils of cycles

Let t_1, t_2 be two different cycles. Then the set $(t_1, t_2)^\perp = t_1^\perp \cap t_2^\perp$ is called the pencil (orthogonalpencil) of t_1 and t_2. The cycles t_1, t_2 are called carrier-cycles, the cycles of $(t_1, t_2)^\perp$, however, pencil-cycles.

Theorem 16

Let $(t_1, t_2)^\perp$ be a pencil and $\mu \in \mathfrak{M}$. Then
$\mu((t_1, t_2)^\perp) = (\mu(t_1), \mu(t_2))^\perp$.

In the further investigations of pencils we frequently use a "transformation for simplification". This simplification method is very important for all the following proofs.
We formulate this method in the next theorem.

Theorem 17

By using suitable cycle-transformations it is possible to transform any pencil to the following pencil:

Carrier-cycles

$$k_o = \{ (Z_1, Z_2) \in \mathcal{Z} \ / \ Z_1\overline{Z}_2 - \overline{Z}_1 Z_2 = 0 \}$$

$$k_1 = \{ (Z_1, Z_2) \in \mathcal{Z} \ / \ Z_1\overline{Z}_1 + m (1 + \varepsilon) Z_1\overline{Z}_2 - m \varepsilon \overline{Z}_1 Z_2 - - c Z_2\overline{Z}_2 = 0 \}$$

with $m \in K$, $c \in K^*$.

Cycles of the pencil $(k_o, k_1)^{\perp}$

$$h_o = \{ (Z_1, Z_2) \in \mathcal{Z} \ / \ Z_1\overline{Z}_2 + \overline{Z}_1 Z_2 + m Z_2\overline{Z}_2 = 0 \}$$

$$h_i = \{ (Z_1, Z_2) \in \mathcal{Z} \ / \ Z_1\overline{Z}_1 - m_i (Z_1\overline{Z}_2 + \overline{Z}_1 Z_2) + + (m_i^2 - c_i) Z_2\overline{Z}_2 = 0 \}$$

with $m_i \in K$, $c_i = m_i^2 + m\, m_i - c$.

Vice versa two cycles $h_i, h_j \in (k_o, k_1)^{\perp}$ with $i \neq j$ produce the pencil $(h_i, h_j)^{\perp}$. Any arbitrarily chosen cycles $h_i, h_j \in (k_o, k_1)^{\perp}$ yield the same pencil. The cycles k_o, k_1 and all the following cycles k_r are elements of this pencil

$$k_r = \{ (Z_1, Z_2) \in \mathcal{Z} \ / \ Z_1\overline{Z}_1 - \overline{M}_r Z_1\overline{Z}_2 - M_r \overline{Z}_1 Z_2 + + (M_r\overline{M}_r - c_r) Z_2\overline{Z}_2 = 0 \}$$

with $M_r \in R$, $c_r \in K^*$, $M_r\overline{M}_r = c_r - c$, $M_r + \overline{M}_r + m = 0$.

Two cycles $k_r, k_s \in (h_i, h_j)^{\perp}$ with $r \neq s$ produce the pencil $(k_r, k_s)^{\perp}$. Any arbitrarily chosen cycles $k_r, k_s \in (h_i, h_j)^{\perp}$ yield the same pencil. Therefore, we have $(k_r, k_s)^{\perp} = (k_o, k_1)^{\perp}$.

The two pencils $(k_r, k_s)^{\perp}$ and $(h_i, h_j)^{\perp}$ are said to be conjugate. Two arbitrarily chosen cycles always determine exactly two such conjugate pencils. These pencils can be transformed to the pencils

$(h_i, h_j)^\perp$, $(k_r, k_s)^\perp$. In all the following proofs it is sufficient to investigate these special pencils.

Theorem 18

There exist pairs of cycles with exactly one (touching cycles) common point and in the case $|K| > 2$ also pairs of cycles with exactly two (intersecting cycles) common points. Pairs of cycles without common points (passing cycles) exist iff, there is a polynom $f(x) = x^2 + mx - c$ with $m \in K$, $c \in K^*$, irreducible in K .

Theorem 19

Let two cycles of a pencil have exactly the point A or exactly the two points A , B in common with $A \neq B$, $A \nparallel B$ or no point at all in common, then this holds also for any other two cycles of the pencil.

Now we define:

A pencil $(t_1, t_2)^\perp$ of cycles with at least two different cycles is called

parabolic, iff $b_1 \cap b_2 = \{A\}$

elliptic, iff $b_1 \cap b_2 = \{A, B\}$ with $A \nparallel B$

hyperbolic, iff $b_1 \cap b_2 = \emptyset$

The point A , respectively the points A and B , sometimes are called carrier-points of the pencil.
With these definitions the pencils are classified completely.

Theorem 20

The conjugate pencil of a given pencil is always of the same class.

Thus, the conjugate of an elliptic pencil is again elliptic. Let A , B be the carrier-points of the one and P , Q the carrier-points of the other pencil, then naturally it follows $A \nparallel B$ and $P \nparallel Q$. But it is very surprising that we further have $A \| Q$, $A \| P$, $B \| Q$, $B \| P$.

Theorem 21

In the finite case with $|K| = q$ every elliptic pencil consists of exactly $q - 1$, every parabolic pencil of exactly q and every hyperbolic pencil of exactly $q + 1$ cycles. There exist exactly $\frac{1}{2}q^2(q+1)^2$ elliptic, exactly $(q+1)^2(q-1)$ parabolic and exactly $\frac{1}{2}q^2(q-1)^2$ hyperbolic pencils.

Theorem 22 (Three-reflection-theorem)

Let b_3, b_2, b_1 be three different cycles of a pencil $(t_1, t_2)^{\perp}$. Then there exists exactly one cycle $b_4 \in (t_1, t_2)^{\perp}$ such that

$$\sigma_{b_3}\,\sigma_{b_2}\,\sigma_{b_1} = \sigma_{b_4} .$$

Proof:

According to theorem 17 we may assume that k_2, k_1, k_o (in this order) are the cycles b_3, b_2, b_1. This means no restriction of the validity of the proof.

$$k_o = \{(z_1, z_2) \in \mathfrak{P} \;/\; z_1\overline{z}_2 - \overline{z}_1 z_2 = 0\}$$

$$k_r = \{(z_1, z_2) \in \mathfrak{P} \;/\; z_1\overline{z}_1 - \overline{M}_r z_1\overline{z}_2 - M_r\overline{z}_1 z_2 + (M_r\overline{M}_r - c_r)\, z_2\overline{z}_2 = 0\}$$

with $r \in \{1,2\}$, $M_r\overline{M}_r = c_r - c$, $M_r + \overline{M}_r + m = 0$, $M_1 \neq M_2$.

Here are the equations of the reflections in these cycles:

$$\sigma_{k_o}\begin{cases} z_1' = \overline{z}_1 \\ z_2' = \overline{z}_2 \end{cases} \qquad \sigma_{k_r}\begin{cases} z_1' = M_r\overline{z}_1 + (c_r - M_r\overline{M}_r)\overline{z}_2 = M_r\overline{z}_1 + c\,\overline{z}_2 \\ z_2' = \overline{z}_1 - \overline{M}_r\overline{z}_2 \end{cases}$$

The product of these three reflections is

$$\sigma_{k_2}\,\sigma_{k_1}\,\sigma_{k_o}\quad\begin{cases} z_1' = (c + \overline{M}_1 M_2)\overline{z}_1 + c\,(M_2 - M_1)\,\overline{z}_2 \\ z_2' = (\overline{M}_1 - \overline{M}_2)\overline{z}_1 + (c + M_1\overline{M}_2)\overline{z}_2 \end{cases}$$

M_r fulfills certain conditions. With these conditions we obtain $N(M_1 - M_2) \neq 0$. Therefore, we can write

$$\begin{cases} z_1' = \dfrac{c + \bar{M}_1 M_2}{\bar{M}_1 - \bar{M}_2}\, \bar{z}_1 + c\, \bar{z}_2 \\[2ex] z_2' = \bar{z}_1 + \dfrac{c + \bar{M}_2 M_1}{\bar{M}_1 - \bar{M}_2}\, \bar{z}_2 \end{cases},$$

Now we define $M_3 = \dfrac{c + \bar{M}_1 M_2}{\bar{M}_1 - \bar{M}_2}$. Then it follows

$$\begin{cases} z_1' = M_3 \bar{z}_1 + c\, \bar{z}_2 \\ z_2' = \bar{z}_1 - \bar{M}_3 \bar{z}_2 \end{cases}$$

These are the equations of the reflection in the cycle

$$k_3 = \{ (z_1, z_2) \in \mathcal{P} \ / \ z_1 \bar{z}_1 - \bar{M}_3 z_1 \bar{z}_2 - M_3 \bar{z}_1 z_2 + (M_3 \bar{M}_3 - c_3)\, z_2 \bar{z}_2 = 0 \}$$

with $c_3 = M_3 \bar{M}_3 + c$.

It is easy to see $M_3 + \bar{M}_3 = -m$ and

$$c_3 = M_3 \bar{M}_3 + c = \frac{c_1 c_2}{N(M_1 - M_2)} \in K^* .$$

This finally means $k_3 \in (h_i, h_j)^{\perp}$.

<u>Theorem 23</u>

<u>In the (K,R)-plane there exist quadrupels of mutually orthogonal cycles. These cycles are elements of a pencil in the case Char $K = 2$ but not in the case Char $K \neq 2$. The product of the reflections in the cycles of such an orthogonal quadrupel always is the identical transformation.</u>

Here is an example of an orthogonal quadrupel in the case Char K ≠ 2:

$$\{(z_1, z_2) \in \mathcal{Z} \,/\, z_1\bar{z}_2 - \bar{z}_1 z_2 = 0\},\ \{(z_1, z_2) \in \mathcal{Z} \,/\, z_1\bar{z}_2 + \bar{z}_1 z_2 = 0\},$$

$$\{(z_1, z_2) \in \mathcal{Z} \,/\, z_1\bar{z}_1 - z_2\bar{z}_2 = 0\},\ \{(z_1, z_2) \in \mathcal{Z} \,/\, z_1\bar{z}_1 + z_2\bar{z}_2 = 0\}.$$

Theorem 24 (Converse of the three-reflection-theorem)

Let b_4, b_3, b_2, b_1 be four cycles, and further let us assume $\sigma_{b_4}\sigma_{b_3}\sigma_{b_2}\sigma_{b_1} = \mathrm{id}$. Then, either these cycles are elements of a pencil or they are mutually orthogonal. The case of an orthogonal quadrupel which does not belong to a pencil (Theorem 23) is only possible, if Char K ≠ 2.

Theorem 25

Every mapping of $\mathcal{L}$ is the product of at most 4 reflections. (In other words: The minimal decomposition-length of products of reflections is at most 4)

The decomposition-length 4 is possible indeed. For instance the spiral similarity of Theorem 9 (with $N(B) \neq 1$ and $B \notin K^*$ has this length. In several steps then it must be shown that every product of 5 reflections can be reduced to one of 3 reflections.

9. Conclusion

The aim of this work was to show that many theorems of the MINKOWSKI-plane can be proved by "counting", by "calculating", by working with "equations" for cycles and transformations - actually by primitive methods.

What remains to be done? We sketch only two themes.

It certainly would be very nice to study the known theorems of classical school geometry (Theorems of PYTHAGORAS, THALES, CEVA, PTOLEMÄUS,...; six point circle of FEUERBACH; line of EULER;...) and then to examine their validity in the MINKOWSKI-plane by using the methods developed in this paper.

Our work - especially the two theorems 11 and 13 - shows another, very important, but also very complicated problem. Is it possible to characterize the known MÖBIUS-MINKOWSKI- and LAGUERRE-planes - in short the BENZ-planes - in a theory of reflections like the "Spiegelungsgeometrie" by BACHMANN? What are the axioms of such a reflection geometry?

10. References

[1] R. ARTZY
A Pascal theorem applied to Minkowski geometry, Journal of geometry 3 (1973), 93 - 105

[2] W. BENZ
(a) Über die Grundlagen der Geometrie der Kreise in der pseudo-euklidischen Geometrie, Journal für Mathematik 232 (1968), 41 - 76
(b) mit W. LEISSNER - H. SCHAEFFER
Kreise, Zykel, Ketten. Zur Geometrie der Algebren. J.Ber. Deutsch. Math. Vereinigung 74 (1972), 107 - 122
(c) Vorlesungen über Geometrie der Algebren, Berlin (1973)

[3] H.J. DIENST
(a) Eine charakteristische Spiegelungseigenschaft hermitescher Quadriken, Journal of geometry 5 (1974), 67 - 81
(b) Schnitt- und Zykelgeometrien, Habilitationsschrift Darmstadt (1975)
(c) Minkowski-Ebenen mit Spiegelungen, Monatshefte der Mathematik 84 (1977), 197 - 208

[4] W. HEISE - H. KARZEL
Symmetrische Minkowski-Ebenen, Journal of geometry 3 (1973), 5 - 20

[5] E.M. SCHROEDER
Gemeinsame Eigenschaften euklidischer, galileischer und minkowskischer Ebenen, Mitteilungen der Math. Gesellschaft Hamburg 10 (1974), 185 - 217

[6] H. SCHAEFFER
Lecture given in Istanbul, to appear

[7] H. ZEITLER
Über (K,L)-Ebenen, Dissertation Kassel (1977)

ON THE SPACE OF RIEMANNIAN METRICS ON SURFACES AND CONTACT MANIFOLDS

David E. Blair

1. We begin by reviewing some of the work of Berger, Bourguignon, Ebin and Marsden on the set of all Riemannian metrics on a manifold. Let M be a compact orientable C^∞ manifold of dimension $n \geq 2$, $\mathcal{D}$ the diffeomorphism group of M and $\mathcal{M}$ the set of all Riemannian metrics on M. We shall be interested in metrics with the same volume element, i.e. if Ω is a non-vanishing n-form on M, let $\mathcal{M}_\Omega$ be the set of Riemannian metrics whose volume element is Ω. Denote by $\mathcal{D}_\Omega$ the diffeomorphisms which preserve Ω. Ebin [4] showed that $\mathcal{M}_\Omega/\mathcal{D}_\Omega$ is homeomorphic to $\mathcal{M}_0/\mathcal{D} \subset \mathcal{M}/\mathcal{D}$ where $\mathcal{M}_0$ denotes the set of all metrics with the same total volume $\int_M \Omega$.

Given $g \in \mathcal{M}$, define a differential operator δ on symmetric covariant tensor fields analogously to the codifferential on differential forms; for a tensor field D of type $(0,2)$ we have $(\delta D)_i = -g^{\ell m}\nabla_m D_{\ell i}$. In [4] Ebin proved a slice theorem for $\mathcal{M}$; given $g \in \mathcal{M}$ there exists a neighborhood U of g in the orbit of g under the action of $\mathcal{D}$ and a submanifold S of $\mathcal{M}$ containing g such that $\mathcal{M}$ is locally diffeomorphic (in the ILH sense) to $U \times S$. Furthermore the tangent space of S at g is the kernel of δ [1]. Thus, to study equivalence classes of metrics in $\mathcal{M}/\mathcal{D}$ it suffices to study curves $g(t)$ in $\mathcal{M}$ for which the deformation tensor field D defined by

$$D_{ij} = \frac{\partial g_{ij}}{\partial t}\Big|_{t=0}$$

satisfies $\delta D = 0$. Finally, note that if $g(t)$ is a curve with Ω the volume element of $g(t)$ for all t, then expressing $g(t)$ in terms of a local $g(0)$-orthonormal basis it is easy to see and well-known that $\mathrm{tr} D = 0$.

In [3] Bourguignon, Ebin and Marsden proved that in dimension > 2, the space of symmetric tensor fields D of type $(0,2)$ with $\delta D = 0$ and $\mathrm{tr} D = 0$ is infinite dimensional and, hence, that $\mathcal{M}_\Omega/\mathcal{D}_\Omega$ is infinite dimensional. For $n = 2$ one could compute the Laplacian ΔD (defined analogously to the Laplacian on forms, Lichnerowiez [5]) and obtain $\Delta D + 2KD = 0$ where K is the Gaussian curvature of $g(0)$. Thus, since Δ is elliptic, $\mathcal{M}_\Omega/\mathcal{D}_\Omega$ is finite dimensional. Taking a different approach, however, we obtain the following result.

Theorem 1.1. If $n = 2$, $\dim(\mathcal{M}_\Omega/\mathcal{D}_\Omega) = 2$.

Proof: Since M is a smooth surface we may choose local isothermal coordinates such that $g(0)$ is given by $ds^2 = \lambda^2(dx^2+dy^2) = \lambda^2 dz d\bar{z}$, $z = x + iy$ where λ is a positive function. Suppose that with respect to the orthonormal basis $\{\frac{1}{\lambda}\frac{\partial}{\partial x}, \frac{1}{\lambda}\frac{\partial}{\partial y}\}$, D is given by the matrix $\begin{pmatrix} v & u \\ u & -v \end{pmatrix}$ and set $f = u + iv$. Then the condition $\delta D = 0$ becomes

$$\frac{\partial f}{\partial \bar{z}} + \left(2i \frac{\partial \ln \lambda}{\partial z}\right) f = 0 . \qquad (1)$$

Since we will be interested in non-zero solutions of this equation, set $f = e^{-h}$; then $\frac{\partial h}{\partial \bar{z}} = 2i \frac{\partial \ln \lambda}{\partial z}$ which has solution

$$h(z) = -\frac{2i}{\pi} \int_{|\zeta|<R} \frac{\partial \ln \lambda/\partial z}{\zeta - z} \, d\xi d\eta \quad \text{where} \quad \zeta = \xi + i\eta \quad \text{and} \quad |\zeta| \leq R$$

is a disk in the domain of the local coordinates. Now, if f_1 and f_2 are solutions of (1), f_1/f_2 is holomorphic and globally defined by the way f_1 and f_2 transform in the overlap of coordinate neighborhoods; but M is compact and hence, f_1/f_2 is a constant. Therefore, any deformation tensor D given by a function f satisfying (1) differs from a fixed one by a multiplicative complex constant and, hence, the dimension of $\mathcal{M}_\Omega/\mathcal{D}_\Omega$ is 2.

2. The author was led to the above discussion by his interest in contact manifolds. Given a contact form η on a contact manifold, consider the set $\mathcal{A}$ of all Riemannian metrics associated to η. If $\mathcal{C}$ denotes the set of strict contact transformations, we shall see that we can form the space $\mathcal{A}/\mathcal{C}$ and ask its dimensionality. In particular we prove that if a 3-dimensional contact manifold admits a K-contact metric, $\mathcal{A}/\mathcal{C}$ is infinite dimensional. We conjecture this in general, however, as $\mathcal{A}/\mathcal{C}$ does not sit nicely in $\mathcal{M}/\mathcal{D}$, we must use a specialized argument.

By a <u>contact manifold</u> we mean a C^∞ manifold M of dimension $2n + 1$ admitting a global 1-form η such that $\eta \wedge (d\eta)^n \neq 0$. Given a contact form η, it is well known that there exists a unique vector field ξ on M such that $\eta(\xi) = 1$ and $d\eta(\xi, X) = 0$; ξ is called the <u>characteristic vector field</u> of the contact structure η.

We introduce associated metrics by reviewing their construction [2,6]. Let k' be any Riemannian metric on M and define a metric k by

$$k(X,Y) = k'(-X + \eta(X)\xi, -Y + \eta(Y)\xi) + \eta(X)\eta(Y) ;$$

k is then a Riemannian metric with respect to which η is the covariant form of ξ. Let $X_1,\ldots,X_{2n},\xi$ be a local k-orthonormal basis and consider the $2n \times 2n$ matrix $(d\eta(X_i,X_j))$; it is non-singular and, hence, can be written uniquely as a product FG where F is orthogonal and G is positive definite symmetric. G defines a Riemannian metric g on M by the matrix $\begin{pmatrix} G & 0 \\ 0 & 1 \end{pmatrix}$, its components being given with respect to $X_1,\ldots,X_{2n},\xi$. Similarly F defines a global tensor field φ of type (1,1) by the matrix $\begin{pmatrix} F & 0 \\ 0 & 0 \end{pmatrix}$; φ satisfies $\varphi^2 = -I + \eta \otimes \xi$ and $\varphi\xi = 0$. g is called an <u>associated metric</u> of η and (η,g) or (φ,ξ,η,g) a <u>contact metric structure</u>. In particular φ and g are so constructed that $d\eta(X,Y) = g(X,\varphi Y)$. The above is equivalent to postulating the existence of tensors φ and g such that $g(X,\xi) = \eta(X)$, $\varphi^2 = -I + \eta \otimes \xi$ and $d\eta(X,Y) = g(X,\varphi Y)$.

A diffeomorphism $\psi : M \to M$ is a <u>contact transformation</u> if $\psi^*\eta = \tau\eta$ for some non-vanishing function τ. We say that ψ is a <u>strict contact transformation</u> if $\tau \equiv 1$.

<u>Lemma 2.1</u>. If ψ is a strict contact transformation and g an associated metric, then ψ^*g is an associated metric.

<u>Proof</u>: It is immediate that $\eta(\psi_*\xi) = 1$ and $d\eta(\psi_*\xi, \psi_*X) = 0$ for all vector fields X and, hence, $\psi_*\xi = \xi$. Define φ^* by

$$(\psi^*g)(X,\varphi^*Y) = d\eta(X,Y)$$

then $\psi_*\varphi^* = \varphi\psi_*$. Applying this twice we have

$$\varphi^{*2} = -I + \eta \otimes \xi$$

and

$$(\psi^*g)(\xi,X) = \eta(X)$$

is trivial. These three equations show that ψ^*g is an associated metric.

Thus, if $\mathcal{A}$ denotes the set of all associated metrics and $\mathcal{C}$ the strict contact transformations $\mathcal{A}/\mathcal{C}$ is well defined. We remark, however, that the converse of the above lemma is not true, i.e. one may have a diffeomorphism ψ such that g and ψ^*g are both associated metrics without ψ being a strict contact transformation. Thus, even though we could map $\mathcal{A}/\mathcal{C}$ into $\mathcal{M}/\mathcal{D}$ by choosing a representative g of an orbit in $\mathcal{A}/\mathcal{C}$ and assigning to it its orbit in $\mathcal{M}/\mathcal{D}$, the map need not be one-to-one.

Given a contact metric structure (φ,ξ,η,g) we define a tensor field h by $h = \frac{1}{2}\mathcal{L}_\xi\varphi$, $\mathcal{L}$ denoting Lie differentiation. h is a symmetric operator on TM and h anti-commutes with φ [2]. It is easy to see that ξ is a Killing vector field with respect to g if and only if $h = 0$. An associated metric with respect to which ξ is Killing is called a K-<u>contact metric</u>. Clearly we always have $h\xi = 0$ and, hence, 0 is an eigenvalue of h. Moreover since h anti-commutes with φ, if λ is an eigenvalue of h so is $-\lambda$.

If ψ is a strict contact transformation, let φ^* be as in the proof of Lemma 2.1 and $h^* = \frac{1}{2}\mathcal{L}_\xi\varphi^*$.

Lemma 2.2. $\psi_* h^* = h\psi_*$. If $\lambda^*(p)$ is an eigenvalue of h^* at $p \in M$, $\lambda^*(p)$ is an eigenvalue of h at $\psi(p)$. If $\dim M = 3$, $\lambda^*(p) = \pm\lambda(\psi(p))$.

Proof: That $\psi_* h^* = h\psi_*$ is a straightforward computation using the definitions of h and h^*. The second statement follows from the first. If $\dim M = 3$, the eigenvalues of h are 0 and $\pm\lambda$; hence, $\lambda^*(p) = \pm\lambda(\psi(p))$.

Recall that on a contact manifold there exist local coordinates (x^i, y^i, z), $i = 1,\ldots,n$ on a neighborhood U such that $d\eta = dz - \sum y^i dx^i$. This is a classical theorem of Darboux and we call such a coordinate neighborhood a Darboux neighborhood on M.

Lemma 2.3. Let U be a Darboux neighborhood on a 3-dimensional contact manifold and $\hat{U}$ the open half ball of radius ϵ_σ lying in U given by $\hat{U} = \{(x,y,z) \mid x^2 + y^2 + z^2 < \epsilon_0^2, x > 0\}$. Let ψ be a strict contact transformation which is not the identity on $\hat{U}$. Then there exists $p \in \hat{U}$ such that $\psi(p) \notin \hat{U}$.

Proof: Suppose $\psi(\hat{U}) \subset \hat{U}$, then since ψ preserves the volume element $\eta \wedge d\eta$, $\psi(\hat{U}) = \hat{U}$. If $(x,y,z) \to (\psi^1, \psi^2, \psi^3)$ denotes the mapping and $\eta = dz - ydx$, $\psi^*\eta = \eta$ becomes

$$\frac{\partial\psi^3}{\partial x} - \psi^2 \frac{\partial\psi^1}{\partial x} = -y ,$$

$$\frac{\partial\psi^3}{\partial y} - \psi^2 \frac{\partial\psi^1}{\partial y} = 0 ,$$

$$\frac{\partial\psi^3}{\partial z} - \psi^2 \frac{\partial\psi^1}{\partial z} = 1 .$$

Moreover $\psi^*d\eta = d\eta$ gives $\frac{\partial\psi^1}{\partial z} = \frac{\partial\psi^2}{\partial z} = 0$. Consequently, $\psi^3(x,y,z) = z + \rho(x,y)$, but since ψ^1 and ψ^2 are independent of z and $\psi(\hat{U}) = \hat{U}$, $\rho = 0$. Therefore $\psi^2 \frac{\partial\psi^1}{\partial x} = y$ and $\psi^2 \frac{\partial\psi^1}{\partial y} = 0$. Now $y \neq 0$ almost everywhere on $\hat{U}$ so we must have $\psi^1(x,y,z) = \sigma(x)$. Again since ψ^1 and ψ^2 are independent of z and $\psi^3(x,y,z) = z$, any point with $x^2 + y^2 = \varepsilon^2 (\varepsilon < \varepsilon_0)$ is mapped to a point with $(\psi^1)^2 + (\psi^2)^2 = \varepsilon^2$. Therefore $\psi^1(x,y,z)$ must be x (or $-x$) and then by $\psi^2 \frac{\partial\psi^1}{\partial x} = y$, $\psi^2(x,y,z) = y$, i.e. ψ is the identity on $\hat{U}$.

Theorem 2.4. Let M be a 3-dimensional contact manifold admitting a K-contact metric g. Then $\mathcal{A}/\mathcal{C}$ is infinite dimensional.

Proof: Choosing a local unit vector field X g-orthogonal to ξ, $\{X,\varphi X,\xi\}$ is a local orthonormal basis. Moreover since g is K-contact it is easy to check that $[\xi,X]$ is collinear with φX and, hence, choosing X non-invariant by the action of ξ we have $[\xi,X] = \alpha\varphi X$ where α is a non-zero function on the domain of X.

Now let U be a Darboux neighborhood lying in the domain of X and $\hat{U}$ as in Lemma 2.3. Let $\hat{F}$ be the set of C^∞ functions on M which are positive on $\hat{U}$ and vanishing outside $\hat{U}$; clearly

$\hat{F}$ is infinite dimensional. For $f \in \hat{F}$ define a new metric g_f by the matrix

$$((g_f)_{ij}) = \begin{pmatrix} 1+f & 0 & 0 \\ 0 & \frac{1}{1+f} & 0 \\ 0 & 0 & 1 \end{pmatrix}, \quad i,j = 1,2,3$$

with respect to the basis $\{X_1 = X, X_2 = \varphi X, X_3 = \xi\}$ on $\hat{U}$ and $g_f = g$ outside $\hat{U}$. Then it is easy to verify that g_f is an associated metric for the contact structure η and that φ_f is given by the matrix

$$\begin{pmatrix} 0 & \frac{-1}{1+f} & 0 \\ 1+f & 0 & 0 \\ 0 & 0 & 0 \end{pmatrix}$$

again with respect to $\{X, \varphi X, \xi\}$. Setting $\beta_f = \alpha(1+f) - \dfrac{\alpha}{1+f}$ we have by straightforward computation

$$(\mathcal{L}_\xi \varphi_f)X = (\xi f)\varphi X - \beta_f X,$$

$$(\mathcal{L}_\xi \varphi_f)\varphi X = \frac{\xi f}{(1+f)^2} X + \beta_f \varphi X.$$

Using these equations we see that the eigenvalues $\pm \lambda_f$ of $h_f = \frac{1}{2} \mathcal{L}_\xi \varphi_f$ satisfy

$$4\lambda_f^2 = \left(\frac{\xi f}{1+f}\right)^2 + \beta_f^2. \qquad (2)$$

To show that $\mathcal{A}/\mathcal{C}$ is infinite dimensional we show that for any two functions $f_1, f_2 \in \hat{F}$, g_{f_1} and g_{f_2} belong to different orbits, i.e. there does not exist a strict contact transformation ψ such that $g_{f_2} = \psi^* g_{f_1}$. Suppose such a ψ exists.

By Lemma 2.3 let $p \in \hat{U}$ such that $\psi(p) \notin \hat{U}$, then by Lemma 2.2, $\lambda_{f_2}(p) = \pm \lambda_{f_1}(\psi(p)) = 0$ since $g_{f_1} = g$ outside $\hat{U}$ and g is K-contact. Equation (2) then gives $\beta_{f_2}(p) = 0$, but $\alpha \neq 0$ on U and, hence, the definition of β_{f_2} gives $f_2(p) = 0$ contradicting the positivity of f_2 on $\hat{U}$.

Remarks. As mentioned before we conjecture this result for contact manifolds of any dimension and not necessarily admitting a K-contact metric.

By Lemma 2.2 it is clear that if g is a K-contact metric, so is $\psi^* g$. Thus, if $\mathcal{K}$ denotes the set of all K-contact metrics, we may consider $\mathcal{K}/\mathcal{C}$. In the spirit of section 1 we conjecture that on a 3-dimensional contact manifold $\mathcal{K}/\mathcal{C}$ is finite dimensional.

References

[1] Berger, M. and D. Ebin, Some decompositions of the space of symmetric tensors on a Riemannian manifold, J. Diff. Geom., 3 (1969) 379-392.

[2] Blair, D. E., Contact Manifolds in Riemannian Geometry, Lecture notes in mathematics, 509, Springer-Verlag, 1976.

[3] Bourguignon, J., D. Ebin and J. Marsden, Sur le noyan des opérateurs pseudo-différentiels à symbole surjectif et non injectif, C. R. Acad. Sc. Paris, 282 (1976) 867-870.

[4] Ebin, D., The manifold of Riemannian metrics, Global Analysis, Proc. Sympos. Pure Math., AMS, 1970, 11-40.

[5] Lichnerowicz, A., Propagateurs et commutateurs en relativité générale, Publ. Sci. I.H.E.S., vol. 10 (1961)

[6] Sasaki, S., On differentiable manifolds with certain structures which are closely related to almost contact structure I, Tôhoku Math. J. 14 (1962) 146-155.

Michigan State University
East Lansing, Michigan 48824

CIRCLES ON SURFACES IN THE EUCLIDEAN 3-SPACE

RICHARD BLUM

Introduction: The subject matter of this paper came to my attention through a book by Melzak [1] in which the following two statements are made: "(a) if S is a complete sufficiently smooth surface containing exactly four circles, through any point of it, then S is a torus; (b) if the number of such circles is ≥ 5, then it is infinite and the surface is a sphere".

Let us define a surface S in the Euclidean 3-space to have the "n-circle property" if through a generic point of S pass n (but not more than n) real and distinct circles lying on S. Then the Melzak statements can be formulated thus:

(a) if S has the 4-circle property it is a torus.

(b) no surface exists with the n-circle property for $4 < n < \infty$.

Whereas the first statement can easily be shown to be false (every image under inversion of the torus must have the same 4-circle property), the second statement is more difficult to prove or disprove. In the present paper it is shown to be false by proving the following theorem:

Let S be the cyclide defined by the equation

$$S: \quad (x^2 + y^2 + z^2)^2 - 2ax^2 - 2by^2 - 2cz^2 + d^2 = 0$$

in the orthogonal cartesian coordinates x,y,z where the real coefficients a,b,c,d satisfy the conditions

$$0 < d < b \leq a \; ; \quad c < d \; .$$

Then S has the n-circle property where

(1) $n = 6$, if $a \neq b$ and $c \neq -d$.

(2) $n = 5$, if $a = b$, $c \neq -d$ or $a \neq b$, $c = -d$,

(3) $n = 4$, if $a = b$ and $c = -d$.

One recognizes readily that the last case represents the torus. In the first half of the second case (i.e.: $a = b$) S is a surface of revolution whose meridian curve is a "spiric line of Perseus" (see [2]). According to the testimony of Proclos (5th century A.D.) it was considered by Perseus in ca. 130 B.C. as the curve of intersection of a general torus with a plane parallel to the axis of the torus. It may be fitting, therefore, to call the surface S in this case the "Perseus Surface".

The basic idea of the proof of the above theorem is to subject S to an inversion whose center P_0 lies on S. The image of S under this inversion will be a cubic surface S_3 and the number of real circles on S passing through P_0 will be exactly equal to the number of real straight lines on S_3, not lying in the plane at infinity. It suffices, therefore, to count the number of such straight lines on S_3 in order to prove the theorem.

In the particular case of the Perseus Surface a simpler and more constructive proof of the theorem is available. This proof makes use of known facts concerning the bitangents of quartic plane curves and seemed to me of sufficient interest in order to present it here as an alternate proof of this particular case.

As for the historical background of this problem, I was able to unearth the following facts: Kummer in 1865 (see [3]) and Darboux in 1880 (see [4[) recognized that the general cyclide has a 10-circle property but nowhere (as far as I could find out) is there a distinction made between real and non-real circles. This is not amazing since in the 19th century algebraic geometry has been done in the complex field. Still the literature of general cyclides in the 19th century is so large that only a thorough check can determine if there were authors who tried to make this distinction. It is worthwhile to note that there seems to be a revival of the interest for cyclides in the last few years (see [5] and [6]).

In what follows the surface S referred to above is denoted by (S_4) and is called a "ringlike cyclide with center", because it has a center of symmetry and, due to the inequality conditions between its coefficients, the connectivity of the torus (i.e. it is of genus one).

I. <u>The ringlike cyclide with center</u>

Such a surface, when referred to a suitable orthogonal cartesian coordinate system (x, y, z) has the equation:

(1) $$(S_4) : r^4 - 2ax^2 - 2by^2 - 2cz^2 + d^2 = 0 \; ; \; r^2 = x^2 + y^2 + z^2 \; ,$$

where the parameters a, b, c, d satisfy the conditions:

(2) $$0 < d < b \leq a \; ; \; c < d \; .$$

We observe: The image of (S_4) under an inversion (I) with respect to a pole $P_\circ(x_\circ, y_\circ, z_\circ)$ lying <u>on</u> (S_4) is a non-ruled circular cubic surface (S_3). Any circle lying on (S_4) and passing through $P_\circ$ will map under (I) into a straight line. The number of such circles must, therefore, equal the number of real straight lines of (S_3), not lying in the plane at infinity.

According to Cayley, Sylvester and J. Steiner (see [7]), every non-ruled cubic surface contains 27 straight lines which are all distinct if the surface has no singular points. (Some of these straight lines may coincide if the surface admits singularities.) Furthermore, let L be any of the 27 straight lines. Then it is intersected by other 10 of the remaining 26 straight lines. These 10 straight lines lie pairwise in 5 planes passing through L . Since any plane through L intersects the surface in L and a conic F , the 5 planes are, therefore, those belonging to the pencil of planes through L for which F degenerates into 2 straight lines.

The equation of (S_3) :

Since $P_\circ(x_\circ, y_\circ, z_\circ)$ lies on (S_4) we have

(3) $r_\circ^4 - 2ax_\circ^2 - 2by_\circ^2 - 2cz_\circ^2 + d^2 = 0$; $r_\circ^2 = x_\circ^2 + y_\circ^2 + z_\circ^2$.

Let the translation (T) and the inversion (I) be defined by:

(T) : $x \to x + x_\circ$; $y \to y + y_\circ$; $z \to z + z_\circ$;

(I) : $x \to 2R^2x/r^2$; $y \to 2R^2y/r^2$; $z \to 2R^2z/r^2$;

where $R\sqrt{2}$ is the radius of the sphere of inversion.
The image of (S_4) under (T) followed by (I) is the circular cubic surface (S_3) whose equation, obtainable by straight forward calculation, is

(4) (S_3) : $r^2(Ax_\circ x + By_\circ y + Cz_\circ z) + R^2(Ax^2 + By^2 + Cz^2)$

$$+ R^2(x_\circ x + y_\circ y + z_\circ z + R^2)^2 = 0 \quad ;$$

where: $2A = r_\circ^2 - a$; $2B = r_\circ^2 - b$; $2C = r_\circ^2 - c$.

General considerations on the 27 straight lines on (S_3) :
From (4) we see that the intersection of (S_3) with the plane at infinity is formed by the straight line (L) and the absolute circle (U) given by the equations:

(L) : $Ax_\circ x + By_\circ y + Cz_\circ z = 0$; $t = 0$;

(U) : $r^2 = 0$; $t = 0$,

where t is the variable of homogenization.

Since (U) is a non-degenerate conic, (S_3) has only one straight line at infinity namely (L) . Of the remaining 26 straight lines on (S_3) , 10 will pass, according to Cayley and Steiner, through (L) and at most 16 through (U) . (Some of these may coincide because of singularities of (S_3) .) But none of the latter can be real since if a real straight line, say M , passes through a non-real point it must also pass through its complex conjugate. But the complex conjugate of a point on (U) lies also on (U), therefore (M) would have to lie in the plane at infinity and, consequently, be identical to (L) which we have excluded.

We have, therefore, the preliminary result:
The maximum number of real straight lines on (S_3) which do not lie in the plane at infinity and, consequently, the maximum number of (real) circles on (S_4) passing through $P_\circ$, is 10, and these 10 straight lines on (S_3) must lie pairwise on 5 planes passing through (L).

The actual number of real straight lines among these 10 can only be found by determining the 5 planes in question and then investigating the nature of their intersection with (S_3).

The 5 planes through (L) for which the conic (F) degenerates:
Consider the pencil of planes through (L). Its equation will be of the form:

$$(5)\qquad p(K)\ :\ Ax_\circ x + By_\circ y + Cz_\circ z + R^2K = 0 \quad ,$$

where K is a parameter. The intersection of $p(K)$ with (S_3) will, of course, be (L) itself and the conic $F(K)$ which is the intersection of $p(K)$ with the quadric $Q(K)$ given by the equation:

$$(6)\quad Q(K)\ :\ (A-K)x^2 + (B-K)y^2 + (C-K)z^2 + (x_\circ x + y_\circ y + z_\circ z + R^2)^2 = 0\ .$$

In order that the conic $F(K)$ degenerate into two straight lines it is necessary and sufficient for the plane $p(K)$ to be tangent to the quadric $Q(K)$ or, in other words, that the pole of $p(K)$ with respect to $Q(K)$ be situated on $p(K)$. It follows from this that in this case the linear system:

$$\begin{aligned}
&(x_\circ^2 + A - K)x + x_\circ y_\circ y + x_\circ z_\circ z + R^2 x_\circ = kAx_\circ\ ;\\
&x_\circ y_\circ x + (y_\circ^2 + B - K)y + y_\circ z_\circ z + R^2 y_\circ = kBy_\circ\ ;\\
&x_\circ z_\circ x + y_\circ z_\circ y + (z_\circ^2 + C-K)z + R^2 z_\circ = kCz_\circ\ ;\\
&R^2 x_\circ x + R^2 y_\circ y + R^2 z_\circ z + R^4 = kR^2K\ ;\\
&Ax_\circ x + By_\circ y + Cz_\circ z + R^2K = 0 \quad ,
\end{aligned}$$

where k is a factor of proportionality, must be consistent. Since this linear system contains 5 equations in the 4 unknowns x, y, z, k the consistency condition is that the determinant of its coefficients vanishes.

I.e.:

$$
(7) \quad \begin{vmatrix}
x_o^2 + A - K & x_o y_o & x_o z_o & x_o & Ax_o \\
x_o y_o & y_o^2 + B - K & y_o z_o & y_o & By_o \\
x_o z_o & y_o z_o & z_o^2 + C - K & z_o & Cz_o \\
x_o & y_o & z_o & 1 & K \\
Ax_o & By_o & Cz_o & K & 0
\end{vmatrix} = 0 \ .
$$

This is an equation of degree 5 in K whose roots, when introduced into the equation of $p(K)$ will yield the 5 planes we are looking for. But this formidable looking equation is actually easy to solve and yields the following roots:

$$K_1 = A \ ; \ K_2 = B \ ; \ K_3 = C \ ; \ K_4 = D \ ; \ K_5 = E \ ,$$

where: $2D = r_o^2 + d \ ; \ 2E = r_o^2 - d$.

<u>The nature of the degenerated conics</u> $F(K_i)$:

In order to solve our problem we must determine those among the $F(K_i)$ $(i = 1, 2, 3, 4, 5)$, which yield real straight lines. This could be done by direct consideration of $F(K_i)$ as the intersection of $p(K_i)$ with $Q(K_i)$ which gives immediately obvious results for K_1 and K_5 but is rather involved for the other cases. An alternate way, which requires some computation, is to determine the invariant $J(K_i)$ of $F(K_i)$ which is, essentially, the discriminant of a homogeneous quadratic form in two variables. The nature of $F(K_i)$ is then determined by the sign of $J(K_i)$ according to the following scheme:

$$
(8) \quad J(K_i) \begin{cases} > 0 \ ; & \text{two non-real straight lines.} \\ = 0 \ ; & \text{one real straight line (counted twice).} \\ < 0 \ ; & \text{two real and distinct straight lines.} \end{cases}
$$

The following values of $J(K_i)$ are obtained:

1) $J(A) = (a - b)(a - c)(a - d)(a + d)$;
2) $J(B) = (b - a)(b - c)(b - d)(b + d)$;
3) $J(C) = (c - a)(c - b)(c - d)(c + d)$;
4) $J(D) = (a + d)(b + d)(c + d)$;
5) $J(E) = (a - d)(b - d)(c - d)$.

As a result of the above values of $J(K_i)$ and taking into consideration the conditions (2), we have:

If all K_i are distinct: $J(A) > 0$; $J(B) < 0$; $J(E) < 0$;

$$\left.\begin{matrix} J(C) < 0 \\ J(D) > 0 \end{matrix}\right\} \text{ for } -d < c \text{ ; and } \left.\begin{matrix} J(C) > 0 \\ J(D) < 0 \end{matrix}\right\} \text{ for } c < -d .$$

$J(K_i)$ vanishes only in two cases, namely:

1) $a = b$ for which $J(A) = J(B) = 0$;
2) $c = -d$ for which $J(C) = J(D) = 0$.

<u>The main theorem</u>:

Applying scheme (8) to these results leads now directly to the:

<u>Theorem</u>: Let n be the number of circles on (S_4) passing through a given point P_o of (S_4), then:

1) <u>If</u> $a \neq b$ <u>and</u> $c \neq -d$: $n = 6$.
2) <u>If either</u> $a = b$ and $c \neq -d$
<u>or</u> $a \neq b$ and $c = -d$: $n = 5$.
3) <u>If</u> $a = b$ <u>and</u> $c = -d$: $n = 4$.

One recognizes immediately that if $a = b$ (S_4) is a surface of revolution and that if in addition $c = -d$ (S_4) is a torus.

The above considerations remain, of course, valid when $d^2 \leq 0$ (with appropriate modifications of conditions (2)) but do not lead in any of the possible cases to an increase in the number n .

Remark: The proof of the above theorem is constructive in the sense that the circles on (S_4) which pass through P_o can be obtained from $F(K_i)$, given by (5) and (6) , subjecting it to the transformations I and T^{-1} taken in this order. I.e., if $J(K_i) \leq 0$ then $T^{-1}(I(F(K_i)))$ is the analytic expression of the corresponding circle(s) on (S_4) . The circles through P_o and lying on (S_4) are, thus, the intersection of (S_4) with the sphere $T^{-1}(I(p(K_i)))$ where $p(K)$ is given by (5).

II. <u>The Perseus Surface</u>.

In the algebraic geometry of plane curves the spiric line of Perseus, mentioned in the introduction, can also be defined as the bicircular quartic with two axes of symmetry (see [2]). By choosing these as the x- and z-axes respectively, its equation can be reduced to:

$$(1) \qquad (\pi) : (x^2 + z^2)^2 - 2ax^2 - 2cz^2 + d^2 = 0 .$$

We obtain for the equation of the Perseus surface generated by (π) when rotated about the z-axis:

$$(2) \qquad (\Pi) : (x^2 + y^2 + z^2)^2 - 2a(x^2 + y^2) - 2cz^2 + d^2 = 0 .$$

In order for (Π) to have the connectivity of the torus, i.e. to be "<u>ring-like</u>", the coefficients a, c, d must satisfy the conditions:

$$(3) \qquad a > d > 0 ; \quad c < d .$$

One recognizes readily that (Π) reduces to the torus for $c = -d$.

We are now ready to formulate the theorem:
<u>The Perseus surface (Π) where coefficients a, c, d are subject to conditions (3) has the n-circle property where</u>:

1) $n = 5$ if $c \neq -d$,
2) $n = 4$ if $c = -d$.

The proof of this theorem is based on the following observations:
1) The curve of intersection of (Π) with a plane tangent to (Π) in two of its points is a quartic curve degenerated into two (not necessarily real and distinct) circles.
2) Every such "bitangent" plane of (Π) corresponds to a bitangent line of (π) given by equation (1) in such a way that a real bitangent line corresponds to real circular sections and a non-real bitangent line to non-real circular sections.
3) Since for $c \neq -d$, (π) is a quartic curve of genus one with two double points (namely the cyclic points) the number of its bitangents is, by virtue of the Plücker equations, $\tau = 8$. For $c = -d$, (π) degenerates into two circles and the number of its bitangents is $\tau = 5$.
4) Due to the fact that (π) is symmetric with respect to the x- and z-axes these bitangents must either be parallel to the axes or pass through the origin.

The equations of the bitangents of (π) can now be obtained without difficulty:
1) The bitangents parallel to the x-axis are:

$$z = \pm \sqrt{\frac{a^2 - d^2}{2(a - c)}}$$

which are always real because of (3). To these corresponds <u>one</u> circle through every point (P) of (Π) namely the parallel circle through (P) on (Π).

2) The bitangents of the form $x = mz$ where

$$m = \pm \sqrt{\frac{d - c}{a - d}}$$

which are, likewise, always real because of (3). To these correspond <u>two</u> circles through every point (P) of (Π).

3) The bitangents parallel to the z-axis for which the following scheme can be established:

$$x = \pm \sqrt{\frac{d^2 - c^2}{2(a - c)}} \quad \begin{cases} \rightarrow \text{real and distinct for } (-d < c < d) \\ \rightarrow \text{real and coinciding for } (c = -d) \\ \rightarrow \text{non-real for } (c < -d) \end{cases}$$

From this follows that there are <u>two</u> real and distinct circles through every point (P) on (Π) if $(-d < c < d)$
one real circle through (P) on (Π) if $c = -d$
and no real circle if $c < -d$.

4) The bitangents of the form $x = mz$ where

$$m = \pm \sqrt{- \frac{c + d}{a + d}} \quad \begin{cases} \rightarrow \text{non-real for } (-d < c < d) \\ \rightarrow \text{real and coinciding for } (c = -d) \\ \rightarrow \text{real and distinct for } (c < -d) \end{cases}$$

From this follows that, corresponding to these bitangents, there is no real circle if $(-d < c < d)$, one real circle through every point (P) on (Π) when $(c = -d)$ which coincides with the circle under 3), and two real and distinct circles through every point (P) on (Π) when $(c < -d)$. Considering all the cases from 1) to 4) listed above, the theorem follows.

<u>Note</u>: The points of contact of a real bitangent need not be real in order that the corresponding circular sections be real. Thus the real bitangents listed under 3) above have real points of contact with (π) for

$$a - \sqrt{a^2 - d^2} \leq c < d$$

but non-real points of contact for

$$-d \leq c < a - \sqrt{a^2 - d^2}$$

whereas the real bitangents listed under 4) above have no real points of contact for

$$c \leq -d \ .$$

Concluding remarks: It seems reasonable to consider the problem discussed here with respect to a general cyclide (the only one which can, according to Darboux, have a multiple circle property and of which the cyclide with center discussed above is a particular case). It can be shown that this entails additional linear terms in the equation (I, 1). The reasoning is then identical as in the case treated above, but the characteristic equation (I, 7) in K which gives the 5 degenerate conics through L does not split anymore, so no explicit expressions for its roots can be obtained. As a result the sign of the invariants corresponding to (8) can not be determined by the methods used above.

I am unaware of any other way of proving a similar theorem with respect to the general cyclide.

I would like to end this paper with the following:

Conjecture: No surface exists with the n-circle property where

$$6 < n < \infty \ .$$

BIBLIOGRAPHY

[1] Z.A. Melzak: Concrete Mathematics. John Wiley (1973), p. 155.

[2] H. Wieleitner: Spezielle ebene Kurven, Leipzig (1908), p. 25-28.

[3] E.E. Kummer: Crelle Journal, Vol 64 (1865), p. 66-76.

[4] G. Darboux: Bull. Sci. Math., 2nd Series, Vol. lV (1880), p. 348-384.

[5] T.F. Banchoff: J. Diff. Geometry, Vol. 4 (1970), p. 193-205.

[6] T.E. Cecil: J. Diff. Geometry, Vol. ll (1976), p. 451-459.

[7] J. Steiner: Collected Works, Vol. ll, p. 651-659.

CLASSES CARACTERISTIQUES PRINCIPALES ET SECONDAIRES

par A. CRUMEYROLLE (Toulouse)

1°) Cochaînes formelles.

Soit ξ un fibré vectoriel topologique, complexe de rang n, au-dessus d'une base V espace topologique paracompact. $\mathcal{U} = (U_\alpha)_{\alpha \in A}$ est un système d'ouverts trivialisants.

Les φ_j, $j = 1, 2, \ldots, k$, sont des éléments de groupe U(n), $k \leqslant n$. Les (ℓ_i) étant les éléments d'une base de $\mathbb{C}^n$, nous définissons :

$$(\overset{k}{\underset{1}{\square}} \varphi_i)(\ell_{\chi_1} \wedge \ell_{\chi_2} \wedge \ldots \wedge \ell_{\chi_k}) = \varepsilon^{i_1 i_2 \ldots i_k}_{\chi_1 \chi_2 \ldots \chi_k} \varphi_1(\ell_{i_1}) \wedge \ldots \wedge \varphi_k(\ell_{i_k}),$$

$$1 \leqslant \chi_1 < \chi_2 \ldots < \chi_k \leqslant n. \ (\text{Si } \varphi_1 = \varphi_2 = \ldots \varphi_k = \varphi \,,\ \overset{k}{\underset{i}{\square}} \varphi_i = k!(\Lambda^k \varphi))$$

La trace de $(\overset{k}{\underset{1}{\square}} \varphi_i)$ est :

$$\mathrm{Tr}(\overset{k}{\underset{i=1}{\square}} \varphi_i) = \varepsilon^{i_1 i_2 \ldots i_k}_{j_1 j_2 \ldots j_k} (\varphi_1)^{j_1}_{i_1} (\varphi_2)^{j_2}_{i_2} \ldots (\varphi_k)^{j_k}_{i_k} \ ; \qquad (1)$$

elle est indépendante de l'ordre choisi pour ordonner les φ_i.

Si $x \in (U_\alpha \cap U_\beta) \neq \emptyset$, on peut introduire les fonctions de transition $u_{\alpha\beta}$, à valeur dans U(n) avec :

$$u_{\alpha\beta}(x) = \exp(2\pi i \ a_{\alpha\beta}(x)),$$

$a_{\alpha\beta}(x)$ est hermitienne ; x étant fixé une telle écriture est toujours possible car l'exponentielle est surjective dans le groupe U(n) convexe, compact. On peut supposer qu'un tel choix a été fait pour tout x, pour l'instant sans autre considération (on peut choisir $a_{\beta\alpha}(x) = -a_{\alpha\beta}(x)$) : c'est pourquoi nous appelons q-cochaîne formelle l'assignation à $x \in U_{\alpha_o} \cap U_{\alpha_1} \ldots \cap U_{\alpha_q}$, (cet ouvert sera noté encore

$U_{\alpha_o \alpha_1 \ldots \alpha_q}$) de la valeur réelle en x de :

$$c_{\alpha_o \alpha_1 \ldots \alpha_q} = \mathrm{Tr}(a_{\alpha_o \alpha_1} \square a_{\alpha_1 \alpha_2} \square \cdots \square a_{\alpha_{\ell-1} \alpha_\ell} \square \sigma_{\alpha_\ell, \alpha_{\ell+3}, \alpha_{\ell+4}} \square$$
$$\cdots \square \sigma_{\alpha_{q-2}, \alpha_{q-1}, \alpha_q}) \quad (2)$$

(avec $\sigma_{\alpha_o \alpha_1 \alpha_2} = a_{\alpha_o \alpha_1} + a_{\alpha_1 \alpha_2} - a_{\alpha_o \alpha_2}$, noté encore $\delta(a_{\alpha_o \alpha_1})$), ou de toute expression de ce type obtenue par composition au sens ci-dessus d'éléments $a_{\alpha \beta}$, $\sigma_{\alpha \beta \gamma}$, dans un ordre quelconque. On notera que $c_{\alpha_o \alpha_1 \ldots \alpha_q}$ est identiquement nulle si le nombre de facteurs pour $\square$ dépasse n.

L'ensemble de ces cochaînes s'inclut naturellement dans une algèbre sur $\mathbb{R}$ avec le cup-produit comme loi multiplicative, notée $\Sigma(\{U_\alpha\}, \{a_{\alpha \beta}\}, \xi)$, en abrégé Σ.

On définit un cobord formel δ en posant, avec un abus léger de notation pour les indices :

$$\delta(c_{o1\ldots q}) = c_{\hat{o}1\ldots q+1} - c_{o\hat{1}2\ldots,q+1} + \cdots (-1)^{q+1} c_{o1\ldots q, \widehat{q+1}} \quad (3)$$

où $\frown$ indique un indice manquant.

Il est immédiat que $\delta(c_{o12}) = 0$, d'où l'on déduit par récurrence que $\delta(c_{o1\ldots 2k}) = 0$, lorsque

$$c_{o12\ldots 2k} = \mathrm{Tr}(\sigma_{o12} \square \sigma_{234} \square \cdots \square \sigma_{2k-2, 2k-1, 2k}) \quad (4)$$

Nous appellerons $S(\{U_\alpha\}, \{a_{\alpha \beta}\}, \xi)$, en abrégé S, la sous-algèbre de Σ engendrée par des cochaînes de la forme (4), k arbitraire. On a donc :

<u>Lemme 1.</u> Tout élément de S est un cocycle formel.

<u>Lemme 2.</u> Tout élément de S est le cobord formel d'un élément de $\Sigma - S$.

Il suffit de remarquer que $\sigma_{o12\ldots 2k}$ est le cobord formel de :

$$\underline{g_{01,\dots 2k-1} = \mathrm{Tr}(a_{01} \,\square\, \sigma_{123} \,\square\, \dots \,\square\, \sigma_{2k-3,\ 2k-1,\ 2k-1})} \qquad (5)$$

<u>DEFINITION</u> : Nous appelons <u>cocycle secondaire formel</u> tout élément <u>u</u> de la sous-algèbre Σ_1 de Σ engendrée par les cochaînes de la forme (5), tel que $\delta u = 0$.

Par exemple il peut arriver que pour un fibré ξ et un cocycle $a_{\alpha\beta}$, $c_{01\dots 2k}$ soit nul, alors (5) est un cocycle secondaire formel.

On peut itérer cette définition, et envisager les cochaînes de la forme :

$$\ell_{01\dots 2k-2} = \mathrm{Tr}(a_{01} \,\square\, a_{12} \,\square\, \sigma_{234} \,\dots\dots\, \square\, \sigma_{2k-4,2k-3,2k-2}).$$

dont le cobord est un élément nul de Σ_1 et parler de <u>cocycle tertiaire</u> formel... etc...

On obtiendra ainsi une suite finie d'espaces de cocycles principaux, secondaires, tertiaires... etc...

Pour un fibré donné, concrètement (2) pourra s'interpréter comme une cochaîne à valeurs dans un préfaisceau de sections. Selon la nature de ce préfaisceau on pourra obtenir éventuellement des éléments de la cohomologie de V à partir de certaines cochaînes de (Σ). Cela n'aura d'intérêt que si ces éléments ne sont pas identiquement nuls. Les éléments de S définiront des classes principales, ceux de Σ_1 des classes secondaires... etc...

2°) <u>Classes caractéristiques principales et secondaires des fibrés complètement réductibles</u>.

ξ étant un fibré vectoriel complexe de base V, il existe un fibré complexe ξ_1 de base V_1, complètement réductible et une application continue $f = V_1 \to V$ avec $\xi_1 = f^*(\xi)$, f^* étant en cohomologie entière ou réelle une application injective (splitting map).

Il est donc pertinent d'étudier d'abord le cas d'un fibré ξ complètement réductible. Les (U_α), $u_{\alpha\beta}$, $a_{\alpha\beta}$, $\sigma_{\alpha\beta\gamma}$ seront donc relatifs à un tel fibré à groupe structural réduit au tore complexe. $a_{\alpha\beta}(x)$ réel est continu et peut être choisi modulo une matrice diagonale à coefficients entiers.

On établit alors aisément [3] :

Proposition 1 : Les éléments de S sont des cocycles à valeurs entières et définissent des classes caractéristiques principales de ξ (classes de Chern). (4) définit, modulo un facteur constant, la $k^{\text{ième}}$ classe de Chern de ξ.

Les éléments de Σ-S sont des cochaînes à valeurs réelles (terme à préciser ultérieurement), mais en général ne sont point des cocycles.

Lemme 3. Si pour un cocycle $(u_{\alpha\beta})$ à valeurs dans le tore $T^n_{\mathbb{C}}$ complexe, on peut envisager un élément de S nul, c'est aussi possible pour tout autre cocycle cohomologue dans $T^n_{\mathbb{C}}$.

En effet si : $u'_{\alpha\beta} = \exp(-2\pi i\lambda_\alpha)\, u_{\alpha\beta} \exp(2\pi i\lambda_\beta)$

$$u'_{\alpha\beta} = \exp 2\pi i(\lambda_\beta + a_{\alpha\beta} - \lambda_\alpha)$$

on peut choisir : $a'_{\alpha\beta} = \lambda_\beta + a_{\alpha\beta} - \lambda_\alpha$, de sorte que $\sigma'_{\alpha\beta\gamma} = \sigma_{\alpha\beta\gamma}$ (ou $a'_{\alpha\beta} = \lambda_\beta + a_{\alpha\beta} - \lambda_\alpha + \eta_{\alpha\beta}$, $\eta_{\alpha\beta} \in Z$, avec $\delta\eta_{\alpha\beta} = 0$).

Mais on observera que si $c_{o12\ldots 2k}$ est encore nul $g_{o1\ldots 2k-1}$ donné par (5) a varié d'un cobord : $\mathrm{Tr}(\lambda_1 - \lambda_o \,\square\, \sigma_{123} \,\square \ldots\, \square \ldots)$. Si donc on considère $(u_{\alpha\beta})$, modulo une équivalence on doit considérer le cocycle $g_{o1\ldots 2k-1}$, modulo un cobord à valeurs réelles.

Lemme 4. Si $(\mathrm{Tr} \underset{k}{\square} \sigma_{\alpha\beta\gamma})$ est nul, le cocycle $\mathrm{Tr}(a_{\alpha\beta} \underset{k-1}{\square} \sigma_{\beta\gamma\delta})$ définit, modulo un cobord, un cocycle entier.

En effet soit $\mathrm{Tr}(\underset{k}{\square} \sigma_{\alpha\beta\gamma})$, cocycle entier. On considère la suite exacte :

$$0 \to \mathbb{Z} \to \tilde{\mathbb{R}}_C \xrightarrow{\exp 2\pi i} C^* \to 1$$ ($\tilde{\mathbb{R}}_C$ faisceau des germes des fonctions localement continues) qui pour la suite de cohomologie donne un isomorphisme de $H^q(X, \mathbb{Z})$ sur $H^{q-1}(X, \mathbb{C}^*)$.

Au cocycle $\mathrm{Tr} \underset{k}{\square} \sigma_{\alpha\beta\gamma}$ correspond le cocycle : $\exp 2\pi i[\mathrm{Tr}(a_{\alpha\beta} \underset{k-1}{\square} \sigma_{\beta\gamma\delta})]$. Si $\mathrm{Tr}(\underset{k}{\square} \sigma_{\alpha\beta\gamma})$ est nul, le cocycle à valeurs dans $\mathbb{C}^*$ est un cobord donc $\mathrm{Tr}(a_{\alpha\beta} \underset{k-1}{\square} \sigma_{\beta\gamma\delta})$ est modulo un cobord, un cocycle entier.

<u>Exemples de classes secondaires associées à un fibré complètement réductible</u>.

a) <u>Si $\mathrm{Tr}(\sigma_{\alpha\beta\gamma}) = 0$ pour le système des $(a_{\alpha\beta})$</u>, $c_1(\xi)$ est nul, le cocycle des $(u_{\alpha\beta}) = \exp(2\pi i \, \mathrm{Tr}(a_{\alpha\beta}))$ est trivial, de sorte que l'on peut écrire :

$$\exp(2\pi i \, \mathrm{Tr}(a_{\alpha\beta})) = \exp 2\pi i(b_\beta - b_\alpha)$$

$$\mathrm{Tr}(a_{\alpha\beta}) = b_\beta - b_\alpha + \eta_{\alpha\beta} \, , \quad \eta_{\alpha\beta} \in \mathbb{Z}$$

avec $\delta\eta_{\alpha\beta} = 0$ puisque $\delta \, \mathrm{Tr}(a_{\alpha\beta}) = 0$.
Les $(\eta_{\alpha\beta})$ définissent donc un élément de $H^1(V, \mathbb{Z})$. Si $b_\alpha = \mathrm{Tr}(u_\alpha)$,
$a_{\alpha\beta} - u_\beta + u_\alpha = a'_{\alpha\beta}$, $\exp(2\pi i \, a'_{\alpha\beta}) = u'_{\alpha\beta}$.

$$\eta_{\alpha\beta} = \mathrm{Tr}(a'_{\alpha\beta}).$$

Le fibré principal de cocycle $\mathrm{Tr}(a'_{\alpha\beta})$ est un revêtement de V. Si $V = \mathbb{C}^*$ on peut interpréter $H^1(V, \mathbb{Z})$ classe secondaire comme la "surface de Riemann" de logarithme.
Plus généralement cet élément est lié à l'indice de Maslov.

<u>On a donc un cocycle $u'_{\alpha\beta}$, avec $\mathrm{Tr}(\sigma'_{\alpha\beta\gamma}) = 0$ et $\mathrm{Tr}(a'_{\alpha\beta})$ définit un élément de $H^1(V, Z)$</u>.

b) Supposons que : $\mathrm{Tr}(\underset{k}{\square} \sigma_{\alpha\beta\gamma}) = 0$ pour les $(a_{\alpha\beta})$. Si

$$u'_{\alpha\beta} = (\exp 2\pi i\, \lambda_\alpha^{-1}) \exp(2\pi i\, a_{\alpha\beta}) \exp(2\pi i\, \lambda_\beta)$$

$$= \exp 2\pi i(\lambda_\beta - \lambda_\alpha + a_{\alpha\beta}) \text{ (cocycle cohomologue à } (u_{\alpha\beta})).$$

Posons : $a'_{\alpha\beta} = \lambda_\beta - \lambda_\alpha + a_{\alpha\beta}$, alors $\mathrm{Tr}(\underset{k}{\square}\, \sigma'_{\alpha\beta\gamma}) = 0$, pour les $(a'_{\alpha\beta})$

$a''_{\alpha\beta} = a'_{\alpha\beta} + \eta_{\alpha\beta}$, avec $\eta_{\alpha\beta}$ matrice diagonale à coefficients entiers satisfait $\mathrm{Tr}\, \underset{k}{\square}\, \sigma''_{\alpha\beta\gamma} = 0$ pour les $(a''_{\alpha\beta})$ si $\delta\eta_{\alpha\beta} = 0$.

$$\mathrm{Tr}((a''_{\alpha\beta} - a'_{\alpha\beta}) \underset{k-1}{\square} \sigma_{\alpha\beta\gamma}) , \quad (\sigma_{\alpha\beta\gamma} = \sigma'_{\alpha\beta\gamma} = \sigma''_{\alpha\beta\gamma})$$

est un cocycle à valeurs entières.

On dira donc de ce cocycle qu'il définit une classe caractéristique secondaire à valeurs entières. Pour le définir on a postulé que $\mathrm{Tr}(\underset{k}{\square}\, \sigma_{\alpha\beta\gamma}) = 0$ et introduit un fibré élément de $H^1(X, \mathbb{Z}^n)$ d'ailleurs arbitraire.

On pourrait seulement postuler que pour le même recouvrement (U_α), $\sigma'_{\alpha\beta\gamma} = \sigma''_{\alpha\beta\gamma}$ et $\mathrm{Tr}[(a'_{\alpha\beta} - a''_{\alpha\beta}) \underset{k-1}{\square}\, \sigma'_{\beta\gamma\delta}]$ est un cocycle à valeurs réelles, localement constantes (à valeurs dans le faisceau trivial $\tilde{\mathbb{R}}$).

c) Supposons que $\mathrm{Tr}(\underset{k}{\square}\, \sigma_{\alpha\beta\gamma}) = 0$ pour un recouvrement $\mathcal{U} = (U_\alpha)$, avec les coefficients $(a_{\alpha\beta})$, (5) étant de plus réel, constant sur chaque composante connexe des $U_{o,1,\ldots 2k-1}$. Alors $g_{o1,\ldots 2k-1}$ donné par (5) définit un élément de $H^*(\mathcal{U}, \tilde{\mathbb{R}})$, $\tilde{\mathbb{R}}$ étant le faisceau constant des nombres réels.

Cette situation peut se présenter, en particulier, si les fonctions de transitions sont localement constantes, ou dans le cas différentiable si le groupe d'holonomie est nul.

Si $\mathcal{U}$ est contractile simple on obtient un élément de $H^*(V, \tilde{\mathbb{R}})$.

Observons que si on change le cocycle $(u_{\alpha\beta})$ en tout autre

$u'_{\alpha\beta}$, équivalent dans le tore $T^n_{\mathbb{C}}$, selon le lemme 3 on peut encore envisager $\mathrm{Tr}(\underset{k}{\square}\, \sigma'_{\alpha\beta\gamma}) = 0$, mais la condition imposée à (5) ne se conserve pas ; il en serait cependant ainsi si on travaillait dans un sous-groupe discret.

3°) Le cas différentiable (fibrés complètement réductibles).

V est maintenant une variété paracompacte, réelle, de dimension finie, différentiable C^∞. On peut supposer $\mathcal{U}$ contractile, simple.

Cherchons à expliciter l'élément de la cohomologie de De Rham associée à une classe caractéristique.

On considère la suite exacte de faisceaux :

$$0 \longrightarrow \widetilde{\mathbb{R}} \xrightarrow{i} \widetilde{\mathcal{a}}_o \xrightarrow{d_o} \widetilde{\mathcal{a}}_1 \xrightarrow{d_1} \widetilde{\mathcal{a}}_2 \longrightarrow \ldots . \widetilde{\mathcal{a}}_p \xrightarrow{d_p} \ldots \qquad (6).$$

où i est l'inclusion, $\widetilde{\mathcal{a}}_p(V)$ le faisceau des germes des formes différentielles $\mathcal{a}_p$ de degré p, d_p la différentielle extérieure restreinte à $\widetilde{\mathcal{a}}_p$. On utilise la suite d'isomorphismes :

$$\check{H}^{2k}(V, \ker d_o) \underset{\delta_1^{2k-1}}{\simeq} \check{H}^{2k-1}(V, \ker d_1) \underset{\delta_2^{2k-2}}{\simeq} \ldots . \simeq \underset{\delta^1_{2k-1}}{\check{H}^1} (V, \ker d_{2k-1}) \simeq \frac{Z^{2k}(V)}{d\mathcal{a}^{2k-1}(V)} \qquad (7)$$

($\check{H}^\lambda$ est la cohomologie de Čech à valeurs dans un faisceau, on n'a pas explicité l'inclusion du faisceau $\widetilde{\mathbf{Z}}$ dans $\widetilde{\mathbb{R}}$) venant de la suite exacte de faisceaux :

$$0 \to \ker d_\alpha \subseteq \widetilde{\mathcal{a}}_\alpha \xrightarrow{d\alpha} \ker d_{\alpha+1} \to 0.$$

En ce qui concerne les classes principales, si on part de (4) on lui associe :

$$T_r(a_{o_1} \square\ \sigma_{123} \square \ldots .\ \sigma_{2k-3,\, 2k-2,\, 2k-1}), \text{ puis}$$

$$T_r(da_{o_1} \square\ \sigma_{123} \quad \ldots\ldots\ldots\ldots\ldots\ldots\ldots)$$

da_{o_1} s'écrit : $-\frac{1}{2\pi i}(\omega_o - \omega_1)$, ω étant une connexion unitaire à matrice diagonale (adaptée à la réduction complète), de formes respectives ω_o et ω_1 au-dessus de U_o et U_1. Appliquant à nouveau le cobord on trouve :

$$- \frac{1}{2\pi i} \quad T_r(\Omega \sqcap \sigma_{123} \dots \sqcap \sigma_{2k-3,\ 2k-2,\ 2k-1}), \text{ avec } d\omega = \Omega, \text{ puis}$$

en recommençant on aboutit à :

$$(- \frac{1}{2\pi i})^k \; k! \; \mathrm{Tr}\ (\Omega \wedge \Omega \wedge \dots \Omega) \quad (8), \text{ avec } k \text{ facteurs.}$$

Proposition 2 - La $k^{\text{ième}}$ classe de Chern $c_k(\xi)$, considérée comme élément de $\check{H}^{2k}(X, \widetilde{\mathbb{R}})$, est représentée par :

$$\frac{1}{k!} \; \underline{c_{o12\dots 2k}} = \frac{1}{k!} \quad \mathrm{Tr}(\underset{k}{\sqcap} \; \sigma_{\alpha\beta\gamma}) \qquad \text{(formule 4) (ou}$$

$(\frac{-1}{2\pi i})^k \; \mathrm{Tr}(\wedge^k \Omega))$ en cohomologie de De Rham.

Nous avions donné ce résultat dans [3].

Remarque : On peut bâtir à partir de là une approche complète de la construction des classes de Chern pour les fibrés non complètement réductibles qui ne nécessite dans le cas différentiable que le théorème de "splitting map" du (2°), avec en outre dans le cas topologique la propriété suivante :

Etant donné ξ fibré de rang n au-dessus de V, si $f^*(\xi) = \overset{n}{\underset{1}{\oplus}} \xi_i$, il existe $u \in H^*(V, \mathbb{Z})$ tel que $c(\overset{n}{\underset{1}{\oplus}} \xi_i) = f^*(u)$.

Revenons maintenant aux fibrés complètement réductibles et plaçons nous d'abord dans les hypothèses de l'exemple (c) du 2°). Il est loisible de supposer que le recouvrement $\mathcal{U} = \underset{\alpha}{\cup}(U_\alpha)$ est contractile simple, la cohomologie de $\mathcal{U}$ est alors identique à celle de V selon un théorème de Leray. Au cocycle $\mathrm{Tr}(a_{\alpha\beta} \underset{k-1}{\sqcap} \sigma_{\beta\gamma\sigma}) \in \check{Z}^{2k-1}$ (V, ker d_o), on doit appliquer $(\delta_1^{2k-2})^{-1}$ (cf.(7)), ce qui conduit à $\mathrm{Tr}(b_\beta \underset{k-1}{\sqcap} \sigma_{\beta\gamma\delta})$, puis à $\mathrm{Tr}(db_\beta \underset{k-1}{\sqcap} \sigma_{\beta\gamma\delta})$ les b_α étant des fonctions au-dessus des U_α, telles que

$$\mathrm{Tr}(db_\beta - db_\alpha) \underset{k-1}{\sqcap} \sigma_{\beta\gamma\delta}) = \mathrm{Tr}(da_{\alpha\beta} \underset{k-1}{\sqcap} \sigma_{\beta\gamma\delta}) = 0 ;$$

$(\delta_1^{k-2})^{-1}$ conduit au représentant : $\mathrm{Tr}(db \underset{k-1}{\sqcap} \sigma_{\beta\gamma\delta})$, où (db est une forme locale, les db_α sont seulement définis, mais par passage à la trace, on a une expression qui a valeur globale). Itérant on obtient :

$$(10) : \left(\frac{-1}{2\pi i}\right)^{k-1} \mathrm{Tr}(db \underset{k-1}{\sqcap} \Omega),$$ Ω étant la courbure d'une connexion adaptée quelconque (on a $\Omega = d\omega$ et $d\Omega = 0$).

La $p^{\text{ième}}$ cohomologie de De Rham $H^p(V)$ n'est autre que $H^p(\Gamma(\widetilde{\mathcal{A}}(V)))$, où $\Gamma(\widetilde{\mathcal{A}}(V))$ est le complexe des cochaînes $\Gamma(\widetilde{\mathcal{A}}_p(V))$, ($\widetilde{\mathcal{A}}_p(V)$ faisceau des germes des formes différentielles de degré p, $\Gamma(\widetilde{\mathcal{A}}(V))$ module des sections).

Au-dessus de $U_{o1\ldots 2k-1}$ contractile on peut envisager θ forme locale, ω_α forme de connexion adaptée au-dessus de U_α, et $\omega_\alpha - \theta = db_\alpha$, modulo une forme exacte (on rappelle que $\omega_\beta - \omega_\alpha = - 2\pi i \, da_{\alpha\beta}$) . Nous avons obtenu :

<u>Proposition 3</u> - <u>La classe caractéristique secondaire représentée en cohomologie de Čech par</u> $g = \mathrm{Tr}(a_{\alpha\beta} \underset{k-1}{\sqcap} \sigma_{\beta\gamma\delta})$ <u>quand</u> $\mathrm{Tr}(\underset{k}{\sqcap} \sigma_{\alpha\beta\gamma}) = 0$, <u>et g localement constante, réelle, est représentée en cohomologie de De Rham</u> $H^{2k-1}(\Gamma(\widetilde{\mathcal{A}}(V)))$ <u>par l'élément</u>

$$\left(\frac{-1}{2\pi i}\right)^{k-1} \widetilde{\mathrm{Tr}}((\omega-\theta) \underset{k-1}{\sqcap} \Omega) \quad (11),$$ <u>du faisceau des germes des sections d'un préfaisceau déterminé par le système de tous les ouverts contractiles</u> $\mathcal{O}$, <u>d'une connexion adaptée</u> ω, <u>de courbure</u> Ω <u>et d'une forme locale</u> θ <u>telle que</u> $d\omega = d\theta = \Omega$, <u>au-dessus de</u> $\mathcal{O}$. [+]

<u>Remarque 1</u> : L'élément (11) peut être défini à l'aide de toute connexion adaptée ; cela résulte de la démonstration, mais on peut aussi voir que si on pose $\omega_t = \omega + t\eta$($\eta$ forme à valeurs dans

[+] On pourra conférer sur ce point [2 bis], pages 52-53, (11) conduit à un élément de $\mathcal{A}_p(V)$ qui par image réciproque donne sur le fibré principal une forme exacte.

Hom (ξ,ξ) à matrice diagonale), $\Omega_t = \Omega + td\eta$, en prenant $\theta_t = \theta + t\eta$ (11) se reproduit modulo une forme exacte.

<u>Remarque 2</u> : S'il est possible d'envisager $\mathrm{Tr}(\underset{k}{\sqcap}\Omega) = 0$ pour une certaine forme de courbure alors :

(12) $\left(\frac{-1}{2\pi i}\right)^{k-1} \widetilde{\mathrm{Tr}}(\omega \underset{k-1}{\sqcap} \Omega)$ définit une forme fermée sur V.

Si par exemple le fibré ξ admet une connexion plate, alors (11) définit un élément nul en cohomologie; s'il est possible de choisir une autre connexion ω' avec $\mathrm{Tr}(\underset{k}{\sqcap}\Omega') = 0$, alors $\widetilde{\mathrm{Tr}}(\omega' \sqcap \Omega')$ définit un élément de la cohomologie de De Rham. Cette situation est celle qui se présente dans la définition de la classe dite de Godbillon-Vey [1,5], ω' est la connexion basique.

Posant $H(t) = \mathrm{Tr}(\omega_t \underset{k-1}{\sqcap} \Omega_t)$ avec $\omega_t = \omega_o + t\eta$, modulo une forme exacte, on obtient :

$$\mathrm{Tr}(\omega_t \sqcap d\eta \underset{k-2}{\sqcap} \Omega_t) = \mathrm{Tr}\,(\eta \underset{k-1}{\bar{\sqcap}} \Omega_t) \quad \text{et}$$

$$H(1) - H(0) = k\int_o^1 \mathrm{Tr}(\eta \underset{k-1}{\sqcap} \Omega_t)dt \qquad (13) \qquad (\Omega_t = \Omega_o + td\eta)$$

Il est bien connu d'ailleurs que :

$$\mathrm{Tr}(\underset{k}{\sqcap}\Omega_1) - \mathrm{Tr}(\underset{k}{\sqcap}\Omega_o) = k\, d\int_o^1 \mathrm{Tr}(\eta \underset{k-1}{\sqcap} \Omega_t)dt \qquad (14)$$

et si les traces figurant au premier membre de (14) sont nulles, l'intégrale du deuxième membre de (13) définit une classe appelée "exotique" dans la littérature antérieure.

<u>4°</u>) <u>Les classes à valeurs dans $\mathbb{Z}_q$. Classes de Stiefel-Whitney.</u>

Au lieu d'envisager $u_{\alpha\beta}(x) = \exp\,(2\pi i\, a_{\alpha\beta}(x))$ comme au 1°) il est loisible d'introduire

$$u_{\alpha\beta}(x) = \exp\left(\frac{2\pi i}{q}\, a_{\alpha\beta}(x)\right), \quad q \in \mathbb{Z}^*$$

et les résultats donnés dans le cas topologique s'adaptent sans peine.

Si nous prenons maintenant un fibré vectoriel réel, de base paracompacte, il existe un "splitting map", pour la cohomologie à valeurs dans $\mathbb{Z}_2$. Cela nous conduit à envisager q = 2 et à construire pour un fibré réel totalement réductible ξ :

$$c'_{\alpha_o\alpha_1\ldots\alpha_k} = \mathrm{Tr}\,(b_{\alpha_o\alpha_1} \square\ b_{\alpha_1\alpha_2} \square \ldots \square\ b_{\alpha_{k-1},\alpha_k}) \qquad (15)$$

$O(n, \mathbb{R}) \subset U(n, \mathbb{R})$, où les $b_{\alpha\beta}$ sont des matrices diagonales à coefficients entiers ($b_{\alpha\beta} = a_{\alpha\beta}$ et q = 2).

On cherche à interpréter (15) comme cocycle (k+1)-aire.

Avec une simplification dans les indices on a :

$$c'_{o1\ldots k} = \sum_{i_1, i_2 \ldots i_k} b_{o1}^{i_1}\ b_{12}^{i_2} \ldots b_{k-1,k}^{i_k} \qquad (16)$$

avec somme étendue aux arrangements de 1, 2,.., k à k. Pour éviter plus loin la difficulté de diviser par k! nous remplaçons $c'_{o1\ldots k}$ par:

$$d'_{o1\ldots k} = \sum_{i_1<i_2\ldots<i_k} b_{o1}^{i_1}\ b_{12}^{i_2} \ldots b_{k-1,k}^{i_k} \qquad (17)$$

dont nous prendrons la réduction mod 2 (red), la somme étant étendue aux combinaisons de 1,..., n, k à k.

Pour k = 1, la cochaîne red $(d'_{o1..k}) = d_{o1\ldots k}$ est de manière évidente un cocycle à valeurs dans $\mathbb{Z}_2$, on peut donc raisonner par récurrence sur k, ce qui est aisé et laissé au lecteur, pour établir cette propriété pour tout k.

Il est alors immédiat de vérifier que les cocycles $d_{o1\ldots k}$ satisfont aux axiomes de définition des classes de Stiefel-Whitney pour k = 0, 1,...r, si r est le rang du fibré ξ.

De la suite $(i_1, i_2 \ldots i_k)$ on déduit par permutation un élément $(\ell_1, \ell_2, \ldots \ell_k)$ et une vérification directe montre que :

$$\mathrm{red} \sum_{i_1<i_2<i_k} b_{o1}^{\ell_1}\ b_{12}^{\ell_2}, \ldots b_{k-1,k}^{\ell_k} \qquad s(i_k) = \ell_k \quad s \in S_k$$

définit aussi bien la $k^{\text{ième}}$ classe de S.W. $\omega_k(\xi)$.

Il existe donc k! sommes telles que (17) pouvant définir le $k^{\text{ième}}$ cocycle de S.W.

On adoptera donc la méthode de définition suivante :

$\frac{1}{k}$ $c'_{01\ldots k}$ est une cochaîne à valeurs dans $\mathbb{Q}$ dont on peut prendre la réduction mod 2 :

$s_{01\ldots k} = \mathrm{red}(\frac{1}{k!}\, c'_{01\ldots k})$ est un cocycle à valeurs dans $\mathbb{Q}_{/2\mathbb{Z}}$ et définit un élément de $H^k(X, \frac{\mathbb{Q}}{2\mathbb{Z}})$.

L'inclusion naturelle $Z_2 = \mathbb{Z}_{/2\mathbb{Z}} \to \mathbb{Q}_{/2\mathbb{Z}}$ passe à une application en cohomologie. Ainsi :

<u>Proposition 4</u> : <u>Le cocycle $\frac{1}{k!}$ $\mathrm{Tr}(b_{\alpha_o\alpha_1} \square b_{\alpha_1\alpha_2} \square \cdots \square b_{\alpha_{k-1},\alpha_k})$ à valeurs dans $\mathbb{Q}_{/2\mathbb{Z}}$ a pour image naturelle par inclusion $\mathbb{Z}_{/2\mathbb{Z}} \to \mathbb{Q}_{/2\mathbb{Z}}$ la $k^{\text{ième}}$ classe de Stiefel-Whitney</u>.

<u>Remarque</u> : $\mathbb{Q}_{/2\mathbb{Z}}$ n'a pas de structure multiplicative venant par quotient de celle de $\mathbb{Q}$ ($2\mathbb{Z}$ n'est-pas un idéal dans $\mathbb{Q}$!) de sorte que l'on ne pourra construire $w(\xi)$ à valeurs dans $\mathbb{Q}_{/2\mathbb{Z}}$, par cup-produit).

La méthode signalée pour la construction des classes de Chern des fibrés quelconques s'adapte immédiatement, dans le cas topologique.

REFERENCES

1. <u>R. BOTT</u> - Lectures on characteristic classes and foliations. Lecture Notes 279 (Springer) (1972).

2. <u>H. CARTAN</u> - Colloque topologie - Bruxelles, 1950 - Masson, Paris (1951).

2 bis : <u>CHERN et SIMONS</u>. Characteristic forms and geometric invariants. Annals of math. Janvier 1974. p.48-69.

3. A. CRUMEYROLLE - Comptes-Rendus Acad. Sc. Paris. t.285-1967, pp. 461-463.

4. C. GODBILLON - Séminaire Bourbaki, 1971-72. n°421.

5. C. GODBILLON et J. VEY - Comptes-Rendus Acad. Sc. Paris, 1971, pp. 92-95.

6. F. KAMBER et Ph. TONDEUR : Foliated bundles and characteristic classes. Lect. Notes 493 (Springer) (1975).

7. F. KAMBER et Ph. TONDEUR : G-foliations and their characteristic classes. Bulletin of. A.M.S. Vol 84, n° 6. nov. 1978. (ces deux articles comportent une bibliographie extrêmement abondante)

8. C. LAZAROV et J. Pasternak - Secondary characteristic classes for riemannien fibrations. J. Diffe. Geometry 11 (1976).

9. D. LEHMANN - Classes caractéristiques exotiques... Ann. Inst. Fourier 24 (1974).

10. P. MOLINO - Comptes-Rendus Acad. Sciences Paris 272 (1971) 779-781 et 1376-1378. Propriétés cohomologiques et propriétés topologiques des feuilletages à connexion transverse projetable. Topology Vol 12, n° 4. nov. 1973.

11. I. VAISMAN - Cohomology and diff. forms (Marcel Dekker (1973))

12. K. YAMATO - Examples of foliations with non trivial exotic characteristic classes (Osaka J. Math 12 (1975) (401-417).

DEFORMATION THEORY AND STABILITY FOR HOLOMORPHIC FOLIATIONS (*)

by

T. Duchamp
Department of Mathematics
The University of Utah
Salt Lake City, Utah 84112 U.S.A.

and

M. Kalka
Department of Mathematics
The Johns Hopkins University
Baltimore, Maryland 21218 U.S.A.

Introduction

A holomorphic foliation on a compact manifold M^n is a foliation given by local submersions $f_\alpha : U_\alpha \to \mathbb{R}^{2q}$ which patch together via local maps $\phi :_{\alpha\beta} : \mathbb{R}^{2q} \to \mathbb{R}^{2q}$ which are local biholomorphisms when $\mathbb{R}^{2q}$ is identified with $\mathbb{C}^q$. We outline here our study of various deformation and stability properties of such foliations. Details will appear in [1] and [2].

Associated to a holomorphic foliation F is a structure sheaf Θ_F, the sheaf of germs of F-invariant, holomorphic vector fields on the normal bundle of F; these are lifts, via the submersions f_α, of local holomorphic vector fields on $\mathbb{C}^q$. Our main tool is a resolution of this sheaf by an elliptic complex. In the case of C^∞-foliations similar resolutions appear in the work of Hamilton [4], Heitsch [5], Kamber-Tondeur [6], Mostow [8] and Vaisman [11]; and when M is complex and the maps f_α are holomorphic Heitsch [5] has constructed a resolution of Θ_F. We use this resolution to extend Kuranishi's Theorem [7] (on the existence of a locally complete, finite dimensional analytic family of complex structures close to a given complex structure) to holomorphic foliations for which there exists a transverse C^∞-foliation. Next we consider various notions of stability and indicate the proofs of the following two theorems. First if $H^1(M,\Theta_F) = 0$ then every holomorphic foliations near F is equivalent to F via a C^∞-diffeomorphism of M. This is true even if F is not Hausdorff and without assuming the existence of a foliation transverse to F. To state the second result we let $\underset{\sim}{Q}_F$ be the sheaf of germs of F-invariant C^∞-sections of the

(*) Research partially supported by NSF grants MCS78-01826 and MCS78-00399 respectively.

normal bundle of F. There is an inclusion $\Theta_F \hookrightarrow Q_F$ which induces a map in cohomology $\iota_*: H^1(M,\Theta_F) \to H^1(M,Q_F)$. We show that if F is a Riemannian foliation and if there is a foliation transverse to F which is given by a closed locally decomposable form then the triviality of ι_* implies that every holomorphic foliation near F is conjugate to F as a C^∞-foliation.

Elliptic Complexes Associated with a Holomorphic Foliation

Let M be a compact manifold and let F be a holomorphic foliation on M. If we let L denote the tangent bundle to F, then the normal bundle $Q \equiv TM/L$ carries an almost complex structure J, obtained by lifting the standard almost complex structure on $\mathbb{C}^q$ to Q via the local submersions f_α. The complex structure on Q induces a splitting of the complexified normal bundle in the standard way

$$Q^{\mathbb{C}} = Q^{(1,0)} \oplus Q^{(0,1)} .$$

The integrabitity conditions for the almost complex structure are given in the following version of the complex Frobenius Theorem [9], [3].

Theorem: Let $E \subset T_M^{\mathbb{C}}$ be a subbundle of complex codimension q with $E + \overline{E} = T_M^{\mathbb{C}}$. Let $Q^{(1,0)} \equiv T_{M/E}^{\mathbb{C}}$. Then the following are equivelant

(*i*) $[E, E] \subseteq E$

(*ii*) $dQ^{(1,0)*} \wedge \underset{\sim}{\Omega}_M^s \subset Q^{(1,0)*} \wedge \underset{\sim}{\Omega}_M^{s+1}$ for all $s \geqslant 0$

(*iii*) $Q^{(1,0)}$ arises from a codimension - 2q holomorphic foliation as above and E is the kernel of the quotient map $\pi^{(1,0)}; T_M^{\mathbb{C}} \to Q^{(1,0)}$.

We have used the notation $\underset{\sim}{B}$ for the sheaf of germs of sections of the vector bundle B and Ω_M^s for the $s^{\underline{th}}$ exterior power of the complexified cotangent bundle of M.

This theorem allows us to describle the collection of foliations near F as a subset of $\mathrm{Hom}_{\mathbb{C}}(E, Q^{(1,0)})$ as follows. Choose a splitting $\rho\colon Q \to T_M$ of the exact sequence

(1) $0 \to L \to T_M \to Q \to 0$

This induces in an obvious way a splitting

(2) $$0 \to E \xrightarrow{\iota_{\mathbb{C}}} T_M^{\mathbb{C}} \underset{\rho^{(1,0)}}{\overset{\pi^{(1,0)}}{\rightleftarrows}} Q^{(1,0)} \to 0 .$$

This splitting permits us to define a one-one correspondence between $\mathrm{Hom}_{\mathbb{C}}(E,Q^{(1,0)})$ and distributions (in the sense of the Frobenius Theorem) near E. For $\phi \in \mathrm{Hom}_{\mathbb{C}}(E,Q^{(1,0)})$ let

$$E_\phi \subseteq T_M^{\mathbb{C}} = \{\iota_\phi(X) = \iota_{\mathbb{C}}(X) + \rho^{(1,0)}\cdot\phi(X) \mid X \in E\} .$$

It is easy to see that $E_\phi + \overline{E_\phi} = T_M^{\mathbb{C}}$ and hence that the complex Frobenius Theorem gives a one-one correspondence between holomorphic foliations near F and the set

$$\mathrm{Fol}(F) = \{\phi \in \mathrm{Hom}_{\mathbb{C}}(E,Q^{(1,0)})\colon [E_\phi, E_\phi] \subseteq E_\phi\} .$$

The topology we place on $\mathrm{Fol}(F)$ is that induced by the C^∞-topology on $\mathrm{Hom}_{\mathbb{C}}(E,Q^{(1,0)})$.

The Frobenius Theorem also allows us to construct an elliptic complex (E^{*s},d_E). Here $E^{*s} = \Lambda^s(E^*)$ and d_E is the operator defined as follows. For $s = 0$, d_E is the composition

$$E^{*0} \equiv \Omega_M^0 \xrightarrow{d} \Omega_M^1 \xrightarrow{\iota_{\mathbb{C}}^*} E^{*1}$$

and for $s \geq 1$, d_E is the unique operator making the following diagram commute

$$\begin{array}{ccccccc} 0 \to \underset{\sim}{Q}^{(1,0)*} \wedge \underset{\sim}{\Omega}_M^{s-1} & \to & \underset{\sim}{\Omega}_M^{s} & \xrightarrow{\iota_E^*} & \underset{\sim}{E}^{*s} \\ \downarrow d & & \downarrow d & & \downarrow d_E \\ 0 \to \underset{\sim}{Q}^{(1,0)*} \wedge \underset{\sim}{\Omega}_M^{s} & \to & \underset{\sim}{\Omega}_M^{s+1} & \xrightarrow{\iota_E^*} & \underset{\sim}{E}^{*(s+1)} \end{array}$$

Let $\mathcal{O}_F \subset C_M^\infty$ be the subsheaf of germs of C^∞ functions which are lifts, via the submersions f_α, of holomorphic functions on $\mathbb{C}^q$.

<u>Lemma 1</u>: The sequence

$$0 \to \mathcal{O}_F \to C_M^\infty \xrightarrow{d_E} \underset{\sim}{E}^{*1} \xrightarrow{d_E} \underset{\sim}{E}^{*2} \to \cdots$$

is a resolution of $\mathcal{O}_F$.

<u>Lemma 2</u>: The complex $(E^*; d_E)$ is elliptic.

The proofs of these lemmas are straightforward after writing d_E in a coördinate system (x,z) adapted to the foliation.

We now define the sheaf Θ_F of germs of locally constant (along leaves of F) holomorphic vector fields as the lifts via the submersions f_α of local holomorphic vector fields on $\mathbb{C}^q$. To define a resolution of Θ_F set

$$\underset{\sim}{E}_Q^{*s} = \underset{\sim}{E}^{*s} \otimes_{\mathcal{O}_F} \Theta_F = \underset{\sim}{E}^{*s} \otimes_{\mathbb{C}} \underset{\sim}{Q}^{(1,0)}$$

and let $d_Q: \underset{\sim}{E}_Q^{*s} \to \underset{\sim}{E}_Q^{*s+1}$ be defined as $d_E \otimes \mathrm{id}$. By lemmas 1 and 2 the complex $(\underset{\sim}{E}_Q^{*s}, d_Q)$ is a resolution of Θ_F and it is elliptic. Using the theory of elliptic complexes the following theorem is evident.

Theorem 3: The cohomology groups $H^s(M, \mathcal{O}_F)$ and $H^s(M, \Theta_F)$ are finite dimensional.

The Kuranishi Family of a Holomorphic Foliation

We wish to characterize $\mathrm{Fol}(F)$ as the kernel of a nonlinear partial differential operator $D: \mathrm{Hom}_{\mathbb{C}}(E, Q^{(1,0)}) \to \mathrm{Hom}_{\mathbb{C}}(\Lambda^2 E, Q^{(1,0)})$ which, by analogy with the deformation theory of complex manifolds, is of the form $D = d_Q - [\ ,\ ]_Q$. Here $[\ ,\ ]_Q: E_Q^{*r} \times \underset{\sim}{E}_Q^{*s} \to \underset{\sim}{E}_Q^{*(r+s)}$ is bilinear and satisfies the following identities:

(*i*) $[\phi, \psi]_Q = (-1)^{rs}[\psi, \phi]_Q$

(*ii*) $d_Q [\phi, \psi]_Q = [d_Q \phi, \psi]_Q + (-1)^r [\phi, d_Q \psi]_Q$

(*iii*) $(-1)^{st}[\phi,[\psi,\tau]_Q]_Q + (-1)^{rs}[\psi,[\tau,\phi]_Q]_Q + (-1)^{rt}[\tau,[\phi,\psi]_Q]_Q = 0$

for $\phi \in \underset{\sim}{E}_Q^{*r}$, $\tau \in \underset{\sim}{E}_Q^{*s}$, $\tau \in \underset{\sim}{E}_Q^{*t}$. Unfortunatly, we do not see how to do this in general. However, if the splitting (2) is induced by a foliation $F^\perp$ transverse to F, we can proceed in analogy with the case of a complex manifold. Choosing coordinates adapted to the local product structure (x,z) and letting

$$\phi = \phi^\alpha_{JB}\, dx^J \wedge d\bar{z}^B \otimes \frac{\partial}{\partial z^\alpha} \in \underset{\sim}{E}_Q^{*r}$$

and

$$\psi = \psi^\alpha_{KG}\, dx^K \wedge d\bar{z}^G \otimes \frac{\partial}{\partial z^\alpha} \in \underset{\sim}{E}_Q^{*s}$$

the bracket $[\phi,\psi]_Q$ is defined by the equation

$$[\phi,\psi]_Q = \frac{1}{2r!s!}\left(\phi^{\gamma}_{JB}\frac{\partial\psi^{\alpha}_{KG}}{\partial z^{\gamma}} + (-1)^{rs+1}\psi^{\gamma}_{KG}\frac{\partial\phi^{\alpha}}{\partial z^{\gamma}}\right) dx^J \wedge d\bar{z}^B \wedge dx^K \wedge d\bar{z}^G \otimes \frac{\partial}{\partial z^{\alpha}}$$

where summation and multi-index conventions have been employed.

<u>Proposition 4</u>. $\phi \in \mathrm{Hom}_{\mathbb{C}}(E,Q^{(1,0)})$ defines a holomorphic foliation if and only if $D\phi = d_Q\phi - [\phi,\phi]_Q = 0$.

We are now ready to state our main theorem.

<u>Theorem 5</u>: Let F be a holomorphic foliation on a compact C^{∞}-manifold M and let $F^{\perp}$ be a C^{∞}-foliation transverse to F . Then there is a local analytic subset $B \subseteq H^1(M, \Theta_F)$ and a holomorphic map

$$B \to \mathrm{Fol}(F) \subseteq \mathrm{Hom}_{\mathbb{C}}(E,Q^{(1,0)}) : \quad t \to F_t$$

which defines a locally complete family of holomorphic foliations in the sense that if F' is a holomorphic foliation sufficiently close to F then F' is conjugate to some F_t via a diffeomorphism close to the identity. Furthermore, given a Riemannian metric respecting the local product structure on M induced by F and $F^{\perp}$ this diffeomorphism can be unambiguously defined.

We give a brief outline of the construction of B since we will need it in our discussion of stability. Using Hodge theory we can, upon choosing a Riemannian metric on M, find differential operators $G_Q\colon \Gamma(E_Q^{*s}) \to \Gamma(E_Q^{*s})$, $s \geq 0$ and $H_Q\colon \Gamma(E_Q^{*s}) \to H^s(M, \Theta_*)$ with the property that

$$I = H_Q + (d_Q d_Q^* + d_Q^* d_Q) \circ G = H_Q + \Delta_Q \circ G \ .$$

It can be shown that every holomorphic foliation F' close to F is conjugate to one defined by a section ϕ of $\mathrm{Hom}_{\mathbb{C}}(E, Q^{(1,0)})$ which satisfies

$$d_Q\phi - [\phi,\phi]_Q = 0$$
$$d_Q^*\phi = 0 \ .$$

Now for $\phi_0 \in H^1(M,\Theta_F)$, which we identify with $\mathrm{Ker}\Delta_Q$, there is a unique solution $\phi = \phi(\phi_0)$ of the equation

$$\phi = \phi_0 + d_Q^* G_Q[\phi,\phi]_Q$$

depending holomorphically on ϕ_0. It can then be shown, as in [7], that the local analytic set B mentioned above is

$$B = \{\phi_0\ H^1(M,\Theta_F) \mid H_Q[\phi(\phi_0),\ \phi(\phi_0)] = 0\}$$

and the map $B \to \mathrm{Fol}(F)$ is given by $\phi_0 \to \phi(\phi_0)$.

Stability Theorems

A foliation F is called holomorphically stable if there is a neighborhood N of F so that for each foliation $F' \in N$ then there is a diffeomorphism of M sending leaves of F' to leaves of F and respecting the associated complex structures.

It is not difficult to show that F is holomorphically stable if $H^1(M,\Theta_F) = 0$, with no assumption on the existence of a transversal foliation. The prrof is as in [4], only one can use a standard implicit function theorem since the d_Q-complex is elliptic.

A more refined notion of stability is given in the following definition.

Defintion: The holomorphic foliation F is said to be infinitesimally stable under holomorphic deformations if the map $\iota_*: H^1(M,\Theta_F) \to H^1(M,Q_F)$ induced by the inclusion $\Theta_F \to Q_F$ is trivial and F is said to be stable under holomorphic deformations if every holomorphic foliation F' sufficiently close to F is conjugate to F via a diffeomorphism of M sending leaves of F' to leaves of F but not necessarily respecting complex structures.

Recall that an $SL(p)$-foliation is a foliation whose tangential distribution is given as the kernel of a closed locally decomposable p-form on M, and that a foliation is called Hermitian if it is holomorphic and Riemannian.

Theorem 6: Let F be a real codimension-2q Hermitian foliation on an M-dimensional compact manifold M and suppose there is an SL(p) SL(p) - foliation transverse to F, where $p = n - 2q$. If F is infinitesimally stable under holomorphic deformations, then it is stable under holomorphic deformations.

To describe the proof of this theorem we need the notion of a baselike form, due to Reinhart [10]. A form $\phi \in E_Q^{*s}$ is called baselike if in coordinates (x,z) adapted to the local product structure $\phi = \phi_B^\alpha(z)d\bar{z}^B \otimes \frac{\partial}{\partial z^\alpha}$.

Proposition 7: Infinitesmal stablity under holomorphic deformations is equivelant to the condition that every class in $H^1(M,\Theta_F)$ has a baselike representative.

Theorem 6 is then proved by showing that under the conditions of the theorem every harmonic element is baselike and futher that if $\phi_0 \in \operatorname{Ker} \Delta_Q$ is baselike and of small norm then so is the unique solution of the equation $\phi = \phi_0 + d_Q^* G_Q[\phi,\phi]_Q$. If ϕ baselike then the foliation defined by E_ϕ is same as the foliation defined by E_0 , from a C^∞ viewpoint, which completes the proof of the theorem.

In order to compute the map ι_* we make the following definition.

Definition: A Hausdorff holomorphic foliation F on a compact manifold M is called a cohomology product foliation if there are forms $\omega_1,\ldots,\omega_r$ on M with $d_Q\omega_i \in \Gamma(Q^{(1,0)*} \wedge T_M^{*\mathbb{C}})$ for $i=1,\ldots,r$ such that for each leaf $N \to M$ of F and each integer $1 \leqslant s \leqslant \dim N$ there is an isomorphism $H^s(N,R) \cong \wedge^s(j^*\omega_1,\ldots,j^*\omega_r)_n$ of the real cohomology of N with the exterior algebra on the free graded commutative algebra generated by $j^*\omega_1,\ldots,j^*\omega_r$ truncated at $n = \dim N$.

If we denote the space of global baselike forms by A^*_b , and the restriction of d_Q to A^*_b by d_b then the following version of the Leray-Hirsch theorum holds.

Theorem 8: Let F be a Hausdorff, holomorphic, cohomology product foliation. Then there is a commutative diagram

$$\begin{array}{ccc} H^1(M,\Theta_F) & \xrightarrow{\iota_*} & H^1(M,Q_F) \\ \cong\uparrow & & \cong\uparrow \\ H^1(A^*_b,d_b) \oplus \Gamma(\Theta_F) \otimes \Lambda^1(\omega_1,\ldots,\omega_r)_n & \xrightarrow{j_*} & \Gamma(Q_F) \otimes \Lambda^1(\omega_1,\ldots,\omega_r)_n \end{array} .$$

Hence F is infinitesimally stable under holomorphic deformations if and only if $\Gamma(\Theta_F) = 0$. Here j_* is the map induced by projection onto the second factor composed with the inclusion ι .

We conclude with an example. One can show [3] that the construction of the Kobayashi metric on a complex manifold can be extended to yield a pseudo-metric on the leaf space of a holomorphic foliation. If the pseudo-metric is non-degenerate the foliation is called hyperbolic.

Lemma 9: If F is a hyperbolic foliation on a compact manifold M , and if there is a C^∞-foliation transverse to F then $\Gamma(M,\Theta_F) = 0$.

Theorem 10: Let F be a holomorphic foliation on a compact manifold M . Suppose that there is an $SL(p)$-foliation transverse to F , and that F is a cohomology product foliation. Then if F is hyperbolic it follows that F is stable under holomorphic deformations.

Let $\pi: M \to N$ be a G-principal bundle and let F be the foliation of M by the fibres of π. If N is complex F has a holomorphic structure induced by the complex structure on N. Suppose that $\omega: TM \to \underline{\underline{g}}$, is a connection form whose curvature lies in $\Gamma(Q^{(1,0)*} \wedge T_M^{*\mathbb{C}} \otimes \underline{\underline{g}})$. Then the induced map $\omega^*: \Lambda^{\bullet}\underline{\underline{g}}^* \to \Gamma(\Lambda^{\bullet}T_M^{*\mathbb{C}})$ takes closed left invariant forms on G to forms whose differentials lie in $\Gamma(Q^{(1,0)*} \wedge \Lambda^{\bullet}T_M^{*\mathbb{C}})$. Hence F is a cohomology product foliation. If ω is flat then the horizontal distribution defines an $SL(p)$-foliation transverse to F. Hence we have the following theorem.

Theorem 11: Let F be given as above. Then F is infinitesimally stable under holomorphic deformations if and only if $\Gamma(N,\Theta_N] = 0$. Further if ω is flat and $\Gamma(N,\Theta_N) = 0$ then F is stable under holomorphic deformations.

Bibliography

1. T. Duchamp and M. Kalka, "Stability Theorems for Holomorphic Foliations," preprint.

2. ______, "Deformation Theory for Holomorphic Foliations," J. Diff. Geom., to appear.

3. ______, "Holomorphic Foliations and the Kobayashi Metric," Proc. A.M.S. 67 (1977) p. 117-122.

4. R. S. Hamilton, "Deformation Theory for Foliations, " preprint.

5. J. Heitsch, "A Cohomology for Foliated Manifolds," Comm. Math. Helv. 50 (1975) p. 197-218.

6. F. Kamber - P. Tondeur, Invariant Differential Operators and the Cohomology of Lie Algebra Sheaves, Memoirs A.M.S. Vol. 68 (2).

7. M. Kuranishi, "New Proof for the Existence of Locally Complete Families of Complex Analytic Structures," Proceedings of the Conference on Complex Analysis, Minneapolis, Springer (1965) p. 142-154.

8. M. Mostow, Continuous Cohomology of Spaces with Two Topologies, Memoirs A.M.S. Vol 7 (1976) .

9. L. Nirenberg, "A Complex Frobenius Theorem," Seminar on Analytic Functions, Institute for Advanced Study, Princeton, (1957) p. 172-189.

10. B. Reinhart, "Harmonic Integrals on Almost Product Manifolds," Trans. A.M.S. 88 (1958) p.243-275.

11. I. Vaisman, "Variétés Riemannienne Feuilletées," Czechoslovak Math. J. 21 (1971) p. 46-75.

VANISHING THEOREMS AND STABILITY OF COMPLEX ANALYTIC FOLIATIONS

by

Joan Girbau

Facultat de Ciènces
Universitat Autònoma de Barcelona
Bellaterra (Barcelona) Spain

I.Vaisman formulated in [7] a vanishing theorem implying a rigidity result for some complex analytic foliate structures, but there was a sign error in the proof of that theorem. In this communication we shall give a new vanishing theorem implying a rigidity result for some complex foliations .

I wish to express my gratitude to Professor I. Vaisman for several useful suggestions.

§1. A vaninshing theorem

In the whole paper we shall refering to a compact complex manifold of complex dimension $n + m$ endowed with a complex analytic foliation $\mathcal{F}$ of complex dimension n . For this configuration, everything which is locally constant on the leaves of $\mathcal{F}$ will be labelled foliate and the notation c.a.f. will mean complex analytic foliate. We suppose that $\mathcal{F}$ is defined by an adapted atlas (U, z^a, z^u) (index convention: $a,b... = 1,...,n; u,v... = n+1,...,n+m$) where z^a = const. define the leaves. Let us suppose that we have a Hermitian bundle-like metric on M given locally by $g = g_{a\bar{b}}\, dz^a\, \overline{dz^b} + g_{u\bar{v}}\, \theta^u\, \overline{\theta^v}$

We shall design by $\left\{ \frac{\partial}{\partial z^a} , Z_a \right\}$ the dual base of $\left\{ dz^a, \theta^u \right\}$. We shall say that g is a <u>bundle-like pseudo-Kähler metric</u> iff $dw' = 0$ where

$$w' = \sqrt{-1}\; g_{a\bar{b}}\; dz^a \wedge \overline{dz^b}$$

(which gives a globally defined form).

Let $E \longrightarrow M$ be a c.a.f. vector bundle with r-dimensional fibres endowed with a <u>foliate</u> Hermitian metric (index convention: A,B... = 1,...,r). We have the usual Laplace operator Δ''_E acting on E-valued forms. Let us desing by $H_b^{p,q}(M,E)$ the space of E-valued <u>base-like</u> (p,q)-forms φ such that $\Delta''_E\, \varphi = 0$. Let us design by $\Omega_b^p(E)$ the sheaf of germs of holomorphic E-valued <u>base-like</u> (p,0)-forms.

Let $E^* \longrightarrow M$ be the dual bundle. Denote by $p: P(E^*) \longrightarrow M$ the projective bundle defined by the 1-dimensional subspaces of the fibres of E^* . $P(E^*)$ carries a complex analytic foliation of codimension $n+r-1$ whose leaves are covering spaces of the leaves of $\mathcal{F}$ on M . Let $p^*(E)$ be the pull-back of E by p on $P(E^*)$ and let S be the subbundle of $p^*(E)$ consisting of the pairs (y_z, u_z) $(z \in M,\ y_z \in E_z^*,\ u_z \in E_z)$ for which $y_z(u_z) = 0$. Then the quotient bundle $Q(E) = p^*(E)/S$ is a c.a.f. line bundle over $P(E^*)$. Le Potier's isomorphism [6] can be generalized to this situation and we have

$$H^q(M, \Omega_b^p(E)) \cong H^q(P(E^*), \Omega_b^p(Q(E)))$$

We define the <u>canonical bundle</u> of M to be the line bundle obtained by the products $U_\alpha \times C$ identifying $(p,z) \in U_\alpha \times C$ to

$$\left(p, \frac{\partial(z_\rho^1 \ldots z_\rho^n)}{\partial(z_\alpha^1 \ldots z_\alpha^n)}\right) \quad U_\rho \times C ,$$

$p \in U_\alpha \cap U_\beta$. We shall denote this bundle by $K(M)$. ($K(M)$ is c.a.f.). In order to enunciate our first main theorem we give the following definition. We shall say that the couple (M,E) satisfies the condition $C(p,q)$ if we have

$$H^q(M, \Omega_b^p(E)) = H^q(D^{p,\cdot}(E), d'')$$

where $D^{p,q}(E)$ denotes the space of base-like E-valued (p,q)-forms.

Theorem 1.- Let M be a compact complex manifold of dimension $n+m$ with a complex analytic foliation $\mathcal{F}$ of codimension n endowed with a bundle-like pseudo-Kähler metric g . Let $E \longrightarrow M$ be a c.a.f. r-dimensional vector bundle with a foliate Hermitian metric h. Let (Θ_A^B) be the curvature matrix of the connexion associated to h and (Ω_a^b) the transversal part of the curvature matrix associated to g . Let $R_{a\bar{b}}$ be the Ricci tensor corresponding to this connexion. Suppose that

$$Q(t,X) = (r+1)\, h_{AC}\, \Theta_B^C(Z_a, \bar{Z}_b)\, X^a\, \overline{X^b} + (R_{a\bar{b}} - \sum_A \Theta_A^A(Z_a, \bar{Z}_b))\, X^a\, \overline{X^b}\, h(t,t) > 0$$

for any local section t of E and any transversal vector field $X = \sum X^a Z_a$, $X \neq 0$. Suppose that for a given $q > 0$ the couple $(P(E^*)$, $K(P(E^*)^* \otimes Q(E))$ satisfies the condition $C(n+r-1,q)$. Then $H^q(M,\Omega_b^0(E)) = 0$.

The proof is based on the Le Potier's isomorphism and a vanishing theorem for line bundles.

In order to know when the condition $C(p,q)$ is fulfilled the following proposition is useful.

Proposition 1.- Let $\alpha_s^p(E)$ be the sheaf of germs of C^∞ E-valued foliate forms of foliate type $(s,p,0,0)$. If $H^1(M, \alpha_s^0(E))=0$, the couple (M,E) satisfies the condition $C(s,1)$.

§ 2. Stability

Let Θ be the sheaf of germs of vector fields on M such that the corresponding infinitessimal transformations preserve the foliation $\mathcal{F}$ .Let Ψ be the sheaf of germs of c.a.f. cross-sections of the transverse (orthogonal) bundle. $\mathcal{F}$ is called Kodaira-Spencer stable if $H^1(M,\Theta)=0$ (This stability was studied in [5]). $\mathcal{F}$ is called Duchamp-Kalka stable if $H^1(M,\Psi)=0$.This stability has been studied in [1] .

We can ennunciate our second main theorem.

Theorem 2.-Let M be a compact complex manifold of complex dimension n+m with a complex analytic foliation $\mathcal{F}$ of codimension n,endowed with a bundle-like pseudo-Kähler metric g.Suppose that

(a) The leaves of $\mathcal{F}$ are closed subsets of M and the first De Rham cohomology group of a generic leaf is trivial.

(b) The transversal part of the second connexion corresponding to g has positive bisectional holomorphic curvature.

Then $\mathcal{F}$ is Duchamp-Kalka stable.

The proof is based on theorem 1 and proposition 1.

§ 3. Examples

Example 1 .- Proposition 5 of [7] is true and can be obtained as corollary of the theorem 2.

Example 2 .- Let V be the projective space $P_n(C)$.Let $p:P(T(V)) \longrightarrow V$ be the projectivization of the tangent bundle T(V) of complex type (1,0) . Let M be the manifold P(T(V)) with the foliation $\mathcal{F}$ given by the fibres of that fibre-bundle.As an application of theorem 2 we can prove that $\mathcal{F}$ is Duchamp-Kalka stable.Let Φ be the sheaf of germs of holomorphic cross-sections of the tangent bundle of the leaves.Using a vanishing theorem of Bochner-Lichnerowicz [5] we can prove that $H^1(M,\Phi)=0$.This fact with the Duchamp-Kalka stability implies that $\mathcal{F}$ is also Kodaira-Spencer stable.

References

[1] T.Duchamp-M.Kalka,Stability Theorems for Holomorphic Foliations. Preprint University of Utah.

[2] J.Girbau,Fibrés semi-positifs et semi-négatifs sur une variété Kählerienne compacte.Annali di Mat.Pura ed Appl. 101 (1974) 171-183.

[3] R.S.Hamilton,Deformation theory for foliations.Preprint.

[4] K.Kodaira-D.C.Spencer,Multifoliate structures.Ann.of Math. 74 (1961) 52-100.

[5] A.Lichnerowicz,Variétés kählériennes et première classe de Chern . J.of Diff.Geom.1 (1967) 195-223.

[6] J.Le Potier, Annulation de la cohomologie à valeurs dans un fibré vectoriel holomorphe positif de rang quelconque.Math.Ann. 218 (1975) 35-53.

[7] I.Vaisman,A class of complex analytic foliate manifolds with rigid structure.J.of Diff.Geom. 12 (1977) 119-131.

POWER SERIES EXPANSIONS, DIFFERENTIAL GEOMETRY OF GEODESIC SPHERES AND TUBES, AND MEAN-VALUE THEOREMS

A. Gray and L. Vanhecke

In this note we shall describe some interesting power series associated with Riemannian manifolds. The basic problem associated with these power series is to express the coefficients in terms of geometric data such as curvature.

1. VOLUMES OF GEODESIC BALLS. Let (M,g) be an n-dimensional Riemannian manifold of class C^{ω} and let $m \in M$. Let $r > 0$ be so small that the exponential map $\exp_m$ is defined on a ball of radius r in the tangent space $T_m(M)$. We denote by $S_m(r)$ the volume of the *geodesic sphere*

$$G_m(r) = \{p \in M \mid d(m,p) = r\}$$

and by $V_m(r)$ the volume of the *geodesic ball*

$$B_m(r) = \{p \in M \mid d(m,p) \leqslant r\}.$$

In [GV1] we were concerned with the following problem: *To what extent do the functions* $V_m(r)$ *determine the Riemannian geometry of* M ? In particular we made the following conjecture:

(I) *Suppose* $V_m(r) = \omega r^n$ *for all* $m \in M$ *and all sufficiently small* $r > 0$. *Then* M *is flat.*

(Here ω = the volume of the unit ball in $\mathbb{R}^n$.) In [GV1] we formulated four other conjectures by comparing the volume functions $V_m(r)$ to those of the nonflat rank one symmetric spaces. In what follows we call these spaces the *model spaces*. These conjectures can be summarized as follows: Let M' be a model space with volume function $V'_m(r)$. Suppose that M is a Riemannian manifold with adapted holonomy group and with the property that $V_m(r) = V'_m(r)$ for all $m \in M$ and all sufficiently small r . Then M is locally isometric to M' .

Our method for attacking these conjectures was to use the power series expansion for $V_m(r)$. In [GV1] we determined the first four nonzero terms using the theory of normal coordinate vector fields and power series expansions of tensor fields on M as developed in [GR1]. The complete power series expansions for the volumes of geodesic balls in the model spaces may be derived by using Jacobi vector fields [GR1]. See also [H].

The conjectures for the quaternionic projective space and for the Cayley plane are true. For the other model spaces we answered the question affirmatively in many important special cases; however, the general problem remains open. For example, we proved in [GV1]:

THEOREM 1. Let M be a Riemannian manifold satisfying the assumptions of conjecture (I). Then M is flat in the following cases:

a. $\dim M \leqslant 3$;

b. M is Einsteinian, or more generally, M has nonnegative or nonpositive Ricci curvature;

c. M is conformally flat;

d. M is a product of surfaces;

e. M is the product of symmetric spaces of classical type;

f. M is a compact oriented four-dimensional manifold whose Euler characteristic and signature satisfy $\chi(M) \geqslant -\frac{3}{2}|\tau(M)|$.

2. THE SECOND FUNDAMENTAL FORM OF GEODESIC SPHERES. In [CV2] a power series expansion for the second fundamental form of $G_m(r)$ is given. Then, using the Gauss equation, it is possible to obtain a power series expansion for the integral of the scalar curvature of the geodesic spheres. Combining this with the expansion for $V_m(r)$ one obtains the following result:

THEOREM 2. Let M be a Riemannian manifold with adapted holonomy group such that for all $m \in M$ and all sufficiently small r, $G_m(r)$ has the same spectrum as the spheres in one of the model spaces. Then M is locally isometric to the model space.

We refer to [CV1,2] for other results related to the study of $G_m(r)$ as a submanifold of M.

3. MEAN-VALUE THEOREMS. In [GW] a power series expansion for the *mean value*

$$M_m(r,f) = \frac{\int_{G_m(r)} f * 1}{S_m(r)}$$

of an integrable real-valued function on M was found. It is possible to characterize certain Riemannian manifolds by the mean-values of harmonic functions. For example, Willmore [WI] proved

THEOREM 3. Let M be a Riemannian manifold such that for all $m \in M$ and all sufficiently small r

$$M_m(r,f) = f(m)$$

for all harmonic functions f. Then M is a harmonic space. The converse is also true.

See [WI] and [BS, pp.159-160].

Using the power series expansion for $M_m(r,f)$ Gray and Willmore [GW] generalized theorem 3 as follows:

THEOREM 4. Let M be a Riemannian manifold. Then, M is an Einstein space if and only if

$$M_m(r,f) = f(m) + O(r^6)$$

for all $m \in M$, and all sufficiently small r and all harmonic functions f.

The formula obtained for $M_m(r,f)$ is a generalization of the Pizzetti formula [CH], [P1,2]. In order to obtain this generalization the authors introduce some differential operators that are generalizations of powers of the Laplacian. The properties of these operators make it possible to compute one additional term in the power series expansion for $V_m(r)$ when M is a surface. This was done in [GV2]. There we formulated the following problem:

Is it possible that for all $m \in M$ *and all sufficiently small* r, *the function* $V_m(r)$ *is a polynomial other than* πr^2 ?

We know of no examples of surfaces other than E^2 for which $V_m(r)$ is a polynomial, but we have been unable to prove that such surfaces do not exist. However in [GV2] we did prove the following partial result.

THEOREM 5. Let M be a 2-dimensional Riemannian manifold of class C^ω and such that for all $m \in M$ and all sufficiently small r, $V_m(r)$ is a polynomial of degree $\leqslant 4$. Then M is locally flat.

To prove this theorem it was necessary to use the first five terms in the power series for $V_m(r)$. Further in [GV2] relations between the volume expansions and the isoperimetric inequality were given.

4. VOLUMES OF TUBES. The theory of normal coordinate vector fields and power series expansions is generalized in [GV3,4] for *Fermi coordinates* and *Fermi vector fields*. This is in particular useful to study volume functions of tubes about submanifolds.

Let P be a connected embedded submanifold of a Riemannian manifold M; we assume P, M and the embedding of class C^ω. Further we suppose that P has compact closure. By a *tube of radius* r (always supposed to be sufficiently small) about P we will mean the set

$$T(P,r) = \{\exp_m(x) \mid m \in P, x \in T_m^{\perp}(P), \|x\| \leq r\}.$$

Further we denote by $V_P(r)$ the n-dimensional volume of $T(P,r)$.

We determined in [GV4] the first three nonzero terms in the power series for $V_P(r)$. Further, using a relation between the Fermi vector fields and the Jacobi vector fields (see [GR2]) we determined the complete formulas for $V_P(r)$ when M is a flat space or a nonflat rank one symmetric space.

In [GV3] we generalized a theorem of Weyl for tubes about curves:

THEOREM 6. Let σ be a curve in a Euclidean space or a rank one symmetric space. Then the volume $V_\sigma(r)$ of a tube of radius r about σ in M is independent of the embedding; it depends only on r and the length of σ.

In fact Weyl proved this theorem for submanifolds P of arbitrary dimension in a Euclidean space or a sphere S^n [WE]. We proved in [GV4] that such a result is no longer true for the rank 1 symmetric spaces. For example, the volume of a tube about a surface P (real dimension 2) can be different for holomorphic and nonholomorphic embeddings in $\mathbb{C}P^n$.

Further in [GV3] we considered for tubes about curves conjectures analogous to conjecture (I). In contrast to the situation for geodesic spheres, it is possible to give complete affirmative answers for tubes. For example we proved

THEOREM 7. Let M be an n-dimensional Riemannian manifold with the property that for all small r and all sufficiently short geodesics we have

$$V_\sigma(r) = \frac{(\pi r^2)^{\frac{n-1}{2}}}{(\frac{n-1}{2})!} L(\sigma),$$

where $L(\sigma)$ denotes the length of σ. Then M is flat.

Similar theorems characterizing Einstein and rank 1 symmetric spaces are also given in [GV3].

The difference between the cases of geodesic balls and tubes can be explained intuitively as follows. Knowledge of the volumes of small geodesic balls yields information about the curvatures at each point of M. However, knowledge of the volumes of small tubes is stronger, because it yields curvature information in each direction at each point. This additional information turns out to be sufficient to obtain affirmative answers.

5. GENERALIZATIONS OF STEINER'S FORMULA. When P is an orientable hypersurface of an orientable Riemannian manifold M, one may consider the set

$$P_r = \{p \in T(P,r) \mid d(p,P) = r\}.$$

P_r has two components P_r^+ and P_r^-. These are the hypersurfaces parallel to P. By computing the volume functions $V_P^{\pm}(r)$ for the half-tubes, one obtains a refinement of the function $V_P(r)$ for a tube.

In [AGV] the terms of order less than or equal 5 in the power series expansion for $V_P^{\pm}(r)$ are given as well as the complete formulas when M is Euclidean space or a simply connected rank one symmetric space. By doing this we obtain generalizations of the Steiner formula in E^n and S^n [AL], [ST].

6. COMPARISON THEOREMS. Intuitively one believes that for Riemannian manifolds of nonnegative sectional curvature one has

$$V_m(r) \leqslant \omega r^n.$$

Bishop [BI] proved this result for all r less than the distance from m to its cut locus under the assumptions that M is complete and has nonnegative Ricci curvature. Furthermore, using the power series expansion for $V_m(r)$ it is easy to see that the inequality holds for sufficiently small r under the weaker assumption that the scalar curvature of M is positive at m.

What is the situation for submanifolds? By considering the power series for $V_P(r)$ it is shown in [GV4] that

$$V_P(r) \leqslant \frac{(\pi r^2)^{\frac{n-q}{2}}}{(\frac{n-q}{2})!}$$

for succiciently small r provided that the sectional curvature K^M of M is positive on P. Here $n = \dim M$ and $q = \dim P$.

There is also a global inequality for $V_P(r)$ corresponding to Bishop's inequality provided that $K^M \geqslant 0$ everywhere on M [GR2]. This inequality is a simultaneous generalization of Bishop's inequality and the Weyl tube formula.

Let R^P and R^M denote the curvature operators of P and M respectively.

THEOREM 8. Suppose $r > 0$ is not larger than the distance between P and its nearest focal point.

(i) If $K^M \geqslant 0$, then

$$V_P(r) \leqslant \frac{(\pi r^2)^{\frac{n-q}{2}}}{(\frac{n-q}{2})!} \sum_{c=0}^{[\frac{q}{2}]} \frac{k_{2c}(R^P - R^M) r^{2c}}{(n-q+2)(n-q+4)\dots(n-q+2c)}.$$

(ii) If $K^M \leqslant 0$, then "$\leqslant$" is replaced by "$\geqslant$" in this inequality.

The k_{2c} are integrals of polynomial expressions involving curvature closely related to the Gauss-Bonnet integrands. In fact let R_0 be a tensor field on M

of the same type as the curvature tensor field. Then

$$k_0(R_0) = \text{volume of } P \;,\quad k_2(R_0) = \frac{1}{2}\int_P \tau(R_0)\,dm \;,$$

$$k_4(R_0) = \frac{1}{8}\int_P \{\tau(R_0)^2 - 4\|\rho(R_0)\|^2 + \|R_0\|^2\}\,dm.$$

Here $\tau(R_0)$ and $\rho(R_0)$ denote the scalar and Rici curvatures of M. Moreover if $q = \dim P$ is even and P is compact one has

$$k_q(R^P) = (2\pi)^{q/2}\chi(P)$$

where $\chi(P)$ is the Euler characteristic of P.

REFERENCES

[AGV] E. ABBENA, A. GRAY & L. VANHECKE, "Steiner's formula for the volume of a parallel hypersurface in a Riemannian manifold," preprint I.H.E.S., Bures-sur-Yvette, 1979.

[AL] C.B. ALLENDOERFER, "Steiner's formula on a general S^{n+1}," Bull. Amer. Math. Soc. 54 (1958), 128-135.

[BI] R. BISHOP, "A relation between volume, mean curvature, and diameter," Amer. Math. Soc. Notices 10 (1963), 364.

[BS] A.L. BESSE, Manifolds all of whose geodesics are closed, Ergebnisse der Mathematik, vol. 93, Springer-Verlag, 1978.

[CH] R. COURANT & D. HILBERT, Methods of mathematical physics, vol. 2, Interscience, 1962.

[CV1] B.-Y. CHEN & L. VANHECKE, "Total curvatures of geodesic spheres," Arch. Math.(Basel) 32 (1979), 404-411.

[CV2] B.-Y. CHEN & L. VANHECKE, "Differential geometry of geodesic spheres," to appear.

[GR1] A. GRAY, The volume of a small geodesic ball in a Riemannian manifold," Michigan Math. J. 20 (1973), 329-344.

[GR2] A. GRAY, "Comparison theorems for volumes of tubes," to appear.

[GV1] A. GRAY & L. VANHECKE, "Riemannian geometry as determined by the volumes of small geodesic balls," Acta Math. 142 (1979), 157-198.

[GV2] A. GRAY & L. VANHECKE, "Oppervlakten van geodetische cirkels op oppervlakken," to appear in Med. Konink. Acad. Wetensch. Lett. Schone Kunst. België Kl. Wetensch..

[GV3] A. GRAY & L. VANHECKE, "The volume of tubes about curves in a Riemannian manifold," to appear.

[GV4] A. GRAY & L. VANHECKE, "The volume of tubes in a Riemannian manifold," to appear.

[GW] A. GRAY & T.J. WILLMORE, "Mean-value theorems for Riemannian manifolds," to appear.

[H] S. HELGASON, "The Radon transform on Euclidean spaces, compact two-point homogeneous spaces and Grassmann manifolds," Acta Math. 113 (1965), 153-180.

[P1] P. PIZZETTI, "Sulla media dei valori che una funzione dei punti dello spazio assume alla superficie di una sfera," Atti R. Accad. Rend. Cl. Sci. Fis. Mat. Natur. ser. 5, 18 (1909), 182-185.

[P2] P. PIZZETTI, "Sull significato geometrica del secundo parametro differenziale di una funzione sopra una superficie qualunque," Atti R. Accad. Rend. Cl. Sci. Fis. Mat. Natur. ser. 5, 18 (1909), 309-316.

[ST] J. STEINER, "Über parallele Flächen," Monatsbericht der Akademie der Wissenschaften zu Berlin (1840), 114-118. Also Werke, vol. 2 (1882), 171-176.

[WE] H. WEYL, "On the volumes of tubes," Amer. J. Math. 61 (1939), 461-472.

[WI] T.J. WILLMORE, "Mean-value theorems in harmonic Riemannian spaces," J. London Math. Soc. 25 (1950), 54-57.

Department of Mathematics
University of Maryland
College Park, Maryland 20742, U.S.A.

Departement Wiskunde
Katholieke Universiteit Leuven
Celestijnenlaan 200B
B-3030 LEUVEN (Belgium)

ON DISTANCE-DECREASING COLLINEATIONS

by Zvi Har'El

§1. *Introduction.* Collineations are defined in Eisenhart's "Non Riemannian Geometry" as "transformations of points of an affinely connected manifold into points of the manifold such that paths are transformed into paths". The same notion is evidently applicable when considering distinct manifolds as domain and range.

The search for distance decreasing maps was initiated by Ahlfors (1938) who proved that a holomorphic map of the hyperbolic unit disc (with Gaussian curvature $\equiv -1$) into a Hermitian surface whose Gaussian curvature is bounded above by -1, is distance-decreasing. This has been generalized to other classes of maps with a distance-decreasing property being achieved if the range is more curved (negatively) than the domain.

In this contribution, we describe a similar phenomenon in the class of collineation, and illustrate its global nature. Detailed proofs will appear in the paper "Projective mappings and distortion theorems", to be published shortly in the Journal of Differential Geometry.

§2. *Projective maps.* A *projective map*, or a *collineation*, of one manifold into another (both endowed with fixed symmetric affine connections) is a map $f : M \to M'$ which preserves paths. That is, for every path γ in M, $f\circ\gamma$ is a path in M'. We recall that a *path* is a reparametrized geodesic, i.e., a curve $\gamma : I \to M$ for which there exists a reparametrization (*"an affine parameter"*) $\varphi : I \to \mathbb{R}$ such that $\gamma\circ\varphi^{-1}$ has a parallel velocity vector field. It is well known that *bijective collineations* of the real projective space $\mathbb{R}P^n$ are induced by linear transformations of an (n+1)-dimensional Euclidean space, and hence are diffeomorphisms. But, this is not the case even for S^n, the universal covering of $\mathbb{R}P^n$, as the bijection $f : S^n \to S^n$ given by

$$f(x) = x \quad \text{for } x \text{ rational} \;, \; f(x) = -x \quad \text{otherwise},$$

preserves great circles but is not even continuous ($x \in S^n$ is considered rational if such are all its coordinates with respect to the standard embedding of S^n in $\mathbb{R}^{n+1}$. See Brickell's (1974) paper reviewed in MR56#6563).Hence, we assume in the remainder of this paper that collineations are smooth and that bijective collineations - also called *projective transformations* — are diffeomorphisms.

The situation most-investigated classically is where the domain and range coincide point-wise, but have separate symmetric affine connections, ∇ and ∇' respectively. The identification map of (M,∇) onto (M,∇') is projective if and only if there exists a smooth 1-form σ on M with the following property: For any two vector fields X,Y on M, $\nabla'_X Y - \nabla_X Y = \sigma(X)Y + \sigma(Y)X$. In that case, we say that ∇ and ∇' are *projectively related.* This may be generalized as follows: Let M_f be the dense open submanifold of M on which rank f attains its maximum ($M_f = M$ if f has a constant rank).

Proposition: *Let* $f : (M,\nabla) \to (M',\nabla')$ *be a smooth map, the connections* ∇,∇' *being* *symmetric. If* f *is projective then there exists a smooth* 1*-form* σ *on* M_f *such that*

(1) $$\nabla'_X f_* Y - f_* \nabla_X Y = \sigma(X)\, f_* Y + \sigma(Y)\, f_* X.$$

Conversely, if (1) *holds with* σ *defined on* M, f *is projective.*

We omit the details of the proof. Nevertheless we remark, that the defining formula for σ is

(2) $$\nabla'_D\, f_*\dot\gamma = 2\sigma(\dot\gamma) f_*\dot\gamma$$

where γ is an arbitrary geodesic, $\dot\gamma$ is its velocity vector field, and ∇'_D is ∇'-differentiation with respect to the parameter along the path $f\circ\gamma$ in M'. This fails to define $\sigma(\dot\gamma)$ if $f_*\dot\gamma$ vanishes, i.e., on the kernel of f_*. This observation explains why we need ker f_* to be a smooth distribution, i.e., rank f to be constant — or otherwise we must restrict to M_f. It is unknown to the author whether there exist collineations with a non-constant rank.

In 1931, Whitehead defined a *projective parameter* on a geodesic $\gamma : I \to M$ as a solution $p : I \to \mathbb{R}$ of the Schwarzian differential equation

(3) $$Sp \equiv D\left(\frac{D^2 p}{Dp}\right) - \frac{1}{2}\left(\frac{D^2 p}{Dp}\right)^2 = \frac{2}{n-1} \operatorname{Ric}(\dot\gamma,\dot\gamma),$$

where n = dim M and Ric is the Ricci tensor. It is classically known that projective parameters are defined up to a fractional linear transformation $\tilde p = \frac{ap+b}{cp+d}$ $(a,b,c,d \in \mathbb{R})$, and a projective change of the symmetric connection preserves both paths and their projective parameters. This property holds for general bijective collineations as well, as may be shown from the formula (valid in M_f as a consequence of (1)):

(4) $$f^*\operatorname{Ric}' = \operatorname{Ric} - d\sigma - (n-1)(\nabla\sigma - \sigma\otimes\sigma).$$

We remark that the clue to the proof is the following observation: If φ is an affine parameter for the path $f\circ\gamma$, one may readily compute from (2) that $2\sigma(\dot\gamma) = D^2\varphi / D\varphi$ and hence $(\nabla\sigma - \sigma\otimes\sigma)(\dot\gamma,\dot\gamma) = \frac{1}{2} S\varphi)$.

Examples: (a) Hyperbolic space H^n: In the Poincaré model, paths are circles orthogonal to the rim, and a convenient projective parameter is p = tanhs, where s is the hyperbolic arc length. This parameter has the range (-1,1), and any other parameter is obtained by a fractional linear transformation. Note that in the Klein model, paths are (segments of) straight lines, and p may be taken as Euclidean arc length.
(b) Using central projection, one can map S^n (minus the equator) or $\mathbb{R}P^n$ (minus the points at infinity) on $\mathbb{R}^n$ (the tangent hyperplane at the pole). This is a bijective collineation, hence the Euclidean arc length in $\mathbb{R}^n$ induces projective parameters in S^n and $\mathbb{R}P^n$, with the range $(-\infty, \infty)$.

As we are going to see below, formula (4) is crucial in proving distance- or volume-decreasing properties of collineation. We have mentioned that the validity of (4) in the bijective case is responsible for the preservation of the projective parameters defined by (3). To control the general case, we define *strong collineations* (or *strongly projective maps*) as collineations which preserve the projective parameters.

Proposition: Let $f : M \to M'$ *be a strong collineation of manifolds with symmetric affine connections. Then, for any* $v \in TM_f$

$$(f^*Ric)(v,v) = Ric(v,v) - (n-1)(\nabla\sigma - \sigma\otimes\sigma)(v,v). \tag{5}$$

Note, that if rank f is not constant, one might prefer to replace $(\nabla\sigma - \sigma\otimes\sigma)(v,v)$ by $\frac{1}{2} S\varphi|_0$, where φ is an affine parameter for the path $t \mapsto (f\circ\exp)(tv)$, and hence interpret (5) for any v with $f_*v \neq 0$.

§3. *Distortion theorems.* Having defined collineations for arbitrary manifolds with symmetric affine connections, we now consider Riemannian manifolds, wishing to investigate how collineations distort volume and distance. The main results are as follows:

Theorem 1: Let $f : M \to M'$ *be a collineation of* n*-dimensional Riemannian manifolds,* M *being complete. If the Ricci curvature of* M *is bounded below by a constant* $-A$, *and the Ricci curvature of* M' *is bounded above by a constant* $-B < 0$, *then either* f *is totally degenerate, or* $A > 0$ *and* f *is volume decreasing up to a constant* $(A/B)^{n/2}$.

Theorem 2: Let $f : M \to M'$ *be a strong collineation of Riemannian manifolds,* M *being complete. If the Ricci curvature of* M *is bounded below by a constant* $-A$, *and the Ricci curvature of* M' *is bounded above by a constant* $-B < 0$, *then either* f *is constant, or* $A > 0$ *and* f *is distance decreasing up to a constant* $(A/B)^{1/2}$.

The sketch of the proof follows: Denoting $u = \left(\frac{f^* \mathrm{vol}'}{\mathrm{vol}}\right)^2$, the squared ratio of volume elements, one uses (4) and the curvature condition

$$Ric \geq -A \ , \ Ric' \leq -B < 0 \tag{6}$$

to show that

$$-Bu^{1/n} \geq -A -(n-1) \frac{\Delta u}{2n(n+1)u}$$

where Δ is the Laplace-Beltrami operator on M. We then apply the Yau-Omori maximum principle in the following form: For a complete Riemannian manifold M whose Ricci curvature is bounded below, a nonnegative smooth function u on M, and arbitrary positive constants α, δ, there exists a sequence of points $\{p_\nu\}$ on M such that

$$\lim u(p_\nu) = \sup u,$$

$$\lim(u(p_\nu) + \delta)^{-1-2\alpha}(\Delta u)(p_\nu) \leq 0.$$

We find that

$$\lim(u(p_\nu) + \delta)^{-1-2\alpha} u(p_\nu)(u(p_\nu))^{1/n} - (A(B)) \leq 0$$

and the conclusion follows for $\alpha < \frac{1}{2n}$, which forces sup u to be finite, and hence either $u \equiv 0$ or $0 < \sup u \leq (A/B)^n$.

In a similar manner, denoting $u = \frac{f^*ds'^2}{ds^2}$, the ratio of squared elements of arc length along a fixed geodesic, we use (5) and the curvature condition (6) to deduce

$$-Bu \geq -A -(n-1) \frac{D^2u}{u} .$$

The conclusion follows using Yau-Omori principle with $\alpha < \frac{1}{2}$.

As bijective collineations and their inverses are necessarily strong, we get:

Corollary: A projective transformation of negatively curved complete Einstein space is an isometry.

Examples: We remark that the most important conditions in both theorems are the curvature condition (6) and the completeness of the domain. We suspect that the former may be sometimes relaxed, as the following example shows: Let $f : H^2 \to \mathbb{R}^2$ be the inclusion map of the Klein model of the Hyperbolic unit disc (with Gaussian curvature $\equiv -1$) into the Euclidean plane. This is a collineation with a range whose Gaussian curvature vanishes identically; nevertheless, it is distance decreasing (with maximum ratio achieved at the origin).

On the other hand, the completeness is crucial for the validity of our distortion theorems, thus proving their global nature. To illustrate that, consider $g : D \to H^2$, the identification map of the Euclidean open unit disc onto the Klein model mentioned above. Here (6) is satisfied clearly; but, being the inverse of f above, g is clearly distance increasing (with the ratio becoming infinitely large on the rim). The reason is the incompleteness of D as an open submanifold of the Euclidean plane $\mathbb{R}^2$.

Department of Mathematics
Technion — Israel Institute of Technology
Haifa 32000, Israel

ON A PARAMETRIX FORM IN A CERTAIN V-SUBMERSION

Haruo Kitahara

For a foliated riemannian manifold M, the complex

$$0 \longrightarrow \Lambda^{0,0}(M) \xrightarrow{d''} \Lambda^{0,1}(M) \xrightarrow{d''} \cdots \xrightarrow{d''} \Lambda^{0,s}(M) \xrightarrow{d''} \cdots$$

is not elliptic. But if M is compact, we may define the Laplacian $\square''$ acting on $\Lambda^{0,s}(M)$ and discuss the harmonic integrals (cf. [6]).

Recently, S. Zucker [15] gives estimates for the parametrix form for the Laplacian on a complete, non-compact riemannian manifold.

For a foliated riemannian manifold M whose leaves are closed, we may construct a V-fibre bundle whose total space is M (cf. [4], [7]). It seems for us to be natural to regard M as a V-submersion over a V-manifold with isolated singularities (cf. [10], [11], [14]).

In this note, we define a riemannian V-submersion and some differential geometric properties. Moreover, we discuss a parametrix form for the Laplacian on a riemannian V-manifold with isolated singularities by means of differential geometric properties of a foliated riemannian manifold.

The parametrix form on a compact riemannian V-manifold has been obtained by W. J. Baily, Jr. [1], then riemannian V-manifolds in which we are interested are "complete" non-compact.

Our methods are owing to S. Zucker [15].

We shall be in C^{∞}-category. Manifolds are, topologically, paracompact, connected, Hausdorff spaces.

The author wishes to extend his hearty thanks to Mr. S. Yorozu who read critically the manuscript.

1. V-manifolds. We recall here briefly the definition of an n dimensional V-manifold (cf. [1], [4], [9]). By a C^∞-local uniformazing system (l.u.s.) $\{\tilde{U}, G, \varphi\}$ for an open subset U of a paracompact, connected, Hausdorff space B, we mean a collection of the following objects:

(1.1.1) $\tilde{U}$: a connected open set in $\mathbb{R}^n$.

(1.1.2) G : a finite group of C^∞-automorphisms of $\tilde{U}$.

(1.1.3) φ : a continuous map of $\tilde{U}$ onto U such that $\varphi \circ \sigma = \varphi$ for all $\sigma \in G$, and φ induces a homeomorphism of $\tilde{U}/G$ onto U.

Let $\{U, G, \varphi\}$ and $\{\hat{U}', G', \varphi'\}$ be l.u.s.'s for U and U' respectively such that $U \subset U'$. By an injection of $\{\tilde{U}, G, \varphi\}$ into $\{\tilde{U}', G', \varphi'\}$ we mean a diffeomorphism λ of $\tilde{U}$ into $\tilde{U}'$ such that there exists an isomorphism η of G into G' satisfying $\lambda \circ \sigma = \eta(\sigma) \circ \lambda$ and $\varphi = \varphi' \circ \lambda$.

An n dimensional C^∞-V-manifold consists of a paracompact, connected, Hausdorff space B and a family $\mathcal{F}$ of C^∞-l.u.s.'s for open subsets of B satisfying the following conditions:

(1.2.1) If $\{\hat{U}, G, \varphi\}$, $\{\hat{U}', G', \varphi'\} \in \mathcal{F}$ and $\varphi(U) \subset \varphi'(U')$, then there exists an injection of $\{\hat{U}, G, \varphi\}$ into $\{\hat{U}', G', \varphi'\}$.

(1.2.2) The open sets U, for which there exists a l.u.s. $\{\hat{U}, G, \varphi\} \in \mathcal{F}$, form a basis of open sets in B.

Let B be a C^∞-V-manifold. For any $b \in B$, we take a l.u.s. $\{\tilde{U}, G, \varphi\} \in \mathcal{F}$ such that $b \in \varphi(\tilde{U})$ and choose a $\tilde{b} \in \tilde{U}$ such that $\varphi(\tilde{b}) = b$. Then the structure of the isotropy subgroup $G_{\tilde{b}}$ of G at $\tilde{b}$ is independent of the choice of $\tilde{U}$ and $\tilde{b}$, and is uniquely determined by b . We call $G_{\tilde{b}}$ the "isotropy group at b ". An ordinary manifold is nothing other than a C^∞-V-manifold for which the isotropy group of each point reduces to the unit group. Let S be the set of all "singular points" of B , i.e. the points of B with non-trivial isotropy groups. Let $b \in S$ and $\tilde{b}$ and $G_{\tilde{b}}$ as above. Then, taking a suitable coordinate system about $\tilde{b}$, $G_{\tilde{b}}$ becomes a finite group of linear transformations. Hence, S is expressed locally by a finite union of linear submanifolds of $\tilde{U}$.

Let $(B_1, \mathcal{F}_1)$ and $(B_2, \mathcal{F}_2)$ be C^∞-V-manifolds. By a C^∞-V-manifold map, we mean a collection of maps $\{h_{\tilde{U}_1}\}$ ($\{\tilde{U}_1, G_1, \varphi_1\} \in \mathcal{F}_1$) satisfying the following conditions:

(1.3.1) There is a correspondence $\{\tilde{U}_1, G_1, \varphi_1\} \longrightarrow \{\tilde{U}_2, G_2, \varphi_2\}$ of $\mathcal{F}_1$ into $\mathcal{F}_2$ such that for any $\{\tilde{U}_1, G_1, \varphi_1\} \in \mathcal{F}_1$ there exists a C^∞-map $h_{\tilde{U}_1}$ of $\tilde{U}_1$ into $\tilde{U}_2$.

(1.3.2) Let $\{\tilde{U}_1, G_1, \varphi_1\}$, $\{\tilde{U}_1', G_1', \varphi_1'\} \in \mathcal{F}_1$ and $\{\tilde{U}_2, G_2, \varphi_2\}$, $\{\tilde{U}_2', G_2', \varphi_2'\} \in \mathcal{F}_2$ be the corresponding l.u.s.'s (in the sense of (1.1)) and $\varphi_1(\tilde{U}_1) \subset \varphi_1'(\tilde{U}_1')$. Then for any injection λ_1 of $\{\tilde{U}_1, G_1, \varphi_1\}$ into $\{\tilde{U}_1', G_1', \varphi_1'\}$ there exists an injection λ_2 of $\{\tilde{U}_2, G_2, \varphi_2\}$ into $\{\tilde{U}_2', G_2', \varphi_2'\}$ such that $\lambda_2 \circ h_{\tilde{U}_1} = h_{\tilde{U}_1'} \circ \lambda_1$.

It follows from (1.3) that there exists uniquely a continuous map h of B_1 into B_2 such that for any $\{\widetilde{U}_1, G_1, \varphi_1\} \in \mathcal{F}_1$, and for the corresponding $\{\widetilde{U}_2, G_2, \varphi_2\} \in \mathcal{F}_2$, $\varphi_2 \circ h_{\widetilde{U}_1} = h \circ \varphi_1$. We call h a C^∞-map of B_1 into B_2 defined by a C^∞-V-manifold map $h = \{h_{\widetilde{U}_1}\}$. In particular, $\mathbb{R}$ (the reals) is regarded as a C^∞-V-manifold defined by a single l.u.s. $\{\mathbb{R}, \{1\}, \{1\}\}$, then a C^∞-function on a C^∞-V-manifold B is defined as a C^∞-V-manifold map $B \longrightarrow \mathbb{R}$.

Let $(B, \mathcal{F})$ be an n dimensional C^∞-V-manifold. Suppose that $\widetilde{U}$ ($\{\widetilde{U}, G, \varphi\} \in \mathcal{F}$) is contained in $\mathbb{R}^n$ and fix a coordinate system ($x^1, \cdots, x^n$) on each $\widetilde{U}$. For any injection λ of $\{\widetilde{U}, G, \varphi\}$ into $\{\widetilde{U}', G', \varphi'\}$,

$$g_\lambda(b) := \frac{\partial x'^\alpha \circ \lambda}{\partial x^\beta} \quad \text{(the Jacobian matrix of } \lambda \text{ at b)}$$

is a C^∞-map of $\widetilde{U}$ into $GL(n:\mathbb{R})$, and the system $\{g_\lambda\}$ defines a V-bundle ($T(B), B, \pi, \mathbb{R}^n, GL(n:\mathbb{R})$) whose "fibre" $\pi^{-1}(b)$ ($b \in B$) is not always a vector space. This V-bundle is called the tangent vector bundle over B.

Let $b \in \varphi(\widetilde{U})$, $\{\widetilde{U}, G, \varphi\} \in \mathcal{F}$ and choose any $\widetilde{b} \in \widetilde{U}$ such that $\varphi(\widetilde{b}) = b$. Then $\pi^{-1}(b) \cong \mathbb{R}^n / \{g_\sigma(\widetilde{b}) \mid \sigma \in G_{\widetilde{b}}\}$. $\widetilde{b} \times \mathbb{R}^n$ may be identified with $T_{\widetilde{b}}$ (the tangent space to $\widetilde{U}$ at $\widetilde{b}$) by the correspondence:

$$\widetilde{b} \times \begin{pmatrix} a^1 \\ a^2 \\ \vdots \\ a^n \end{pmatrix} \longmapsto X := \sum_{\alpha=1}^{n} a^\alpha \partial/\partial x^\alpha.$$

Then, denoting by $T_{\widetilde{b}}^{G_{\widetilde{b}}}$ the linear subspace of $T_{\widetilde{b}}$ consisting of all $G_{\widetilde{b}}$-invariant vectors, $\pi^{-1}(b)$ contains a vector space T_b $:= \varphi*(T_{\widetilde{b}}^{G_{\widetilde{b}}})$ which is independent of the choice of $\widetilde{U}$ and $\widetilde{b}$. An element of T_b is called a tangent vector to B at b. A cross-section X of the tangent vector bundle $T(B)$ is called a vector field over B. $\widetilde{X}_{\widetilde{U}}$ being a G-invariant cross-section (i.e. a G-invariant vector field over $\widetilde{U}$), we have $\widetilde{X}_{\widetilde{U}}(\tilde{b}) \in T_{\tilde{b}}^{G}\tilde{b}$ and so $X(b) \in T_b$. Hence, $X(b)$ being a tangent vector at b for any $b \in B$, the set of all vector fields forms a vector space.

Let $(B,\mathscr{F})$ be a C^∞-riemannian V-manifold with a riemannian metric $(\ ,\ )$. By definition, to give a riemannian metric $(\ ,\ )$ on $(B,\mathscr{F})$ is to give a riemannian metric $(\ ,\ )_{\widetilde{U}}$ on each $\widetilde{U}$ ($\{\widetilde{U}, G, \varphi\} \in \mathscr{F}$) such that for any injection λ of $\{\widetilde{U}, G, \varphi\}$ into $\{\widetilde{U}', G', \varphi'\}$, $(\widetilde{X}, \widetilde{Y})_{\widetilde{U}} = (\lambda(\widetilde{X}), \lambda(\widetilde{Y}))_{\widetilde{U}'}$, where $\widetilde{X}, \widetilde{Y}$ are any vector fields on $\widetilde{U}$, and $\lambda(\widetilde{X}), \lambda(\widetilde{Y})$ corresponding vector fields on $\lambda(\widetilde{U}) \subset \widetilde{U}'$. In particular, each $(\ ,\ )_{\widetilde{U}}$ is G-invariant. The C^∞-riemannian V-manifold $(B,\mathscr{F})$ has the "Levi-Civita connection" ∇, that is, ∇ is defined by a G-invariant riemannian connection $\nabla_{\widetilde{U}}$ on each $\widetilde{U}$ ($\{\widetilde{U}, G, \varphi\} \in \mathscr{F}$) such that for any injection λ of $\{\widetilde{U}, G, \varphi\}$ into $\{\widetilde{U}', G', \varphi'\}$, $\lambda(\nabla_{\widetilde{U}\ \widetilde{X}}\widetilde{Y}) = \nabla_{\widetilde{U}'\ \lambda(\widetilde{X})}\lambda(\widetilde{Y})$. Then the curvature tensor $R_B := \{R_{\widetilde{U}}\}$ of ∇ is defined by

$$(R_{\widetilde{U}})(\widetilde{X},\widetilde{Y})\widetilde{Z} := \nabla_{\widetilde{U}\ \widetilde{X}}\nabla_{\widetilde{U}\ \widetilde{Y}}\widetilde{Z} - \nabla_{\widetilde{U}\ \widetilde{Y}}\nabla_{\widetilde{U}\ \widetilde{X}}\widetilde{Z} - \nabla_{\widetilde{U}\ [\widetilde{X},\widetilde{Y}]}\widetilde{Z}$$

for vector fields $\widetilde{X}, \widetilde{Y}, \widetilde{Z}$ on each $\widetilde{U}$. By the standard arguments R_B is well-defined on B.

Let $(B,\mathscr{F})$ be an n dimensional C^∞-riemannian V-manifold. For each $s = 0, 1, 2, \cdots, n$, a differential s-form ϕ on B is defined by a G-invariant s-form $\phi_{\widetilde{U}}$ on each $\widetilde{U}$ ($\{\widetilde{U}, G, \varphi\} \in \mathscr{F}$) such that $\phi_{\widehat{U}} = \phi_{\widetilde{U}'} \circ \lambda$ for any injection λ of $\{\widetilde{U}, G, \varphi\}$ into $\{\widetilde{U}', G', \varphi'\}$. Let $\Lambda^s(B)$ be the vector space of all differential s-forms on B. By the local expression we may define the exterior product $\wedge$ and the exterior differentiation d on $\wedge^s(B)$. And for any $\phi \in \wedge^s(B)$, we may define the (n-s)-form $*\phi \in \wedge^{n-s}(B)$ by $\phi_{\tilde{U}} \wedge * \phi_{\tilde{U}} := (\phi_{\tilde{U}}, \phi_{\tilde{U}})_{\tilde{U}}$. Then, if ϕ is differentiable, we mwy define the codifferential operator δ by $\delta\phi := (-1)^{ns+n+1} * d * \phi$. If either $\phi \in \wedge^s(B)$ or $\psi \in \wedge^{s+1}(B)$ has a compact support, we may define the global inner product $\langle \ , \ \rangle$ and $\langle d\phi, \psi \rangle = \langle \phi, \delta\psi \rangle$. The operator $\square := d\delta + \delta d$ acting on $\wedge(B) := \sum_{s \geqq 0} \wedge^s(B)$ is called the Laplacian.

2. Foliations with bundle-like metrics. Hereafter, adopt the indices and summation convention as following:

$$1 \leqq A, B, C, \cdots \leqq m \ , \ 1 \leqq i, j, k, \cdots \leqq r \ ,$$

$$r+1 \leqq \alpha, \beta, \gamma, \cdots \leqq m \ (\text{ or } \ 1 \leqq \alpha, \beta, \gamma, \cdots \leqq n) \ .$$

Let M be an m dimensional riemannian manifold with a riemannian metric $(\ , \ )$ and E a foliation on M of codimension n ($n := m - r > 0$). Let $E^{\perp}$ be the transversal (or normal) plane field to E with respect to $(\ , \ )$ of dimension r . Then there exist flat charts $\{(U;(x^i,x^\alpha))\}$ with respect to E such that E is defined by $dx^\alpha = 0$. On a flat chart $(U;(x^i,x^\alpha))$ there exist 1-forms w^i and vector fields v_α such that $\{w^i, dx^\alpha\}$ and $\{\partial/\partial x^i, v_\alpha\}$ form dual bases for the cotangent space and the tangent space at each point in U respectively. Then we may get

$$w^i = dx^i + A^i_\alpha \, dx^\alpha \quad \text{and} \quad v_\alpha = \partial/\partial x^\alpha - A^i_\alpha \, \partial/\partial x^i \ .$$

We may choose A^i_α such that $(\partial/\partial x^i, v_\alpha) = 0$, and so the riemannian metric $(\ , \ )$ has the local expression

$$ds^2 = g_{ij}(x^k,x^\gamma) \, w^i \cdot w^j + g_{\alpha\beta}(x^k,x^\gamma) \, dx^\alpha \cdot dx^\beta$$

where $g_{ij}(x^k,x^\gamma) := (\partial/\partial x^i, \partial/\partial x^j)$ and $g_{\alpha\beta}(x^k,x^\gamma) := (v_\alpha , v_\beta)$.

In particular, if the metric (,) has the local expression

$$ds^2 = g_{ij}(x^k, x^\gamma)\, w^i \cdot w^j + g_{\alpha\beta}(x^\gamma)\, dx^\alpha \cdot dx^\beta \quad ,$$

the metric (,) is called a bundle-like metric compatible with the foliation E .

Hereafter suppose that M has a bundle-like metric compatible with the foliation E .

According to I. Vaisman (cf. [4], [12], [13]) we define the second connection D on M induced from (,) as following:

(2.1.1) $$D_{\partial/\partial x^i} \partial/\partial x^j = \Gamma^k_{ij}\, \partial/\partial x^k \ , \quad D_{V_\alpha} \partial/\partial x^j = \Gamma^k_{\alpha j}\, \partial/\partial x^k \ ,$$

$$D_{\partial/\partial x^i} V_\beta = 0 \ , \qquad D_{V_\alpha} V_\beta = \Gamma^\gamma_{\alpha\beta} V_\gamma \ ,$$

(2.1.2) $$(\partial/\partial x^i)(\partial/\partial x^j, \partial/\partial x^k) = (D_{\partial/\partial x^i} \partial/\partial x^j, \partial/\partial x^k) + (\partial/\partial x^j, D_{\partial/\partial x^i} \partial/\partial x^k) \ ,$$

$$V_\alpha (V_\beta, V_\gamma) = (D_{V_\alpha} V_\beta, V_\gamma) + (V_\beta, D_{V_\alpha} V_\gamma) \ ,$$

(2.1.3) $$T(\partial/\partial x^i, \partial/\partial x^j) = 0 \ , \quad T(\partial/\partial x^i, V_\beta) = 0 \ ,$$

$$T(V_\alpha, \partial/\partial x^j) = 0 \ , \quad T(V_\alpha, V_\beta) = T^k_{\alpha\beta}\, \partial/\partial x^k \ ,$$

where T denotes the torsion tensor of D , that is, for any vector fields X, Y on M , $T(X,Y) := D_X Y - D_Y X - [X,Y]$. Note that the torsion T of D doesn't always vanish. Then we get

(2.2.1) $$\Gamma^k_{ij} = \Gamma^k_{ji} = \frac{1}{2} g^{\ell k} (\partial g_{\ell j}/\partial x^i + \partial g_{i\ell}/\partial x^j - \partial g_{ij}/\partial x^\ell) \ ,$$

(2.2.2) $$\Gamma^\gamma_{\alpha\beta} = \Gamma^\gamma_{\beta\alpha} = \frac{1}{2} g^{\varepsilon\gamma} (V_\alpha(g_{\varepsilon\beta}) + V_\beta(g_{\alpha\varepsilon}) - V_\varepsilon(g_{\alpha\beta})) \ ,$$

(2.2.3) $$\Gamma^k_{\alpha j} = \partial A^k_\alpha / \partial x^j \ .$$

And, we get

(2.3.1) $R^{i}_{\alpha AB} = 0$, $R^{\alpha}_{iAB} = 0$, $R^{\alpha}_{\beta k \ell} = 0$, $R^{\alpha}_{\beta k \gamma} = 0$,

(2.3.2) $R^{i}_{jk\ell} = \partial \Gamma^{i}_{\ell j}/\partial x^{k} - \partial \Gamma^{i}_{kj}/\partial x^{\ell} + \Gamma^{h}_{\ell j}\Gamma^{i}_{kh} - \Gamma^{h}_{kj}\Gamma^{i}_{\ell h}$,

(2.3.3) $R^{\alpha}_{\beta\gamma\delta} = v_{\gamma}(\Gamma^{\alpha}_{\delta\beta}) - v_{\delta}(\Gamma^{\alpha}_{\gamma\beta}) + \Gamma^{\varepsilon}_{\delta\beta}\Gamma^{\alpha}_{\gamma\varepsilon} - \Gamma^{\varepsilon}_{\gamma\beta}\Gamma^{\alpha}_{\delta\varepsilon}$,

where R denotes the curvature tensor of D , that is, for any vector fields X, Y, Z on M ,

$$R(X,Y)Z := D_X D_Y Z - D_Y D_X Z - D_{[X,Y]}Z \quad ,$$

and, in local expression

$$R(X_A, X_B)X_C := R^{D}_{CAB}\, X_D \ , \quad X_A := \partial/\partial x^{i} \quad \text{or} \quad v_{\alpha} \ .$$

Hereafter we denote $R^{\perp}(\cdot,\cdot)$ the linear transformation of the sections $\Gamma(E^{\perp})$ of $E^{\perp}$ into itself.

A differentiable curve $c(t) : [0,1] \longrightarrow M$ is called to be transversal if $\dot{c}(t) \in \Gamma(E^{\perp})$ where $\dot{c}(t)$ denotes the differential with respect to the parameter t . Let $c(t)$ be a transversal curve in M . Then, taking its local expression $c(t) = (c^{i}(t), c^{\alpha}(t))$ we get

$$\begin{aligned} \dot{c}(t) &= \dot{c}^{i}(t)\,\partial/\partial x^{i} + \dot{c}^{\alpha}(t)\,\partial/\partial x^{\alpha} \\ &= (\ \dot{c}^{i}(t) + A^{i}_{\alpha}\,\dot{c}^{\alpha}(t)\ )\,\partial/\partial x^{i} + \dot{c}^{\alpha}(t)\, v_{\alpha} \\ &= \dot{c}^{\alpha}(t)\, v_{\alpha} \quad (\text{ by the transversality of } c(t)\ ). \end{aligned}$$

Hence, a transversal curve $c(t)$ is called a geodesic if

(2.4) $$D_{\dot{c}(t)}\,\dot{c}(t) = \left(\frac{d^2 c^{\alpha}}{dt^2} + \Gamma^{\alpha}_{\beta\gamma}\frac{dc^{\beta}}{dt}\frac{dc^{\gamma}}{dt}\right) v_{\alpha} := 0 \quad .$$

Suppose that for any $p \in M$ and any $X \in E^{\perp}_{p}$ we have the transversal geodesic $c(t) : [0,1] \longrightarrow M$ with the initial vector X , that is,

$$\begin{cases} \dfrac{d^2c^\alpha}{dt^2} + \Gamma^\alpha_{\beta\gamma}\dfrac{dc^\beta}{dt}\dfrac{dc^\gamma}{dt} = 0 \quad , \quad \left.\dfrac{dc^\alpha}{dt}\right|_{t=0} = X^\alpha \quad , \\ \\ \dfrac{dc^i}{dt} + A^i_\alpha \dfrac{dc^\alpha}{dt} = 0 \quad . \end{cases}$$

Then, we define a map $\exp^\perp_p$ of $E^\perp_p$ into M by

(2.5) $\qquad \exp^\perp_p X := c(1)$

which is called the transversal exponential map at p. Note that if M is complete in the bundle-like metric, $\exp^\perp_p tX$ may be defined on $(-\infty,+\infty)$. In fact, a transversal geodesic with respect to the Levi-Civita connection induced from the bundle-like metric coincides with respect to the second connection (cf. [5]).

For any $p \in M$, let $S(p)$ be the slice through p defined by $x^\alpha = 0$ in a flat chart $(U;(x^i,x^\alpha))$ around p. For any $q \in S(p)$, if we choose a suitable neighborhood U_q of the vector $(0,\cdots,0)$ in $E^\perp_q$, $\exp^\perp_q : U_q \longrightarrow \exp^\perp_q(U_q) \cap U$ is a diffeomorphism, and so, there exists a unique vector $u \in U_q$ such that $\exp^\perp_q u = y$ for any $y \in \exp^\perp_q(U_q)$. We correspond y with (y^α) where $u = y^\alpha v_\alpha|_q$. Then for any $q' \in S(p)$, letting $u' := y^\alpha v_\alpha|_{q'}$, $\exp^\perp_q u$ and $\exp^\perp_{q'} u'$ are contained in same slice through y. Hence we may identify the zero section of $E^\perp|_{S(p)}$ with $S(p)$. Therefore, taking a suitable neighborhood W of the zero section of $E^\perp|_{S(p)}$, we may regard $\exp^\perp$ as a diffeomorphism of W to a flat chart $(U;(x^i,x^\alpha))$. $(U;(x^i,x^\alpha))$ is called a transversal normal flat chart, and any $x \in U$ is written as $x = (x^i,x^\alpha)$ where $c(t) = \exp^\perp_q tx^\alpha v_\alpha|_q$ is a geodesic orthogonal to $S(p)$ at $q \in S(p)$ and $(x^i,0)$ are the coordinates of $c \cap S(p)$. (x^i,x^α) are called the transversally normal flat coordinates on U.

Let $c(t)$ be a transversal geodesic in M and Y a transversal Jacobi field along $c(t)$, that is, Y is a transversal vector field along $c(t)$ satisfying the equation:

$$(2.6) \qquad D_{\dot{c}} D_{\dot{c}} Y + R(Y,\dot{c})\dot{c} = 0 .$$

In fact, let $\{ e_1, \cdots, e_n = \dot{c} \}$ be a parallel orthonormal base of $\Gamma(E^{\perp})$ along the transversal geodesic $c(t)$ and $Y(t) = y^{\alpha}(t)e_{\alpha}$, $y^{\alpha}(t)$ are real valued functions. Then $D_{\dot{c}}Y = \dot{y}^{\alpha}e_{\alpha}$ and $D_{\dot{c}}D_{\dot{c}}Y = \ddot{y}^{\alpha}e_{\alpha}$. Hence, Y is a transversal Jacobi field if and only if

$$\ddot{y}^{\alpha} + y^{\beta} R^{\alpha}_{n\beta n} = 0 .$$

From the uniqueness theorem for solutions of the second order differential equations, the transversal Jacobi field in the above sense is well defined.

Let $K : [a,b] \times]-\varepsilon,\varepsilon[\longrightarrow M$ ($\varepsilon > 0$) be a variation of a transversal geodesic $c = c_0$ such that $K(s,t)$ is a transversal curve for each s and each t respectively. We have

$$L(c_t) := \int_a^b [E(s,t)]^{1/2} ds$$

where $E(s,t) := (K_*(\partial/\partial s), K_*(\partial/\partial s))_{K(s,t)}$. Suppose that, for all $t \in]-\varepsilon,\varepsilon[$, $K(a,t) = p$ and $K(b,t) = p'$. Then, from the minimality of c_0, we have

$$(dL(c_t)/dt)_{t=0} = 0 .$$

Now, we calculate

$$dL(c_t)/dt = \int_a^b \partial E^{1/2}/\partial t \, ds = \frac{1}{2} \int_a^b E^{-1/2} \partial E/\partial t \, ds .$$

For $t = 0$, $E(s,t) = 1$. Then we have

$$(dL(c_t)/dt)_{t=0} = \frac{1}{2} \int_a^b (\partial E/\partial t)_{t=0} \, ds .$$

However, letting $X := K_*(\partial/\partial s)$ and $Y := K_*(\partial/\partial t)$, we have

$$\partial E/\partial t = 2\,\partial(\,Y,\,X\,)/\partial s + 2\,(\,T(Y,X),\,X\,)$$

$$= 2\,\partial(\,Y,\,X\,)/\partial s \qquad (\text{ by } (2.1.3)\,).$$

Hence, we have

$$(\,dL(c_t)/dt\,)_{t=0} = (\,Y(0),\,\dot{c}\,)\Big|_a^b - \int_a^b (\,Y(0),\,D_{\dot{c}}\dot{c}\,)\,ds$$

$$= -\int_a^b (\,Y(0),\,D_{\dot{c}}\dot{c}\,)\,ds$$

where $Y(0) := (\,Y\,)_{t=0}$. Therefore, we have "the first variation formula":

$$(\,dL(c_t)/dt\,)_{t=0} = -\int_a^b (\,Y(0),\,D_{\dot{c}}\dot{c}\,)\,ds\,.$$

Next, suppose that $K(a,t)$ and $K(b,t)$ are transversal geodesics. We calculate the second variation $(\,d^2L(c_t)/dt^2\,)_{t=0}$:

$$\partial^2 E/\partial t^2 = 2\,\partial(\,D_{\partial/\partial t}X,\,X\,)/\partial t$$

$$= 2\,(\,D_{\partial/\partial t}D_{\partial/\partial t}X,\,X\,) + 2\,(\,D_{\partial/\partial t}X,\,D_{\partial/\partial t}X\,)$$

$$= 2\,(\,D_{\partial/\partial t}D_{\partial/\partial s}Y,\,X\,) + 2\,(\,D_{\partial/\partial s}Y,\,D_{\partial/\partial s}Y\,)$$

$$(\text{ by } (2.1.1) \text{ and } (2.1.3)\,)$$

$$= 2\,(\,D_{\partial/\partial s}D_{\partial/\partial t}Y + R(Y,X)Y,\,X\,)$$

$$+ 2\,(\,D_{\partial/\partial s}Y,\,D_{\partial/\partial s}Y\,)\quad .$$

However, we have

$$(\,D_{\partial/\partial s}D_{\partial/\partial t}Y,\,X\,) = \partial(\,D_{\partial/\partial t}Y,\,X\,)/\partial s$$

$$-\,(\,D_{\partial/\partial t}Y,\,D_{\partial/\partial s}X\,)$$

and $(\,D_{\partial/\partial s}X\,)_{t=0} = 0$ for c_0 is a transversal geodesic. Then we have

$$(d^2L(c_t)/dt^2)_{t=0}$$

$$= (D_{\partial/\partial t}Y, X)\Big|_a^b$$

$$+ \int_a^b [(D_{\partial/\partial s}Y, D_{\partial/\partial s}Y) - (R(Y,X)X, Y)$$

$$- (X, D_{\partial/\partial s}Y)]_{t=0}\, ds .$$

Moreover $(D_{\partial/\partial t}Y)_{s=a} = (D_{\partial/\partial t}Y)_{s=b} = 0$ for $K(a,t)$ and $K(b,t)$ are supposed to be transversal geodesics. Therefore we have "the second variation formula":

$$(d^2L(c_t)/dt^2)_{t=0}$$

$$= \int_a^b [(D_{\partial/\partial s}Y, D_{\partial/\partial s}Y) - (R(Y,X)X, Y)$$

$$- (X, D_{\partial/\partial s}Y)]_{t=0}\, ds .$$

3. The (0,s)-forms on a foliated manifold with a bundle-like metric. On an m dimensional foliated manifold M with a foliation E, we may have a differential form into components as following: Any p-form ϕ may be expressed locally as

$$\sum_{\substack{i_1<\cdots<i_t \\ \alpha_1<\cdots<\alpha_s}} \sum_{t+s=p} \phi_{i_1\cdots i_t\alpha_1\cdots\alpha_s}(x^k, x^\gamma)\, w^{i_1}\wedge\cdots\wedge w^{i_t}\wedge dx^{\alpha_1}\wedge\cdots\wedge dx^{\alpha_s} .$$

We define $\pi_{t,s}\phi$ to be the sum of all those terms with a fixed t and s. As under change of flat charts $[\{dx^\alpha\}]$ goes into $[\{dx^{*\alpha}\}]$ and $[\{w^i\}]$ goes into $[\{w^{*i}\}]$, this operator $\pi_{t,s}$ is independent of the choice of flat charts. Here $[\{\cdot\}]$ denotes the vector space generated by the set $\{ \cdot \}$. $\pi_{t,s}\phi$ is called the component of type (t,s) of ϕ. The type decomposition of forms induces a type decomposition of the exterior derivative d by the rule

$$(\Pi_{u,v}\, d\,)\phi = \Sigma_{t,s} \Pi_{u+t,v+s}\, d\, \Pi_{t,s}\phi \quad .$$

Let $\Pi_{1,0}d = d'$ and $\Pi_{0,1}d = d''$. In general, there will be a component $\Pi_{-1,2}d$; as we are interested only in forms of type $(0,s)$, we shall not introduce a notion for this component.

Proposition 3.1 (cf. [6]). If ϕ is of type $(0,s)$ then $d\phi = d'\phi + d''\phi$. Moreover, $d'\phi = 0$ if and only if ϕ depends only upon (x^α) in the sense that locally

$$\phi = \sum_{\alpha_1<\cdots<\alpha_s} \phi_{\alpha_1\cdots\alpha_s}(x^\alpha)\, dx^{\alpha_1}\wedge\cdots\wedge dx^{\alpha_s} \quad , \tag{3.1}$$

which is called a basic (0,s)-form.

Let $\Lambda^{0,s}(M)$ be the space of all basic (0,s)-forms. Restricted to $\Lambda^{0,*}(M) := \Sigma_{s\geq 0}\Lambda^{0,s}(M)$, $d''^2 = d^2 = 0$.

Hereafter we suppose that M has a bundle-like metric compatible with E .

B. L. Reinhart [6] introduced the $*''$-operation on $\Lambda^{0,s}(M)$ and defined by

$$(*''\phi)_{\beta_1\cdots\beta_{n-s}} = \sum_{\alpha_1<\cdots<\alpha_s} \delta\binom{1\cdots\cdots\cdots\cdots n}{\alpha_1\cdots\alpha_s\beta_1\cdots\beta_{n-s}}(\det(g_{\alpha\beta}))^{1/2} \tag{3.2}$$

$$g^{\alpha_1\nu_1}\cdots g^{\alpha_s\nu_s}\phi_{\nu_1\cdots\nu_s} \quad .$$

According to B. L. Reinhart [6], we may define a L_2-inner product on the space $L_2^{0,s}(M)$ of all global square-summable real basic (0,s)-forms on M by

$$\langle\phi,\psi\rangle := \int_M \phi\wedge *''\psi\, dx^1\wedge\cdots\wedge dx^r \tag{3.3}$$

(cf. [3]). The differential operator d'' maps $\Lambda^{0,s}(M)$ into

$\Lambda^{0,s+1}(M)$. We define $\delta'' : \Lambda^{0,s}(M) \longrightarrow \Lambda^{0,s-1}(M)$ by

$$\delta''\phi := (-1)^{ns+n+1} *'' d'' *'' \phi \quad .$$

Then, we have

$$\langle d''\phi , \psi \rangle = \langle \phi , \delta''\psi \rangle$$

for $\phi \in L_2^{0,s}(M)$, $\psi \in L_2^{0,s+1}(M)$. We define the Laplacian acting on $\Lambda^{0,*}(M)$ by

(3.4) $$\Box'' := d''\delta'' + \delta'' d'' \quad .$$

We express the operators d'' , δ'' and $\Box''$ in terms of the second connection D : For any $\phi = \sum_{\alpha_1 < \cdots < \alpha_s} \phi_{\alpha_1 \cdots \alpha_s}(x^\alpha)\, dx^{\alpha_1} \wedge \cdots \wedge dx^{\alpha_s}$ $\in \Lambda^{0,s}(M)$,

(3.5.1) $$(d''\phi)_{\beta_1 \cdots \beta_{s+1}} = \sum_{\nu=1}^{s+1} (-1)^{\nu-1} D_{\beta_\nu} \phi_{\beta_1 \cdots \hat{\beta}_\nu \cdots \beta_{s+1}}$$

(3.5.2) $$(\delta''\phi)_{\beta_1 \cdots \beta_{s-1}} = - D^\gamma \phi_{\gamma \beta_1 \cdots \beta_{s-1}}$$

(3.5.3) $$\begin{aligned}(\Box''\phi)_{\alpha_1 \cdots \alpha_s} &= - D^\alpha D_\alpha \phi_{\alpha_1 \cdots \alpha_s} \\ &+ \sum_{k=1}^{s} (-1)^k R^{\gamma \cdot \alpha \cdot}{}_{\cdot \alpha \cdot \alpha_k} \phi_{\gamma \alpha_1 \cdots \hat{\alpha}_k \cdots \alpha_s} \\ &+ 2 \sum_{h<k} (-1)^{h+k} R^{\gamma \cdot \alpha \cdot}{}_{\cdot \alpha_h \cdot \alpha_k} \phi_{\alpha \gamma \alpha_1 \cdots \hat{\alpha}_h \cdots \hat{\alpha}_k \cdots \alpha_s} \quad .\end{aligned}$$

4. Riemannian V-submersions. Let M be an m dimensional riemannian manifold with a riemannian metric $(\ ,\ )_M$ and $(B,\mathcal{F})$ an n dimensional riemannian V-manifold with a riemannian metric $(\ ,\ )_B$. By a riemannian V-submersion $\pi : M \longrightarrow B$ we mean a C^∞-V-manifold map $\pi := \{\pi_{\hat{U}}\}$ satisfying the following condition:

(4.1.1) For any $p \in M$, $(\pi_{\tilde{U}})_* : T_p \longrightarrow T_{\tilde{b}}$ ($b = \pi(p)$ and $b = \varphi(\tilde{b})$) is of maximal rank and an isometry.

(4.1.2) For any $b \in B$, $\pi^{-1}(b)$ is a connected (closed) r dimensional manifold ($r = m - n$). $\pi^{-1}(b)$ is called a fibre over b.

Note that the topology of $\pi^{-1}(b)$ doesn't always coincide with the induced one from M.

Hereafter we suppose that B is an n dimensional riemannian V-manifold with isolated singularities.

$B^\circ := \{ b' \in B \mid G_{\tilde{b}'} = \{1\} \}$ is open dense in B. For any $b' \in B^\circ$, all $\pi^{-1}(b')$ are homeomorphic and we denote F its typical one. If we define an action of $G_{\tilde{b}}$ on $\tilde{U}' \times F$ by

$$\begin{aligned} \tilde{U}' \times F &\longrightarrow \tilde{U}' \times F \\ (x, f) &\longmapsto (\sigma^{-1}x, \sigma^{-1}f) \end{aligned} .$$

$\pi^{-1}(b)$ is homeomorphic to $F/G_{\tilde{b}}$ (cf. [7]). Therefore M is a foliated manifold by fibres.

For any $b \in B$, choose a flat chart $(\tilde{U}; (x^i, x^\alpha))$ around any $p \in \pi^{-1}(b)$ such that $x^\alpha = 0$ define fibres, and a l.u.s. $\{ \tilde{U}', G', \varphi' \}$ such that $\pi_{\tilde{U}}(\tilde{U}) \subset \tilde{U}'$ and $\varphi' = (y^\alpha)$ is a coordinates on $\tilde{U}'$, then we have $y^\alpha = y^\alpha(x^k, x^\gamma)$.

Now, we shall prove that $(\ ,\ )_M$ is a bundle-like metric, so we use some notations in §2. $(\ ,\ )_M$ is of form

$$ds^2 = g_{ij}(x^k, x^\gamma)\, w^i \cdot w^j + g_{\alpha\beta}(x^k, x^\gamma)\, dx^\alpha \cdot dx^\beta .$$

As $(\pi_{\tilde{U}})_*$ is an isometry, we have

$$(X, Y)_M = ((\pi_{\tilde{U}})_* X, (\pi_{\tilde{U}})_* Y)_B$$

for any vector fields X, Y on M, and

$$(\pi_{\hat{U}})_*(v_\alpha) = (\pi_{\hat{U}})_*(\partial/\partial x^\alpha) - (A^i_\alpha \circ \pi_{\hat{U}}) \circ (\pi_{\hat{U}})_*(\partial/\partial x^i)$$

$$= (\pi_{\hat{U}})_*(\partial/\partial x^\alpha)$$

$$= (\partial y^\beta/\partial x^\alpha)(\partial/\partial y^\beta) .$$

Then we have

$$g_{\alpha\beta}(x^k,x^\gamma) := (\ v_\alpha ,\ v_\beta\ )_M = h_{\delta\theta} \frac{\partial y^\delta}{\partial x^\alpha}\frac{\partial y^\theta}{\partial x^\beta}$$

where $h_{\delta\theta} := (\ \partial/\partial y^\delta,\ \partial/\partial y^\theta\ )_B$. However, from

$$0 = (\pi_{\hat{U}})_*(\partial/\partial x^i) = (\partial y^\alpha/\partial x^i)(\partial/\partial y^\alpha) ,$$

we have $\partial y^\alpha/\partial x^i = 0$, that is, $y^\alpha = y^\alpha(x^\gamma)$. Then we have

$$\partial g_{\alpha\beta}(x^k,x^\gamma)/\partial x^i = \frac{\partial h_{\delta\theta}}{\partial y^\varepsilon}\frac{\partial y^\varepsilon}{\partial x^i}\frac{\partial y^\delta}{\partial x^\alpha}\frac{\partial y^\theta}{\partial x^\beta} + h_{\delta\theta}\frac{\partial^2 y^\delta}{\partial x^i\partial x^\alpha}\frac{\partial y^\theta}{\partial x^\beta} + h_{\delta\theta}\frac{\partial y^\delta}{\partial x^\alpha}\frac{\partial^2 y^\theta}{\partial x^i\partial x^\beta}$$

$$= 0 .$$

Then, $g_{\alpha\beta}(x^k,x^\gamma)$ are functions of any variables x^γ . Hence, we have

$$ds^2 = g_{ij}(x^k,x^\gamma)\ w^i\cdot w^j + g_{\alpha\beta}(x^\gamma)\ dx^\alpha\cdot dx^\beta$$

where $g_{\alpha\beta}(x^\gamma) := h_{\delta\theta}(\pi(x^\gamma))\ \partial y^\delta/\partial x^\alpha\cdot\partial y^\theta/\partial x^\beta$. Therefore we have the following:

Theorem 4.1. M _has a bundle-like metric compatible with the foliation by fibres._

M has the second connection D and B the Levi-Civita connection ∇_B induced from $(\ ,\ )_B$. Then we have

$$\pi_*(D_X Y^\perp) = \nabla_{B\ \pi_*(X)}(\pi_*(Y^\perp))$$

for any vector field X and any transversal vector field $Y^\perp$ on M . And $R^\perp$ on M corresponds to the curvature tensor of $_B$ via π_* .

We may define geodesics in any $\widetilde{U}'$ ($\{\widehat{U}', G, \varphi\} \in \mathcal{F}$) (cf. [1]). Choose a suitable flat chart U and a corresponding l.u.s. $\{\widehat{U}', G, \varphi\} \in \mathcal{F}$, then we may define the exponential map $\exp_B$ in U' and we have $\exp_p^{\perp} = \exp_{B,b}$ ($p \in \pi^{-1}(b) \cap U$). Note that if M is complete, $\exp^{\perp}$ may be globally defined, but $\exp_B$ is not always globally defined.

5. Parametrix forms. Let M be an m dimensional complete riemannian manifold with a riemannian metric (,) (or g) and $(B, \mathcal{F})$ an n dimensional riemannian C^∞-V-manifold with isolated singularities. Let $\pi : M \longrightarrow B$ be a riemannian V-submersion.

Proposition 5.1. $\Lambda^{0,s}(M) \cong \Lambda^s(B)$ (_isomorphic as vector spaces via the isometry_ π_*).

Proposition 5.2. _The following relations hold between_ $\Lambda^{0,s}(M)$ _and_ $\Lambda^s(B)$:

$$d''\pi^* = \pi^* d , \quad *''\pi^* = \pi^* * , \quad \delta''\pi^* = \pi^*\delta ,$$
$$\Box''\pi^* = \pi^*\Box .$$

In particular, $\phi \in \Lambda^s(B)$ _is_ $\Box$_-harmonic if and only if_ $\pi^*\phi \in \Lambda^{0,s}(M)$ _is_ $\Box''$_-harmonic._

For any $(x,y) \in B \times B$, we define $A(x,y) \in \Lambda^0(B \times B)$ by

$$A(x,y) := -\frac{1}{2}\,[\, r(\pi^{-1}(x), \pi^{-1}(y))\,]^2$$

where $r(\pi^{-1}(x), \pi^{-1}(y))$ is the transversal geodesic distance from $\pi^{-1}(x)$ to $\pi^{-1}(y)$. The value of r is well-defined (cf. [6]). Then we may regard $r(x,y)$ as the "geodesic distance" from x to y . Locally, letting $(U; (\tilde{x}^i, \tilde{x}^\alpha))$ be a flat chart on M , we have

$$r(x,y) = \int_0^1 [\, g_{\alpha\beta}(\hat{x}^\gamma) \frac{d\tilde{x}^\alpha}{dt} \frac{d\tilde{x}^\beta}{dt} \,]^{1/2} \, dt$$

where $c(t) = (\, \tilde{x}^\alpha(t) \,)$, $0 \leqq t \leqq 1$, is a transversal geodesic to the leaf $\pi^{-1}(x)$ at $c(0) \in \pi^{-1}(x)$ and $c(1) \in \pi^{-1}(y)$. As $\partial A/\partial \hat{x}^i = \partial A/\partial \tilde{y}^i = 0$, we may regard $A(x,y)$ as an element of $\Lambda^{0,0}(M \times M)$. Now $d_{\hat{x}} A = (\partial A/\partial \tilde{x}^i)\, d\hat{x}^i + (\partial A/\partial \tilde{x}^\alpha)\, d\hat{x}^\alpha = (v_\alpha(A))\, d\hat{x}^\alpha$. Then we have $d_{\tilde{x}} A = d''_{\tilde{x}} A$ and $d_{\hat{x}} d_{\hat{y}} A = d''_{\tilde{x}} d''_{\tilde{y}} A$. Hence we have

$$\begin{aligned} d''_{\tilde{x}} d''_{\tilde{y}} A &= (\, v_\alpha(v_\beta(A)) \,)\, d\hat{x}^\alpha \cdot d\tilde{y}^\beta \\ &= g_{\alpha\beta}(\hat{x}^\gamma)\, d\hat{x}^\alpha \cdot d\tilde{y}^\beta + O(r) \end{aligned}$$

(for the definition of $O(r)$, see [8]).

We define

$$\alpha_s(x,y) := \frac{1}{s!} (\, d''_{\tilde{x}} d''_{\tilde{y}} A(x,y) \,)^s \quad .$$

Then we have

$$\alpha_s(x,y) = \frac{1}{(s!)^2} g_{\alpha_1 \cdots \alpha_s, \beta_1 \cdots \beta_s}(\hat{x})\, d\hat{x}^{\alpha_1} \wedge \cdots \wedge d\hat{x}^{\alpha_s} \cdot d\tilde{y}^{\beta_1} \wedge \cdots \wedge d\tilde{y}^{\beta_s}$$

$$+ O(r)$$

where $g_{\alpha_1 \cdots \alpha_s, \beta_1 \cdots \beta_s} = g_{\alpha_1\gamma_1} \cdots g_{\alpha_s\gamma_s} \delta^{\gamma_1 \cdots \gamma_s}_{\beta_1 \cdots \beta_s}$. Hence, regarding $\phi \in \Lambda^s(B)$ as $\phi \in \Lambda^{0,s}(M)$, since $\phi^{\beta_1 \cdots \beta_s}(\hat{y}) = \phi^{\beta_1 \cdots \beta_s}(\hat{x}) + O(r)$ and $\frac{1}{s!} g_{\alpha_1 \cdots \alpha_s, \beta_1 \cdots \beta_s}(\hat{x})\, \phi^{\beta_1 \cdots \beta_s}(\hat{x}) = \phi_{\alpha_1 \cdots \alpha_s}(\hat{x})$, we have

$$\alpha_s(x,y) \wedge *''_{\tilde{y}}\, \phi(y) = \phi(x) \cdot dV''_{\tilde{y}} + O(r)$$

where $dV''_{\tilde{y}} := (\, \det(g_{\alpha\beta}(\hat{y})) \,)^{1/2}\, d\tilde{y}^{r+1} \wedge \cdots \wedge d\tilde{y}^m$. Therefore we have following:

Proposition 5.3.

$$\alpha_s(x,y) \wedge *''_{\hat{y}} \phi(y) \wedge d\tilde{y}^1 \wedge \cdots \wedge d\tilde{y}^r = \phi(x) \cdot dV_{\hat{y}} + O(r)$$

<u>where</u> $dV_{\hat{y}} = dV''_{\hat{y}} \wedge d\tilde{y}^1 \wedge \cdots \wedge d\tilde{y}^r$.

Let ΔB be the diagonal set of $B \times B$. Then ΔB corresponds to the set $\Delta M := \{ (\tilde{x},\tilde{y}) \in M \times M \mid \pi(\hat{x}) = \pi(\tilde{y}) \}$. Let W be a neighborhood of ΔB (ΔM). We define a function $\sigma(x,y)$ on $B \times B$ ($M \times M$) as follows:

(5.1.1) $\quad 0 \leqq \sigma(x,y) \leqq 1$.

(5.1.2) $\quad$ The support of σ is contained in W .

(5.1.3) $\quad \sigma(x,y) = \sigma(y,x)$.

(5.1.4) $$\sigma(x,y) = \begin{cases} 1 & \text{on some neighborhood of } \Delta B \ (\Delta M) \\ 0 & \text{outside of } W \end{cases} .$$

Explicitely, for a suitable $\varepsilon > 0$, we take a function μ : $0 \leqq \mu \leqq 1$ on $\mathbb{R}$, $\mu(t) = 1$ ($0 \leqq t \leqq \varepsilon$), $\mu(t) = 0$ ($2\varepsilon < t$). And $\sigma(x,y) := \mu(r(\pi^{-1}(x), \pi^{-1}(y)))$.

Hereafter we often omit the tilde " ~ " in $\hat{x} \in M$ corresponding to $x \in B$ unless confusion.

We define the parametrix form of $\square$ or $\square''$ by

$$w(x,y) := \begin{cases} \dfrac{r^{2-n}(x,y)}{(n-2)S_n} \sigma(x,y) \alpha_s(x,y) & (n > 2) \\[2ex] -\dfrac{\log(r(x,y))}{S_2} \sigma(x,y) \alpha_s(x,y) & (n = 2) \end{cases}$$

where S_n denotes the volume of n-sphere. $w(x,y)$ is a double $(0,s)$-form on M .

Proposition 5.4. <u>For</u> $x \neq y$,

$$q(x,y) := - \square''_{\hat{y}} w(x,y) = O(r^{2-n}) \quad [O(\log r) \ \underline{\text{if}} \ n = 2].$$

Proof. We put $w(x,y) = r^{2-n} a(x,y)$. Let $f \in \Lambda^{0,0}(M)$, that is, locally $\partial f/\partial y^i = 0$ in a flat chart $(\tilde{U}; (\hat{y}^i, \tilde{y}^\alpha))$. Then we have $\Box'' f = - D^\alpha D_\alpha f$. Let $f \in \Lambda^{0,0}(M)$ and $\phi \in \Lambda^{0,s}(M)$. We put $f\phi = \sum_{\alpha_1 < \cdots < \alpha_s} f \phi_{\alpha_1 \cdots \alpha_s} d\hat{y}^{\alpha_1} \wedge \cdots \wedge d\hat{y}^{\alpha_s}$. Then we have

$$(D^\alpha D_\alpha (f\phi))_{\alpha_1 \cdots \alpha_s} = f D^\alpha D_\alpha \phi_{\alpha_1 \cdots \alpha_s} + 2(D^\alpha f)(D_\alpha \phi_{\alpha_1 \cdots \alpha_s}) - (\Box'' f) \phi_{\alpha_1 \cdots \alpha_s} .$$

And, we have

$$(\Box''(f\phi))_{\alpha_1 \cdots \alpha_s} = f (\Box'' \phi)_{\alpha_1 \cdots \alpha_s} - 2(D^\alpha f)(D_\alpha \phi_{\alpha_1 \cdots \alpha_s}) - (\Box'' f) \phi_{\alpha_1 \cdots \alpha_s} .$$

As $\partial(r^{2-n}(x,y))/\partial \tilde{y}^i = 0$, we have

$$\text{(5.2)} \quad \Box''_{\hat{y}} w(x,y) = r^{2-n}(x,y) \Box''_{\hat{y}} a(x,y) + a(x,y) \Box''_{\hat{y}}(r^{2-n}(x,y)) - 2b$$

here $b_{\alpha_1 \cdots \alpha_s} = D^\alpha(r^{2-n}) D_\alpha a_{\alpha_1 \cdots \alpha_s}$. Hence, we have

$$\Box''_{\hat{y}} w(x,y) = O(r^{2-n}) ,$$

for each term of the right hand side of (5.2) is $O(r^{2-n})$. ∎

For any $x \in B$, we take $\tilde{x} \in \pi^{-1}(x)$. For any vector $Y \in E^{\perp}_{\tilde{x}}$, let γ_Y be a transversal geodesic with the initial point $\tilde{x}$ and the initial vector Y , $P_{Y,0}$ a transversal parallel translation with respect to D from Y to 0 along γ_Y . We define a variation K of $\gamma_{E(Y)}$, $E(Y) := Y/\|Y\|$ by

$$K(r,t) := \exp^{\perp}_{\tilde{x}} r(E(Y) + tZ/\|Y\|)$$

for any $Z \in E^{\perp}_{\tilde{x}}$, and

$$\tilde{Z}(r) := (\partial K/\partial t \ (r,0))^{\perp} .$$

Then we have

(5.3) $$D_{E(Y)}\tilde{Z}(r) = D_{\partial/\partial r}(\partial K/\partial t\ (r,0))^{\perp}$$
$$= (\ D_{\partial/\partial r}\ \partial K/\partial t\ (r,0)\ )^{\perp}$$
$$= (\ D_{\partial/\partial t}\ \partial K/\partial r\ (r,0) + T(K_*(\partial/\partial r),K_*(\partial/\partial t))(r,0)\ )^{\perp}$$
$$= (\ D_{\partial/\partial t}\ \partial K/\partial r\ (r,0)\ )^{\perp}\ .$$

$$D_{E(Y)}D_{E(Y)}\tilde{Z}(r) = D_{\partial/\partial r}\ D_{\partial/\partial r}(\partial K/\partial t\ (r,0))^{\perp}$$
$$= [(\ D_{\partial/\partial t}\ D_{\partial/\partial r}\ \partial K/\partial r$$
$$+ R(K_*(\partial/\partial r),K_*(\partial/\partial t))K_*(\partial/\partial r)\ )(r,0)$$
$$+ D_{\partial/\partial r}\ T(K_*(\partial/\partial r),K_*(\partial/\partial t))(r,0)\ ]^{\perp}$$
$$= (\ R(E(Y),Z(r))E(Y)\ )^{\perp}$$

(for $\gamma_{E(Y)}$ is a transversal geodesic).

Therefore we have

(5.4) $$D_Y D_Y\ \tilde{Z}(r) + R^{\perp}(\tilde{Z}(r),Y)Y = 0\ ,\ \tilde{Z}(0) = 0\ ,\ (\ D_Y\ \tilde{Z}\ )(0) = Z\ .$$

Hence we may define

(5.5) $$V(r) := P_{rY,0}\ \tilde{Z}(r)$$
$$Z(r) := P_{rY,0}[\ T_{rY}\ \exp^{\perp}_{\tilde{X}}\ Z\ ]^{\perp} = \frac{1}{r}\ V(r)\ .$$

Proposition 5.5.

$$Z(r) = Z + r\int_0^1 (1-t)(\ P_{rtY,0}R^{\perp}\ )(Y,V)Y\ dt\ .$$

Proof. We have $V(0) = 0$, $(\ D_Y\ V\ )(0) = Z$ and

$$(\ D_Y\ D_Y\ V\ )(r) = P_{rY,0}[\ D_Y\ D_Y\ \tilde{Z}\ ]$$
$$= P_{rY,0}[\ R^{\perp}(Y,\tilde{Z})Y\ ]$$
$$= (\ P_{rY,0}R^{\perp}\ )(Y,V)Y\ .$$

By Taylor's expansion theorem, we have

$$Z(r) = \frac{1}{r}\ V(r)$$

$$= Z + \frac{1}{r}\int_0^1 (r-s)(\ D_Y D_Y V\ )(s)\ ds$$

$$= Z + r\int_0^1 (1-t)(\ D_Y D_Y V\ )(rt)\ dt$$

$$= Z + r\int_0^1 (1-t)(\ P_{rtY,0}R^{\perp}\ )(Y,V)Y\ dt\ . \ \blacksquare$$

For any $x, y \in B$, choose a vector $Y \in E^{\perp}_{\tilde{x}}$ such that γ_Y is the transversal geodesic from $\tilde{x}$ to $\tilde{y}$. $R^{\perp}$ is called uniformly bounded if $P_{rY,0}R^{\perp}$ ($0 \leq r \leq 1$) is bounded. And R_B is called uniformly bounded if the corresponding $R^{\perp}$ is uniformly bounded.

Corollary 5.6. _If_ $R^{\perp}$ _is uniformly bounded on_ M , _there exist positive real numbers_ C _and_ r_o _such that_ $\|(\ T_Y \exp^{\perp}_{\tilde{x}}\ )^{\perp}\| \leq C$ _and_ $\|[(\ T_Y \exp^{\perp}_{\tilde{x}}\ )^{\perp}]^{-1}\| \leq C$ _for all_ $\tilde{x} \in \pi^{-1}(x)$, _all_ $x \in B$ _and all_ $Y \in B^{\perp}(0,r_o)$ _where_ $B^{\perp}(0,r_o)$ _denotes the open transversal ball in_ $E^{\perp}_{\tilde{x}}$ _of radius_ r_o _of the origin_ 0 .

Proof. We check only the first statement, for the second is, then, obvious. Take any $\tilde{x} \in \pi^{-1}(x)$, $x \in B$ and let $\{ Z_\alpha \}$ be an orthonormal basis of $E^{\perp}_{\tilde{x}}$. We put $A(x,y) = P_{Y,0}[\ T_Y \exp^{\perp}_{\tilde{x}}\ ]^{\perp}$. Then the matrix entries of A with respect to $\{ Z_\alpha \}$ are given by

$$a_{\alpha\beta}(x,rY) := (\ A(x,rY)Z_\alpha,\ Z_\beta\ )$$

$$= (\ Z_\alpha,\ Z_\beta\ ) + r\int_0^1 (1-t)(\ (P_{rtY,0}R^{\perp})(Y,V_\alpha)Y,\ Z_\beta\ )\ dt$$

$$= \delta_{\alpha\beta} + r^2\int_0^1 (1-t)(\ (P_{rtY,0}R^{\perp})(Y,Z_\alpha)Y,\ Z_\beta\ )\ dt\ .$$

And $E^{\perp}_{\tilde{x}}$ is isometric to $E^{\perp}_{\tilde{x}'}$ for any $\tilde{x}, \tilde{x}' \in \pi^{-1}(x)$. Therefore, we have

$$\|(\ T_Y \exp^{\perp}_{\tilde{x}}\ )^{\perp}\| \leq C$$

or all $\tilde{x} \in \pi^{-1}(x)$, all $x \in B$ and all $Y \in B^{\perp}(0,r_o)$. ∎

If the conclusion of Corollary 5.6 holds on M , we say that M has a uniformly bounded transversal exponential map.

The notation

$$\phi(x,y) = UO(\psi(r))$$

is read " $\phi(x,y)$ is uniformly $O(\psi(r))$ " if there exist $C > 0$ and $r_0 > 0$ such that

$$|\phi(x,y)| < C\,|\psi(r)|$$

on $N^*(\Delta B, r_o) := \{ (x,y) \in B \times B \mid 0 < r(x,y) < r_o \}$.

Proposition 5.7. *If* $R^{\perp}$ *is bounded,*

$$P_{Y,0}(\, T_Y \exp^{\perp}_{\tilde{x}}\,)^{\perp} - I = UO(r^2) \tag{5.6}$$

where I *denotes the identity map.*

We suppose that M has a uniformly bounded transversal exponential map and the volume of M is finite.

By proposition 5.1, we may introduce a L_2-inner product on $L_2^*(B) := \sum_{s \geqq 0} L_2^s(B)$ by

$$\langle \phi , \psi \rangle := \langle \pi^*\phi , \pi^*\psi \rangle \quad .$$

Then we have following:

Proposition 5.8. $L_2^{0,s}(M) \cong L_2^s(B)$ (*isomorphic as Hilbert spaces via* π^*).

We define the integral operators Ω and Q as follows:

$$(\Omega\phi)(x) := \int_M w(x,y) \wedge *''_{\hat{y}}\, \phi(y) \wedge d\hat{y}^1 \wedge \cdots \wedge d\hat{y}^r$$

$$(Q\phi)(x) := \int_M q(x,y) \wedge *''_{\tilde{y}}\, \phi(y) \wedge d\hat{y}^1 \wedge \cdots \wedge d\tilde{y}^r \quad .$$

Then a formal calculation implies that

$$\Omega \Box'' = I - Q \quad .$$

Let $k(x,y)$ be a double s-form on B supported in $\overline{N^*(\Delta B, r_0)}$ and $k(x,y) = UO(r^{d-n})$. Then we may regard $k(x,y)$ as a double basic $(0,s)$-form on M supported in $\overline{N^*(\Delta M, r_0)}$. For $\phi \in \Lambda^{0,s}(M)$, we define

$$K\phi(x) := \int_M k(x,y) \wedge *''_{\tilde{y}}\, \phi(y) \wedge d\tilde{y}^1 \wedge \cdots \wedge d\hat{y}^r \quad ,$$

$$|K\phi(x)|^2 := \sum_{\alpha_1 < \cdots < \alpha_s} (K\phi)_{\alpha_1 \cdots \alpha_s}(x)\,(K\phi)^{\alpha_1 \cdots \alpha_s}(x) \quad ,$$

$$|k(x,y)|^2 := \sum_{\substack{\alpha_1 < \cdots < \alpha_s \\ \beta_1 < \cdots < \beta_s}} k_{\alpha_1 \cdots \alpha_s, \beta_1 \cdots \beta_s}(x,y)\, k^{\alpha_1 \cdots \alpha_s, \beta_1 \cdots \beta_s}(x,y) \quad ,$$

$$\| K\phi \|^2 := \langle K\phi(x), K\phi(x) \rangle$$

$$= \int_M |K\phi(x)|^2 \, dV_{\tilde{x}} \quad .$$

Then we have the estimate:

$$|K\phi(x)|^2 \leqq \binom{n}{s} \left(\int_M |k(x,y)|\, |\phi(y)|\, dV_{\hat{y}} \right)^2$$

$$\leqq \binom{n}{s} \left(\int_M |k(x,y)|\, |\phi(y)|^2\, dV_{\tilde{y}} \right) \left(\int_M |k(x,y)|\, dV_{\tilde{y}} \right) .$$

Letting $\mu_x := \sup_{x \in B} \int_M |k(x,y)|\, dV_{\tilde{y}}$ and $\mu_y := \sup_{y \in B} \int_M |k(x,y)|\, dV_{\tilde{x}}$, we have

$$\| K\phi \|^2 \leqq \binom{n}{s} \mu_x \mu_y \| \phi \|^2 \quad .$$

If $d > 0$, then we have

$$\mu_x \leqq \int_M r^{d-n}\, dV_{\tilde{y}} < \infty \quad \text{and} \quad \mu_y < \infty \quad .$$

Therefore K is a bounded operator.

Let $k := \sum_{i,j=1}^{\infty} k_{ij}$ where each k_{ij} is supported in $U'_i \times U'_j$ $(\{\tilde{U}'_i, G_i, \varphi_i\} \in \mathcal{F})$, and $K := \sum_{i,j=1}^{\infty} K_{ij}$ a corresponding decomposition with k. Rearranging the indices i and j, we may put $k := \sum_{i=1}^{\infty} k_i$ and $K := \sum_{i=1}^{\infty} K_i$. Therefore, if $d > [\frac{n}{2}]$, K is a compact operator, for the series is norm convergent and each K_i is a compact operator.

If K_1 and K_2 are integral operators with kernels k_1 and k_2 respectively, $K_1 \circ K_2$ is the operator with kernel

$$k(x,y) := \int_M k_1(x,z) \wedge *''_{\tilde{z}}\, k_2(z,y) \wedge d\tilde{z}^1 \wedge \cdots \wedge d\tilde{z}^r .$$

If $k_1 = UO(r^{d_1-n})$ and $k_2 = UO(r^{d_2-n})$, with $d_1, d_2 > 0$, then $k = UO(r^{d_1+d_2-n})$ if $d_1 + d_2 - n < 0$. In fact,

$$|k(x,y)| \leqq \int_M |k_1(x,z)| \cdot |k_2(z,y)|\, dV_{\tilde{z}}$$

$$\leqq C \int_M r(x,z)^{d_1-n}\, r(z,y)^{d_2-n}\, dV_{\tilde{z}} \qquad (\ C : \text{constant}\ )$$

$$\leqq C' \int_{|z| \leqq r_0} |z|^{d_1-n}\, |y-z|^{d_2-n}\, dV_{\tilde{z}} \qquad (\ C' : \text{constant}\ )$$

(by Corollary 5.6)

If $r = r(x,y)$,

$$\int_{|z| \leqq 2r} |z|^{d_1-n}\, |y-z|^{d_2-n}\, dV_{\tilde{z}}$$

$$= r^{d_1+d_2-n} \int_{|u| \leqq 2} |u|^{d_1-n}\, |y'-u|^{d_2-n}\, dV_{\tilde{u}}$$

$$\leqq C\, r^{d_1+d_2-n} \qquad (\ C : \text{constant},\ \ y'^{\alpha} = \frac{1}{r} y^{\alpha}\ )$$

and

$$\int_{2r < |z| \leqq r_0} |z|^{d_1 - n} |y-z|^{d_2 - n} dV_{\tilde{z}}$$

$$\leqq C \int_{2r < |z| \leqq r_0} |z|^{d_1 + d_2 - n} dV_{\tilde{z}}$$

$$\leqq C' (r_0^{d_1 + d_2 - n} - r^{d_1 + d_2 - n})$$

(C, C' : constants).

Then we have $|k(x,y)| \leqq C r^{d_1 + d_2 - n}$ (C : constant).

Therefore we will prove that w and q are both $UO(r^{2-n})$ (UO(log r) if n = 2).

Now, $d''_{\tilde{y}} A = v_\alpha(A)\, d\tilde{y}^\alpha = - r(x,y)\, v_\alpha(r(x,y))\, d\tilde{y}^\alpha$. Letting $X^\alpha = - r(x,y)\, g^{\alpha\beta}(y)\, v_\beta(r(x,y))$, we have

$$\alpha(x,y) = d''_{\tilde{x}}\, d''_{\tilde{y}}\, A(x,y)$$

$$= - v_\alpha(r(x,y)) v_\beta(r(x,y)) d\tilde{x}^\alpha \cdot d\tilde{y}^\beta$$

$$- r(x,y) (v_\alpha(v_\beta(r(x,y)))) d\hat{x}^\alpha \cdot d\hat{y}^\beta$$

and

$$d''_{\tilde{x}}\, X^\alpha = - g^{\alpha\beta}(\hat{y}) v_\theta(r(x,y)) v_\beta(r(x,y)) d\tilde{x}^\theta$$

$$- g^{\alpha\beta}(\hat{y})\, r(x,y)\, (v_\theta(v_\beta(r(x,y)))) d\hat{x}^\theta$$

$$- r(x,y)\, v_\beta(r(x,y))\, (d''_{\tilde{x}}\, g^{\alpha\beta}(\hat{y})) .$$

Then we have

$$\alpha(x,y) = g_{\alpha\beta}(\hat{y})\, d''_{\tilde{x}}\, X^\alpha \cdot d\hat{y}^\beta .$$

And let $\{e_\alpha\}$ be an orthonormal frame in $E^\perp_{\tilde{y}}$ and $X = X^\alpha e_\alpha$, we have

(5.7) $$\alpha(x,y) = d''_{\tilde{x}}\, X^\alpha \cdot e^*_\alpha$$

where $\{e^*_\alpha\}$ is the dual basis to $\{e_\alpha\}$. Then we have

(5.8) $\alpha(x,y) = \sum_{\alpha}((T^*_X \exp^{\perp}_{\tilde{y}})^{\perp})^{-1}\ e^*_{\alpha}\cdot e^*_{\alpha}$

and, by Corollary 5.6,

$$\alpha(x,y) = UO(1) .$$

Therfore, by the defintion of w , we have following:

Theorem 5.9. _If_ M _has a uniformly bounded transversal exponential map, then_

$$w = UO(r^{2-n}) \quad [\ UO(\ \log r\ )\ \text{if}\ n = 2\ ] .$$

Theorem 5.10. _If_ R_B _is uniformly bounded, then_

$$w = UO(r^{2-n}) \quad [\ UO(\ \log r\ )\ \text{if}\ n = 2\ ]$$

where R_B _denotes the curvature tensor of_ ∇_B _on_ B .

Corollary 5.11. _Under the same hypothesis,_ Ω _is a bounded operator_.

The calculations for $q = -\Box''_{\tilde{y}}\, w$ are much more involved. We need to know the behavior of $\Box''$ under exterior multiplication. We use the formula for $\Box''$ in terms of the covariant derivatives: For example, letting $\theta = \phi\wedge\psi$, $\phi\in\Lambda^{0,p}(M)$, $\psi\in\Lambda^{0,q}(M)$, the second term of the rigth hand side of (3.5.3),

$$\tilde{R}^{\perp}(\theta)_{\alpha_1\cdots\alpha_{p+q}} := -\sum_{k=1}^{p+q} R^{\gamma\cdot\alpha\cdot}{}_{\cdot\alpha\cdot\alpha_k}\,\theta_{\alpha_1\cdots\gamma\cdots\alpha_{p+q}}$$

$$= -\sum_{k=1}^{p+q}\sum_{\tau^{-1}(k)\leq p} R^{\gamma\cdot\alpha\cdot}{}_{\cdot\alpha\cdot\alpha_k}\,(\mathrm{sgn}\tau)\,\phi_{\alpha_{\tau(1)}\cdots\gamma\cdots\alpha_{\tau(p)}}\,\psi_{\alpha_{\tau(p+1)}\cdots\alpha_{\tau(p+q)}}$$

$$-\sum_{k=1}^{p+q}\sum_{\tau^{-1}(k)>p} R^{\gamma\cdot\alpha\cdot}{}_{\cdot\alpha\cdot\alpha_k}\,(\mathrm{sgn}\tau)\,\phi_{\alpha_{\tau(1)}\cdots\alpha_{\tau(p)}}\,\psi_{\alpha_{\tau(p+1)}\cdots\gamma\cdots\alpha_{\tau(p+q)}}$$

$$= (\hat{R}^{\perp}(\phi)\wedge\psi)_{\alpha_1\cdots\alpha_{p+q}} + (\phi\wedge\hat{R}^{\perp}(\psi))_{\alpha_1\cdots\alpha_{p+q}} .$$

Then we have

$$(5.9) \qquad \square''(\phi\wedge\psi) = \square''\phi\wedge\psi + \phi\wedge\square''\psi - 2\langle D\phi \mid D\psi\rangle + \mathcal{E}(\phi,\psi)$$

where $\mathcal{E}$ is a bilinear form derivable from the curvature tensor $R^{\perp}$, vanishing when either ϕ or ψ is a 0-form, and $\langle \ \mid \ \rangle$ the partial contraction of the 1-forms arising from D.

Remark. If $R^{\perp}$ is bounded, $\mathcal{E}$ is bounded.

If we put $w = F(r)\,\sigma(r)\,\alpha_s$, we have

$$\begin{aligned} -q &= \square''_y(F(r)\,\sigma(r)\,\alpha_s) \\ &= (\square''_y F)\sigma\alpha_s + F(\square''_y\sigma)\alpha_s + F\sigma(\square''_y\alpha_s) \\ &\quad - 2\langle D_yF \mid D_y\sigma\rangle\alpha_s - 2F\langle D_y\sigma \mid D_y\alpha_s\rangle - 2\sigma\langle D_yF \mid D_y\alpha_s\rangle . \end{aligned}$$

Let $\gamma(s)$ be a transversal geodesic from x to y and $e_1 := \dot{\gamma}(r)$ ($r = r(x,y)$). We choose $\{e_1, \cdots, e_n\}$ so that $\{e_1, \cdots, e_n\}$ forms an orthonormal frame in $E_y^{\perp}$ and take transversal geodesics $\gamma_u(t)$ parameterized by arc-length t from y with initial vectors e_u ($u = 2, \cdots, n$). Then a function $F \in \Lambda^{0,0}(M)$ satisfies

$$\square''_y F(y) = -\frac{d^2}{ds^2}(F\circ\gamma)(r) - \sum_{u=2}^{n}\frac{d^2}{dt^2}(F\circ\gamma_u)(0) .$$

By the orthogonality of e_u to γ at y, e_u may be realized the family $\{C_t\}$ parameterized by t of transversal geodesics with the initial point x and $C_0 = \gamma$. We put

$$F(\gamma_u(t)) = \varphi(\text{length } C_t) .$$

Then we have

$$\begin{aligned} \frac{d^2}{dt^2}(F\circ\gamma_u)(0) &= \frac{d^2\varphi}{ds^2}(r)\left(\frac{d}{dt}(\text{length } C_t)(0)\right)^2 \\ &\quad + \frac{d\varphi}{ds}(r)\frac{d^2}{dt^2}(\text{length } C_t)(0) . \end{aligned}$$

However, $\frac{d}{dt}$(length C_t)(0) = 0 and by the second variation formula,

$$\frac{d^2}{dt^2}(\text{ length } C_t)(0) = (Z_u(r), D_{\dot\gamma} Z_u(r))$$

where each Z_u is the transversal Jacobi field along γ with $Z_u(0) = 0$ and $Z_u(r) = e_u$. Now, letting

$$\begin{aligned} \Theta &:= \det((T_Y \exp_x^{\perp})^{\perp}) \\ &= \det((T_Y \exp_x^{\perp}(e_1))^{\perp} \wedge \cdots \wedge (T_Y \exp_x^{\perp}(e_n))^{\perp}) , \end{aligned}$$

we have (cf. [2])

$$(5.10) \qquad \Box_Y'' F(r) = - \frac{d^2F}{dr^2} - \frac{dF}{dr}\left(\frac{n-1}{r} + \frac{1}{\Theta}\frac{\partial \Theta}{\partial r} \right) .$$

Proposition 5.12. <u>If</u> $R^{\perp}$ <u>is uniformly bounded</u>,

$$\frac{1}{\Theta}\frac{\partial \Theta}{\partial r} = UO(r) .$$

Proof. We note that $\Theta = \det(a_{\alpha\beta})$. As $a_{\alpha\beta} - \delta_{\alpha\beta} = UO(r^2)$, $\Theta = UO(1)$. Then, expanding Θ term by term, it suffices to check that for $Y \in E_x^{\perp}$ with $\|Y\| = 1$,

$$\frac{\partial a_{\alpha\beta}}{\partial r} = (D_Y Z_\alpha(r), Z_\beta) = UO(r)$$

or that $D_Y Z(r) = UO(r)$ for any $Z \in E_x^{\perp}$ (cf. (5.5)). We have

$$V(r) = rZ + r^2 \int_0^1 (1-t)(P_{rtY,0} R^{\perp})(Y,V)Y\, dt$$

$$D_Y V(r) = Z + r \int_0^1 (P_{rtY,0} R^{\perp})(Y,V)Y\, dt .$$

Then we have

$$\begin{aligned} D_Y Z(r) &= - \frac{1}{r^2} (V(r) - r D_Y V(r)) \\ &= \int_0^1 t\, (P_{rtY,0} R^{\perp})(Y,V)Y\, dt \end{aligned}$$

$$= r \int_0^1 t^2 \, (P_{rtY,0} R^{\perp})(Y,Z)Y \, dt$$

$$= UO(r) \quad . \qquad \blacksquare$$

Corollary 5.13. <u>Under the same hypotheses,</u> $\square_Y'' F(r) = UO(r^{2-n})$ <u>and</u> $\square_Y'' \sigma(r) = UO(1)$.

In fact, we have $\square_Y'' F(r) = C \frac{1}{\Theta} \frac{\partial \Theta}{\partial r} r^{1-n}$, and for estimating $\square_Y'' \sigma(r)$, we note that $d\sigma/dr = 0$ in a neighborhood of $\triangle M$.

Proposition 5.14. <u>If</u> $R^{\perp}$, $DR^{\perp}$ <u>and</u> $D^2 R^{\perp}$ <u>are bounded,</u> $D_Y \alpha_s = UO(r)$ <u>and</u> $\square_Y'' \alpha_s = UO(1)$.

Proof. It suffices to prove for $s = 1$. We have

$$\alpha(x,y) = \Sigma_{\alpha}((T_X^* \exp_y^{\perp})^{\perp})^{-1} e^*_{\alpha} \cdot e^*_{\alpha} \quad .$$

Choose $\{e_{\alpha}\}$ at y to be the parallel translate of a fixed basis $\{e^{\circ}_{\alpha}\}$ of $E_x^{\perp}$. Then we have

$$\alpha(e^{\circ}_{\alpha}) = \Sigma_{\beta}(e^*_{\beta} , ((T_X \exp_y^{\perp})^{\perp})^{-1} e^{\circ}_{\alpha}) e^*_{\beta}$$

$$= \Sigma_{\beta}(e^*_{\beta} , (P_{X,0}(T_X \exp_y^{\perp})^{\perp})^{-1} P_{X,0} e^{\circ}_{\alpha}) e^*_{\beta}$$

$$=: \Sigma_{\beta} b_{\alpha\beta}(y,x) e^*_{\beta}$$

where $B := (b_{\alpha\beta}) = A^{-1}$. Hence, we have

$$D_Y \alpha(e^{\circ}_{\alpha}) = \Sigma_{\beta} d_Y'' b_{\alpha\beta}(y,x) e^*_{\beta} + \Sigma_{\beta} b_{\alpha\beta}(y,x) D_Y e^*_{\beta} \quad .$$

To prove $D_Y \alpha = UO(r)$, it suffices to prove that $d_Y'' b_{\alpha\beta}(y,x) = UO(r)$, or $d_X'' a_{\alpha\beta}(x,y) = UO(r)$, and $D_Y e_{\alpha} = UO(r)$.

Now, letting $Y = r E(Y)$ along γ_Y , we have

$$\frac{d}{dt} P_{tY,0}(D_{e_{\beta}} e_{\alpha}) = P_{tY,0}(D_Y D_{e_{\beta}} e_{\alpha})$$

$$= P_{tY,0}(R^{\perp}(Y,\widetilde{e}_{\beta}) e_{\alpha})$$

where $\widetilde{e}_{\beta}$ is the unique vector field which agrees with e_{β} at y .

Then we have

$$P_{Y,0}(\ D_{e_\beta} e_\alpha\ ) = \int_0^1 t\ P_{tY,0}(\ R^\perp(Y,\tilde{e}_\beta)e_\alpha\ )\ dt\ .$$

Hence, we have $D_y e_\alpha = UO(r)$.

For the fixed y , we use a transversal normal flat chart centered at y . As x varies, we take $\{Z_\alpha\}$ to be a basis parallel translated from y . We have, for $X = X^\alpha e_\alpha \in E_y^\perp$,

$$a_{\alpha\beta}(x,y) = \delta_{\alpha\beta} + r^2 \int_0^1 s\ (\ (P_{(1-s)X,0}R^\perp)(\hat{E},\tilde{Z}_\alpha)\hat{E},\ \tilde{Z}_\beta\ )\ ds$$

where $\hat{E} := P_{(1-s)X,0}E(Y)$ and $\tilde{Z}_\alpha := P_{(1-s)X,0}Z_\alpha$. Let $\bar{E}$ denote the unit radial vector field $\partial/\partial r$. If $V \in E_x^\perp$, we have

$$d''_{x,V}\ a_{\alpha\beta} = 2r\ (V,\bar{E}) \int_0^1 s\ (\ (P_{(1-s)X,0}R^\perp)(\hat{E},\tilde{Z}_\alpha)\hat{E},\ \tilde{Z}_\beta\ )\ ds$$

$$+ r^2 \int_0^1 s[\ (\ (d''_{x,V}(P_{(1-s)X,0}R^\perp))(\hat{E},\tilde{Z}_\alpha)\hat{E},\ \tilde{Z}_\beta\ )$$

$$+ (\ (P_{(1-s)X,0}R^\perp)(d''_{x,V}\hat{E},\tilde{Z}_\alpha)\hat{E},\ \tilde{Z}_\beta\ )$$

$$+ (\ (P_{(1-s)X,0}R^\perp)(\hat{E},d''_{x,V}\tilde{Z}_\alpha)\hat{E},\ \tilde{Z}_\beta\ )$$

$$+ (\ (P_{(1-s)X,0}R^\perp)(\hat{E},\tilde{Z}_\alpha)d''_{x,V}\hat{E},\ \tilde{Z}_\beta\ )$$

$$+ (\ (P_{(1-s)X,0}R^\perp)(\hat{E},\tilde{Z}_\alpha)\hat{E},\ d''_{x,V}\tilde{Z}_\beta\ )\ ]\ ds$$

where $d''_{x,V}$ denotes the covariant derivative in the direction V at x . Letting $\tilde{V}_\alpha := P_{x,y}V_\alpha$, we may write

$$\tilde{Z}_\alpha(x,y) = Z_\alpha + r \int_0^1 s(P_{(1-s)X,0}R^\perp)(\hat{E},\tilde{V}_\alpha)\hat{E}\ ds\ .$$

As $\tilde{V}_\alpha$ satisfies the differential equation

$$D_{\partial/\partial s}\, D_{\partial/\partial s}\, \tilde{V}_\alpha(s) = (P_{sE(Y),0}R^\perp)(\hat{E},\tilde{V}_\alpha)\hat{E}\ ,$$

$$D_{\partial/\partial s}\, \tilde{V}_\alpha(r) = Z_\alpha\ ,\ \tilde{V}_\alpha(r) = 0\ .$$

We may apply the theorem of solution of differential equations depending on parameters and initial data. Then we have $d''_{x,V}\tilde{V}_\alpha = UO(1)$, and so, $d''_{x,V}\tilde{Z}_\alpha = UO(r^{-1})$. Using a transversal geodesic polar coordinate at y, we have $d''_{x,V}\hat{E} = UO(r^{-1})$. Hence, we have $d''_{x,V} a_{\alpha\beta} = UO(r)$. Therefore, we have $D_y\alpha = UO(r)$.

Now, we calculate $\square''_y\alpha$. We have

$$\square''_y\alpha = \square''_y d''_x d''_y A = d''_x d''_y \left(n + r \frac{1}{\Theta}\frac{\partial\Theta}{\partial r} \right) .$$

As $r \frac{1}{\Theta}\frac{\partial\Theta}{\partial r}$ is comprised of terms $a_{\alpha\beta}$ and $r(\partial a_{\alpha\beta}/\partial r)$, each of which is expressed with integrals of the same sort, it suffices to prove that $d''_y a_{\alpha\beta} = UO(r)$ and $d''_x d''_y a_{\alpha\beta} = UO(1)$. If $W \in E^\perp_y$, then we have

$$\begin{aligned} d''_{y,W} a_{\alpha\beta} = 2r\,(W,\bar{E}) \int_0^1 (1-t)\,(\,(P_{tY,0}R^\perp)(\hat{E},Z_\alpha)\hat{E},\ Z_\beta\,)\,dt \\ + r^2 \int_0^1 (1-t)\,[\,(\,(d''_{y,W}(P_{tY,0}R^\perp)(\hat{E},Z_\alpha)\hat{E},\ Z_\beta\,) \\ + (\,(P_{tY,0}R^\perp)(d''_{y,W}\hat{E},Z_\alpha)\hat{E},\ Z_\beta\,) \\ + (\,(P_{tY,0}R^\perp)(\hat{E},d''_{y,W}Z_\alpha)\hat{E},\ Z_\beta\,) \\ + (\,(P_{tY,0}R^\perp)(\hat{E},Z_\alpha)d''_{y,W}\hat{E},\ Z_\beta\,) \\ + (\,(P_{tY,0}R^\perp)(\hat{E},Z_\alpha)\hat{E},\ d''_{y,W}Z_\beta\,)\,]\,dt \end{aligned}$$

Again, as Z_α may be expressed with $\tilde{V}_\alpha$, then $d''_{y,W} Z_\alpha = UO(r^{-1})$, and other terms may be controlled by aboves. Then we have $d''_{y,W} a_{\alpha\beta} = UO(r)$.

Before taking $d''_{x,V}\, d''_{y,W}\, a_{\alpha\beta}(x,y)$ for $V \in E^\perp_x$, we use $\{Z_\alpha\}$ parallel translated from the fixed point y, and translate all tensors from x to y. The radial vector field $\partial/\partial r$ at x becomes $-\partial/\partial r$ at y. We have terms of a new kind only in the second derivative terms: $d''_{x,V}\, d''_{y,W}(P_{tY,0} R^{\perp})$. $d''_{x,V}\, P_{x,y}\, d''_{y,W}\, \hat{E}$ and $d''_x\, d''_y\, Z_\alpha$ are treated as above. Note that we have taken $D^2 R^{\perp}$ in one of the terms of $d''_x\, d''_y\, a_{\alpha\beta}$. ∎

Theorem 5.15. *If* $R^{\perp}$, $DR^{\perp}$ *and* $D^2 R^{\perp}$ *are bounded on* M, *then*

$$q = UO(r^{2-n}) \qquad [\ UO(\log r) \text{ if } n = 2\ ].$$

Theorem 5.16. *If* R_B, $\nabla_B R_B$ and $\nabla^2_B R_B$ are bounded on B, *then*

$$q = UO(r^{2-n}) \qquad [\ UO(\log r) \text{ if } n = 2\ ]$$

where R_B *denotes the curvature tensor of* ∇_B *on* B.

Corollary 5.17. *Under the same hypotheses,* $\mathcal{Q}$ *is a bounded operator on* $L^*_2(B)$ *and* $\mathcal{Q}^k$ *is compact if* $k > \frac{n}{4}$ *and the volume of* B *is finite.*

BIBLIOGRAPHY

[1] W. J. Baily, Jr., The decomposition theorem for V-manifolds, Amer. J. Math. 78(1956) 862-883.

[2] M. Berger, P. Gauduchon, E. Mazet, Le spectre d'une variete riemannienne. Lecture Notes in Math. 194, Springer-Verlag, Berlin-Heidelberg-New York, 1971.

[3] H. Kitahara, Remarks on square-integrable basic cohomology spaces on a foliated riemannian manifold, to appear Kodai Math. J.

[4] H. Kitahara, S. Yorozu, A formula for the normal part of the Laplace-Beltrami operator on the foliated manifold, Pacific J. Math. 69(1977) 425-432.

[5] B. L. Reinhart, Foliated manifolds with bundle-like metrics, Ann. of Math. 66(1959) 119-132.

[6] B. L. Reinhart, Harmonic integrals on foliated manifolds, Amer. J. Math. 81(1959) 529-536.

[7] B. L. Reinhart, Closed metric foliations, Michigan Math. J. 8(1961) 7-9.

[8] G. de Rham, Variétés différentiables, 3rd ed., Hermann, Paris, 1960.

[9] I. Satake, The Gauss-Bonnet theorem for V-manifolds, J. Math. Soc. Japan 9(1957) 464-492.

[10] Ta. Takahashi, On the trajectories of the structure vector fields of the Sasakian three dimensional spheres, Sci. Rep. Hirosaki Univ. 23(1976) 57-60.

[11] Ta. Takahashi, Deformations of Sasakian structures and its application to the Briskorn manifolds, Tohoku Math. J. 30(1978) 37-43.

[12] I. Vaisman, Variétés riemanniennes feuilletées, Czechosl. Math. J. 21(1971) 46-75.

[13] I. Vaisman, Cohomology and differential forms, Marcel Dekker, Inc., New York, 1973.

[14] S. Yorozu, Notes on square-integrable cohomology spaces on certain foliated manifolds, to appear Trans. Amer. Math. Soc.

[15] S. Zucker, Estimates for the classical parametrix for the Laplacian, Manuscripta Math. 24(1978) 9-29.

Department of Mathematics
College of Liberal Arts
Kanazawa University
Kanazawa 920, Japan

W. Klingenberg

Stable and unstable motions on surfaces

We begin with some basic concepts and results. As a general reference see [Kl 1] to which we also refer as LCG.

By a surface we mean a compact 2-dimensional manifold M with a riemannian metric g. Instead of (M,g) we also simply write M. If we only want to consider the underlying structure of a differentiable 2-dimensional manifold we say: differentiable surface.

Denote by T_1M the total space of the unit tangent bundle over M, $\tau: T_1M \longrightarrow M$ shall be the projection mapping. A motion on M is an orbit $\{\phi_t X_o, t \in \mathbb{R}\}$, of the geodesic flow on T_1M. I.e., $\tau\phi_t X_o$ is the geodesic $c(t)$ with initial direction $\dot{c}(0) = X_o$ and $\phi_t X_o = \dot{c}(t)$. Thus, $\phi_t X_o$ describes the position and the velocity at the time t of a point which moves on M without exterior forces and which started from $p_o = \tau X_o \in M$ with initial velocity X_o.

We call an orbit $\{\phi_t X_o\}$ periodic if $\phi_\omega X_o = X_o$, for some $\omega > 0$. In general we take the smallest positive ω with this property. ω then is called the prime period of the periodic orbit. The projection $c(t) = \tau\phi_t X_o$ of such an orbit into M yields a closed geodesic of length ω.

On T_1M we can define a riemannian metric $\bar{g}$ as follows: Represent a tangent vector ξ to T_1M by a curve $s \in]-\varepsilon,\varepsilon[\longmapsto X(s) \in T_1M$ with $X'(0) = \xi$. Put $\tau X(s) = c(s)$. Define

$$\bar{g}(\xi,\xi) = g(c'(0),c'(0)) + g(\nabla X(0),\nabla X(0))$$

where $\nabla X(s)$ is the covariant derivative of $X(s)$. The right hand side is independent of the choice of $X(s)$ such that $X'(0) = \xi$: $c'(0) = D\tau.\xi$ and also $\nabla X(0)$ depends only on $X'(0) = \xi$ and $c'(0)$.

Denote by $\bar{d}(\ ,\ )$ the distance on T_1M derived from the riemannian metric $\bar{g}$.

We now can define stability and instability for a given motion $\{\phi_t X_0\}$: $\{\phi_t X_0\}$ is called <u>totally unstable</u> if, for all sufficiently small $\varepsilon > 0$, the following is true: For every perturbation X_0' of the initial velocity X_0 with $X_0' \notin \{\phi_t X_0\}$, the orbit $\{\phi_t X_0'\}$ leaves the ε-neighbourhood of $\{\phi_t X_0\}$, i.e., $\bar{d}(\phi_t X_0, \phi_t X_0')$ is not bounded by ε, for all $t \in \mathbb{R}$.

On the other hand, we call $\{\phi_t X_0\}$ <u>stable</u> if, for every $\varepsilon > 0$, every sufficiently small perturbation X_0' of X_0 (i.e., $\bar{d}(X_0, X_0') <$ some positive $\eta = \eta(\varepsilon)$ the orbit $\{\phi_t X_0'\}$ stays in the ε-neighbourhood of $\{\phi_t X_0\}$:

$$\bar{d}(\phi_t X_0, \phi_t X_0') < \varepsilon, \quad \text{all } t \in \mathbb{R}.$$

Note that "totally unstable" and "stable" are not formal contra positions of each other. Still, they constitute generically the only two possibilities for the periodic orbits:

<u>Theorem 1</u>. For a residual set G^* in the space $G = GM$ of all riemannian metrics on a closed differentiable surface

M, a periodic orbit either is totally unstable or else, it is stable.

Proof. We define G^* to consist of those riemannian metrics g for which the periodic orbits on $T_1(M,g)$ either are hyperbolic or else, of twist type. These are properties of the (non-linear) Poincaré map $\mathcal{P}_{X_0}$, associated to a periodic orbit $\{\phi_t X_0;\ 0 \le t \le \omega\}$:

$\mathcal{P}_{X_0}$ hyperbolic means that the linear Poincaré map $D\mathcal{P}_{X_0}$ has eigenvalues λ with $|\lambda| \neq 1$.

$\mathcal{P}_{X_0}$ being of twist type is a property of the third order jet $D^3\mathcal{P}_{X_0}$ of $\mathcal{P}_{X_0}$: First of all, it means that the eigenvalues λ of $D\mathcal{P}_{X_0}$ satisfy $|\lambda| = 1$, but that they are not 3rd or 4th roots of unity. Under this hypothesis, $\mathcal{P}_{X_0}$ may in appropriate coordinates be represented by

$$\begin{pmatrix} x \\ y \end{pmatrix} \longmapsto \begin{pmatrix} x \cos \phi(x,y) - y \sin \phi(x,y) \\ x \sin \phi(x,y) + y \cos \phi(x,y) \end{pmatrix}$$

with $\phi(x,y) = \alpha + \beta(x^2+y^2)$ plus terms of order > 3. Twist type now means: $\beta \neq 0$.

That G^* is a residual set is proved in LCG.

For $\{\phi_t X_0,\ 0 \le t \le \omega\}$ hyperbolic there exist injective immersions - the so-called (strong) stable and unstable manifolds through X_0 - :

$$W_s, W_u\colon (\mathbb{R},0) \longrightarrow (T_1 M, X_0)$$

such that, for some a,b, $0 < a \leq 1$, $0 < b$,

$$a\,\bar{d}(X_0, W_s(x))e^{-bt} \leq \bar{d}(\phi_t X_0, \phi_t W_s(x)), \quad \text{for} \quad t \leq 0$$

$$a\,\bar{d}(X_0, W_u(x))e^{-bt} \leq \bar{d}(\phi_t X_0, \phi_t W_u(x)), \quad \text{for} \quad t \geq 0$$

at least, as long as $|x|$ is sufficiently small and for those $t \leq 0$ (in the first equation) or $t \geq 0$ (in the second equation) for which the left hand sides remain below some $\varepsilon > 0$.

These estimates therefore show that $\phi_t W_s(x)$, $|x| \neq 0$, leaves every sufficiently small neighbourhood of $\{\phi_t X_0\}$, for $t \longrightarrow -\infty$, whereas $\phi_t W_u(x)$, $|x| \neq 0$, does the same for $t \longrightarrow +\infty$. An arbitrary element X_0', sufficiently near X_0 and not on the orbit $\{\phi_t W_s(x)\}$ of some $W_s(x)$, will move away from $\{\phi_t X_0\}$ for $t \longrightarrow +\infty$.

Let now $\{\phi_t X_0, 0 \leq t \leq \omega\}$ be of twist type. It then is a result of Kolmogorov -Arnold-Moser, cf. LCG, that $\{\phi_t X_0\}$ is stable: In every neighbourhood of this orbit there exist ϕ_t -invariant 2-dimensional tori. The interior of such a torus, i.e., a solid torus, constitutes a neighbourhood from which no orbit can escape. □

We now ask for the existence of stable and unstable periodic orbits on (M,g) for a metric g in the set G^*. For the case of negative Gauss curvature K we have the following result, due to Hadamard and E. Hopf, cf. LCG :

<u>Theorem 2</u>. Assume that (M,g) has negative curvature K. Then all periodic orbits are hyperbolic, i.e., totally unstable. They are dense in T_1M. More precisely, the set $\operatorname{Per} T_1M$ of $X_0 \in T_1M$ for which $\{\phi_t X_0\}$ is periodic is dense in T_1M. However, $\operatorname{Per} T_1M$ has measure 0. □

<u>Note</u>. For $K < 0$ the periodic orbits are (modulo the choice of the initial point) in one to one correspondence with the conjugacy classes of the fundamental group $\pi_1 M$ of M. $\pi_1 M$ has exponential growth which makes it relatively easy to establish the existence of infinitely many prime periodic orbits. That $\operatorname{Per} T_1M$ is even dense is a general phenomenon for flows with a hyperbolic structure and with a flow-invariant volume element - properties which hold for $\phi_t: T_1M \longrightarrow T_1M$ if $K < 0$, c.f. LCG.

If M permits a riemannian metric with $K < 0$, it must have genus > 1. Here we may restrict ourselves to orientable surfaces - if M is not orientable, replace it by its 2-fold orientable covering and project the periodic orbits on this covering down into M.

Consider now surfaces M with $K > 0$. M being orientable this is equivalent to considering a convex surface. In particular, genus $M = 0$. Surfaces for which we do not require that the curvature K has constant sign will be touched upon briefly at the end.

Theorem 3. Let M be a convex surface. Then there exist infinitely many prime periodic orbits on T_1M.

Let now the metric g on M belong to the set G^*. Then infinitely many of these orbits are hyperbolic.

Presumably, there also are infinitely many stable ones. The existence of a single stable periodic orbit would imply this.

At least, if $\min K : \max K \geq 1/4$ then the shortest periodic orbit is stable. The same is true if M permits a free isometric $\mathbb{Z}_2$-action, i.e., if M is the 2-fold covering of the projective plane with a metric of positive curvature.

Proof. The existence of infinitely many prime closed geodesics on M was proved in LCG. For a different proof see [Kl 3].

For $g \in G^*$, a closed geodesic of even index on (M,g) must be hyperbolic. And there do exist infinitely prime ones with even index, cf. [Kl 2], [Kl 3].

That the shortest periodic orbit on a structure (M,g) with $\min K : \max K \geq 1/4$, $g \in G$, is stable is due to Thorbergsson [Th]. The last statement is contained in [LCG]. □

Poincaré [Po] claimed the existence of a non-hyperbolic closed geodesic without self-intersections on every convex surface. His proof, however, seems incomplete. He uses the classification of generic analytic bifurcations for periodic orbits which may occur when we consider a general 1-parameter family

g_τ, $0 \leq \tau \leq 1$, of metrics on the differentiable 2-sphere S^2.

Start e.g. with $M_0 = (S^2, g_0)$ an ellipsoid with three different axes. Then the shortest ellipse represents a non-hyperbolic periodic orbit. If we consider a family $M_\tau = (S^2, g_\tau)$ of surfaces, it may happen that a non-hyperbolic closed geodesic c_τ on M_τ turns into an hyperbolic one. Generically, there are two ways in which this may occur. Let τ_0, $0 < \tau_0 < 1$ be the value where this bifurcation takes place.

(i) If τ increases over $\tau = \tau_0$ then there is being "born" near the double covering c_τ^2 of c_τ a non-hyperbolic closed geodesic c'_τ, $\tau > \tau_0$.

(ii) As $\tau < \tau_0$ approaches τ_0, near c_τ^2 there exists a hyperbolic geodesic c'_τ, $\tau < \tau_0$, which melts into c_τ^2 for $\tau \geq \tau_0$.

To prove the existence of a non-hyperbolic closed geodesic on $M = (S^2, g)$ for every $g \in G$ one might attempt to join $g = g_1$ to g_0 by a path g_τ, $0 \leq \tau \leq 1$, in G with metrics g_τ for which only generic bifurcations take place - at least for the shortest non-hyperbolic geodesic. We believe that in this way it is likely to be true - also for the case of the torus:

For every metric g on a surface M of genus 0 or 1 there exists a non-hyperbolic closed geodesic.

Note that for surfaces of genus > 1 this is false.

We think it to be possible, however, that the first non-hyperbolic closed geodesic c on such a surface (M,g) occurs only very "late", if we order the closed geodesics on (M,g) according to their length: I.e., it might not be possible to give an universal upper bound for the ratio $L(c): L(c_0)$ where c_0 is the shortest closed geodesic and c the shortest non-hyperbolic one. Nor might this be possible for the number of hyperbolic closed geodesics of length $< L(c)$.

References:

[Kl 1] Klingenberg, W: Lectures on Closed Geodesics. Grundlehren der Math. Wissenschaften Bd. 230, Berlin-Heidelberg-New York: Springer 1978

[Kl 2] Klingenberg, W: Über den Index geschlossener Geodätischer auf Flächen. Nagoya Math. I. 69, 107-116 (1978).

[Kl 3] Klingenberg, W: Closed Geodesics on Surfaces of Genus 0. Ann. Scuola Norm. Pisa (1979)

[Po] Poincaré, H.: Sur les lignes géodésiques des surfaces convexes. Trans. Amer. Soc. 6, 237 - 274 (1906).

[Th] Thorbergsson, G: Non-hyperbolic Closed Geodesics. To appear in Math. Scand. (1979).

W. Klingenberg
Mathematisches Institut
der Universität Bonn
Wegelerstraße 10
D-5300 Bonn 1

Vector Fields and Generalized Vector Fields on Fibered Manifolds

Yvette Kosmann-Schwarzbach

The object of this paper is to provide the geometric background for the study of the symmetries and the generalized symmetries of systems of partial differential equations. This field of research currently engages the interest of both mathematicians and mathematical physicists because symmetries lead to either a reduction in the number of variables, or to a reduction in the order of the equations to be solved, while, as a consequence of Noether's theorem, they yield conservation laws which are of intrinsic physical interest.

The systematic study of the *symmetries of systems of partial differential equations*, *i.e*., of the transformations of the equations' dependent and independent variables into others which leave the set of solutions invariant, was introduced by Sophus Lie in the 1870s (see *e.g*. [28], pp. 176-87) when he devised methods that determined the infinitesimal generators of one-parameter groups of symmetries, or infinitesimal symmetries. This area of study was renewed as a result of G. Birkhoff's observation [4] that the use of the symmetries of an equation generalized the *similarity methods* which had long been known in hydrodynamics. Lie's methods were subsequently applied and generalized by numerous authors and books have been published on the subject by Ovsjannikov and by Bluman and Cole.

Generalized symmetries were proposed by Bessel-Hagen in 1921 [3] and have recently been "rediscovered" [2] and studied [10], [1], [22] because of their fundamental importance in the integration of several nonlinear equations, in the theory of completely integrable Hamiltonian systems as well as in the theory of Bäcklund transformations [22], [19].

In geometric language, symmetries of differential operators from a fibered manifold F to a fibered manifold F' are automorphisms of F with respect to which the operator has an equivariance property; if F is a vector bundle such an automorphism need not be linear on each fiber of F. Thus it is natural to study the group of automorphisms of a fibered manifold and the corresponding Lie algebra of infinitesimal automorphisms. An infinitesimal automorphism of a fibered manifold is a projectable vector field; it defines, by means of the Lie derivation, a differential operator of order 1 on the sections of the fibered manifold, more precisely, a differential section operator with values in the vertical tangent bundle (see (1.3)). These two aspects of the same object, 'vector field' and 'differential operator' are essential in the applications to mathematical physics where it is not always clear which point of view is being adopted. Thus it is appropriate to study the mapping defined by the Lie derivation from the set of projectable vector fields into the set of differential operators, and to define the vertical bracket of differential operators (definition (1.18), proposition (1.24)) for which this mapping becomes a Lie algebra homomorphism. The particular case of the Lie algebra of all (linear or nonlinear) infinitesimal automorphisms of a vector bundle is somewhat simpler because the technicality of the differential section operators disappears. (But the general case clearly shows the role of the vertical tangent bundle and of the linearized, or vertical, operators.) There is a real simplification if one considers only the linear infinitesimal automorphisms of a vector bundle (§2) because the first-order quasi-scalar linear differential operators are very easy to describe (proposition (2.2)). The definitions and the proofs in the general case (§1) were modelled after this simpler case which could be read as an introduction.

In §3 we indicate the relationship between Lie group actions and Lie algebra actions on fibered manifolds. These correspond to nonlinear representations of Lie groups and of Lie algebras respectively.

With a view to the applications in the theory of the infinitesimal symmetries of differential operators, we define (§4) the Lie derivative of a differential operator with respect to a pair of projectable vector fields. Actually, we define the Lie derivative for somewhat more general objects, the generalized differential operators between fibered manifolds.

The second part of this paper (§§ 5, 6 and 7) deals with the properties of generalized vector fields on fibered manifolds. The k-vector fields, for $k \geq 1$, generalize the projectable vector fields. Each k-vector field X canonically defines a differential operator of order k on the sections of the fibered manifold, more precisely, a differential section operator of order k with values in the vertical tangent bundle, which is called the Lie derivation with respect to X. <u>Any</u> differential operator of order $k \geq 1$ is the Lie derivation operator for some k-vector field; moreover there exists a unique vertical k-vector field having the prescribed Lie derivation operator. We prove that the set of generalized vector fields (k-vector fields for $k \geq 0$) is a filtered Lie algebra under the vertical bracket defined by (1.18) and (5.9).

We define (§6) the flow of a k-vector field as the solution of an evolution equation. In the analytic case we prove an exponential formula, a series expansion of the flow in terms of the successive Lie derivations.

In §7, the generalization of §4, we define the Lie derivative of a differential operator between fibered manifolds with respect to a pair of generalized vector fields. We prove that the vanishing of the Lie derivative is a necessary and sufficient condition for the equivariance of the operator under the flows of the generalized vector fields. This is the essential property relating symmetries and infinitesimal symmetries. We note that in this generalized situation the infinitesimal criterion is somehow more fundamental because the existence of flows is not guaranteed.

In the last paragraph (§8) we summarize the consequences of the preceding results for the determination of the symmetries of systems of partial differential equations.

1. THE LIE ALGEBRA OF INFINITESIMAL AUTOMORPHISMS OF A FIBERED MANIFOLD.

Let $\pi: F \to M$ be a C^∞ locally trivial fibered manifold over a second countable manifold M. We denote by $\Gamma(F)$ the set of C^∞ sections of $\pi: F \to M$, and by $C^\infty(M)$ the vector space of C^∞ real functions on M. Let $p: VF \to F$ be the <u>vertical bundle</u> of F, <u>i.e.</u> the vector subbundle of the tangent bundle of F whose fibers are the tangent spaces to the fibers of $\pi: F \to M$. All manifolds and maps are assumed to be of class C^∞. When no confusion can arise, we denote a fibered manifold and its total space by the same symbol. If μ is a map, $T\mu$ denotes its tangent mapping. We say that a differential operator is of order r if it involves the derivatives of order at most r of its arguments.

(1.1) <u>Definition</u>. <u>An infinitesimal automorphism of</u> $\pi: F \to M$ <u>is a projectable vector field</u> X <u>on</u> F.

It is clear that a vector field X on F is an infinitesimal automorphism of F if and only if X generates a <u>flow of local automorphisms</u> of the fibered manifold F, whence the terminology.

Given an infinitesimal automorphism X of F we denote by X_M the vector field on M obtained by <u>projection</u>. Conversely, an infinitesimal automorphism of F which projects onto a given vector field X_M on M is called a <u>lifting</u> of X_M to F.

The set of infinitesimal automorphisms of F is a <u>Lie algebra</u> with respect to the opposite of the usual Lie bracket of vector fields. We denote this Lie algebra by $\mathcal{A}(F)$. (The reason for considering the opposite of the usual bracket will appear later; see §3).

Let μ_t be the flow of an infinitesimal automorphism X of F, and let $(\mu_t)_M$ be the flow of X_M. For a section ψ of F, we set

$$(1.1) \qquad \mu_t.\psi = \mu_t \circ \psi \circ (\mu_t)_M^{-1}.$$

We now describe the Lie derivative of sections of F with respect to an infinitesimal automorphism X of F. For each section ψ of F and for each point x in M we consider the tangent vector to F at $\psi(x)$ defined by

$$(1.2) \qquad (L(X)\psi)(x) = \frac{d}{dt}(\mu_t.\psi)(x)|_{t=0}.$$

It is clear that $(L(X)\psi)(x)$ is vertical because for each t, $(\mu_t.\psi)(x)$ is in the fiber of F over x. Thus $L(X)\psi$ is a section of the vector bundle ψ^*VF over M induced by ψ from the vector bundle p: $VF \to F$. It is called the Lie derivative of the section ψ with respect to X. The definition of the Lie derivative in the case of a fibered manifold can be found in [21], [36] 4.3 or [14] §4. We chose here the opposite of the quantity defined by the last two authors; we shall justify this choice in §3. Our choices in previous papers have not been consistent and some corrections have to be made. The expressions in local coordinates of the vector fields must be changed into their opposite in [16] p. 955, lines 4, 6 and 18, and in [19] p. 396, lines 2 and 21, and p. 397, lines 15, 18 and 19; in addition there is a misprint in definition (2.2), p. 396, where $(X.u)_x$ should be the difference $X(j_x^k u) - (Tu)_x(X_M)_x$ and not the sum.

We recall the definition given by K. Uhlenbeck (see [34] §4) of a differential section operator D of order r from a fibered manifold π: $F \to M$ to a fibered manifold p: $G \to F$; for every section ψ of F, $D\psi$ is a section of ψ^*G which, at each point x in M, depends only on the r-jet of ψ at x. From this definition it follows that

(1.3) Proposition. The Lie derivation of sections $\psi \to L(X)\psi$ with respect to an infinitesimal automorphism X of F is a first-order differential section operator from F to VF.

If π: $F \to M$ is a vector bundle, $L(X)\psi$ can be identified with a section of F itself, and we recover the usual notion of the Lie derivative. In the case of a vector bundle, more generally any differential section operator from F to VF can be identified with a differential operator on F, i.e. from F to itself.

The fundamental properties of the Lie derivative of a section are

(1.4) $$\frac{d}{dt}(\mu_t \cdot \psi) = L(X)(\mu_t \cdot \psi),$$

(1.5) $$\frac{d}{dt}(\mu_t \cdot \psi) = (T\mu_t)(L(X)\psi).$$

This last property implies that ψ is invariant with respect to the flow of X if and only if its Lie derivative with respect to X vanishes (see [27], §24).

Let (x^i, y^α) be local coordinates on a trivialized open set of F. Latin indices range from 1 to n, the dimension of M, and Greek indices range from 1 to d, the dimension of the fiber of F. (In applications to the theory of partial differential equations, n is the number of independent variables and d is the number of unknown functions.) We shall sometimes denote (x^i) by x and (y^α) by y. We denote by ∂_i the partial derivation with respect to x^i and by ∂_α the partial derivation with respect to y^α. We use the summation convention. An infinitesimal automorphism X of F can be written in local coordinates as $X = X^i\partial_i + Y^\alpha\partial_\alpha$, where X^i are local functions of the x^j's alone, and Y^α are local functions of all the variables. Then $X_M = X^i\partial_i$.

(1.6) Lemma. <u>If, in local coordinates</u>, $X = X^i\partial_i + Y^\alpha\partial_\alpha$, <u>then</u> $(L(X)\psi)^\alpha(x) = -X^i(x)(\partial_i\psi^\alpha)(x) + Y^\alpha(x,\psi(x))$.

<u>Proof</u>. For (x^i, y^α) in F,

$$\mu_t(x^i, y^\alpha) = (x^i + tX^i(x) + O(t^2),\ y^\alpha + tY^\alpha(x,y) + O(t^2)).$$

Therefore,

$$(\mu_t \cdot \psi)^\alpha(x) = (x^i, \psi^\alpha(x^j - tX^j(x) + O(t^2)) + tY^\alpha(x,\psi(x)) + O(t^2))$$

and

$$\frac{d}{dt}(\mu_t \cdot \psi)^\alpha(x)\big|_{t=0} = -X^i(x)(\partial_i\psi^\alpha)(x) + Y^\alpha(x,\psi(x)).$$

We shall now determine which differential section operators are Lie derivations. Let us introduce at this point several definitions which will be used further on in this paper. They are generalizations of notions introduced by Palais in [35].

Linearized operators. Let D be a differential section operator of order r from π: $F \to M$ to p: $G \to F$. Let ψ be a section of F, let ϕ be a section of ψ^*VF (a vertical vector field on F along ψ), and let x be a point in M. There exists a local one-parameter family ψ^t of local sections of F such that in a neighborhood of x, $\psi^0 = \psi$ and $\left.\frac{d\psi^t}{dt}\right|_{t=0} = \phi$. We set

$$(1.7) \qquad \mathcal{V}D(\psi, \phi)(x) = \frac{d}{dt}(D\psi^t)(x)\Big|_{t=0}$$

since the right-hand side depends only on ψ and ϕ and not on the choice of ψ^t. $\mathcal{V}D(\psi,\phi)(x)$ is a tangent vector to G at $(D\psi)(x)$. This vector projects, by means of Tp, onto $\phi(x)$ which is tangent to F at $\psi(x)$, and it projects by means of $T\pi \circ Tp$ onto 0, which is tangent to M at x. Let us denote by V_MG the vertical bundle of the fibered manifold $\pi \circ p$: $G \to M$. Then $\mathcal{V}D(\psi,\phi)(x)$ is in the fiber of V_MG at $(D\psi)(x)$. It is not hard to see that $\mathcal{V}D(\psi,.)$ is a linear differential operator of order r from ψ^*VF to $(D\psi)^*V_MG$.

We shall now examine two special cases which as a matter of fact are the only ones that arise in practice. i) G is the vertical bundle VF of F. We show that the operator $\mathcal{V}D(\psi,.)$ canonically defines a differential operator $VD(\psi,.)$ from ψ^*VF to itself. Let y be in F and let z and v be elements of the fiber of VF over y. We extend v into a local vertical vector field on F which we denote by the same letter. We set $y(t) = (\exp tv)y$. Then $y(0) = y$ and $\left.\frac{dy(t)}{dt}\right|_{t=0} = v$. We also define $z(t) = T(\exp tv)_y z$, and we set $i(z,v) = \left.\frac{dz(t)}{dt}\right|_{t=0}$. The vector $i(z,v)$ is well-defined, tangent to VF at z (because $z(0) = z$), and such that $Tp(i(z,v)) = v$ (because $p(z(t)) = y(t)$). Thus if we consider

$$(1.8) \qquad VD(\psi,\phi)(x) = \mathcal{V}D(\psi,\phi)(x) - i((D\psi)(x),\phi(x)),$$

we see that it is a vector tangent to VF at $(D\psi)(x)$ whose projection by means of Tp is 0. In fact $VD(\psi,\phi)(x)$ is a vector tangent at $(D\psi)(x)$ to the fiber of p: $VF \to F$ over $\psi(x)$. Since this fiber is a vector space, we can identify the vector $VD(\psi,\phi)(x)$ with an element of the fiber. In other words, $VD(\psi,\phi)$ is a section of ψ^*VF and $VD(\psi,.)$ is a linear differential operator of order r from ψ^*VF to itself which is called the <u>linearized operator</u> of D at ψ. ii) G is of the form π^*F', where F' is a fibered manifold over M. Then D is identified with a <u>differential operator</u> from F to F' (see [34]). The operator $\mathcal{V}D(\psi,.)$ is then identified with a differential operator from ψ^*VF to $(D\psi)^*VF'$, which we denote by $VD(\psi,.)$. It is the linearized operator of D at ψ

as defined by Palais ([35], theorem 17.2). If moreover, F and F' are vector bundles, the linearized operator of D at ψ can be identified with a linear differential operator from F to F', which we shall denote again by $VD(\psi,.)$. If the operator D itself is linear, then $VD(\psi,.)$ is independent of ψ and is equal to D.

We note that if F is a vector bundle, a differential section operator D from F to VF can be identified with a differential operator from F to itself, and that both definitions of VD coincide.

We have previously called the operator VD the variation of D [19] following Lax [25]. It can also be called the Gâteaux derivative of D (see [29], [30]) or, for obvious reasons, the vertical operator of D (see [20]).

We summarize our discussion.

(1.9) Proposition. The linearized operator of a differential section operator of order r from F to VF at ψ is a linear differential operator $VD(\psi,.)$ of order r from ψ^*VF to itself. If F is a vector bundle, $VD(\psi,.)$ is identified with a linear differential operator of order r from F to itself and VD is a differential operator from F × F to F, defined by

$$VD(\psi, \phi) = \frac{d}{dt}(D(\psi + t\phi))\Big|_{t=0}. \tag{1.10}$$

To clarify matters we write the expression in local coordinates of the linearized operator of a differential section operator of order r. We denote by z^A the local coordinates in the fiber of G, where A and the capital Latin indices range from 1 to the dimension of the fiber of G. The multi-index $I(r)$ ranges over all n-multi-indices $(a_1,\ldots,a_n)$ of length r, i.e. such that $\sum_{p=1}^{n} a_p = r$.

$$(D\psi)^A(x) = D^A(x^i,\psi^\alpha(x),\partial_j\psi^\beta(x),\ldots,\partial_{I(r)}\psi^\beta(x)),$$

where each D^A is a function of the variables $(x^i,y^\alpha,y^\alpha_i,\ldots,y^\alpha_{I(r)})$. Then

$$VD(\psi,\phi)(x) = (x^i,\psi^\alpha(x),D^A,0,\phi^\beta(x),E^B)$$

with

(1.11) $$E^B = \frac{\partial D^B}{\partial y^\alpha}\phi^\alpha(x) + \frac{\partial D^B}{\partial y_j^\alpha}\partial_j\phi^\alpha(x) + \dots + \frac{\partial D^B}{\partial y_{I(r)}^\alpha}\partial_{I(r)}\phi^\alpha(x)$$

all the derivatives of D^B being evaluated at the point $(x^i, \psi^\gamma(x), \partial_m\psi^\gamma(x), \dots, \partial_{I(r)}\psi^\gamma(x))$. Also $i(z,v) = (y^\alpha, z^\alpha, v^\alpha, 0)$. So, in the case where $G = VF$,

$$VD(\psi,\phi)(x) = (x^i, \psi^\alpha(x), E^\beta).$$

If F is a vector bundle, $VD(\psi,\phi)(x)$ is identified with (E^β).

Example. Let $F = R^2 \times R$ be the trivial vector bundle with base manifold R^2 and fiber R. Let (x,t) be the coordinates on the base and let ψ be a section of F. Let $D\psi = \partial_t\psi + 3\psi\partial_x\psi + (\partial_x\psi)^2$. Then $VD(\psi,\phi) = \partial_t\psi + 3\partial_x\psi\phi + 3\psi\partial_x\phi + 2\partial_x\psi\partial_x\phi$.

Quasi-scalar operators. Quasi-scalar differential section operators are differential section operators whose linearized operators are quasi-scalar. We first recall the definition of linear quasi-scalar differential operators ([35], definition 19.33).

(1.12) Definition. A linear differential operator D on a vector bundle F is quasi-scalar if, for all x in M, the symbol of D, evaluated on a cotangent vector to M at x, is the multiplication by a scalar in the fiber of F over x.

In general the local coefficients of the first-order terms of a first-order linear differential operator are local matrix-valued functions; such an operator is quasi-scalar if its matrices are scalar multiples of the identity. For this reason, in [15] a first-order quasi-scalar linear differential operator was called an operator with scalar symbol.

(1.13) Definition. A differential section operator D from the fibered manifold F to VF is said to be quasi-scalar if, for each ψ in $\Gamma(F)$, the linearized operator $VD(\psi,.)$ is a quasi-scalar linear differential operator on ψ^*VF, and its symbol is independent of ψ.

For a careful definition of the symbol of a nonlinear differential operator as the collection of the symbols of its linearized operators see [35], 17.7. In short, a quasi-scalar differential section operator is an operator with constant scalar symbol. If F is a vector bundle and if D is linear, definition (1.13) reduces to definition (1.12).

In local coordinates, the condition for a first-order differential section operator D to be quasi-scalar can be written:

$$\frac{\partial D^\beta}{\partial y_j^\alpha}(x^i, y^\gamma, y_m^\gamma) = -X^j(x)\delta_\alpha^\beta,$$

where $X^j(x)$, for each $j = 1,2,\ldots,n$, is a local function on M.

We shall use the following property of quasi-scalar first-order operators.

(1.14) Proposition. A first-order differential operator D from the fibered manifold F to VF is quasi-scalar if and only if there exists a vector field X_M on M such that the differential section operator from F to its tangent bundle defined by $\psi \to D\psi + (T\psi)(X_M)$ is of order 0.

X_M is then unique and is called the projection of D onto M.

Proof. The proof is straightforward, using local coordinates. Let X_M^i be the components of X_M. If $(D\psi)^\beta = -X_M^i \partial_i \psi^\beta + (D_0\psi)^\beta$ where D_0 is of order 0, then D is quasi-scalar. Conversely, assume that D is quasi-scalar. The functions X^j are the components of a vector field X_M. Moreover

$$(D\psi + (T\psi)(X_M))^\beta = D^\beta(x^i, \psi^\alpha(x), \partial_j\psi^\alpha(x)) + X^i\partial_i\psi^\beta,$$

and this quantity's partial derivatives with respect to y_i^α vanish, thus it determines a differential section operator of order 0.

We denote by $\mathcal{B}(F)$ the set of quasi-scalar first-order differential section operators from F to VF. In local coordinates, an element D of $\mathcal{B}(F)$ is of the form $(D\psi)^\alpha(x) = -X^i(x)\partial_i\psi^\alpha(x) + Y^\alpha(x^j, \psi^\beta(x))$, and the corresponding linearized operators have the form

$$VD(\psi,\phi)^\alpha(x) = -X^i(x)\partial_i\phi^\alpha(x) + \frac{\partial Y^\alpha}{\partial y^\beta}(x^j, \psi^\gamma(x))\phi^\beta(x).$$

If F is a vector bundle, we note that a quasi-scalar differential section operator from F to VF is identified with a differential operator on F, and that this operator is necessarily quasi-linear. We are now able to prove that $\mathcal{A}(F)$ and $\mathcal{B}(F)$ are isomorphic.

(1.15) Proposition. *L* is a one-to-one mapping from the set of infinitesimal automorphisms of F onto the set of quasi-scalar first-order differential section operators from F to VF.

We first restate a fundamental lemma. (See [36](4.5.1) and [14] (22).)

(1.16) Lemma. For each ψ in $\Gamma(F)$, $L(X)\psi = X \circ \psi - (T\psi)(X_M)$.

Proof. Given x in M, $(T\psi)_x$ maps the tangent space to M at x into the tangent space to F at $\psi(x)$; thus the right-hand side of the equality is a tangent vector to F at $\psi(x)$; it is in fact vertical because both $(T\psi)_x(X_M)_x$ and $X_{\psi(x)}$ project onto $(X_M)_x$. Moreover,

$$(L(X)\psi)(x) = \frac{d}{dt}(\mu_t\cdot\psi)(x)\Big|_{t=0} = \frac{d}{dt}(\mu_t(\psi(\mu_{tM}^{-1}x)))\Big|_{t=0} = X_{\psi(x)} - (T\psi)_x(X_M)_x,$$

thus proving the lemma.

From this lemma it is clear that *L* maps *A*(F) into *B*(F), since $(X\circ\psi)(x)$ depends only on the 0-jet of ψ at x. To show that *L* is onto and one-to-one we construct a reciprocal mapping. Let D be in *B*(F) and let X_M be its projection onto M (proposition (1.14)). We define a vector field X on F in the following way. Let y be in the fiber of F over x, and let ψ be a section of F such that $\psi(x) = y$; then

$$X_y = (D\psi)_x + (T\psi)_x(X_M)_x.$$

It is clear that since D is quasi-scalar, X_y depends exclusively on y and is independent of the choice of ψ. Moreover, X is a projectable vector field whose projection is X_M because $(D\psi)_x$ is a vertical vector for all x. Thus to D in *B*(F) we have associated an infinitesimal automorphism X of F and, by construction, $L(X) = D$. If D is of the form $L(X)$, then the associated vector field is X itself. Thus *L* is one-to-one from *A*(F) onto *B*(F). We have also proved the following:

(1.17) Corollary. A differential section operator from F to VF is the Lie derivation with respect to a lifting of a vector field X_M on M to an infinitesimal automorphism of the fibered manifold F if and only if it is a quasi-scalar first-order differential section operator with projection X_M.

The Lie algebra structure of *B*(F). The set *B*(F) of quasi-scalar first-order differential operators from F to VF is in a natural way a vector space, because p: $VF \to F$ is a vector bundle. It is easy to see that the mapping L is an isomorphism of vector spaces from $\mathcal{A}(F)$ onto $\mathcal{B}(F)$. Moreover there exists a unique Lie bracket on $\mathcal{B}(F)$ which we denote by $[\ ,\ ]_V$ such that L becomes a Lie algebra homomorphism from the Lie algebra $\mathcal{A}(F)$ of projectable vector fields on F equipped with the opposite of the usual bracket onto $\mathcal{B}(F)$. We shall construct a Lie bracket on the vector space of all differential section operators from F to VF, whose restriction to $\mathcal{B}(F)$ has the required property. We call this bracket the vertical bracket, and we denote it also by $[,]_V$.

Let D_1 and D_2 be differential section operators of order r_1 and r_2 from F to VF. We set, for a section ψ of F,

$$(VD_1 \circ D_2)\psi = VD_1(\psi, D_2\psi);$$

thus $VD_1 \circ D_2$ is a differential section operator from F to VF. We now define a differential section operator from F to VF, denoted by $[D_1,D_2]_V$, by the formula:

$$[D_1,D_2]_V = VD_1 \circ D_2 - VD_2 \circ D_1. \tag{1.18}$$

We can write more explicitly, for ψ in $\Gamma(F)$,

$$[D_1,D_2]_V\psi = VD_1(\psi,D_2\psi) - VD_2(\psi,D_1\psi).$$

In order to prove Jacobi's identity in proposition (1.24) below, we need to define the second linearized operator V^2D of a differential section operator D from F to VD. We set

$$\begin{aligned} &V^2D(a,b,c,d) = \frac{d}{dt}VD(a^t,b^t)\Big|_{t=0} \quad \text{where} \\ &a^0 = a,\ b^0 = b,\ \frac{da^t}{dt}\Big|_{t=0} = c,\ \frac{db^t}{dt}\Big|_{t=0} = d. \end{aligned} \tag{1.19}$$

(Here a^t is a local one-parameter family of local sections of F and b^t for each t is a section of $(a^t)^*VF$, and we make the usual identifications.)

(1.20) Lemma. Let $A^{t,s}$ be a local two-parameter family of local sections of F such that

$$A^{0,0} = a,\ \frac{\partial A^{0,s}}{\partial s}\Big|_{s=0} = b,\ \frac{\partial A^{t,0}}{\partial t}\Big|_{t=0} = c,\ \frac{\partial^2 A^{t,s}}{\partial t\partial s}\Big|_{\substack{t=0\\s=0}} = d.$$

Then

$$V^2D(a,b,c,d) = \frac{\partial^2}{\partial t \partial s} D(A^{t,s}) \Big|_{\substack{t=0 \\ s=0}}.$$

This lemma follows from the definitions. It has the following consequence:

(1.21) $$V^2D(a,b,c,d) = V^2D(a,c,b,d).$$

From the definition (1.19) we see that

(1.22) $$V^2D(a,b,c,d) = V^2D(a,b,c,0) + VD(a,d).$$

We can set $VVD(a,b,c) = V^2D(a,b,c,0)$. By (1.21) VVD is symmetric in the last two arguments. Also VVD has the following property:

(1.23) Lemma. Let D_1, D_2, D_3 be three differential section operators from F to VF. Then for a section ψ of F,

$$(V(VD_1 \circ D_2))(\psi, D_3\psi) = VVD_1(\psi, D_2\psi, D_3\psi) + VD_1(\psi, VD_2(\psi, D_3\psi)).$$

Proof. By a straightforward application of the definitions we obtain

$$V(VD_1 \circ D_2)(\psi, D_3\psi) = V^2D_1(\psi, D_2\psi, D_3\psi, VD_2(\psi, D_3\psi)).$$

Then we apply (1.22) and the definition of VVD_1.

We are now in a position to prove that $[,]_V$ is a Lie bracket:

(1.24) Proposition. $[,]_V$ is a Lie bracket on the vector space of differential section operators from F to VF.

Proof. The antisymmetry of $[D_1, D_2]_V$ is obvious. Its bilinearity is proved as a result of the following properties for differential section operators D_1, D_2, D_3:

$$V(D_1 + D_2) = VD_1 + VD_2,$$

$$VD_1 \circ (D_2 + D_3) = VD_1 \circ D_2 + VD_1 \circ D_3.$$

(This last fact follows from the linearity of $VD_1(\psi, .)$.)

Using lemma (1.23) the proof of Jacobi's identity is a routine computation, which we outline. By definition of the vertical bracket

$$[[D_1,D_2]_V,D_3]_V = V(VD_1 \circ D_2 - VD_2 \circ D_1) \circ D_3$$
$$- VD_3 \circ (VD_1 \circ D_2 - VD_2 \circ D_1).$$

Thus for ψ in $\Gamma(F)$, we obtain

$$[[D_1,D_2]_V,D_3]_V\psi = V(VD_1 \circ D_2)(\psi,D_3\psi) - V(VD_2 \circ D_1)(\psi,D_3\psi)$$
$$- VD_3(\psi,VD_1(\psi,D_2\psi)) + VD_3(\psi,VD_2(\psi,D_1\psi)).$$

Each of the first two terms can be written as a sum of two terms by lemma (1.23). The above expression is then a sum of 6 terms, to which we add the 12 terms obtained by a circular permutation on D_1, D_2, D_3. Most terms appear in pairs with opposite signs and those which do not, cancel by virtue of the symmetry property of VVD.

The vertical bracket preserves the filtration by the order because if D_1 and D_2 are of order r_1 and r_2 respectively, their vertical bracket is of order $r_1 + r_2$. (See § 5.)

(1.25) <u>Proposition</u>. <u>For any vector fields</u> X_1 <u>and</u> X_2 <u>in</u> $A(F)$,

$$L[X_1,X_2] = [L(X_1),L(X_2)]_V$$

where [,] is the opposite of the usual bracket on $A(F)$.

<u>Proof</u>. An easy proof uses local coordinates. Let $X_1 = X_1^i\partial_i + Y_1^\alpha\partial_\alpha$ and let $X_2 = X_2^i\partial_i + Y_2^\alpha\partial_\alpha$. Then

$$[X_1,X_2] = -(X_1^j\partial_jX_2^i - X_2^j\partial_jX_1^i)\partial_i$$
$$- (X_1^i\partial_iY_2^\alpha - X_2^i\partial_iY_1^\alpha + Y_1^\beta\partial_\beta Y_2^\alpha - Y_2^\beta\partial_\beta Y_1^\alpha)\partial_\alpha.$$

On the other hand,

$$VL(X_1)(\psi,L(X_2)\psi)^\alpha = -X_1^i\partial_i(-X_2^j\partial_j\psi^\alpha + Y_2^\alpha(x,\psi))$$
$$+ \partial_\beta Y_1^\alpha(x,\psi)(-X_2^j\partial_j\psi^\beta + Y_2^\beta(x,\psi)),$$

and

$$VL(X_1)(\psi,L(X_2)\psi)^\alpha - VL(X_2)(\psi,L(X_1)\psi)^\alpha = (X_1^i\partial_iX_2^j - X_2^i\partial_iX_1^j)\partial_j\psi^\alpha$$
$$- X_1^i\partial_iY_2^\alpha(x,\psi) + X_2^i\partial_iY_1^\alpha(x,\psi) + Y_2^\beta(x,\psi)\partial_\beta Y_1^\alpha(x,\psi)$$

$$- Y_1^\beta(x,\psi)\partial_\beta Y_2^\alpha(x,\psi).$$

(The terms $X_1^i\partial_\beta Y_2^\alpha\partial_i\psi^\beta - X_2^i\partial_\beta Y_1^\alpha\partial_i\psi^\beta$ appear twice with opposite signs and cancel.)

Thus $L[X_1,X_2]\psi = (VL(X_1) \circ L(X_2) - VL(X_2) \circ L(X_1))\psi$.

(1.26) Proposition. $\mathcal{B}(F)$ is closed under the Lie bracket $[,]_V$ and L is a Lie algebra homomorphism from $A(F)$ onto $\mathcal{B}(F)$.

The proposition follows from proposition (1.25) and from the fact that L is a one-to-one mapping from $A(F)$ onto $\mathcal{B}(F)$. Moreover, if D_1 and D_2 have projections X_{1M} and X_{2M}, then $[D_1,D_2]_V$ has projection $[X_{1M},X_{2M}]$.

Note that it is not true in general that the vertical bracket $[,]_V$ of two differential section operators of order 1 is again of order 1. But it becomes true if both operators are quasi-scalar. This is a justification for the term quasi-scalar since this property generalizes a property of scalar differential operators.

Remark. Kolar gave an expression for the Lie derivative of a section of a fibered manifold with respect to the bracket of two projectable vector fields, using the first prolongation of the vector fields ([14] formula (26)). It is not hard to see (for instance using local coordinates) that the expression given by a 'strong difference' in Kolar's paper is exactly the value on a section of F of the vertical bracket we have defined, so that our result (proposition (1.25)) agrees with his.

If F is a vector bundle, the differential section operators from F to VF are identified with the differential operators on F, and in this case formula (1.18) defines a Lie bracket on the vector space of differential operators on F, which we call the vertical bracket. The Lie derivative of a section with respect to the opposite of the usual bracket of two projectable vector fields is the vertical bracket of the Lie derivatives with respect to each vector field. The vertical bracket reduces to the ordinary Lie bracket or commutator of differential operators only on the subspace of linear differential operators, i.e., when each operator is equal to its linearized operator.

2. THE LIE ALGEBRA OF LINEAR INFINITESIMAL AUTOMORPHISMS OF A VECTOR BUNDLE.

Let π: $F \to M$ be a real (or complex) vector bundle. The set $\Gamma(F)$

of sections of F is a real (or complex) vector space.

For a vector bundle F, we distinguish between automorphisms of F (considered as a fibered manifold) and linear automorphisms of F (those automorphisms of F which are linear on each fiber). In this paragraph, we consider the latter.

(2.1) Definition. A linear infinitesimal automorphism of π: $F \to M$ is an infinitesimal automorphism of F which generates a flow of local linear automorphisms of F.

The set of linear infinitesimal automorphisms of F is a Lie subalgebra of $A(F)$ with respect to the opposite of the usual bracket of vector fields. We denote this Lie algebra by $A_{lin}(F)$.

The Lie derivation associated with a linear infinitesimal automorphism of F is a linear first-order differential operator on F. Conversely a linear first-order differential operator on F is the Lie derivation with respect to a linear infinitesimal automorphism of F only if it is quasi-scalar.

In local coordinates, a linear infinitesimal automorphism is of the form $X = X^i \partial_i + k^\alpha_\beta y^\beta \partial_\alpha$, where X^i and k^α_β are local functions of the x^i's alone. According to lemma (1.6), the corresponding Lie derivatives are expressed by $(L(X)\psi)^\alpha = -X^i \partial_i \psi^\alpha + k^\alpha_\beta \psi^\beta$. We now formulate a simple condition which is equivalent to quasi-scalarity for linear first-order differential operators on F.

(2.2) Proposition. A linear differential operator D on F is first-order and quasi-scalar if and only if there exists a derivation D_M of $C^\infty(M)$ satisfying

$$D(f\psi) = f(D\psi) + (D_M f)\psi$$

for all ψ in $\Gamma(F)$ and all f in $C^\infty(M)$.

For the proof see [15] proposition 8. In [15] a linear differential operator satisfying this condition was called a derivative operator with respect to D_M. This term was chosen for the following reason: there is a one-to-one correspondence between vector fields on M and derivations of $C^\infty(M)$, and similarly there is a one-to-one correspondence between linear projectable vector fields on F and derivative operators on F. More precisely, a differential operator on F is the Lie derivation with respect to a linear lifting of X_M if and only if it is a derivative operator with respect to the derivation X_M of $C^\infty(M)$.

If F is a trivial vector bundle with one-dimensional fiber, the derivative operators of F are the generalized derivations of Miller ([31] chapters 1 and 8).

On the vector space of linear differential operators on F the vertical bracket defined by (1.18) reduces to the usual commutator:

$$[D,D']_V = [D,D'] = D \circ D' - D' \circ D.$$

The vector space of linear quasi-scalar first-order differential operators on F is a Lie subalgebra $\mathcal{B}_{lin}(F)$ of the Lie algebra of all linear differential operators on F equipped with the usual commutator.

(2.3) Proposition. The mapping L: $X \to L(X)$ is an isomorphism from the Lie algebra $A_{lin}(F)$ (equipped with the opposite of the usual bracket) onto the Lie algebra $\mathcal{B}_{lin}(F)$ (equipped with the usual commutator of differential operators).

This result is a direct consequence of proposition (1.26). It can also be proved directly, avoiding the complications arising from the linearized operators that are involved in the general case.

Local Lie algebra structures. We now assume that the vector bundle F is such that the vector fields on M admit canonical liftings to F, *i.e.* there exists a homomorphism λ from the Lie algebra $A(M)$ of vector fields on M equipped with the opposite of the usual bracket into the Lie algebra $A_{lin}(F)$, such that $T\pi \circ \lambda$ is the identity. For each X in $A_{lin}(F)$, we set $X_M = T\pi(X)$ and $k_X = X - \lambda(X_M)$. Then k_X is a vertical linear vector field on F which can be identified with a section of $F \otimes F^*$, where F^* denotes the dual of F. Examples of vector bundles which admit liftings are the trivial vector bundles and all the vector bundles associated with the principal frame bundle of F or its prolongations. The case of the tangent bundle of F has been considered by Lecomte [26].

The opposite of the usual Lie bracket on $A_{lin}(F)$ defines a Lie bracket on the vector space of sections of $TM \oplus (F \otimes F^*)$, giving rise to a local Lie algebra structure on this vector bundle in the sense of Kirillov [13]. (When Shiga had previously introduced the concept [37] he had called it a Lie algebra structure over the manifold M.) Let (X_{1M},k_1) and (X_{2M},k_2) be two sections of $TM \oplus (F \otimes F^*)$. A computation in local coordinates using the isomorphism between $A_{lin}(F)$ and $\mathcal{B}_{lin}(F)$ shows that

(2.4) $[(X_{1M},k_1),(X_{2M},k_2)] = ([X_{1M},X_{2M}],\ L(X_{1M})k_2 - L(X_{2M})k_1 + [k_1,k_2])$

where [,] denotes the opposite of the usual bracket on the Lie algebra of vector fields on M, and $L(X_M)k$ denotes the Lie derivative of k with respect to $\lambda(X_M)$. The local Lie algebra structure defined by formula (2.4) coincides with the structure defined by Lecomte in the case of the tangent bundle [26]. Unless d = 1, this structure is different from the local Lie algebra structure defined on trivial vector bundles by Shiga ([37], example 3, p. 340), although both reduce to the bracket of two scalar first-order linear differential operators in the case where the fiber of F is one-dimensional.

The Lie derivation of sections is a differential representation of this local Lie algebra into the space of sections of F.

3. LIE GROUP ACTIONS AND LIE ALGEBRA ACTIONS ON FIBERED MANIFOLDS.

Let F be a fibered manifold as in paragraph 1. Let G be a Lie group. An action of G on F is a homomorphism $\mu \to \mu_F$ from G into the group of automorphisms of the fibered manifold F which is smooth, i.e., such that $(\mu,y) \to \mu_F y$ is a smooth mapping from G × F to F. Any group action on F defines a representation R of G into the set $\Gamma(F)$ by

(3.1) $$R(\mu)\psi = \mu_F \circ \psi \circ \mu_M^{-1}$$

where μ_M is the projection of μ_F onto M (cf. (1.1), where $R(\mu_t)\psi$ was denoted by $\mu_t \cdot \psi$).

The Lie algebra $A(F)$ of projectable vector fields on F is a topological Lie algebra for the usual topology of uniform convergence of vector fields and their derivatives on compact sets; it is a closed subalgebra of the topological Lie algebra of all vector fields on F, and in fact it is a Fréchet space. It follows that the Lie algebra $B(F)$ of quasi-scalar first-order differential section operators from F to VF is naturally endowed with a topological Lie algebra structure by means of its canonical isomorphism with $A(F)$.

(3.2) Definition. Let g be a topological Lie algebra. An action of g on the fibered manifold F is a continuous homomorphism $X \to X_F$ from g into $A(F)$, equipped with the opposite of the usual bracket.

In particular, if g is finite-dimensional, an action of g on F is merely a homomorphism from g into $A(F)$. Any Lie algebra action gives rise to a mapping θ from g to $B(F)$ defined by

(3.3) $$\theta(X) = L(X_F).$$

$\theta(X)$ is called the Lie derivation of sections of F associated with the Lie algebra action $X \to X_F$. Proposition (1.25) yields the following corollary:

(3.4) Proposition. If $X \to X_F$ is an action of g on F, the mapping $\theta: X \to \theta(X) = L(X_F)$ is a Lie algebra homomorphism from g into $B(F)$.

Note that when F is a vector bundle, θ is a representation of g into $\Gamma(F)$ by differential operators of order 1.

(3.5) Proposition. Let $\mu \to \mu_F$ be an action of a Lie group G on the fibered manifold F. For each X in the Lie algebra g of G and each y in F we set

(3.6) $$(X_F)_y = \frac{d}{dt}((\exp tX)_F y)\Big|_{t=0}.$$

Then each X_F is in $A(F)$ and the mapping $X \to X_F$ is an action of g on F. Moreover θ is the differential of R, i.e. for each x in M,

(3.7) $$(\theta(X)\psi)(x) = \frac{d}{dt}(R(\exp tX)\psi)(x)\Big|_{t=0}.$$

Proof. It is clear that for each X in g, X_F is in $A(F)$, and that the mapping $X \to X_F$ is linear. In addition we must prove that, for two elements X_1 and X_2 of g,

(3.8) $$[X_1,X_2]_F = [X_{1F},X_{2F}],$$

the right-hand side being the opposite of the usual bracket on $A(F)$. This follows from the facts that

$$[X_1,X_2] = \frac{\partial}{\partial t}\frac{\partial}{\partial s}(\exp(tX_1)\exp(sX_2)\exp(tX_1)^{-1})\Big|_{\substack{t=0\\s=0}}$$

and that $\mu \to \mu_F$ is a group homomorphism such that $(\mu,y) \to \mu_F y$ is smooth.

We note that our definitions yield $(\exp tX)_F = \exp(tX_F)$. Thus we can write

$$\theta(X)\psi = L(X_F)\psi = \frac{d}{dt}((\exp(tX_F).\psi)\Big|_{t=0} = \frac{d}{dt}((\exp tX)_F.\psi)\Big|_{t=0}$$
$$= \frac{d}{dt}(R(\exp tX)\psi)\Big|_{t=0},$$

thus proving (3.7).

Remark. We shall sometimes call the Lie algebra action $X \to X_F$ the differential of the Lie group action $\mu \to \mu_F$. The motive for definition (3.2) should now be clear: while the differential of a Lie group action to the right is a homomorphism from the Lie algebra of the Lie group into the Lie algebra of vector fields equipped with the usual bracket (see for instance Nomizu [32], §6, p. 16), Lie group actions to the left give rise to antihomomorphisms, i.e., to homomorphisms into the Lie algebra we denoted by $A(F)$. The motive for definition (1.2) of the Lie derivation of sections should also be clear: we obtain a Lie algebra representation into the space of sections with respect to the usual commutator of differential operators (or rather the generalization of the commutator to the nonlinear case which is the vertical bracket); we shall see in paragraphs 5 and 6 that the "differential operator" aspect turns out to be more fundamental than the "vector field" aspect in generalizations, and this will justify our choice.

If F is a vector bundle we can single out among the group actions those which are linear, i.e., such that the group acts by linear automorphisms of F. Linear group actions give rise, by means of formula (3.1), to linear representations of the group into $\Gamma(F)$. It is clear that the differential of a linear Lie group action is a linear Lie algebra action in the sense that the image of g is in the Lie subalbebra $A_{lin}(F)$ of linear infinitesimal automorphisms of F. A linear Lie algebra action $X \to X_F$ gives rise to a linear representation of the Lie algebra g into $\Gamma(F)$ by means of $\theta(X) = L(X_F)$, where each $\theta(X)$ is a linear differential operator of order 1 on $\Gamma(F)$.

But if we do not assume that $\mu \to \mu_F$ (resp. $X \to X_F$) is linear, we obtain more general representations of the Lie group G (resp. the Lie algebra g) into $\Gamma(F)$ which are actually formal nonlinear representations of G (resp. g) into $\Gamma(F)$ in the sense of Flato et al. [7]. In fact the differential operators on F can be expanded into formal power series of n-linear mappings from $(\Gamma(F))^n$ $(n \geq 0)$ into $\Gamma(F)$; moreover the ordinary composition law of the mappings $R(\mu)$ coincides with the restriction of the group law on formal power series and, what is less obvious, the bracket $[,]_V$ which we have defined in (1.18) is exactly the restriction to differential operators of the bracket $[,]_*$ that Flato et al. ([7], p. 406) defined on power series. (See [20].)

4. LIE DERIVATIVES OF DIFFERENTIAL OPERATORS.

In this paragraph we shall define the Lie derivatives of differential operators with respect to pairs of infinitesimal automorphisms. In applications to the theory of partial differential equations an explicit condition for the equivariance of a differential operator is furnished by the vanishing of its Lie derivatives. Although in practice the fibered manifolds on which the operators act are vector bundles or even products of vector spaces, we shall still treat the case of fibered manifolds for the sake of completeness and because, when the fiber is curved, the introduction of the linearized operators becomes more natural. We shall consider as well the case of the generalized differential operators (defined below) because of their intrinsic geometric interest. Base-preserving differential operators appear in the classical problems but they are not essentially simpler than the more general operators which we shall discuss. When u is a linear differential operator between vector bundles, it can be considered to be a section of a vector bundle, so the definition of its Lie derivative follows from paragraphs 1 and 2, but when u is nonlinear we need the method indicated in the present paragraph.

Let F and F' be two fibered manifolds over base manifolds M and M'. A generalized differential operator of order r ($r \geq 0$) from F to F' is a morphism u from J^rF, the bundle of r-jets of F, to F', which projects onto a diffeomorphism u_M from M onto M'. The diffeomorphism u_M is called the projection of u. Generalized differential operators have been discussed in the literature, *e.g.*, by Kupershmidt [23]. Such an operator of order r from F to F' can also be viewed as a differential operator of order r from F to u_M^*F', or as a mapping from $\Gamma(F)$ to $\Gamma(F')$ such that, for ψ in $\Gamma(F)$ and x' in M', $(u\psi)(x')$ depends only on the r-jet of ψ at $u_M^{-1}x'$. The set of generalized differential operators of order r from F to F' will be denoted by $\mathrm{diff}^r(F,F')$. Any morphism u from F to F' which projects onto a diffeomorphism from M onto M' defines an element $\hat{u}$ of $\mathrm{diff}^0(F,F')$ by

$$\hat{u}\psi = u \circ \psi \circ u_M^{-1}. \tag{4.1}$$

We shall often write $u.\psi$ as in (1.1) or even $u\psi$ for $\hat{u}\psi$, thus denoting by the same letter the morphism u and the generalized differential operator of order 0 it defines. We set $\mathrm{diff}(F,F') = \bigcup_{r\geq 0} \mathrm{diff}^r(F,F')$.

If F and F' are vector bundles, we denote by $\mathrm{diff}^r_{\mathrm{lin}}(F,F')$ the set of generalized linear differential operators of order r from F to F', *i.e.*, of linear morphisms from J^rF to F' which project onto a diffeomorphism from M onto M'. We set $\mathrm{diff}_{\mathrm{lin}}(F,F') = \bigcup_{r\geq 0} \mathrm{diff}^r_{\mathrm{lin}}(F,F')$.

We shall need the following definition when we consider the Lie derivative of a generalized differential operator, in particular of a morphism of fibered manifolds:

(4.2) Definition. Let $p: G \to F'$ be a fibered manifold over F'. Let u be a generalized differential operator of order r from F to F'. A differential section u-operator of order r from F to G associates to a section ψ of F a section of $(u\psi)^*G$ which, at each point x' in M', depends only on the r-jet of ψ at $u_M^{-1}x'$.

If D is a differential section u-operator and if ψ is a section of F, then $D\psi$ is a section of G over M' whose value at $x' \in M'$ is in the fiber of G over $(u\psi)(x')$ and depends only on the r-jet of ψ at $u_M^{-1}x'$. It is clear that if $M = M'$, $F = F'$, and if u is the identity operator, we recover the notion of a differential section operator from F to G. The set of all differential section u-operators of order r from F to G will be denoted by $\mathrm{diff}_u^r(F,G)$, and we set $\mathrm{diff}_u(F,G) = \bigcup_{r \geq 0} \mathrm{diff}_u^r(F,G)$. These are vector spaces when G is a vector bundle over F'. We shall consider mainly the case of $G = VF'$. If F' itself is a vector bundle over M', then a differential section u-operator from F to VF' is just a generalized differential operator from F to F' with projection u_M.

Lie derivative of a generalized differential operator. Let u be in $\mathrm{diff}^r(F,F')$ with projection u_M. Let X be an infinitesimal automorphism of F with projection X_M, and let X' be an infinitesimal automorphism of F' with projection $X'_{M'}$. Let μ_t, μ_{tM}, μ'_t, $\mu'_{tM'}$ be respectively, the flows of X, X_M, X', $X'_{M'}$. We define $u^{(t)} = \hat{\mu}'_t \circ u \circ \hat{\mu}_t^{-1}$. It is a generalized differential operator of order k from F to F' with projection $\mu'_{tM'} \circ u_M \circ \mu_{tM}^{-1}$ such that $u^{(t)}\psi = \mu'_t . u(\mu_t^{-1} . \psi)$. (The expression $\mu . \psi$ was defined in (1.1).)

(4.3) Definition. The Lie derivative of a generalized differential operator u of order r from F to F' with respect to the pair of infinitesimal automorphisms (X,X') of F and F' is the differential section u-operator $L(X,X')u$ of order $r + 1$ from F to VF' defined by

$$(L(X,X')u)\psi(x') = \frac{d}{dt}(u^{(t)}\psi)(x')\Big|_{t=0} \tag{4.4}$$

for ψ in $\Gamma(F)$ and x' in M'.

We shall now list certain cases of particular interest:

If $M = M'$ and u is a differential operator of order r from F to F', the Lie derivative $L(X,X')u$ of u is a differential section u-operator of order r+1 from F to VF', i.e., for each point x in M, $(L(X,X')u)(x)$ is a vertical tangent vector to F' at the point $(u\psi)(x)$ which only depends upon the (r+1)-jet of ψ at x.

If F' is a vector bundle over M', the Lie derivative of a generalized differential operator of order r from F to F' is a generalized differential operator of order r+1 from F to F'. If (i) F and F' are vector bundles, (ii) $X \in A_{\text{lin}}(F)$ and $X' \in A_{\text{lin}}(F')$, and (iii) $u \in \text{diff}^r_{\text{lin}}(F,F')$, then $L(X,X')u \in \text{diff}^{r+1}_{\text{lin}}(F,F')$.

In [17] and [18] we studied the Lie derivative of a morphism u from F to F' which projects onto a diffeomorphism from M onto M', which derivative is a differential section u-operator of order 1 from F to VF'. We note that Lie derivatives of generalized differential operators appear as special cases of this situation, when F itself is the bundle of r-jets of a fibered manifold. The Lie derivatives of sections of F' also appear as special cases when $F = M = M'$.

The essential property of the Lie derivative of a generalized differential operator is that its vanishing constitutes an infinitesimal criterion for the equivariance of the operator. More precisely, we say that u is equivariant with respect to (μ_t,μ'_t) if $u^{(t)} = u$ for all t, i.e., $\hat{\mu}'_t \circ u = u \circ \hat{\mu}_t$ for all t. Then,

(4.5) Proposition. *Let μ_t (resp. μ'_t) be a one-parameter group of automorphisms of the fibered manifold F (resp. F') with infinitesimal generator X (resp. X'). A necessary and sufficient condition for u to be equivariant with respect to the pair (μ_t,μ'_t) is that $L(X,X')u = 0$.*

For the proof, see theorem (2.2) of [18].

Computation of $L(X,X')u$. We shall give a formula for the Lie derivative of a generalized differential operator in terms of the Lie derivation of sections. To state the formula in its full generality we need to extend the definition of the linearized operator given in §1 to the case of a generalized differential operator u from F to F': the linearized operator of u at ψ is a linear generalized differential operator $Vu(\psi,.)$ from ψ^*VF to $(u\psi)^*VF'$. (The definition is copied after (1.7) and (1.8).) If F and F' are vector bundles, the linearized operator of u at ψ can be identified with a linear generalized operator from F to F'. If D is a differential section operator from F to VF, and if u is in $\mathrm{diff}^r(F,F')$, we denote by $Vu \circ D$ the differential section u-operator from F to VF' defined by $(Vu \circ D)\psi = Vu(\psi, D\psi)$.

(4.6) Proposition. For any u in $\mathrm{diff}^r(F,F')$ and any X in $A(F)$, X' in $A(F')$, $L(X,X')u = L(X') \circ u - Vu \circ L(X)$.

Proof. Both sides of the equality map sections of F into sections of $(u\psi)^*VF'$. Moreover

$$
\begin{aligned}
(L(X,X')u)\psi &= \frac{d}{dt}\mu'_t . u(\mu_t^{-1}.\psi)\Big|_{t=0} \\
&= \lim_{t\to 0} \frac{1}{t}(\mu'_t . u(\mu_t^{-1}.\psi) - u(\mu_t^{-1}.\psi)) + \lim_{t\to 0} \frac{1}{t}(u(\mu_t^{-1}.\psi) - u\psi) \\
&= L(X')(u\psi) - Vu(\psi, L(x)\psi).
\end{aligned}
$$

Whenever F and F' are vector bundles, proposition (4.6) expresses the equality of two generalized differential operators from F to F'.

Lie derivatives with respect to brackets. We first remark that the set $h(F,F')$ of mappings L which associate to u in diff(F,F') Lu in $\mathrm{diff}_u(F,VF')$, is a Lie algebra under the vertical bracket: it is a vector space in a natural way, the linearized $VL(u,.)$ of L at u is defined by

$$
VL(u,v) = \frac{d}{dt}(Lu^t)\Big|_{t=0}, \quad \text{where} \quad u^0 = v \quad \text{and} \quad \frac{du^t}{dt}\Big|_{t=0} = v,
$$

and is a linear map from $\mathrm{diff}_u(F,VF')$ to itself, and $[L_1,L_2]_V = VL_1 \circ L_2 - VL_2 \circ L_1$ defines a Lie bracket. (The proof is analogous to that of proposition (1.24).) We shall prove that the Lie derivation of generalized differential operators under a pair of Lie brackets $([X_1,X_2],[X'_1,X'_2])$ is the vertical bracket of the Lie derivations $L(X_1,X'_1)$ and $L(X_2,X'_2)$, which bracket we have just defined.

From proposition (4.6), we obtain, for ψ in $\Gamma(F)$,

$$
\begin{aligned}
\frac{d}{dt}(L(X,X')u^t)\psi\Big|_{t=0} &= \frac{d}{dt}L(X')(u^t\psi)\Big|_{t=0} - \frac{d}{dt}Vu^t(\psi, L(X)\psi)\Big|_{t=0} \\
&= VL(X')(u\psi, v\psi) - Vv(\psi, L(X)\psi).
\end{aligned}
$$

Whence

$$
(4.7) \qquad VL(X,X')(u,v)\psi = VL(X')(u\psi, v\psi) - Vv(\psi, L(X)\psi).
$$

Definition (1.19) of the second linearized operator and properties

(1.21) and (1.22) carry over to the case of generalized differential operators. We obtain, for a section ψ of F and a section ϕ of ψ^*VF,

$$(4.8)\quad V(L(X,X')u)(\psi,\phi) = VL(X')(u\psi,Vu(\psi,\phi)) - V^2u(\psi,L(X)\psi,\phi,VL(X)(\psi,\phi)).$$

We now state and prove the generalization of proposition (1.25).

(4.9) <u>Proposition</u>. <u>For any</u> u <u>in</u> $\mathrm{diff}^r(F,F')$, X_1, X_2 <u>in</u> $A(F)$, X_1', X_2' in $A(F')$.

$$[L(X_1,X_1'),L(X_2,X_2')]_V u = L([X_1,X_2],[X_1',X_2'])u,$$

<u>where</u> [,] <u>denotes the opposite of the usual bracket on</u> $A(F)$ <u>and</u> $A(F')$.

<u>Proof</u>. We apply the left-hand side of this equality to a section ψ of F.

$$([L(X_1,X_1'),L(X_2,X_2')]_V u)\psi = VL(X_1,X_1')(u,L(X_2,X_2')u)\psi - VL(X_2,X_2')(u,L(X_1,X_1')u)\psi.$$

By formula (4.7) we transform this expression to:

$$VL(X_1')(u\psi,(L(X_2,X_2')u)\psi) - V(L(X_2,X_2')u)(\psi,L(X_1)\psi) - VL(X_2')(u\psi,(L(X_1,X_1')u)\psi) + V(L(X_1,X_1')u)(\psi,L(X_2)\psi).$$

By proposition (4.4) and the linearity of VL(u,v) in v, we have:

$$VL(X_1')(u\psi,(L(X_2,X_2')u)\psi) = VL(X_1')(u\psi,L(X_2')u\psi) - VL(X_1')(u\psi,Vu(\psi,L(X_2)\psi)).$$

By formula (4.8) and the preceding calculations, we obtain:

$$\begin{aligned}([L(X_1,X_1'),L(X_2,X_2')]_V u)\psi &= VL(X_1')(u\psi,L(X_2')u\psi) - VL(X_1')(u\psi,Vu(\psi,L(X_2)\psi))\\ &- VL(X_2')(u\psi,L(X_1')u\psi) + VL(X_2')(u\psi,Vu(\psi,L(X_1)\psi))\\ &- VL(X_2')(u\psi,Vu(\psi,L(X_1)\psi))\\ &+ V^2u(\psi,L(X_1)\psi,L(X_2)\psi,VL(X_2)(\psi,L(X_1)\psi))\\ &+ VL(X_2')(u\psi,Vu(\psi,L(X_2)\psi))\\ &- V^2u(\psi,L(X_2)\psi,L(X_1)\psi,VL(X_1)(\psi,L(X_2)\psi)).\end{aligned}$$

Canceling terms and applying the generalized versions of (1.21) and (1.22) and proposition (1.25), we finally obtain:

$$\begin{aligned}([L(X_1,X_1'),L(X_2,X_2')]_V u)\psi &= VL(X_1')(u\psi,L(X_2')u\psi) - VL(X_2')(u\psi,L(X_1')u\psi)\\ &\quad + Vu(\psi,VL(X_2)(\psi,L(X_1)\psi)) - Vu(\psi,VL(X_1)(\psi,L(X_2)\psi))\\ &= L[X_1',X_2'](u\psi) - Vu(\psi,L[X_1,X_2]\psi)\\ &= (L([X_1,X_2],[X_1',X_2'])u)\psi,\end{aligned}$$

thereby proving the proposition.

Proposition (4.9) states that L: $(X,X') \in A(F) \times A(F') \to L(X,X') \in h(F,F')$ is a Lie algebra homomorphism. It is actually an injective homomorphism. If F' is a <u>vector bundle</u>, an element L of $h(F,F')$ is a mapping from diff(F,F') to itself such that u and Lu project onto the same diffeomorphism u_M from M to M'. The image of $A(F) \times A(F')$ under L is a Lie subalgebra of $h(F,F')$ of mappings of degree 1 from diff(F,F') to itself. We could further clarify its structure by defining the analogue of the quasi-scalar differential operators of §1. If F and F' are both vector bundles, the image of $A_{lin}(F) \times A_{lin}(F')$ under L is a Lie subalgebra of $h(F,F')$ on which the vertical bracket reduces to the usual commutator; its elements can be defined by purely algebraic properties, which generalize those of the derivative operators of proposition (2.2).

<u>Local Lie algebra structure</u>. Let F and F' be two vector bundles over the same manifold M and such that the vector fields on M admit canonical liftings to F and F'. (See the end of §2.) The opposite of the usual bracket on $A_{lin}(F) \times A_{lin}(F')$ defines a local Lie algebra structure on the vector bundle $TM \oplus (F \otimes F^*) \oplus TM \oplus (F' \otimes F'^*)$. The explicit formula is easily deduced from (2.4). The Lie derivation of linear differential operators from F to F' is a differential representation of this local Lie algebra.

<u>Lie algebra actions and nonlinear representations.</u> Let g be a topological Lie algebra acting on fibered manifolds F and F' by $X \in g \to X_F \in A(F)$ and $X \in g \to X'_{F'} \in A(F')$. The Lie derivative $\underline{\theta}(X)u$ of a generalized differential operator u from F to F' with respect to X is the differential section u-operator from F to VF' defined by

$$\underline{\theta}(X)u = L(X_F,X'_{F'})u. \tag{4.10}$$

Setting $\theta(X) = L(X_F)$ and $\theta'(X) = L(X'_{F'})$ as in definition (3.3), we obtain from proposition (4.6),

$$(4.11) \qquad \underline{\theta}(X)u = \theta'(X) \circ u - Vu \circ \theta(X).$$

As a corollary of proposition (4.9), we see that $\underline{\theta}$ is a Lie algebra homomorphism from g into $h(F,F')$, i.e., for X_1 and X_2 in g,

$$(4.12) \qquad \underline{\theta}[X_1,X_2] = [\underline{\theta}(X_1),\underline{\theta}(X_2)]_V.$$

If F' is a vector bundle, $\underline{\theta}$ defines a representation, in general nonlinear, of g into any invariant vector space contained in diff(F,F'). In particular, if F = F', the restriction of $\underline{\theta}$ to the vector space of differential operators from F to F' defines a representation of g which is linear if and only if g acts linearly on F and F'. These remarks correct a statement we made in [18], (2.16): diff(F,F') itself is not a vector space, even when F' is a vector bundle, but some of its subsets, such as the set of generalized differential operators which project onto a given diffeomorphism, constitute vector spaces; to obtain a representation of g we must restrict $\underline{\theta}$ to such a vector space. It would be interesting to determine upon which subspaces g acts irreducibly.

5. THE LIE ALGEBRA OF GENERALIZED VECTOR FIELDS.

Because the geometric theory which we have just exposed has applications to the study of systems of partial differential equations, it is desirable to treat the generalized infinitesimal symmetries which have been introduced more recently in an analogous fashion. Here we study in some detail the geometric properties of the generalized vector fields on a fibered manifold.

The notion of k-vector field was introduced by R. Hermann [8], §7, [9] and developed by H. H. Johnson [11], [12]. We adapt Johnson's definition given in [11] to the case of a fibered manifold π: $F \to M$. For every nonnegative integer k, we denote by ρ^k: $J^kF \to F$ the fibered manifold of k-jets of F, ρ^k being the target projection. We denote by $T\pi$ the differential of π which is the projection of the tangent space TF of F onto the tangent space of M. We denote by p: $TF \to F$ the tangent bundle of F.

(5.1) Definition. A k-vector field on the fibered manifold $\pi: F \to M$ is a mapping X form J^kF to TF such that:

(i) X is a base-preserving morphism from the fibered manifold $\rho^k: J^kF \to F$ into the fibered manifold $p: TF \to F$, and

(ii) for each pair (z,z') of elements of J^kF having the same source,

$$T\pi(X_z) = T\pi(X_{z'}).$$

Assumption (i) simply says that the value of X at $j_x^k\psi$, where ψ is a local section of F and x is a point in M, is a tangent vector to F at the point $\psi(x)$. Assumption (ii) means that the projection of $X(j_x^k\psi)$ depends only on x and not on the choice of ψ, and implies that every k-vector field X on F projects onto a vector field X_M on M. A 0-vector field on F is a vector field on F (in the ordinary sense) projectable onto M, i.e., an infinitesimal automorphism of the fibered manifold M. With the usual identifications, any k-vector field is also a k'-vector field for $k' \geq k$.

We recover the notion of k-vector field as defined by Johnson in [11] as the particular case where F is a trivial bundle $M \times M' \to M$. Thus our definition actually generalizes Johnson's.

Let $A^k(F)$ be the vector space of all k-vector fields on F, for $k \geq 0$. In particular, $A^0(F)$ is the vector space $A(F)$ of §1. Let $\mathcal{D}(F) = \bigcup_{k \geq 0} A^k(F)$. An element of $\mathcal{D}(F)$ is called a generalized vector field on F.

If F is a vector bundle, it is natural to define a linear (resp. affine) k-vector field as a k-vector field which is a linear (resp. affine) morphism of vector bundles from $\rho^k: J^kF \to F$ into $p: TF \to F$. The linear and affine k-vector fields play a particular role in the theory of generalized symmetries. (See for example [10], p. 440.)

Expression in local coordinates. Let (x^i, y^α) be local coordinates on a trivialized open set of F as in §1. We can choose local coordinates on J^kF, $(x^i, y^\alpha, y_i^\alpha, \ldots, y_{I(k)}^\alpha)$, where I(k) ranges over all n-multi-indices of length k, i.e., $I(k) = (a_1, a_2, \ldots, a_n)$ and $\sum_{p=1}^{n} a_p = k$. A k-vector field on F is written locally

(5.2) $$X = X^i \partial_i + Y^\alpha \partial_\alpha,$$

where the X^i's are functions of the x^j's alone and the Y^α's are functions of all the coordinates $(x^i, y^\beta, y_i^\beta, \ldots, y_{I(k)}^\beta)$ on J^kF.

For an affine k-vector field X, each Y^α is an affine function of the variables $y^\beta, y_i^\beta, \ldots, y_{I(k)}^\beta$, that is:

$$Y^\alpha(x^i, y^\beta, y_i^\beta, \ldots, y_{I(k)}^\beta) = q^\alpha(x) + q_\beta^\alpha(x) y^\beta + q_\beta^{\alpha i}(x) y_i^\beta + \ldots + q_\beta^{\alpha I(k)}(x) y_{I(k)}^\beta.$$

<u>The Lie derivation of sections with respect to a k-vector field</u>. To each k-vector field X on F there corresponds a differential section operator from F into the vertical bundle VF of F. For each section ψ of F, and for each point x of M, we set

$$(5.3) \qquad (L(X)\psi)_x = X(j_x^k\psi) - (T\psi)_x(X_M)_x,$$

where T denotes the differential of the mapping ψ. The difference of the two terms in the right-hand side is well defined in the tangent vector space to F at $\psi(x)$, and it is clear that it is a vertical tangent vector, both terms having the same projection $(X_M)_x$ on M. Thus to each section ψ of F, $L(X)$ associates $L(X)\psi$, a vertical tangent vector field to F along ψ.

In the case k = 0, (5.3) reduces to the Lie derivative of sections with respect to a projectable vector field (see lemma (1.16)): $L(X)$ is a differential section operator of order 1 from F to VF, and whenever X is not 0, $L(X)$ is actually of order 1. For $k \geq 1$, we obtain a differential section operator from F to VF which is of order k. See below for a more detailed study.

If F is a vector bundle, $L(X)$ can be identified with a differential operator from F to itself, which obviously generalizes the classical Lie derivative.

<u>The Lie derivative in local coordinates</u>. Let $X = X^i\partial_i + Y^\alpha\partial_\alpha$, be a k-vector field ($k \geq 0$) on F. By the definition of $T\psi$ it is easy to see that

$$(T\psi)_x(X_M)_x = X^i(x)\partial_i + (X^i(x)\partial_i\psi^\alpha(x))\partial_\alpha.$$

Moreover

$$X(j_x^k\psi) = X^i(x)\partial_i + Y^\alpha(x, \psi^\beta(x), \partial_i\psi^\beta(x), \ldots, \partial_{I(k)}\psi^\beta(x))\partial_\alpha.$$

Thus $(L(X)\psi)_x$ is the vertical tangent vector at the point $\psi(x)$,

$$(L(X)\psi)_x = [-X^i(x)\partial_i\psi^\alpha(x) + Y^\alpha(x,\psi^\beta(x),\partial_i\psi^\beta(x),\dots,\partial_{I(k)}\psi^\beta(x))]\partial_\alpha,$$

and we obtain

$$(5.4) \quad (L(X)\psi)^\alpha_x = -X^i(x)\partial_i\psi^\alpha(x) + Y^\alpha(x,\psi^\beta(x),\partial_i\psi^\beta(x),\dots,\partial^\beta_{I(k)}(x)).$$

We shall sometimes write $X.\psi$ instead of $L(X)\psi$. We shall also write somewhat improperly

$$(X.\psi)^\alpha = -X^i\partial_i\psi^\alpha + Y^\alpha(x,\psi,\partial_i\psi,\dots,\partial_{I(k)}\psi).$$

It is clear from these formulas that, when F is a vector bundle, if X is a k-vector field with $k \geq 1$, then $L(X)$ acts on the sections of F as a differential operator of order k. These formulas are those assumed <u>a priori</u> in the physics literature [2], [1] and in [33], except that we have assumed that the X^i's depend only on x, but this is not a loss of generality as we shall see presently.

<u>Normal form of a k-vector field</u>. We now interpret lemma 2 of Kumei [22] and a remark by Olver [33], p. 16.

(5.6) <u>Proposition</u>. <u>If</u> $k \geq 1$, <u>for each</u> k-<u>vector field</u> X <u>on</u> F, <u>there exists a unique vertical</u> k-<u>vector field</u> X <u>on</u> F <u>such that</u> $L(\tilde{X}) = L(x)$.

In fact, set $\tilde{X}(j^k_x\psi) = X(j^k_x\psi) - (T\psi)_x(X_M)_x$. It is clear that $\tilde{X}$ is vertical and that $L(\tilde{X}) = L(X)$. Conversely if X is vertical and satisfies $L(\tilde{X}) = L(X)$, then by (5.3), $\tilde{X}$ has to satisfy $X(j^k_x\psi) - (T\psi)_x(X_M)_x = \tilde{X}(j^k_x\psi)$, whence the unicity of $\tilde{X}$.

If X is a 0-vector field, there exists a unique vertical 1-vector field $\tilde{X}$ such that $L(\tilde{X}) = L(X)$. Indeed, we set $\tilde{X}(j^1_x\psi) = X_{\psi(x)} - (T\psi)_x(X_M)_x$.

For a k-vector field X ($k \geq 0$) we call $\tilde{X}$ the <u>normal form</u> of X. This is Olver's term in [33]. In local coordinates, if $X = X^i\partial_i + Y^\alpha\partial_\alpha$, then $\tilde{X} = Z^\alpha\partial_\alpha$ with

$$Z^\alpha = Y^\alpha - X^i y^\alpha_i,$$

and for each section ψ of F,

$$(L(X)\psi)^{\alpha} = (L(\tilde{X})\psi)^{\alpha} = -X^{i}\partial_{i}\psi^{\alpha} + Y^{\alpha}(x,\psi,\partial_{i}\psi,\ldots,\partial_{I(k)}\psi).$$

We denote by v^k the restriction of the mapping $X \to \tilde{X}$ to the vector space $A^k(F)$ of k-vector fields on F. If $k = 0$, v^0 is a one-to-one mapping from $A(F)$ onto a proper subspace of the space of vertical 1-vector fields. Locally the vertical 1-vector fields which are the normal form of a 0-vector field can be written $(X^i y^{\alpha}_i - Y^{\alpha})\partial_{\alpha}$.

If $k \geq 1$, v^k maps $A^k(F)$ onto the vector space $V^k(F)$ of all vertical k-vector fields. The kernel of v^k is the subspace of $V^1(F)$ consisting of the 1-vector fields X such that for all sections ψ, $X(j^1_x\psi) = (T\psi)_x(X_M)_x$. The kernel is locally generated over the ring of functions on M by the n 1-vector fields $\partial_i + y^{\alpha}_i\partial_{\alpha}$. Any k-vector field X can be written in a unique way as the sum of a vertical k-vector field $\tilde{X}$ and a 1-vector field $\hat{X}$ with standard form 0. In local coordinates, if $X = X^i\partial_i + Y^{\alpha}\partial_{\alpha}$, then $\tilde{X} = (Y^{\alpha} - X^i y^{\alpha}_i)\partial_{\alpha}$ and $\hat{X} = X^i(\partial_i + y^{\alpha}_i\partial_{\alpha})$.

We shall now translate the above results into propositions concerning the differential operators of order k on F.

(5.7) Proposition. For all $k \geq 0$, the restriction of the mapping L to the vector space $V^k(F)$ of vertical k-vector fields on F is a canonical isomorphism from $V^k(F)$ onto the vector space of all differential section operators of order k from F to VF.

Proof. The mapping L is obviously linear. Given a differential section operator D of order k from F to VF, for each point x in M, $(D\psi)(x)$ is an element in the vertical tangent space of F at $\psi(x)$ which depends only on the k-jet of ψ at x. We denote it by $X^D(j^k_x\psi)$. It is clear that X^D is the unique vertical k-vector field on F such that $L(X^D) = D$ since, according to (5.3), for a vertical k-vector field X, $(L(X)\psi)_x = X(j^k_x\psi)$.

Assume that F is a vector bundle. The proposition then asserts that the most general differential operator of order k ($k \geq 0$) on F is obtained as the Lie derivation with respect to a vertical k-vector field. We note that to affine (resp. linear) vertical k-vector fields there correspond affine (resp. linear) differential operators of order k. When $k = 0$, the proposition expresses the fact that a vertical vector field on F is a base-preserving morphism from F into itself. When $k = 1$, the proposition implies that any first-order differential operator on F is the Lie derivation of sections of F with respect to a certain (unique) vertical 1-vector field. We know from proposition (1.15) that such an operator, if and only if it is quasi-scalar, is also the

Lie derivation with respect to an ordinary vector field (which is not vertical if the operator is not of order 0).

(5.8) Proposition. If $k \geq 1$, the restriction of L to $A^k(F)$ is a canonical projection from the vector space $A^k(F)$ of k-vector fields on F onto the vector space of all differential section operators of order k from F to VF. The kernel of this projection is the space of 1-vector fields with standard form 0.

The proposition follows from proposition (5.7), the property $L \circ v^k = L$, and the preceding study of the kernel of v^k. We note that proposition (1.15) settles the case where $k = 0$, and it should be compared to proposition (5.8). We summarize the results for a vector bundle F. If $k = 0$, L maps the space $A(F)$ of projectable vector fields injectively into (but not onto) the space of differential operators of order 1; its image is the space of quasi-scalar first-order differential operators. For generalized vector fields with $k \geq 1$, the situation is very different from that of the classical case $k = 0$ because L maps $A^k(F)$ onto the space of all differential operators of order k, but it is not injective.

Because of proposition (5.6), in all applications the k-vector fields considered ($k \geq 1$) will be assumed to be vertical, i.e., of the form $X = Y^\alpha(x, y^\beta, y_i^\beta, \ldots, y_{I(k)}^\beta)\partial_\alpha$ corresponding to differential operators of order k defined by $(X.\psi)^\alpha = Y^\alpha(x, \psi^\beta, \partial_i\psi^\beta, \ldots, \partial_{I(k)}\psi^\beta)$.

We say that a generalized vector field X is equivalent to 0 if and only if $\tilde{X} = 0$. Two generalized vector fields are equivalent if and only if their difference is equivalent to 0. (The same notion was introduced for Lie-Bäcklund operators by Anderson and Ibragimov, [1], p. 56.)

If F is a vector bundle of rank 1 -- the case of one scalar unknown function in the theory of partial differential equations --, every affine 1-vector field on F is equivalent to a 0-vector field.

The Lie algebra of generalized vector fields. The space $\mathcal{D}(F) = \bigcup_{k \geq 0} A^k(F)$ of generalized vector fields is, in a natural way, a filtered vector space. We wish to define a Lie bracket on $\mathcal{D}(F)$ that will make $\mathcal{D}(F)$ a filtered Lie algebra and that will extend the opposite of the usual bracket of projectable vector fields.

We have already proved (proposition (1.24)) that the vertical bracket $[,]_V$ defined by formula (1.18) is a Lie bracket on the vector space of all differential section operators from F to VF. By proposition (5.7) the vector space $\mathcal{V}(F)$ of the vertical generalized vector fields on F is isomorphic under L to the vector space of differential section operators from F to VF, and therefore the vertical bracket defines a unique Lie bracket on $\mathcal{V}(F)$ such that L becomes a Lie algebra isomorphism. We define a Lie bracket on $\mathcal{D}(F)$ by considering $\mathcal{D}(F)$ as the direct sum of the Lie algebra $\mathcal{V}(F)$ and the kernel of v^1, considered as an Abelian Lie algebra. In other words, we set, for two generalized vector fields X_1 and X_2,

$$[X_1,X_2] = [\tilde{X}_1,\tilde{X}_2] \tag{5.9}$$

where $[\tilde{X}_1,\tilde{X}_2]$ is defined by the requirement $L[\tilde{X}_1,\tilde{X}_2] = [L(\tilde{X}_1),L(\tilde{X}_2)]_V$, and we call $[X_1,X_2]$ the vertical bracket of X_1 and X_2.

Let X_1 be a k_1-vector field and X_2 a k_2-vector field. Without loss of generality we assume that $k_1 \leq k_2$ and that X_1 and X_2 are vertical. In local coordinates $X_1 = Y_1^\alpha\partial_\alpha$ and $X_2^\alpha = Y_2^\alpha\partial_\alpha$. Then

$$[L(X_1),L(X_2)]_V\psi^\alpha = VY_1(\psi,Y_2^\beta(\psi)) - VY_2^\alpha(\psi,Y_1^\beta(\psi)). \tag{5.10}$$

Taking into account the local expressions for the operators VY_1^α and VY_2^α (formula (1.11)), we obtain

$$\begin{aligned}[L(X_1),L(X_2)]_V\psi^\alpha = {} & -Y_1^\beta \frac{\partial Y_2^\alpha}{\partial y^\beta} + Y_2^\beta \frac{\partial Y_1^\alpha}{\partial y^\beta} - \partial_i Y_1^\beta \frac{\partial Y_2^\alpha}{\partial y_i^\beta} \\ & + \partial_i Y_2^\beta \frac{\partial Y_1^\alpha}{\partial y_i^\beta} - \ldots - \partial_{I(k_1)} Y_1^\beta \frac{\partial Y_2^\alpha}{\partial y_{I(k_1)}^\beta} \\ & + \partial_{I(k_1)} Y_2^\beta \frac{\partial Y_1^\alpha}{\partial y_{I(k_1)}^\beta} - \ldots - \partial_{I(k_2)} Y_1^\beta \frac{\partial Y_2^\alpha}{\partial y_{I(k_2)}^\beta},\end{aligned}$$

where each operator in the sum is evaluated at ψ. Therefore, if $k_1 = k_2 = k$,

$$\begin{aligned}[X_1,X_2] = -\Bigg(& Y_1^\beta \frac{\partial Y_2^\alpha}{\partial y^\beta} - Y_2^\beta \frac{\partial Y_1^\alpha}{\partial y^\beta} + \partial_i Y_1^\beta \frac{\partial Y_2^\alpha}{\partial y_i^\beta} - \partial_i Y_2^\beta \frac{\partial Y_1^\alpha}{\partial y_i^\beta} \\ & + \ldots + \partial_{I(k)} Y_1^\beta \frac{\partial Y_2^\alpha}{\partial y_{I(k)}^\beta} - \partial_{I(k)} Y_2^\beta \frac{\partial Y_1^\alpha}{\partial y_{I(k)}^\beta}\Bigg)\partial_\alpha.\end{aligned} \tag{5.11}$$

If X_1 and X_2 are not in normal form, we obtain the formula for

$[X_1,X_2]$ by replacing Y_1^α and Y_2^α in this formula by $Y_1^\alpha - X_1^i y_i^\alpha$ and $Y_2^\alpha - X_2^i y_i^\alpha$ respectively.

We observe that this is just the opposite of the usual bracket if X_1 and X_2 are projectable vector fields on F in the ordinary sense. By the bilinearity of the bracket just defined it suffices to prove that the two brackets coincide in two particular cases:

i) $X_1 = Y_1^\alpha \partial_\alpha$ and $X_2 = Y_2^\alpha \partial_\alpha$ where $Y_1(\psi)$ and $Y_2(\psi)$ depend only on the 0-jet of ψ. In this case formula (5.11) reduces to

$$[X_1,X_2] = -\left(Y_1^\beta \frac{\partial Y_2^\alpha}{\partial y^\beta} - Y_2^\beta \frac{\partial Y_1^\alpha}{\partial y^\beta} \right)$$

which is indeed the opposite of the usual bracket.

ii) $X_1 = X_1^i \partial_i$ and $X_2 = X_2^i \partial_i$. In this case we have to apply formula (5.11) with $Y_1^\alpha = -X_1^i y_i^\alpha$ and $Y_2^\alpha = -X_2^i y_i^\alpha$. We obtain

$$[X_1,X_2] = -(\partial_i X_1^j X_2^i y_j^\alpha - \partial_i X_2^j X_1^i y_j^\alpha)\partial_\alpha = (X_1^i \partial_i X_2^j - X_2^i \partial_i X_1^j) y_j^\alpha \partial_\alpha$$
$$= -(X_1^i \partial_i X_2^j - X_2^i \partial_i X_1^j)\partial_j,$$

which is again the opposite of the usual expression of the Lie bracket.

We also note that the bracket we have defined above on generalized vector fields by means of the vertical bracket of differential operators coincides with the opposite of the bracket defined by Johnson in [11] definition 3. This is clear both from the expression in local coordinates (5.11) and from lemma 1 and definition 3 of Johnson.

If $L(X_1)$ is of order k_1 and $L(X_2)$ of order k_2, $[L(X_1), L(X_2)]_V$ is of order $k_1 + k_2$. (It is actually of order $k_1 + k_2 - 1$ when the fiber of F is one-dimensional and $k_1 \geq 1$, $k_2 \geq 1$.) Thus the bracket of a k_1-vector field and a k_2-vector field is a $(k_1 + k_2)$-vector field. The vector space $\mathcal{D}(F)$ of generalized vector fields on F is a filtered Lie algebra in the following sense: $\mathcal{D}(F)$ is the union of the vector spaces $A^k(F)$, $k \geq 0$, $A^{k+1} \supset A^k$ for each $k \geq 0$, and $[A^{k_1}, A^{k_2}] \subset A^{k_1+k_2}$ for all $k_1 \geq 0$, $k_2 \geq 0$.

<u>Local Lie algebra structure</u>. Assume that F is a vector bundle such that the vector fields on M admit canonical liftings to F. (See the end of §2.) The linear k-vector fields on F are the sections of the vector bundle $TM \oplus (F \otimes (J^kF)^*) = TM \oplus \mathrm{Hom}(J^kF, F)$. If we consider the infinite-dimensional vector bundle JF of infinite jets of F, we can

form the inductive vector bundle $TM \oplus \mathrm{Hom}(JF,F)$ (see [37], §1 or [24], 1.1). The linear generalized vector fields are the sections of this vector bundle and the vertical bracket defines a local Lie algebra structure on this vector bundle.

The Lie derivation of sections is a differential representation of this local Lie algebra into the space of sections of F.

Lie algebra actions and nonlinear representations. Let $\mathcal{g}$ be a topological Lie algebra. We assume that $\mathcal{D}(F)$ is equipped with the inductive topology defined by the natural topologies of the subspaces $A^k(F)$.

(5.12) Definition. A generalized action of $\mathcal{g}$ on F is a continuous homomorphism from $\mathcal{g}$ into the topological Lie algebra $\mathcal{D}(F)$.

If $\mathcal{g}$ is finite-dimensional, a generalized action of $\mathcal{g}$ on F is simply an assignment $X \in \mathcal{g} \to X_F \in \mathcal{D}(F)$ (the X_F's are differential operators of arbitrary order) such that the vertical bracket $[X_{1F}, X_{2F}]$ corresponds to the Lie bracket $[X_1, X_2]$.

If F is a vector bundle, the generalized linear actions correspond to actions on the sections of F by linear differential operators of arbitrary order. Since in that case the vertical bracket coincides with the usual commutator, we recognize what physicists call (linear) realizations of the Lie algebra $\mathcal{g}$.

Any generalized action of $\mathcal{g}$ on F gives rise to a representation of $\mathcal{g}$ into the space of sections of F. This representation is nonlinear unless the action itself is linear.

6. FLOWS OF GENERALIZED VECTOR FIELDS

Recall from paragraph 1, formula (1.4) that the essential property of the Lie derivative of sections of a fibered manifold F with respect to a projectable vector field is that $\frac{d}{dt}(\mu_t \cdot \psi)\big|_{t=0} = L(X)(\mu_t \cdot \psi)$ where μ_t is the flow of X. We shall presently use this property as the definition of the flow in the case of the generalized vector fields. The transformations μ_t will act on the set of sections, or rather of local sections, of F. But, unlike the situation in the classical case, $k = 0$, they will not arise from automorphisms of F nor even of a higher-order jet bundle of F. This follows from a theorem of Ibragimov and Anderson [10] extending an old result of Bäcklund. More precisely, following R. Hermann [8] and H. H. Johnson [12], we set

(6.1) Definition. Let X be a generalized vector field on a fibered manifold F. Let ψ be a section of F. An integral curve of X through ψ is a mapping Ψ: $(x,t) \to \Psi(x,t)$ from $U \times I$ to F, where U is a nonempty open set in M and I is a nonempty open interval in $\mathbb{R}$ containing 0, such that for each t in I, $x \to \Psi(x,t)$ is a local section of F, denoted by $\mu_t \cdot \psi$, satisfying on U

$$\frac{d}{dt}(\mu_t \cdot \psi) = L(X)(\mu_t \cdot \psi) \tag{6.2}$$

$$\mu_0 \cdot \psi = \psi. \tag{6.3}$$

The mapping $\psi \to \mu_t \cdot \psi$ is called the flow of X.

Thus as Olver ([33], chapter 1) and Kupershmidt ([24], theorem 5.6) have pointed out, finding the flow of a given generalized vector field X is equivalent to solving the system of partial differential evolution equations with initial conditions

$$\begin{cases} \dfrac{\partial \Psi}{\partial t} = L(X)\Psi \\ \Psi(x,0) = \psi(x). \end{cases} \tag{6.4}$$

The unknown function $\Psi(x,t)$ is a function of n+1 variables; in the right-hand side of the equation $L(X)$ is a differential operator of order k (for X a k-vector field) which involves only the derivatives with respect to x and not to t; this is why the system is in evolution form. This system is nonlinear unless X is a linear generalized vector field. Of course existence and unicity of the solution do not hold in general. (For results concerning flows on Banach manifolds, see, e.g., Chernoff and Marsden [5], §3 and the references therein.)

If X is such that the initial-value problem (6.4) has a unique, maximal local solution, the flow of X is a local one-parameter group of mappings from the set of local sections of F into itself, i.e., when both sides are defined, the following relations hold:

$$\mu_{t_1} \cdot (\mu_{t_2} \cdot \psi) = \mu_{t_1+t_2} \cdot \psi \quad \text{and} \quad (\mu_t)^{-1} \cdot \psi = \mu_{-t} \cdot \psi. \tag{6.5}$$

A simple example which can be found in [1] and [33] is the flow of the 1-vector field $X = y_x \partial_y$ where x and y are the coordinates on the trivial vector bundle $\mathbb{R} \times \mathbb{R} \to \mathbb{R}$. System (6.4) in this case is

$$\begin{cases} \dfrac{\partial \Psi}{\partial t} = \dfrac{\partial \Psi}{\partial x}, \\ \Psi(x,0) = \psi(x). \end{cases}$$

The flow of X is therefore given by $(\mu_t \cdot \psi)(x) = \psi(x+t)$. But this example is just the classical case in disguise because X is equivalent to the 0-vector field $X = -\partial_x$. For an example of a flow in the genuinely generalized case, see [19] example (2.4) (but the vector field X should read $X = v\partial_u + u_{xx}\partial_v$.) Another example is furnished by $X = y_{xx}\partial_y$ (see [1], p. 55); finding the flow of X amounts to solving the heat equation; thus in this case there is only a "semi-flow".

We can generalize formula (1.5) as follows:

(6.6) <u>Proposition</u>. <u>If the generalized vector field</u> X <u>admits a flow</u> μ_t, <u>for a local section</u> ψ <u>of</u> F,

$$\frac{d}{dt}(\mu_t \cdot \psi) = (T\mu_t)(L(X)\psi)$$

<u>when both sides are defined</u>.

<u>Proof</u>. We note that

$$\frac{d}{dt}(\mu_t \cdot \psi)(x) = \frac{d}{ds}(\mu_t \cdot \mu_s \cdot \psi)(x)\Big|_{s=0}.$$

Since $(\mu_t \cdot \mu_s \cdot \psi)(x) = \mu_t((\mu_s \cdot \psi)(\mu_{tM}^{-1}x))$, it follows that the above expression is a vertical tangent vector to F at $\mu_t(\psi(\mu_{tM}^{-1}x))$, image of $\frac{d}{ds}(\mu_s \cdot \psi)(\mu_{tM}^{-1}x)\Big|_{s=0}$ under the tangent mapping to μ_t at the point $\psi(\mu_{tM}^{-1}x)$. Whence

$$\frac{d}{dt}(\mu_t \cdot \psi)(x) = (T\mu_t)_{\psi(\mu_{tM}^{-1}x)}\,(L(X)\psi)_{\mu_{tM}^{-1}x}.$$

As a consequence of proposition (6.6) we see that

(6.7) <u>Proposition</u>. <u>A local section</u> ψ <u>of</u> F <u>is invariant under the flow of</u> X <u>if and only if its Lie derivative with respect to</u> X <u>vanishes</u>.

This is the extension of the classical result (see Lichnerowicz [27], §24). It gives an infinitesimal criterion for the invariance of a section, and is therefore useful in the applications.

<u>Exponential formula</u>. We now assume that the manifolds and the maps we consider are analytic, and that F is a vector bundle. We say that a generalized vector field X on F admits an analytic flow if, for each local section ψ of F, for each x in an open set U of M, there exists a

neighborhood Ω of 0 in $\mathcal{R}$, which may depend on ψ and x, such that $(\mu_t\cdot\psi)(x) = \Psi(x,t)$ is an analytic function of t from Ω to the fiber of F over x, a finite-dimensional vector space. By Taylor's formula, an analytic flow satisfies for each x in U and t in Ω,

$$(\mu_t\cdot\psi)(x) = \sum_{n=0}^{\infty} \frac{t^n}{n!} \frac{d^n(\mu_t\cdot\psi)}{dt^n}(x)\bigg|_{t=0}.$$

In order to evaluate the coefficient of t^n in the above series, we set

$$L^{(0)}(X)\psi = \psi,$$
$$L^{(1)}(X)\psi = L(X)\psi,$$

and we define by induction differential operators on F, $L^{(n)}(X)$, satisfying, for $n \geq 2$,

$$L^{(n)}(X) = \nabla L^{(n-1)}(X) \circ L(X),$$

that is, for each ψ in $\Gamma(F)$,

$$L^{(n)}(X)\psi = \nabla L^{(n-1)}(X)(\psi, L(X)\psi).$$

(Each $\nabla L^{(n)}(X)$ is defined by means of formula (1.10).)

(6.8) <u>Lemma</u>. <u>For each</u> $n \geq 0$, <u>the differential operator</u> $L^{(n)}(X)$ <u>satisfies</u>

$$\frac{d^n(\mu_t\cdot\psi)}{dt^n}(x) = (L^{(n)}(X)(\mu_t\cdot\psi))(x)$$

<u>Proof</u>. We prove the lemma by induction. For n = 0, the property is trivial. For n = 1, it follows from (6.2) in the definition of the flow. We assume that the lemma holds for all integers $\leq n-1$. By the definition of $L^{(n)}(X)$, the definition of the linearized operators, and the property $\frac{d}{ds}(\mu_{t+s}\cdot\psi)\big|_{s=0} = \frac{d}{dt}(\mu_t\cdot\psi) = L(X)(\mu_t\cdot\psi)$, we obtain

$$\begin{aligned} L^{(n)}(X)(\mu_t\cdot\psi) &= \nabla L^{(n-1)}(X)(\mu_t\cdot\psi, L(X)(\mu_t\cdot\psi)) \\ &= \frac{d}{ds}L^{(n-1)}(X)(\mu_{t+s}\cdot\psi)\Big|_{s=0} = \frac{d}{dt}L^{(n-1)}(X)(\mu_t\cdot\psi). \end{aligned}$$

By the induction assumption, the property holds for n.

The lemma implies in particular that

$$\frac{d^n(\mu_t\cdot\psi)}{dt^n}(x)\bigg|_{t=0} = (L^{(n)}(X)\psi)(x)$$

and therefore we have proved the following proposition:

(6.9) Proposition. If X is a generalized vector field which admits an analytic flow μ_t, then for each local section of F, x in U, t in Ω

$$(\mu_t\cdot\psi)(x) = \sum_{n=0}^{\infty} \frac{t^n}{n!}(L^{(n)}(X)\psi)(x).$$

A slightly different version of (6.9) was proved by Johnson ([11], theorem 3) who called the resulting formula an "exponential formula". Indeed, since μ_t is the flow of X it is reasonable to call it the exponential of X and to denote it by exp(tX). Proposition (6.9) expresses the equality, for each local section ψ of F and for small t's, of the two local sections of F, $\exp(tX).\psi$ and $\sum_{n=0}^{\infty} \frac{t^n}{n!}L^{(n)}(X)\psi$. This equality can be further summarized by

$$\exp(tX) = \sum_{n=0}^{\infty} \frac{t^n}{n!}L^{(n)}(X).$$

where both sides act on the local sections of F. Thus the operators $L^{(n)}(X)$ play the role of the successive powers of the variable X in the expansion of the operator exp(tX).

We have $L^{(2)}(X) = VL(X) \circ L(X)$. We can express $L^{(3)}(X)$ in terms of the differential operator $V^2L(X)$ on F^4, which was defined in (1.19):

$$\begin{aligned} L^{(3)}(X)\psi = VL^{(2)}(X)(\psi,L(X)\psi) &= \frac{d}{dt} L^{(2)}(X)(\mu_t\cdot\psi)\Big|_{t=0} \\ &= \frac{d}{dt}VL(X)(\mu_t\cdot\psi,L(X)(\mu_t\cdot\psi))\Big|_{t=0} \\ &= V^2L(X)(\psi,L(X)\psi,L(X)\psi,VL(X)(\psi,L(X)\psi)). \end{aligned}$$

In short we can write $L^{(3)}(X) = V^2L(X) \circ VL(X) \circ L(X)$. More generally we could define the differential operator $V^nL(X)$ on F^{2^n} and express $L^{(n)}(X)$ as a composition of $V^{n-1}L(X), \ldots, V^2L(X), VL(X), L(X)$.

Remark. If X is a linear generalized vector field, $L(X)$ is a linear differential operator and $L^{(2)}(X) = L(X) \circ L(X)$; more generally, $L^{(n)}(X)$ is the nth power of the operator $L(X)$. Thus in the linear case we recover a form of the usual exponential formula for linear operators in Banach spaces. The power series expansion in example 4 of [1], p. 55, is a particular case of the exponential formula in the linear case.

Flows of generalized vector fields and Lie-Bäcklund transformations. It follows from proposition (6.9) that, in general, the value of $\mu_t.\psi$ at a point x in M depends on all the successive partial derivatives of ψ at x, that is, on the infinite jet of ψ at x. We are thus led to regard the flow of a generalized vector field X, at least formally, as a mapping of the set of local sections of F into itself, arising from a morphism from the infinite-dimensional vector bundle JF of infinite jets of F into itself. If X is not equivalent to a 0-vecor field the flow of X does not arise from a morphism from a finite-dimensional jet bundle of F into F. This viewpoint is implicit in the work of Anderson, Ibragimov and their collaborators. (See for instance [1], pp. 8, 47, 51.) We interpret some of the definitions found in reference [1] in geometric language, but as elsewhere in this article, we consider only transformations and infinitesimal transformations which are projectable. The bundle F is the trivial vector bundle $R^n \times R^d$. Projectable Lie-Bäcklund operators (pp. 48-50) are infinite prolongations of ∞-vector fields (the analogue of the k-vector fields when the bundle J^kF is replaced by the infinite-dimensional bundle JF). The Lie-Bäcklund equations (11.7) are the defining equations of the flow of a Lie-Bäcklund operator; formulation (12.8) - (12.9) of these equations for the case of a vertical vector field is identical to the evolution equation with initial condition we have considered in (6.4). Lie-Bäcklund tangent transformation groups appear as the one-parameter groups of automorphisms of JF corresponding to flows of Lie-Bäcklund operators.

7. LIE DERIVATIVES OF DIFFERENTIAL OPERATORS WITH RESPECT TO GENERALIZED VECTOR FIELDS

We retain the notations of the preceding paragraphs. Although we treat the case of the generalized differential operators from F to F', we keep in mind that the particular case of differential operators is the most important one for applications, and also that subtantial simplifications occur when F and F' are vector bundles. For a simplified exposition of this question and some examples, see [19].

Since we are not assured in general of the existence of the flow of a generalized vector field, we cannot extend definition (4.3) to the case where X and X' are generalized vector fields. Instead we shall adopt the generalization of proposition (4.6) as the definition of the Lie derivative and we shall show that when X and X' possess flows, (4.5) still holds in this more general case.

(7.1) Definition. The Lie derivative of a generalized differential operator u from F to F' with respect to the pair (X,X'), where X is a k-vector field on F and X' is a k'-vector field on F', is the differential section u-operator $L(X,X')u$ from F to VF' defined by

$$L(X,X')u = L(X') \circ u - Vu \circ L(X). \tag{7.2}$$

If u is of order r, and if a = max(k,k',1), then $L(X,X')u$ is of order r+a.

(7.3 Proposition. Assume that the flows μ_t and μ'_t of X and X' exist For small values of t, we define $u^{(t)}$ (as in §4) by $u^{(t)}\psi = \mu'_t.u(\mu_t^{-1}.\psi)$. The following relations hold:

$$\frac{d}{dt}(u^{(t)}\psi)(x') = ((L(X,X')u^{(t)})\psi)(x') \tag{7.4}$$

and, in particular,

$$\frac{d}{dt}(u^{(t)}\psi)(x')\Big|_{t=0} = ((L(X,X')u)\psi)(x'). \tag{7.5}$$

Moreover

$$\frac{d}{dt}(u^{(t)}\psi)\cdot(x') \tag{7.6}$$

$$= (T\mu'_t)_{(u(\mu_t^{-1}.\psi))(\mu'^{-1}_{tM'}x')}((L(X,X')u(\mu_t^{-1}\circ\psi))(\mu'^{-1}_{tM'}x').$$

(ψ is a local section of F, x' is a point in M' and t is a real number small enough for both sides to be defined.)

Proof. To prove (7.4), we observe that

$$\frac{d}{dt}(u^{(t)}\psi)(x') = \frac{d}{ds}\mu'_{t+s}.(u(\mu_t^{-1}.\psi)(x')\Big|_{s=0} + \frac{d}{ds}\mu'_t.(u(\mu_{t+s}^{-1}.\psi)(x')\Big|_{s=0}$$

and we evaluate each term in this sum:

$$\frac{d}{ds}\mu'_{t+s}.(u(\mu_t^{-1}.\psi)(x')\Big|_{s=0} = \frac{d}{ds}\mu'_s.(\mu'_t\circ u(\mu_t^{-1}\circ\psi))(x')\Big|_{s=0}$$

$$= \frac{d}{ds}\mu'_s.(u^{(t)}\psi)(x')\Big|_{s=0} = L(X')(u^{(t)}\psi)(x')$$

and

$$\frac{d}{ds}\mu'_t.(u(\mu_{t+s}^{-1}.\psi)(x')\Big|_{s=0} = \frac{d}{ds}\mu'_t\circ(u(\mu_t^{-1}.\mu_s^{-1}.\psi))(x')\Big|_{s=0}$$

$$= \frac{d}{ds}(u^{(t)}(\mu_s^{-1}.\psi))(x')\Big|_{s=0} = -Vu^{(t)}(\psi, L(X)\psi)(x').$$

Combining these two expressions, (7.4) follows from definition (7.2) of $L(X,X')u^{(t)}$.

To prove (7.6), we evaluate each term, replacing an expression of the form $\mu.\psi$ by its value $\mu \circ \psi \circ \mu_M^{-1}$. We obtain:

$$\frac{d}{ds}\mu'_t\mu'_s u\mu_t^{-1}\psi(\mu_{tM}u_M^{-1}\mu'^{-1}_{sM'}\mu'^{-1}_{tM'}(x'))\Big|_{s=0}$$

$$= (T\mu'_t)_{u\mu_t^{-1}\psi\mu_{tM}u_M^{-1}\mu'^{-1}_{tM'}x'}\frac{d}{ds}\mu'_s u\mu_t^{-1}\psi(\mu_{tM}u_M^{-1}\mu'^{-1}_{sM'})(\mu'^{-1}_{tM'}(x'))\Big|_{s=0}$$

$$= (T\mu'_t)_{(u(\mu_t^{-1}.\psi))(\mu'^{-1}_{tM'}x')}(L(X')(u(\mu_t^{-1}.\psi)))(\mu'^{-1}_{tM'}x')$$

and

$$\frac{d}{ds}\mu'_t(u\mu_s^{-1}\mu_t^{-1}\psi\mu_{tM}\mu_{sM}u_M^{-1})(\mu'^{-1}_{tM'}x')\Big|_{s=0}$$

$$= (T\mu'_t)_{(u(\mu_t^{-1}.\psi))(\mu'^{-1}_{tM'}x')}\frac{d}{ds}(u\mu_s^{-1}\mu_t^{-1}\psi\mu_{tM}\mu_{sM})(u_M^{-1}(\mu'^{-1}_{tM'}x'))\Big|_{s=0}$$

$$= -(T\mu'_t)_{(u(\mu_t^{-1}.\psi))(\mu'^{-1}_{tM'}x')}(Vu(\mu_t^{-1}.\psi, L(X)(\mu_t^{-1}.\psi))(u_M^{-1}(\mu'^{-1}_{tM'}.x')).$$

Adding these two expressions and comparing with definition (7.2), we obtain (7.6).

Since (7.5) follows from the definition (7.1), it is clear that definition (7.1) is indeed a generalization of definition (4.3).

Using proposition (7.3) it is easy to prove an infinitesimal criterion for the equivariance of an operator:

(7.7) *Proposition.* *Let* X *(resp.* X'*) be a generalized vector field on* F *(resp.* F'*) which admits a flow* μ_t *(resp.* μ'_t*). A generalized differential operator* u *from* F *to* F' *is equivariant with respect to the pair* (μ_t, μ'_t) *if and only if* $L(X,X')u = 0$.

Proof. If u is equivariant with respect to (μ_t,μ'_t), then $u^{(t)} = u$ for all t. Thus $L(X,X')u = 0$ by (7.5). Conversely, if $L(X,X')u = 0$, then by (7.6) $u^{(t)}$ is independent of t. Since $u^{(0)} = u$, this means that $u^{(t)} = u$ for all t, or that u is equivariant with respect to (μ_t,μ'_t).

Proposition (4.5) appears as a particular case of this result.

Exponential formula. We show how proposition (6.9) extends to the case of a pair of generalized vector fields acting on differential operators. Let F and F' be analytic vector bundles over the analytic manifold M. Let u be a differential operator from F to F'. We assume that for each local section ψ of F, for each x in an open set U of M, $(u^{(t)}\psi)(x)$ is an analytic function of t in a neighborhood Ω of 0. We set

$$L^{(0)}(X,X')u = u,$$
$$L^{(1)}(X,X')u = L(X,X')u,$$

and we define by induction $L^{(n)}(X,X')$, satisfying, for $n \geq 2$,

$$L^{(n)}(X,X') = VL^{(n-1)}(X,X') \circ L(X,X'),$$

that is, for each differential operator u,

$$L^{(n)}(X,X')u = VL^{(n-1)}(X,X')(u,L(X,X')u).$$

The following proposition is a consequence of Taylor's formula and of formula (7.4):

(7.8) Proposition. For x in U and t in Ω,

$$(u^{(t)}\psi)(x) = \sum_{n=0}^{\infty} \frac{t^n}{n!}((L^{(n)}(X,X')u)\psi)(x).$$

Proposition (7.8) expresses the equality of the two differential operators $u^{(t)}$ and $\sum_{n=0}^{\infty} \frac{t^n}{n!}L^{(n)}(X,X')u$. We can regard the differential operator $u^{(t)}$ as the image of u under the flow of the pair (X,X'), and we may denote it by $\exp t(X,X')u$. With this notation, proposition (7.8) can be symbolized by the following formal equality between mappings from the vector space of differential operators from F to F' into itself

$$\exp t(X,X') = \sum_{n=0}^{\infty} \frac{t^n}{n!} L^{(n)}(X,X').$$

and this is why we call (7.8) an exponential formula.

We note that formula (7.9) appears as the extension to the nonlinear case of the well-known formula from the theory of Lie groups ([6], 19.11.2.2),

$$\text{(7.9)} \qquad \mathrm{Ad}(\exp(tX)) = \sum_{n=0}^{\infty} \frac{t^n}{n!}(\mathrm{ad}\ X)^n.$$

Indeed if F = F', X = X' and X is linear, for any <u>linear</u> differential operator u,

$$L(X,X)u = L(X) \circ u - u \circ L(X) = (\mathrm{ad}\ L(X))u,$$

and, moreover, $L^{(n)}(X,X) = (L(X,X))^n$. Thus, in this case, when restricted to linear differential operators on F, formula (7.9) reduces to

$$\exp t(X,X) = \sum_{n=0}^{\infty} \frac{t^n}{n!}(\mathrm{ad}\ L(X))^n,$$

which is very similar to (7.9).

8. SYMMETRIES AND GENERALIZED SYMMETRIES OF DIFFERENTIAL OPERATORS

As we have shown in several related publications [15] - [19], the geometric notions of infinitesimal automorphisms, generalized vector fields and Lie derivatives are the tools necessary to understand the infinitesimal symmetries and the generalized infinitesimal symmetries of differential operators, and to obtain algorithmic methods leading to their determination, especially in the nonlinear case. We summarize here some of the results concerning the symmetries of differential operators. They will serve as examples of applications of the theorems proved in this paper. We retain the notations of the preceding paragraphs. We assume that F and F' are vector bundles over the same base manifold M, and that D is a differential operator from F to F'.

<u>Symmetries and infinitesimal symmetries</u>. We say that a one-parameter group μ_t of automorphisms of the fibered manifold F is a group of <u>symmetries</u> of D if there exists a local one-parameter group μ'_t of automorphisms of the fibered manifold F', projecting onto the same diffeomorphisms μ_{tM} of M as μ_t, preserving the zero section of F', and such

that D is equivariant with respect to the pair (μ_t, μ'_t) for all t. This condition is clearly sufficient to ensure the fundamental property that $D\psi = 0$ implies $D(\mu_t \cdot \psi) = 0$, i.e., that the set of local solutions of D is invariant under μ_t.

It follows from proposition (4.5) that an infinitesimal automorphism X of F is the infinitesimal generator of a one-parameter group of symmetries of D if and only if there exists an infinitesimal automorphism X' of F', projecting onto the same vector field X_M on M as X, preserving the zero section of F', and such that $L(X,X')D = 0$. Such an infinitesimal automorphism of F is called an infinitesimal symmetry of D. If X is an infinitesimal symmetry of D, then $D\psi = 0$ implies that $VD(\psi, L(X)\psi) = 0$, i.e., the set of local solutions of D is infinitesimally invariant under X.

We chose the preceding definitions here and in [16] because of the elegant relationship between symmetries and infinitesimal symmetries and because the infinitesimal symmetries of a differential operator, so defined, can actually be determined in a algorithmic way [16]. Although the equivalence with the more classical, a priori weaker, definition of an infinitesimal symmetry is far from being obvious and probably does not hold in general, it holds for interesting and sufficiently general cases.

Generalized infinitesimal symmetries and generalized symmetries. The definition of the generalized infinitesimal symmetries is more basic than that of generalized symmetries. A generalized vector field X on F is a generalized infinitesimal symmetry of D if, for each local section ψ of F, there exists a vertical generalized vector field X' on F' (which may depend on ψ) preserving the zero section of F', and such that $L(X,X')D\psi = 0$. If X is a generalized infinitesimal symmetry of D, then $D\psi = 0$ implies that $VD(\psi, L(X)\psi) = 0$, i.e., the set of local solutions of D is infitesimally invariant under X.

The flow of a generalized infinitesimal symmetry of D is called a local one-parameter group of generalized symmetries of D. The generalized symmetries are not, in general, automorphisms of F.

It follows from proposition (7.7) that if X is a generalized infinitesimal symmetry of D, and if X and X' have flows μ_t and μ'_t, D is equivariant with respect to the pair (μ_t, μ'_t) for small t's. In particular the set of local solutions of D is invariant under μ_t. We therefore obtain, with these definitions, a satisfactory relationship between infinitesimal and 'global' properties. The definition of generalized infinitesimal symmetries proposed in [19], (3.3) is less interesting than the one above because, contrary to what we asserted in

[19], (3.4), it is not obvious that it would imply the 'global' invariance of the set of solutions of the operator under the flow of a generalized infinitesimal symmetry.

Clearly we obtain symmetries and infinitesimal symmetries in the ordinary sense as special cases of the more general objects just defined: the requirement that X' be vertical is not a restriction since any 0-vector field is equivalent to a vertical 1-vector field. The fact that we let X' depend on the section ψ is prompted by examples from the literature, where generalized infinitesimal symmetries are known (e.g., for the sine-Gordon and for the Korteweg-de Vries equations) and satisfy our definition only if we let X' depend on ψ. (See the examples in [19], (3.6) and (3.7).)

Many different names have been ascribed to what we call the generalized infinitesimal symmetries of a differential operator, and to objects with closely related definitions: admissible operators, invariance or covariance infinitesimal transformations, generalized infinitesimal similarities, etc... . We note that our definition coincides with that of a generalized vector field leaving a system of partial differential equations invariant in the sense of Johnson [12]; indeed if D is of order r, VD $\circ$ L(X) is the rth prolongation of X acting on D considered as a function on J^rF; thus theorem 1 of [12] implies, in the analytic case, that the flow of a generalized infinitesimal symmetry of D leaves the set of solutions of D invariant.

In conclusion, we cite two problems for whose solution the geometry developed here seems essential.

An interesting, and not altogether obvious, theorem is that the generalized infinitesimal symmetries of a given differential operator form a Lie algebra under the vertical bracket defined in §§1 and 5 (see [20]). Johnson had proved ([12], §3) a similar result using the bracket he defined by means of prolognations of k-vector fields (which is the opposite of the vertical bracket), but his approach does not apply to generalized infinitesimal symmetries considered as differential operators, which is the most important viewpoint for applications. In [29] and [30], Magri defines the bracket of two nonlinear operators on a vector space by means of Gâteaux derivatives, proves that it is a Lie bracket and applies this fact to expose the Lie algebra structure of the set of symmetries of a differential operator on a trivial vector bundle. The geometric notion of the vertical bracket yields the generalization of Magri's result and unifies the two widely divergent approaches.

Finally, we mention that, in the much needed general theory of

Bäcklund transformations, generalized infinitesimal symmetries play an essential role. We have shown in [19] that what is classically called a Bäcklund transformation between two evolution equations defined by evolution operators P and Q is a differential operator for which the pair (P,Q) constitutes a generalized infinitesimal symmetry. This approach seems to hold promise for applications: as we have shown in §7, conditions concerning generalized infinitesimal symmetries can be expressed by the vanishing of Lie derivatives which can lead to an algorithmic determination of unknown Bäcklund transformations.

REFERENCES

1. R. L. Anderson and N. H. Ibragimov, Lie-Bäcklund transformations in applications, SIAM Studies in Applied Mathematics 1, Philadelphia 1979.
2. R. L. Anderson, S. Kumei and C. E. Wulfman, Generalization of the concept of invariance of differential equations, Phys. Rev. Lett. 28 (1972), 988-991.
3. E. Bessel-Hagen, Über die Erhaltungssätze der Elektrodynamik, Math. Ann. 84 (1921), 258-276.
4. G. Birkhoff, Hydrodynamics, Princeton University Press, Princeton N. J. 1950 (2nd edition 1960).
5. P. R. Chernoff and J. E. Marsden, Properties of infinite dimensional Hamiltonian systems, Lecture Notes in Math. 425, Springer-Verlag, Berlin 1974.
6. J. Dieudonné, Cours d'analyse 4, Gauthiers-Villars, Paris 1971.
7. M. Flato, G. Pinczon and J. Simon, Non linear representations of Lie groups, Ann. sci. Éc. norm. sup. Sér. IV, 10 (1977), 405-418.
8. R. Hermann, E. Cartan's geometric theory of partial differential equations, Advances in Math. 1 (1965), 265-317.
9. R. Hermann, Geometry, Physics and Systems, Marcel Dekker, New York 1973.
10. N. H. Ibragimov and R. L. Anderson, Groups of Lie-Bäcklund transformations, Doklady Akad. Nauk SSSR 227 (1976), 539-542 (Soviet Math., Doklady 17 (1976), 437-441).
11. H. H. Johnson, Bracket and exponential for a new type of vector field, Proc. Amer. math. Soc. 15 (1964), 432-437.
12. __________, A new type of vector field and invariant differential systems, ibid. 675-678.
13. A. A. Kirillov, Local Lie algebras, Uspehi mat. Nauk 31 (1976), 57-76 (Russ. math. Surveys 31 (1976), 55-75).
14. I. Kolář, Fundamental vector fields on associated vector bundles, Časopis Pěst. Mat. 102 (1977), 419-425.
15. Y. Kosmann, On Lie transformation groups and the covariance of differential operators, Differential Geometry and Relativity, M. Cahen and M. Flato eds., Reidel, Dordrecht 1976.
16. Y. Kosmann-Schwarzbach, Sur les transformations de similitude des équations aux dérivées partielles, C. r. Acad. Sci., Paris, 287 Sér. A (1978), 953-956.
17. __________, Dérivées de Lie des morphismes de fibrés, Publ. math. Univ. Paris VII, 3 (1978), 55-72.
18. __________, Infinitesimal conditions for the equivariance of

morphisms of fibered manifolds, Proc. Amer. math. Soc. 77 (1979), 374-380.

19. __________, Generalized symmetries of nonlinear partial differential equations, Lett. math. Phys. 3 (1979), 395-404.

20. __________, The vertical bracket and its applications to symmetries and Bäcklund transformations, to appear.

21. N. Kuiper and K. Yano, On geometric objects and Lie groups of transformations, Nederl. Akad. Wet., Proc., Ser. A 58 (1955), 411-420.

22. S. Kumei, Invariance transformations, invariance group transformations, and invariance groups of the sine-Gordon equations, J. math. Phys. 16 (1975), 2461-2468.

23. B. A. Kupershmidt, On the geometry of jet manifolds, Uspehi mat. Nauk 30 (1975), 211-212.

24. __________, Geometry of jet bundles and the structure of Lagrangian and Hamiltonian formalisms, preprint (1979).

25. P. D. Lax, Integrals of nonlinear equations of evolution and solitary waves, Comm. pure appl. Math. 21 (1968), 467-490.

26. P. Lecomte, Sur l'algèbre de Lie des automorphismes infinitésimaux du fibré tangent, C. r. Acad. Sci., Paris, 288 Sér. A (1979), 661-663.

27. A. Lichnerowicz, Géométrie des groupes de transformations, Dunod, Paris 1958.

28. S. Lie, Gesammelte Abhandlungen, vol. 3, Teubner, Leipzig 1922.

29. F. Magri, An operator approach to Poisson brackets, Ann. of Phys. 99 (1976), 196-228.

30. __________, An operator approach to symmetries, Nuovo Cimento 34 B (1976), 334-344.

31. W. Miller, Jr., Lie theory and special functions, Academic Press, New York 1968.

32. K. Nomizu, Lie groups and differential geometry, Math. Soc. Japan, 1956.

33. P. J. Olver, Symmetry groups and conservation laws in the formal variational calculus, preprint (1978).

34. R. S. Palais, Banach manifolds of fiber bundle sections, Actes Congrès Intern. Math. (Nice 1970), vol. 2, Gauthiers-Villars, Paris 1971, 243-249.

35. __________, Foundations of global non-linear analysis, Benjamin, New York 1968.

36. S. Salvioli, On the theory of geometric objects, J. diff. Geometry 7 (1972), 257-278.

37. K. Shiga, Cohomology of Lie algebras over a manifold, I, J. math.

Soc. Japan 26 (1974), 324-361.

Department of Mathematics
Brooklyn College
City University of New York
Brooklyn, N.Y. 11210

Lie algebras of order 0 on a manifold

by P. Lecomte

1. The infinite dimensional Lie algebras on a manifold have been intensely studied during the last years by many people. Those authors have mostly worked on the Lie algebras of vector fields and in particular the Lie algebra of infinitesimal automorphisms of some geometric structure, except K. Shiga who has defined a more general class of Lie algebras called differential Lie algebras on a manifold.

The problems studied about infinite dimensional Lie algebras on a manifold can be roughly classified as follows :

(i) To study those algebras from a purely algebraic point of view, is it possible to extend the basic tools and results of the finite dimensional case ?

(ii) To study the cohomology of Chevalley of those algebras.

(iii) To study the relations between the algebraic properties of those algebras and the geometric properties of the manifolds on which they are defined ; does the algebraic structure characterize the geometric structure ?

The purpose of this paper is to introduce and to study a large class of infinite dimensional Lie algebras on a manifold - called of order 0 - which has not received much attention although it has many interesting properties. We have started in this way with some particular case and now provide a general setting for the theory that we want to illustrate here.

In contrast to what happens in the theory of Lie algebras of vector fields, the case of algebras of order 0 gives rise to a very rich answer to the above problem (i) ; about problem (iii), we will see that in many case, the algebraic structure of Lie algebras of order 0 characterizes the differentiable structure of the base manifold as in the case of the algebra of the vector fields.

On the other hand, the methods used here are rather different from the basic tools of the study of Lie algebras of vector fields and are more algebraic.

In this paper, we shall restrict ourselves to the study of the group of automorphisms of a Lie algebra of order 0 ; it is a good illustration for the results obtained and the methods used and, moreover, we have no place to describe the whole theory.

2. We now define a Lie algebra of order 0 on a manifold M.

Let L be a finite dimensional Lie algebra and let $E \xrightarrow{p} M$ be a vector bundle over the manifold M, with typical fiber L. We say that E is a vector bundle of type L if it admits a cocycle whose transition maps take their values in the group aut L of automorphisms of L. In this case each fiber of E has a natural Lie algebra structure (defined by the cocycle) so that the space $\Gamma_k(E)$ of global cross-sections of class C_k $(k=0,1,2,\ldots,\infty)$ of E has a natural Lie algebra structure (defined pointwise). The Lie algebras so obtained are called Lie algebras of order 0 over M. We will set $\Gamma(E) = \Gamma_\infty(E)$.

This class of Lie algebras on M is very large. We shall only mention the following examples. Let $F \xrightarrow{q} M$ be any real vector bundle over M. Then E = Hom(F,F) is a vector bundle of type $gl(m,\mathbb{R})$ (m : dimension of any fiber of F) in a natural way. Suppose moreover that F is equipped with some structure η (a metric of signature (a,b) ; a symplectic form). Restricting then Hom(F,F) to its subbundle $E_\eta = Hom_\eta(F,F)$ of linear transformations preserving η, we obtain vector bundles of type so(a,b) or sp(m/2). One can also restrict Hom(F,F) to the subbundle of traceless linear transformations obtaining so a bundle of type $sl(m,\mathbb{R})$. Similar examples are obtained for a complex bundle F.

3. We shall recall some facts about finite dimensional Lie algebras which are needed in the sequel. Let L be such an algebra and denote by $\theta(L)$ the set of linear $T : L \longrightarrow L$ such that

$$T \circ ad(x) = ad(x) \circ T \ , \ \forall \ x \in L \ .$$

The elements of $\theta(L)$ are the intertwining operators of L (for the adjoint representation). Then the following lemma is easily proved.

Lemma 1. <u>The set $\theta(L)$ is an associative algebra with unit. If the center $z(L)$ of L vanishes, then $\theta(L)$ is commutative so that</u>

$$\theta(L) = N(L) \oplus S(L)$$

<u>where</u> $N(L)$ (resp. $S(L)$) <u>denotes the nilpotent</u> (resp. <u>semisimple</u>) <u>part of</u> $\theta(L)$.

Let us now say that L is <u>decomposable</u> if it is the direct sum of two non trivial ideals :

$$L = L_1 \oplus L_2 \quad . \qquad (*)$$

If (*) holds, then we obtain a $T \in \theta(L)$ by setting $T|_{L_i} = i.\mathbb{1}|_{L_i}$ ($i=1,2$, $\mathbb{1}$ = identity on L). Conversely, suppose $T \in \theta(L)$. Then each eigenspace of T is a proper ideal of L so that if T has at least two distinct eigenvalues, then L is decomposable. We see that the properties of $\theta(L)$ and the decomposability of L are closely related. One proves easily the following

Lemma 2. <u>The Lie algebra</u> L <u>admits a decomposition</u> $L_1\oplus\ldots\oplus L_p$ <u>into a direct sum of non trivial and non decomposable ideals</u> L_i ($i \leq p$). <u>If moreover</u> $z(L) = 0$, <u>then this decomposition is unique up to the order of the factors</u>.

4. In order to study the group aut $\Gamma(E)$, it will be convenient to study the decomposability property of $\Gamma(E)$ and hence to compute the algebra $\theta(\Gamma(E))$. This is the purpose of this section.

Since each fiber E_x ($x \in M$) of E is a Lie algebra, we may consider the associative algebras $\theta(E_x)$ ($x \in M$) and construct then in a natural way a vector subbundle $\theta(E)$ of $Hom(E,E)$ (which is of type $\theta(L)$) such that $\theta(E)_x = \theta(E_x)$ for each $x \in M$. One obviously has

$$\Gamma_k(\theta(E)) \subset \theta(\Gamma_k(E))$$

and one can moreover prove that

Theorem 3. <u>If</u> $z(L) = 0$, <u>then</u> $\Gamma_k(\theta(E)) = \theta(\Gamma_k(E))$ <u>for</u> $k=0,1,2,\ldots,\infty$.

This theorem is a basic tool of the theory of Lie algebras of order 0. (The proofs of the propositions we shall give in the rest of the paper are based on it). It should be mentionned that the notion of intertwining operators cannot be used successfully in the study of Lie algebras of vector fields since in this case those operators are trivial.

5. We now study the decomposability of $\Gamma_k(E)$. We first state the following proposition which is an easy consequence of the previous theorem.

Proposition 4. _Let_ $z(L) = 0$. _If_ L _is not decomposable, then_ $\Gamma_k(E)$ _is not decomposable_.

In fact, if $\Gamma_k(E)$ is decomposable, then section 3 and Theorem 3 show that each fiber of E, and hence L, is decomposable.

The converse of Proposition 4 is false. It depends on the topology of M. This problem can be completely solved using some cohomological arguments. We shall only mention here the case of a simply connected manifold M. As a first step we state

Proposition 5. _Let_ $z(L) = 0$ _and let_ M _be simply connected_. _Then_ E _is the direct sum of subbundles_ E_i $(i \leq p)$ _each of which being of non decomposable type_ L_i (see Lemma 2). _This decomposition is unique up to the order of the factor_ E_i.

If $E = \bigoplus_{i=1}^{p} E_i$, where E_i is of non decomposable type L_i , one gets a decomposition $\Gamma_k(E) = \bigoplus_{i=1}^{p} \Gamma_k(E_i)$ of $\Gamma_k(E)$ as a direct sum of non decomposable ideals. Moreover, under the assumptions of Proposition 5, one obtains (uniqueness being guaranteed by Theorem 3)

Proposition 6. _Each_ $\Gamma_k(E)$ $(k=0,1,2,\ldots,\infty)$ admits a _decomposition into non decomposable ideals_. _This decomposition is unique up to the order of the factors_. _Each of them is a Lie algebra of order_ 0 _on_ M.

6. Let $E' \xrightarrow{p'} M'$ be another vector bundle over another manifold M'. Suppose there exists a diffeomorphism $\lambda : M \longrightarrow M'$. We then say that a linear map $\tau : \Gamma(E) \longrightarrow \Gamma(E')$ is a <u>differential operator over</u> λ if it induces a differential operator $\lambda^*\tau : \Gamma(E) \longrightarrow \Gamma(\lambda^*E')$ where λ^*E' denotes the pull-back of E' over λ (roughly, τ becomes a differential operator after having identified M' with M by means of λ). In this case, if ω is the domain of a trivialisation of E and $\lambda(\omega)$ the domain of a trivialisation of E', then the corresponding local form of $\tau(A)$ $(A \in \Gamma(E))$ looks like

$$\tau(A)_x = \sum_{0 \leq |\alpha| \leq s} T_x^\alpha(D_x^\alpha A) \qquad (x \in \omega)$$

where $\alpha = (\alpha^1,\ldots,\alpha^n)$ is a multi-index and the T^α's are smooth matrices.

Suppose now that L' is a finite dimensional Lie algebra and that E' is of type L'. Suppose moreover that $z(L) = 0$, $z(L') = 0$ and that L and L' are both non decomposable and let $\mu : \Gamma(E) \longrightarrow \Gamma(E')$ be an isomorphism of Lie algebras. Then (a priori M and M' are not diffeomorphic)

Theorem 7. <u>The Lie algebras</u> L <u>and</u> L' <u>are isomorphic</u> ; <u>the manifolds</u> M <u>and</u> M' <u>are diffeomorphic</u>. <u>Moreover</u> μ <u>is a differential operator over a diffeomorphism</u> $\lambda : M \longrightarrow M'$, <u>of order</u> $s \leq \dim L-1$ <u>and the</u> T^α's <u>take the form</u>

$$T^\alpha = N_1^{\alpha^1} \circ \ldots \circ N_n^{\alpha^n} \circ \mu_0$$

<u>where</u> $N_1^{\#},\ldots,N_n^{\#} \in C_\infty(\omega, N(L))$ <u>and</u> $\mu_0 \in C_\infty(\omega, \text{aut } L)$.

If in particular $\theta(L) = \mathbb{K}.\mathbb{1}$ ($\mathbb{K} = \mathbb{R}$ or $\mathbb{C}$) and $\theta(L') = \mathbb{K}.\mathbb{1}$, then this implies

Theorem 8. <u>The isomorphism</u> μ <u>is induced in a natural way by an isomorphism</u> $\nu : E \longrightarrow E'$ <u>of vector bundles of type</u> $L \approx L'$.

If we suppose M simply connected, then one can apply section 4 and decompose $\Gamma(E)$ into a direct sum of non decomposable Lie algebras of order 0. Restricting then μ to each factor $\Gamma(E_i)$, we can use the previous results. We obtain in this way the structure of the isomorphisms $\mu : \Gamma(E) \longrightarrow \Gamma(E')$ in the case where $z(L) = 0$, $z(L') = 0$ for possibly decomposable L and L'. The results are easily formutaled.

Going back now to problem (iii) of section 1, we see that

Theorem 9. <u>If</u> $z(L) = 0$ <u>and if</u> L <u>is non decomposable, then the algebraic structure of</u> $\Gamma(E)$ <u>characterizes the differential structure of</u> **M**. <u>Moreover if</u> $\theta(L) = \mathbb{K}.\mathbb{1}$, <u>then it also characterizes the bundle structure of</u> E. <u>If</u> $z(L) = 0$ <u>and</u> **M** <u>is simply connected, then its differential structure is characterized by any Lie algebra of order</u> 0 <u>defined by</u> L.

References

1. I. Amemiya. Lie algebra of vector fields and complex structure. J. of Math. Soc. Japan, vol. 27, n° 4, oct. 1975, p. 545.

2. I. Amemiya, K. Masuda, K. Shiga. Lie algebras of differential operators. Osaka J. of Math., vol. 12, n° 1, April 1975, p. 139.

3. A. Avez, A. Lichnerowicz et A. Diaz-Miranda. Sur l'algèbre des automorphismes infinitésimaux d'une variété symplectique. J. of diff. Géom., vol. 9, n° 1, March 1974, p. 1.

4. A. Koriyama. On Lie algebras of vector fields with invariant submanifolds. Nagoya Math. J., vol. 55, 1974, p. 91.

5. P. Lecomte. Derivations of linear endomorphisms of the tangent bundle. Bull. Soc. Roy. Sc. Liège, 1979, to appear.

6. P. Lecomte. Sur l'algèbre de Lie des automorphismes infinitésimaux du fibré tangent. Comptes Rendus Acad. Sc. Paris, 1979, to appear.

7. A. Lichnerowicz. Fibrés vectoriels, structure unimodulaires exactes et automorphismes infinitésimaux. J. de Math. Pures et Appliquées, vol. 56(2), 1977, p. 183.

8. K. Matzuda. Homomorphisms of the Lie algebras of vector fields. J. of Math. Soc.Japan, vol. 28, n° 3, 1976, p. 506.

9. H. Omori. Infinite dimensional Lie transformation groups. Lecture Notes in Mathematics, 427, Springer-Verlag, 1976.

10. L.E. Pursall, M.E. Shanks. The Lie algebra of a smooth manifold. Proc. Amer. Math. Soc., vol. 5, 1954, p. 468.

11. K. Shiga. Cohomology of Lie algebras over a manifold, I. J. of Math. Soc. Japan, vol. 26, n° 2, 1974, p. 324.

12. K. Shiga. Cohomology of Lie algebras over a manifold, II. J. of Math. Soc. Japan, vol. 26, n° 4, 1974, p. 587.

13. K. Shiga, T. Tsujishita. Differential representation of vector fields. Kodai Math. Sem. Rep., vol. 28, n° 2-3, March 1977, p. 214.

14. F. Takens. Derivations of vector fields. Compositio Mathematica, vol. 26(2), 1973, p. 151.

15. G. Van Calk. The Lie algebra of vector fields of a differential manifold. Thèse de Doctorat, Bruxelles, 1977.

Institut de Mathématique
Université de Liège
Avenue des Tilleuls, 15
B-4000 LIEGE (Belgium)

Introduction à l'étude de certains systèmes différentiels

Paulette LIBERMANN

à Charles Ehresmann

Le but de ce travail, qui ne contient pas de résultats nouveaux, est d'exposer brièvement les notions qui sont à la base de la théorie des systèmes différentiels sur les variétés ; cette théorie, qui généralise celle des équations aux dérivées partielles dans les espaces numériques, a ses applications non seulement en Géométrie Différentielle mais aussi en Mécanique et en Physique théorique.

Les notions exposées sont dûes à C. Ehresmann, précurseur de la Géométrie Différentielle moderne (jets holonomes et semi-holonomes, connexions, pseudogroupes de Lie, groupoïdes différentiables etc).

Cet article était achevé juste avant la disparition de ce grand Mathématicien ; il pourra lui servir d'hommage.

Notre bibliographie, naturellement incomplète, donne des références d'articles où sont étudiés d'une manière plus approfondie les sujets traités dans ce travail ou relatifs à la cohomologie de Spencer ; ces articles contiennent eux-mêmes une bibliographie détaillée.

Nous avons essayé d'introduire les diverses notions de manière naturelle : par exemple le 1-jet d'une forme différentielle ω de degré 1 est un jet semi-holonome d'ordre 2 ; cette forme peut être considérée comme une connexion ; sa courbure obstacle à l'intégrabilité, est la différentielle extérieure $d\omega$.

I - Rappels et notations

Pour toute application $f : A \longrightarrow B$, A est la source de f, f(A) le but de f. Les variétés seront de dimension finie, de classe C^∞ (ainsi que les applications). Pour toute variété M on désigne le fibré tangent et le fibré cotangent par TM et T^*M.

Une surmersion ("fibered manifold") (E,M,π) est une submersion surjective $\pi : E \longrightarrow M$ (c'est-à-dire telle que l'application π soit de rang égal à la dimension de M) ; les submersions sont caractérisées par l'existence de sections locales. Si (E,M,π) et (E',M,π') sont deux submersions, la sous-variété de $E \times E'$ des couples (z,z') tels que $\pi(z) = \pi'(z')$ sera désignée par $E \times_M E'$ ou $E \underset{\pi \times \pi'}{\times} E'$. On désignera par VTE (fibré tangent vertical) l'ensemble des vecteurs tangents aux fibres $E_x = \pi^{-1}(x)$ de la surmersion (E,M,π).

Etant données deux variétés V et W, deux applications f et f' définies dans un voisinage de $x \in V$, à valeurs dans W, sont dites k-équivalentes en x si $f(x) = f'(x)$ et si f et f' s'expriment au moyen de cartes locales au voisinage de x et f(x) par des applications ayant même développement de Taylor à l'ordre k ($k \geq 0$). La classe d'équivalence de f en x s'appelle le k-jet de f en x ; on le note $j^k_x f$; x est appelé la source de $j^k_x f$, f(x) son but. L'ensemble $J^k(V,W)$ des k-jets de V dans W est muni d'une structure de variété ; les applications source $\alpha : J^k(V,W) \longrightarrow V$, but $\beta : J^k(V,W) \longrightarrow W$ (ainsi que $\alpha \times \beta$) sont des submersions. L'application $j^k f : x \rightarrow j^k_x f$ est une section locale de $J^k(V,W)$ pour la projection α. A la composition des applications correspond une composition des jets, d'où la notion de jet inversible (jet de difféomorphisme). Si $k' < k$, on a la surmersion $J^k(V,W) \longrightarrow J^{k'}(V,W)$; un k-jet est inversible s'il détermine un 1-jet inversible (théorème des fonctions inverses).

Si (E,M,π) est une submersion, on désignera par $J_k E$ l'espace des k-jets des sections locales (donc $J_k E \subset J^k(M,E)$) ; la projection $J_k E \longrightarrow J_{k-1} E$ définit une structure de fibré affine ; en particulier le fibré $J_1E \longrightarrow E$ admet pour fibré vectoriel associé le fibré des applications linéaires de TM dans VTE.

L'ensemble $J^k(V,W)$ s'identifie à J_kE où : $E = V \times W$, $M = V$, $\pi = p_1 : V \times W \longrightarrow V$.

II - Systèmes différentiels et prolongements holonomes

Soient V et W deux variétés. Un système différentiel d'ordre k est par définition une sous-variété S_k de la variété $J^k(V,W)$. Une solution (locale) d'un tel système est une application $f : U \subset V \longrightarrow W$ (U ouvert de V) telle que $\forall x \in U$, le jet $j^k_x f$ appartienne à S_k ; l'application $j^k f : U \longrightarrow S_k$ (définie par $j^k f(x) = j^k_x f$) est une section locale de S_k relativement à l'application source $\alpha : S_k \rightarrow V$.

Lorsque V et W sont des espaces numériques, on retrouve les systèmes d'équations aux dérivées partielles.

Le système S_k est dit complètement intégrable si pour tout $X_k \in S_k$, il existe une solution f de S_k satisfaisant la condition $j^k_x f = X_k$ (où $x = \alpha(X_k)$) et $j^k f$ est une section locale de S_k dont l'image contient X_k ; donc :

Propriété 1. Pour que le système S_k soit complètement intégrable, il est nécessaire que la restriction à S_k de l'application source α soit une submersion (et par suite $\alpha(S_k)$ est une sous-variété ouverte de V).

On supposera désormais cette condition réalisée et l'on se ramènera au cas où $\alpha(S_k) = V$.

En posant comme dans I, $E = V \times W$, $M = V$, $\pi = p_1$, on se ramène donc à la définition suivante :

Définition 1. Un système différentiel R_k d'ordre k pour une surmersion (E,M,π) est une sous-variété de J_kE telle que la restriction de la projection source α à R_k soit une surmersion [R_k est une sous-surmersion (fibered submanifold) de $\alpha : J_k(E) \longrightarrow M$]

Supposons R_k complètement intégrable ; soit $X_k \in R_k$ et soit f une solution de R_k telle que $j^k_x f = X_k$; pour tout $r > 0$, le jet $j^{k+r}_x f$ s'identifiant au jet $j^r_x j^k f$, ce jet $j^{k+r}_x f$ appartient à $J_r R_k \cap J_{r+k}E$.

On appelle prolongement d'ordre r de R_k le sous-ensemble

$$R_{k+r} = J_r R_k \cap J_{k+r}E$$

de $J_{k+r}E$.

Donc pour que le système R_k soit complètement intégrable, il est nécessaire que pour tout $r > 0$, l'application $R_{k+r} \longrightarrow R_r$ soit surjective.

Il est à remarquer que R_{k+r} n'est pas nécessairement une sous-variété de $J_{k+r}E$ pour $r > 0$.

Pour obtenir des résultats concernant les systèmes différentiels, on doit faire des hypothèses de régularité supplémentaires.

Si (E,M,π) est une fibration vectorielle et si R_k est un sous-fibré vectoriel de J_kE, on a un système différentiel linéaire dont le maniement est plus commode que celui d'un système non linéaire ; H. Goldschmidt [3a] a étendu aux systèmes non linéaires la théorie de Spencer ; il a défini les systèmes formellement intégrables ; il a démontré que ces systèmes pouvaient se définir de la manière suivante ; un système R_k est formellement intégrable si pour tout $r \geq 0$, il possède les propriétés P_r :

1°) l'application π_{k+r} ; $R_{k+r+1} \longrightarrow R_{k+r}$ est surjective

2°) cette application définit une sous-surmersion de π_{k+r} : $J_{k+r+1}(E)|R_{k+r+1} \longrightarrow R_{k+r}$

Un système complètement intégrable peut ne pas être formellement intégrable (M. Janet a donné des exemples de systèmes différentiels complètement intégrables dont le prolongement n'est pas complètement intégrable) ; d'autre part H. Lewy a donné un exemple de système formellement intégrable sans solutions, (voir [4c]).

Par contre [3a], si les variétés E,M et R_k sont analytiques, alors tout système formellement intégrable est intégrable.

De la théorie de Spencer, on déduit [3a] qu'il existe un nombre entier $k_o > k$ ne dépendant que de n, p = dim E et k, tel que si R_k vérifie la conditions P_r pour $0 \leq r \leq k_o - k$, alors R_k est formellement intégrable, c'est-à-dire il suffit d'étudier un nombre fini de prolongements pour vérifier que R_k est formellement intégrable.

Il est utile de considérer d'autres prolongements que ceux définis par le foncteur J_k (k-jet de section).

Par exemple, on désigne par T^k_nM (resp. T^{*k}_nM) l'espace des k-jets de $\mathbb{R}^n$ dans M (resp. de M dans $\mathbb{R}^n$) de source (resp. but) 0. Pour n = k = 1, on retrouve le fibré tangent TM et le fibré cotangent T^*M.

Si dim M = n, le sous-ensemble $H^k(M)$ de T^k_nM formé des jets inversibles (c'est-à-dire de jets de difféomorphismes locaux) est l'espace des k-repères (pour k = 1, on a l'espace H(M) des repères au sens usuel) ; $H^k(M)$ est un fibré principal, de groupe structural L^k_n (groupe des k-jets inversibles de $\mathbb{R}^n$ dans $\mathbb{R}^n$ de source et but 0). De même $H^{*k}(M)$, sous-ensemble des jets inversibles de T^{*k}_nM est un L^k_n-fibré principal (espace des k-corepères).

Soit (E,M,π) une surmersion ; si n = dim M, on définit le sous-espace $\mathcal{T}^k_nE$ de T^k_nE, image réciproque de $H^k(M)$ par la projection $T^k\pi$: $T^k_nE \longrightarrow T^k_nM$; en particulier si E = M et $\pi = id_M$, $\mathcal{T}^k_nM = H^k(M)$.

On démontre [7d] :

<u>Proposition</u> 1. <u>$\mathcal{T}^k_n E$ est difféomorphe au produit fibré</u> $J_k E \times_M H^k(M)$.

En effet si $\sigma : U \subset \mathbb{R}^n \to E$ est une application telle que $j^k_o\sigma \in \mathcal{T}^k_n E$, en restreignant au besoin la source de σ, il résulte du théorème des fonctions inverses que $\varphi = \pi \circ \sigma$ est un difféomorphisme ; donc $s = \sigma \circ \varphi^{-1}$ est une section locale de E ; la donnée de φ et s détermine σ.

Si (P,M,π) est une fibration principale, de groupe structural G, on démontre [7d] que $(\mathcal{T}^k_n P, M, T^k\pi)$ est une fibration principale de groupe structural $T^k_n(G) \times L^k_n$, alors qu'en général $J_k P$ n'est muni d'une structure de fibré principal, d'où l'utilité du foncteur $\mathcal{T}^k_n$.

Si l'on considère les prolongements $TT^k_p M$ et $T^k_p TM$ (où $p > 0$ est un entier quelconque), on a les projections :

$$\begin{array}{ccc} TT^k_p M & \xrightarrow{T\tau^k_p} & TM \\ \tau \downarrow & & \\ T^k_p M & & \end{array} \qquad \text{et} \qquad \begin{array}{ccc} T^k_p TM & \xrightarrow{T^k_p \tau} & T^k_p M \\ \downarrow \tau^k_p & & \\ TM & & \end{array}$$

<u>Théorème</u> 1. (lemme de Schwarz pour les variétés)

<u>Il existe un difféomorphisme canonique</u>

$$\psi^k_p : TT^k_p M \longrightarrow T^k_p TM$$

<u>tel que</u> (1) $\tau = T^k_p\tau \circ \psi^k_p$, $T\tau^k_p = \tau^k_p \circ \psi^k_p$;

<u>de plus, pour une surmersion</u> (E,M,π), on a :

(2) $\psi^k_n(T\mathcal{T}^k_n E) = \mathcal{T}^k_n TE$;

<u>en particulier</u>

(3) $\psi^k_n(TH^k(M)) = \mathcal{T}^k_n(TM)$

En effet tout élément de $TT^k_p M$ ou de $T^k_p TM$ est le "jet partiel" (cf [2c]) d'une application de $\mathbb{R} \times \mathbb{R}^p$ dans M, d'ordre 1 par rapport à la "variable" $t \in \mathbb{R}$, d'ordre k par rapport à la "variable" $u \in \mathbb{R}^p$. Au moyen d'une carte locale au voisinage de $x \in M$ tout élément $\rho \in (TT^k_p M)_x$ est représenté par le quadruplet $(a,b,c,d) \in \mathbb{R}^n \times L^k_{p,n} \times \mathbb{R}^n \times L^k_{p,n}$ (où $L^k_{p,n}$ = ensemble des k-jets de source et but 0 de $\mathbb{R}^p$ dans $\mathbb{R}^n$). De même tout $\rho' \in (T^k_p TM)_x$ est représenté au moyen de cette carte locale par $(a',b',c',d') \in \mathbb{R}^n \times \mathbb{R}^n \times L^k_{p,n} \times L^k_{p,n}$; ρ est encore représenté par une application $g : \mathbb{R} \times \mathbb{R}^p \longrightarrow \mathbb{R}^n$ et ρ' par une application $g' : \mathbb{R} \times \mathbb{R}^p \longrightarrow \mathbb{R}^n$, g et g' étant des applications polynômes de degré 1 par rapport à t et k par rapport à u. Si l'on pose $g(t,u) = g_t(u)$, d'après le lemme de Schwarz pour les espaces numériques, pour toute différentielle $D^s g_t (s=1,\dots,k)$, on a :

$$\frac{d}{dt}(D^s g_t(u)\Big|_{t=0} = D^s\left(\frac{dg_t}{dt}\right)(u)\Big|_{t=0} .$$

On en déduit qu'à tout (a,b,c,d) représentant $\rho \in (TT^k_p M)_x$, on associe $(a,c,b,d) \in (T^k_p TM)_x$; les conditions (1) sont vérifiées par suite de l'échange de b et c.

En particulier si p = n, $\rho \in TH^k(M)$ est représenté par $(a,b,c,d) \in \mathbb{R}^n \times L^k_n \times \mathbb{R}^n \times L^k_n$ et $\psi^k_n(\rho)$ par $(a,c,b,d) \in \mathbb{R}^n \times \mathbb{R}^n \times L^k_n \times L^k_n$ et $\psi^k_n(\rho) \in \mathcal{T}^k_n(TM)$. Par images réciproques, on en déduit pour toute surmersion la relation (2).

Pour toute fibration principale (P,M,π), de groupe structural G, le fibré tangent s'identifie au produit fibré $TP/_G \times_M P$ où le fibré vectoriel $TP/_G$, de base M est l'espace des vecteurs tangents à P mod. les translations à droite de G.

Plus généralement une surmersion (E,M,π) admet un "parallèlisme fibré" (cf [7f]) s'il existe un fibré vectoriel $T_{red}E$, de base M tel que TE soit isomorphe à $T_{red}E \times_M E$; $T_{red}E$ est le fibré tangent "réduit".

On a alors [7f].

<u>Théorème 2</u>. <u>Si une surmersion</u> (E,M,π) <u>admet un parallèlisme fibré, de fibré tangent réduit</u> $T_{red}E$, <u>alors la surmersion</u> $(\mathcal{T}^k_n E, M, \mathcal{T}^k_n \pi)$ <u>admet un parallèlisme fibré, de fibré tangent réduit</u> $J_q T_{red}E$.

En effet on a la suite de difféomorphismes :

$$T\mathcal{T}^k_n E \Longleftrightarrow \mathcal{T}^k_n TE \Longleftrightarrow J_k TE \times_M H^k(M) \Longleftrightarrow J_k(T_{red}E \times_M E) \times_M H^k(M)$$

$$\Longleftrightarrow J_k T_{red}E \times_M J_k E \times_M H^k(M) \Longleftrightarrow J_k T_{red}E \times_M \mathcal{T}^k_n E.$$

<u>Corollaire</u>. <u>Si</u> (P,M,π) <u>est une fibration principale, alors pour le fibré principal</u> $\mathcal{T}^k_n P$, <u>de base</u> M, <u>on a l'isomorphisme</u>

$$T(\mathcal{T}^k_n P)/_{G_k} \Longleftrightarrow J_k({}^{TP}/_G),$$

<u>où</u> G_k <u>est le groupe structural de</u> $\mathcal{T}^k_n P \longrightarrow M$ (<u>isomorphe à</u> $T^k_n(G) \times L^k_n$). <u>En particulier, on a l'isomorphisme</u> :

$$J_k(TM) \Longleftrightarrow TH^k/_{L^k_n}$$

Le corollaire est immédiat et la dernière relation s'obtient en prenant la fibration principale (M,M,id_M).

Cette dernière propriété avait été démontrée dans [7b] en utilisant des groupes locaux à un paramètre.

Les résultats de ce paragraphe permettront, dans les paragraphes ultérieurs, d'associer à un système différentiel défini par une fibration principale un système linéaire.

Remarque. Par les mêmes méthodes [3a], on démontre qu'il existe un isomorphisme de VTJ_kE sur J_kVTE (où VTE est le fibré vertical tangent à la surmersion (E,M,π)).

On en déduit que si VTE admet un isomorphisme sur le produit fibré

$VT_{red}E \times_M E$ (où $VT_{red}E$ est un fibré vectoriel), alors

VTJ_kE est isomorphe à $J_kVT_{red}E \times_M J_kE$.

Ceci généralise une propriété des fibrés vectoriels ; pour un tel fibré, on a en effet $VTE = E \times_M E$ (et $VT_{red}E = E$).

III - Prolongements semi-holonomes. Application aux systèmes différentiels de type fini.

En plus des jets usuels (que l'on appelle holonomes), C. Ehresmann a introduit les jets non holonomes et semi-holonomes. Ces derniers s'obtiennent en oubliant la condition de symètrie de Schwarz dans les dérivations successives. Les prolongements semi-holonomes (qui contiennent les prolongements holonomes de même ordre) d'un système différentiel sont des sous-variétés ; nous verrons, dans les systèmes de type fini, que l'intégrabilité est équivalente à la coïncidence des prolongements holonomes et semi-holonomes.

Soit (E,M,π) une surmersion ; J_1J_1E est appelé le prolongement non holonome d'ordre 2 de E ; par itération de J_1, on obtient le prolongement non holonome d'ordre k.

On définit le prolongement semi-holonome $\bar{J}_2E \subset J_1J_1E$ de la manière suivante : une section locale $U \subset M \longrightarrow J_1E$ sera dite adaptée en $x \in U$ si $s(x) = j^1_x(\beta.s)$ où β est l'application but $J_1E \longrightarrow E$; le jet j^1_xs est alors appelé semi-holonome ; $\bar{J}_2E$ est l'ensemble de ces jets semi-holonomes ; c'est encore le noyau de la double flèche :

$$\begin{array}{ccc} J_1J_1E & \xrightarrow{J_1} & J_1E \\ \beta \big\downarrow & & \\ J_1E & & \end{array}$$

Remarquons que si s est adaptée en tout $x \in U$, alors s s'écrit

$j^1 f$ et $j^1_x s = j^1_x j^1 f = j^2_x f$; on a un 2-jet holonome.

Par récurrence, on définit $\bar{J}_k E$ comme le noyau de la double flèche :

$$\begin{array}{ccc} J_1\bar{J}_{k-1}E & \xrightarrow{J_1\beta} & J_1\bar{J}_{k-2}E \\ \downarrow \beta & & \\ \bar{J}_{k-1}E & & \end{array}$$

(en remarquant que $\bar{J}_{k-1}E \subset J_1\bar{J}_{k-2}E$).

Les k-jets semi-holonomes se composent ; un k-jet semi-holonome est inversible s'il détermine un 1-jet inversible.

On définit de même les prolongements $\bar{T}^k_p M$, $\bar{T}^{*}_p M$ ainsi que le fibré principal $\bar{H}^k(M)$ dans k-jets semi-holonomes inversibles de $\mathbb{R}^n$ dans M, dont le groupe structural $\bar{L}^k_n$ est le groupe des k-jets semi-holonomes inversibles de $\mathbb{R}^n$ dans $\mathbb{R}^n$; un élément de $\bar{L}^k_n$ peut être représenté par :

$$y^i = \sum(a^i_j x^j + a^i_{j_1 j_2} x^{j_1} \otimes x^{j_2} + \ldots + a^i_{j_1 \ldots j_k} x^{j_1} \otimes \ldots \otimes x^{j_k}) \qquad i=1,\ldots,n,$$

où la matrice a^i_j est inversible.

On définit également, pour toute surmersion (E,M,π), l'espace $\bar{\mathcal{T}}^k_n E$, image réciproque de $\bar{H}^k(M)$; la proposition 1 devient

<u>Proposition</u> 2. $\bar{\mathcal{T}}^k_n E$ <u>est difféomorphisme au produit fibré</u> $\bar{J}_K E \times_M \bar{H}^k(M)$.

Pour toute fibration principale (P,M,π), l'espace $\bar{\mathcal{T}}^k_n P$ est un fibré principal, en particulier $\bar{\mathcal{T}}^k_n H(M)$.

<u>Corollaire</u>. Il existe un difféomorphisme canonique

$\bar{J}_k H(M) \Longleftrightarrow \bar{H}^{k+1}(M)$ (en particulier $J_1 H \Longleftrightarrow \bar{H}^2$)

d'où un difféomorphisme canonique :

$$\bar{L}^{k+1}_n \Longleftrightarrow \bar{T}^k_n(L_n)$$

<u>Preuve</u> : $\mathcal{T}_n H(n)$ étant difféomorphe à $J_1 H(M) \times_M H(M)$, on a la double flèche

$$\begin{array}{ccc} \mathcal{T}_n H(M) & \xrightarrow{p_2} & H(M) \\ \downarrow \beta & & \\ H(M) & & \end{array}$$

et $\bar{H}^2(M)$ est le noyau de cette double flèche ; donc $J_1 H(M)$ est difféomorphe à $\bar{H}^2(M)$

(propriété démontrée dans [14]).

Par récurrence on démontre cette propriété pour k quelconque ; si $M = \mathbb{R}^n$, $\bar{H}^k(M)$ s'identifie à $\mathbb{R}^n \times \bar{L}^k_n$, d'où le difféomorphisme $\bar{L}^{k+1}_n \Longleftrightarrow \bar{T}^k_n(L_n)$.

Remarques - 1°) Le corollaire permet de définir sur $\bar{J}_k H(M)$ une structure de fibré principal ; on peut en déduire sur $J_k H(M)$ une structure de sous-fibré principal.

2°) Pour tout groupe de Lie G, $T^k_n(G)$ (resp. $\bar{T}^k_n(G)$) est muni d'une structure de groupe de Lie (par prolongement de la loi de composition), produit semi-direct $T^k_{n,e}(G) \times G$ (resp. $\bar{T}^k_{n,e}(G) \times G$) où $T^k_{n,e}(G)$ (resp. $\bar{T}^k_{n,e}(G)$) est le groupe des k-jets de $\mathbb{R}^n$ dans G, de but 0 ; mais le difféomorphisme $\bar{T}^k_n(L_n) \Longleftrightarrow \bar{L}^{k+1}_n$ n'est pas un isomorphisme de groupes. Du corollaire, on déduit un difféomorphisme :
$\bar{L}^{k+1}_n / L_n \Longleftrightarrow \bar{T}^k_{n,e}(L_n)$.

Si l'on considère le fibré vectoriel trivial $E_o = M \times \mathbb{R}$, on a

$$J_1 E_o = T^* M \oplus E_o \quad \text{(où } T^*M \text{ est le fibré cotangent à M)}$$

Proposition 3. Il existe un difféomorphisme canonique

$$J_1(T^*M) \Longleftrightarrow \bar{T}^{*2}(M)$$

(où $\bar{T}^{*2}(M)$ est l'espace des 2-jets semi-holonomes de M dans $\mathbb{R}$, de but 0).

En effet soit $\omega : U \subset M \longrightarrow T^*M$ une forme différentielle dans l'ouvert U ; pour tout $x \in U$, par définition, il existe une fonction numérique $f : U' \subset U \longrightarrow \mathbb{R}$ telle que $f(x) = 0$, $j^1_x f = \omega(x)$; l'application $w : U' \longrightarrow T^*M$ est donc x-adaptée et $j^1_x w \in \bar{T}^{*2}(M)$; pour que $j^1_x w \in T^{*2}(M)$, il faut que $w = j^1 f$ (=df).

Par exemple si ω est définie au moyen de coordonnées locales par $\omega = \sum_{i=1}^{n} a_i dx^i$, $j^1_x \omega$ est défini par : $a_1(x), \ldots, a_n(x), \frac{\partial a_i}{\partial x^j}(x)$; pour que le 2-jet soit holonome, il faut et il suffit (si U est simplement connexe) que $\frac{\partial a_i}{\partial x^j} = \frac{\partial a_j}{\partial x^i}$ $(i,j=1,\ldots,n)$.

On peut également définir les prolongements sesquiholonomes à l'ordre k ; par exemple le prolongement sesquiholonome $\check{J}_k E \subset \bar{J}_k E$ est défini pour $k > 2$, comme le noyau de la double flèche

$$\begin{array}{ccc} J_1 J_{k-1} E & \longrightarrow & J_1 J_{k-2} E \\ \downarrow & & \\ J_{k-1} E & & \end{array}$$

Pour $k = 2$, on a $\check{J}_2 E = \bar{J}_2 E$.

La projection $\check{J}_k E \longrightarrow J_{k-1} E$ définit une structure de fibré affine.

Si $R_k \subset J_k E$ est un systèmes différentiel, on définit le prolongement sesquiholonome $\check{R}_{k+1}$ de R_k par $J_1 R_k \cap \check{J}_{k+1}E$. On a $R_{k+1} \subset \check{R}_{k+1}$.

Lemme. Si un système différentiel $R_k \subset J_k E$ est difféomorphe à sa projection R_{k-1} sur $J_{k-1}E$, alors $\check{R}_{k+1}$ est difféomorphe à R_k.

En effet le difféomorphisme $\phi : R_k \to R_{k-1}$ se prolonge en un difféomorphisme $J_1\phi : J_1R_k \to J_1R_{k-1}$ et $\check{R}_{k+1}$ noyau de la double flèche $(J_1\phi, \beta : J_1R_k \to R_k)$ est l'image de R_k par l'injection $\phi^{-1} \circ i$, où i est l'injection $R_k \to J_1R_{k-1}$.

Il en résulte qu'alors l'application $R_{k+1} \to R_k$ est surjective si et seulement si $\check{R}_{k+1}$ et R_{k+1} coïncident. Dans ce cas le système est complètement intégrable ; en effet le système R_k s'exprime, au moyen de coordonnées locales, comme un système d'équations aux dérivées partielles Σ_k de Mayer-Lie c'est-à-dire tel que les dérivées partielles d'ordre k sont fonctions des dérivées d'ordre inférieur ; si l'application $R_{k+1} \to R_k$ est surjective, la condition d'intégrabilité obtenue en dérivant le système Σ_k sont satisfaites et d'après le théorème de Frobenius, le système Σ_k (d'où R_k) est complètement intégrable.

Les résultats précédents se résument de la manière suivante [2d], [7e] :

Théorème 3. Pour qu'un système différentiel R_k, tel que R_k soit difféomorphe à sa projection R_{k-1}, soit complètement intégrable, il faut et il suffit que le prolongement holonome R_{k+1} coïncide avec le prolongement sesquiholonome $\check{R}_{k+1}$.

Plus généralement un système différentiel R_k est dit de type fini [7a] s'il existe un entier r tel que le prolongement R_{k+r} soit une sous-variété de $J_{k+r}E$ et que R_{k+r} soit difféomorphe à R_{k+r-1}. Le système R_{k+r} est du type précédent. Donc pour que R_{k+r} soit complètement intégrable il faut et il suffit que $\check{R}_{k+r+1}$ soit identique à R_{k+r+1}. Pour que, de plus, R_k soit complètement intégrable, il faut et il suffit que l'application $R_{k+r} \to R_k$ soit surjective. Donc dans le cas d'un système de type fini, la théorie des systèmes formellement intégrables est inutile pour étudier l'existence des solutions.

Parmi les systèmes de type fini, figurent les connexions.

Définition. Etant donnée une surmersion (E,M,π), une connexion holonome d'ordre k est une section

$$C_k : J_{k-1}E \longrightarrow J_kE \ ;$$

R_k est ici identique à $C_k(J_{k-1}E)$.

En particulier une connexion d'ordre 1 est un relèvement $C_1 : E \to J_1E$; c'est un champ d'éléments de contact sur E, transverse aux fibres ; on retrouve les connexions au sens usuel. Si E est un fibré vectoriel, on impose à la connexion d'être linéaire (morphisme de fibrés vectoriels).

Le relèvement C_k se prolonge en un relèvement $J_1C_k : J_1J_{k-1}E \longrightarrow J_1J_kE$ et l'application composée $J_1C_k \circ C_k : J_{k-1}E \longrightarrow J_1J_kE$ est à valeurs dans $\check{J}_{k+1}E$ (dans $\bar{J}_2E$ pour k = 1), ce qui justifie encore l'introduction des jets semi-holonomes et sesquiholonomes. On vérifie que $J_1C_k \circ C_k(J_{k-1}E)$ est le prolongement $\check{R}_{k+1}$. Donc d'après le théorème 3, une connexion est intégrable si et seulement si $J_1C_k \circ C_k$ est à valeurs dans $J_{k+1}E$. En utilisant le fait que la projection $\check{J}_{k+1}E \longrightarrow J_kE$ est une fibration affine, on définit la courbure obstacle à l'intégrabilité ; en particulier pour k = 1, on obtient le courbure usuelle.

IV - Groupoïdes différentiables et pseudogroupes

Les groupoïdes considérés ici sont des ensembles (et même des variétés). Un groupoïde ϕ est un ensemble muni d'une loi de composition partielle satisfaisant les axiomes :

1°) Tout $\varphi \in \phi$ admet une unité à droite unique $\alpha(\varphi)$ (telle que $\varphi.\alpha(\varphi) = \varphi$) et une unité à gauche unique $\beta(\varphi)$ (telle que $\beta(\varphi).\varphi = \varphi$).

2°) $\varphi_2\ \varphi_1$ est défini si et seulement si $\alpha(\varphi_2) = \beta(\varphi_1)$.

3°) Si $(\varphi_3\ \varphi_2)\varphi_1$ est défini, il en est de même de $\varphi_3(\varphi_2\varphi_1)$ et $(\varphi_3\varphi_2)\varphi_1 = \varphi_3(\varphi_2\varphi_1)$.

4°) Pour tout φ, il existe φ^{-1} (unique) tel que $\varphi\varphi^{-1} = \beta(\varphi)$, $\varphi^{-1}\varphi = \alpha(\varphi)$

L'ensemble M des unités de ϕ sera appelée base de ϕ. On a les applications $\alpha : \phi \longrightarrow M$, $\beta : \phi \longrightarrow M$, $\alpha \times \beta : \phi \longrightarrow M \times M$.

Le groupoïde est dit transitif si $(\alpha \times \beta)$ est surjectif.

Exemples :

1°) Si ϕ admet une seule unité, la loi de décomposition est partout définie et l'on a un groupe.

2°) Soit H un ensemble, $\phi = H \times H$ est muni d'une structure de groupoïde transitif par la loi de composition : (b',a')(b,a) est défini si et seulement si b = a' et alors le composé est (b',a) ; les éléments de la diagonale Δ_H sont les unités.

3°) Soit $\pi : K \longrightarrow H$ une application surjective ; le produit fibré $K \times_H K$ est muni d'une structure de groupoïde non transitif.

On a de manière évidente la notion de sous-groupoïde ; pour tout groupoïde ϕ, on ne considérera que des sous-groupoïdes ayant même base.

Un groupoïde différentiable ϕ est une variété différentiable munie d'une structure de groupoïde telle que

1°) la base M est une sous-variété de ϕ

2°) les applications $\alpha : \phi \longrightarrow M$, $\beta : \phi \longrightarrow M$ sont différentiables et sont des submersions (ce sont donc des surmersions).

3°) la loi de composition $(\varphi,\varphi') \longrightarrow (\varphi\varphi')$ (qui est définie sur le produit fibré des surmersions α et β donc sur une sous-variété de $\Phi \times \Phi$) est différentiable.

4°) l'application $\varphi \longrightarrow \varphi^{-1}$ est un difféomorphisme de Φ sur lui-même.

On a la notion de sous-groupoïde-variété

Exemples

1°) On retrouve d'abord les 3 exemples précédents en munissant le groupoïde d'une structure de variété (groupe de Lie, produit de variétés, produit fibré de variétés).

2°) Un groupoïde différentiable tel que $\alpha = \beta$ est une somme de groupes de Lie ; notamment on a les fibrés vectoriels.

3°) Un groupoïde de Lie est un groupoïde différentiable Φ tel que l'application $\alpha \times \beta : \Phi \longrightarrow M \times M$ soit une surmersion ; il est donc transitif ; on démontre qu'il est localement trivial [2f], [7d] au sens suivant : pour tout couple $(x,y) \in M \times M$, il existe un voisinage ouvert U de y dans M et une application différentiable $s : U \longrightarrow \Phi$ telle que $\alpha s(y') = x$, $\beta s(y') = y$ pour tout $y' \in U$.

L'image réciproque de la diagonale Δ_M par $\alpha \times \beta$ est un groupoïde, somme de groupes.

Ce sont surtout les groupoïdes de Lie que nous considérerons dans la suite. Par exemple l'ensemble $\Pi^k(M)$ des q-jets inversibles de M dans M est un groupoïde de Lie.

Rappelons qu'un pseudogroupe de transformations différentiables sur une variété M est un ensemble Γ de difféomorphismes locaux vérifiant les axiomes :

1°) Si f appartient à Γ, il en est de même de f^{-1} ; si f et f' de sources U et U' appartiennent à Γ, alors l'application composée $f' \circ f$ (dont la source $f^{-1}[f(U) \cap U']$ est éventuellement vide) appartient à Γ.

2°) Si $U = \bigcup_i U_i$, où chaque U_i est un ouvert de M pour qu'un difféomorphisme f de source U, appartienne à Γ, il faut et il suffit que sa restriction f_i à chaque U_i appartienne à Γ.

3°) L'application identique de M appartient à Γ.

On vérifie que pour tout $k > 0$, l'ensemble $\mathcal{J}^k(\Gamma)$ des k-jets des difféomorphismes appartenant à Γ est muni d'une structure de groupoïde ($\mathcal{J}^k(\Gamma)$ est le groupoïde d'ordre k associé à Γ) : $\mathcal{J}^k(\Gamma)$ est un sous-ensemble (non nécessairement une sous-variété) du groupoïde $\Pi^k(M)$ (associé au pseudogroupe de tous les difféomorphismes de M).

Le pseudogroupe Γ est dit transitif si pour tout couple $(x,y) \in M \times M$, il existe $f \in \Gamma$ tel que $f(x) = y$. Le groupoïde associé $\mathcal{J}^k(\Gamma)$ est alors transitif.

Un sous-groupoïde différentiable Φ de $\Pi^k(M)$ est un système différentiel d'ordre k dont les solutions constituent un pseudogroupe Γ mais Φ ne coïncide avec $\mathcal{J}^k(\Gamma)$ que s'il est complètement intégrable.

Un pseudogroupe de Lie d'ordre k sur une variété M est par définition [2c], [7a] un pseudogroupe Γ de difféomorphismes tel que :

a) $\mathcal{J}^k(\Gamma)$ est un sous-groupoïde différentiable de $\Pi^k(M)$.

b) Γ est complet d'ordre k c'est-à-dire Γ est l'ensemble des solutions de $\mathcal{J}^k(\Gamma)$.

Exemples : 1°) le pseudogroupe Γ des automorphismes locaux analytiques complexes de $\mathbb{C}^n$ (que l'on peut identifier à $\mathbb{R}^{2n}$) est un pseudogroupe de Lie d'ordre 2 ; chaque automorphisme local vérifie les conditions de Cauchy-Riemann.

Le sous-pseudogroupe $\Gamma' \subset \Gamma$ constitué des applications affines est d'ordre 2. Remarquons que $\mathcal{J}^1(\Gamma') = \mathcal{J}^1(\Gamma)$.

2°) le pseudogroupe des difféomorphismes locaux de $\mathbb{R}^n$ de déterminant 1 est d'ordre 1, de même que le pseudogroupe des difféomorphismes locaux de $\mathbb{R}^{2n}$ laissant invariante la 2-forme différentiable extérieure $dx^1 \wedge dy^1 + \ldots + dx^n \wedge dy^n$.

3°) le pseudogroupe des isométries locales d'un espace euclidien.

Un pseudogroupe de Lie est de type fini s'il est défini par un système différentiel de type fini (exemple 3) ; dans le cas contraire, il est de type infini ("groupes infinis" au sens d'E. Cartan) : exemples 1 et 2.

Pour pouvoir obtenir des résultats sur les pseudogroupes de Lie, on doit faire des hypothèses supplémentaires de régularité [4b] : on suppose par exemple que pour tout $s > 0$, $\mathcal{J}^s(\Gamma)$ est un sous-groupoïde différentiable de $\Pi^s(M)$.

Nous désignerons dans la suite par pseudogroupe de Lie transitif un pseudogroupe Γ tel que pour tout $s > 0$, $\mathcal{J}^s(\Gamma)$ est un groupoïde de Lie.

Une équation de Lie non linéaire d'ordre k peut être définie comme la donnée d'un groupoïde de Lie, sous-groupoïde de $\Pi^k(M)$

Une section locale inversible d'un groupoïde différentiable Φ est une section locale pour la surmersion α, telle que $\beta \circ s$ soit un difféomorphisme local de la base M.

L'ensemble Θ des sections inversibles de Φ constitue un pseudogroupe pour la loi de composition suivante : $(s,s') = s''$ où s'' est la section $x \to s'(\beta(x))s(x))$.

On définit ainsi le prolongement holonome $\Phi^k \subset J_k\Phi$ (espace des k-jets de sections inversibles). Par exemple $\Pi^k(M)$ est le prolongement d'ordre k de $\Pi^\circ(M) = M \times M$.

On définit le prolongement semi-holonome $\bar{\Phi}^k \subset \bar{J}_k\Phi$ et le prolongement sesquiholonome $\check{\Phi}^k$. Si Φ est un groupoïde de Lie, il en est de même de Φ^k, $\bar{\Phi}^k$ et $\check{\Phi}^k$; pour tout k' $(0 \leq k' < k)$, on a la projection $\Phi^k \to \Phi^{k'}$ $(\bar{\Phi}^k \to \bar{\Phi}^{k'}, \check{\Phi}^k \to \Phi^{k'})$.

Soit ϕ un groupoïde de Lie, de base M ; l'ensemble $\phi_{x_o} = \alpha^{-1}(x_o)$ est un fibré principal, de base M, de groupe structural $G_{x_o} = (\alpha \times \beta)^{-1}(x_o, x_o)$, de projection β (on montre d'abord que ϕ_{x_o} est un fibré principal "abstrait", puis en utilisant le fait que ϕ est localement trivial, on montre que c'est un fibré principal différentiable [7d]).

Inversement si (P, M, π) est une fibration principale, de groupe structural G, alors l'espace quotient $(P \; P)/\rho$ (où ρ est la relation d'équivalence définie par $(zg, z'g) \sim (z', z)$ pour tout $g \in G$) est muni d'une structure de groupoïde de Lie dont l'ensemble des unités d'identifie à M ; c'est le groupoïde des isomorphismes de fibres sur fibres [2b], [7d] ; le groupoïde associé à $\phi_{x_o} = \alpha^{-1}(x_o)$ est ϕ lui-même

<u>Exemple</u> : sur une variété M, l'ensemble $\pi^k(M)$ des q-jets inversibles de M dans M est un groupoïde de Lie, associé à $H^k(M)$ (espace des repères d'ordre k); $\pi^k_{x_o}(M)$ est l'ensemble des k-jets de source x_o. <u>Les équations de Lie non linéaires</u> sont habituellement définies par <u>la donnée d'un sous-fibré principal de</u> $\pi^k_{x_o}(M)$.

Si P est un fibré principal et ϕ son groupoïde associé, il y a correspondance biunivoque entre automorphismes locaux de ce fibré et sections locales inversibles de ϕ ; on en déduit [7d] que ϕ^k (resp. $\bar{\phi}^k$, $\check{\phi}^k$) est le groupoïde associé à $\mathcal{T}^k_n P$ (resp. $\bar{\mathcal{T}}^k_n P$, $\check{\mathcal{T}}^k_n P$).

La notion de système différentiel d'ordre k s'étend aux groupoïdes de Lie quelconques : un tel système est défini par un sous-groupoïde de Lie Ψ de ϕ^k ; pour qu'il soit complètement intégrable, il est nécessaire que pour tout $q > 0$, la projection $\Psi^q \cap \phi^{q+k} \longrightarrow \Psi$ soit surjective (les solutions de Ψ sont des sections inversibles de ϕ).

Si l'on revient au groupoïde $\pi^k(M)$, prolongement de $\pi^o(M) = M \times M$; il est associé à $H^k(M)$ (qui peut s'écrire $\mathcal{T}^k_n M$) ; on définit $\bar{\pi}^k(M)$ et $\check{\pi}^k(M)$.

Une G-<u>structure d'ordre</u> k (où G est un groupe de Lie, sous-groupe de L^k_n) est un G-sous-fibré principal H_G de $H^k(M)$; cette G-structure est un système différentiel ; elle est dite <u>intégrable</u> si ce système est complètement intégrable c'est-à-dire si pour tout $z \in H_G$ il existe un difféomorphisme f : U (voisinage de 0 dans $\mathbb{R}^n$) $\longrightarrow$ M tel que $j^k_o f = z$ et $j^1_u(f \circ \tau_u) \in H_G$ pour tout $u \in U$ (τ_u est la translation de $\mathbb{R}^n$: $v \longrightarrow v-u$).

Les automorphismes locaux d'une G-structure H_G (difféomorphismes locaux dont le k-jet laisse invariant H_G) sont les solutions du groupoïde ϕ associé à H_G. Si ce système ϕ est complètement intégrable, la G-structure est dite <u>transitive et localement homogène</u> (alors pour tout $(z, z') \in H_G \times H_G$, il existe un difféomorphisme local transformant z en z') ; le pseudogroupe des automorphismes locaux est alors <u>de Lie</u>.

Si H_G est intégrable, il en est de même de Φ (c'est le cas des G-structures considérées dans les exemples 1,2,3 de pseudogroupes de Lie) ; mais Φ peut être intégrable sans que H_G le soit ; par exemple la structure presque complexe sur la sphère S_6 définie par les octaves de Cayley n'est pas intégrable ; par contre cette structure est homogène : S_6 s'identifie à l'espace homogène $\mathbb{G}_2/_{S\,U_3}$ ($\mathbb{G}_2$ groupe simple exceptionnel à 14 paramètres).

Supposons, pour simplifier l'exposé, la G-structure du premier ordre ; on définit le prolongement semi-holonome $\bar{H}_G^{q+1}$ de H_G comme le sous-fibré principal de $\bar{H}^{q+1}(M)$, image de $\bar{J}_q H_G$ par l'isomorphisme $\bar{J}_q H \longrightarrow \bar{H}^{q+1}(M)$; le prolongement holonome est par définition $H_G^{q+1} = \bar{H}_G^{q+1} \cap H^{q+1}(M)$; ce n'est pas nécessairement un fibré principal ; pour que la G-structure soit intégrable, il est nécessaire que pour tout $q > 0$, l'application $H_G^{q+1} \longrightarrow H_G$ soit surjective ; on en déduit [7d], [7e] qu'alors H_G^{q+1} est un fibré principal ; cette condition réalisée, la stucture est dite q-intégrable. L'obstacle à la q-intégrabilité est le <u>tenseur de structure</u> d'ordre q, notion introduite par C. Ehresmann [2a] et D. Bernard [1] pour l'ordre 1 ; pour l'ordre supérieur voir [7d], [4a] . Le point de vue de C. Ehresmann est le suivant : une G-structure est une section globale s du fibré $H/_G$ (de fibres isomorphes à l'espace homogène ${}^{L_n}/_G$) ; le jet $j^1 s$ définit une section globale de ${}^{\bar{H}_2}/_{\bar{G}^2}$ (où $\bar{G}^2$ est le groupe structural de $\bar{H}_G^2$) ; pour que la structure soit 1-intégrable, la section doit vérifier un système différentiel du premier ordre ; de même à l'ordre supérieur (équations <u>admissibles</u> de P. Molino [9]).

On définit de même le tenseur de structure pour les groupoïdes de Lie, sous-groupoïdes de $\Pi(M)$.

Remarque : il existe toujours des sections locales de ${}^H/G$ et donc des G-structures locales ; il existe même des G-structures intégrables dans un ouvert de M (transformées par une carte locale de la G-structure triviale dans $\mathbb{R}^n$) ; le tenseur de structure est donc important du point de vue global. On sait qu'il y a des obstacles topologiques à l'existence d'une section globale de $H/_G$ (par exemple seules les sphères S_2 et S_6 admettent des structures presque complexes).

V - <u>Déplacements infinitésimaux des groupoïdes différentiables et des fibrés principaux</u>.

Pour tout groupoïde différentiable, un <u>déplacement infinitésimal</u> est un vecteur tangent à Φ qui est α-vertical et dont l'origine est une unité de Φ; en notant depl Φ l'ensemble des déplacements infinitésimaux de Φ , on a

$$\text{depl}\,\Phi = M \underset{i \times \pi}{\times} \overset{\alpha}{V} T\Phi$$

où i est l'injection canonique $M \longrightarrow \Phi$ et π la projection sur Φ du fibré tangent α-vertical $V^{\alpha} T\Phi$.

Dans le cas d'un espace homogène $P/_G$ (P groupe de Lie, G sous-groupe de Lie de P), on retrouve les déplacements infinitésimaux de la méthode du repère mobile.

Si Φ est une somme de groupes ($\alpha = \beta$), M est la sous-variété des unités de ces groupes et depl Φ est la réunion de leurs algèbres de Lie ; le fibré vertical VT Φ s'identifie au produit fibré depl $\Phi \times_M \Phi$; c'est le cas notamment des fibrés vectoriels.

Proposition 4 ([7e])

Si Φ est le groupoïde associé à un fibré principal P, il existe un difféomorphisme canonique de depl Φ sur $TP/_G$, espace des vecteurs tangents à P mod. les translations à droite de G.

En effet si Λ est la projection $P \times P \longrightarrow \Phi = {}^{(P \times P)}/_{\rho}$ et γ un chemin $I \longrightarrow \Phi$ tel que $j^1_0 \gamma \in$ depl Φ , alors l'image réciproque par Λ de $\gamma(I)$ est la classe $(\delta(0)G, \delta(I)G)$ où δ est un chemin $I \longrightarrow P$.

On posera depl P = $^{TP}/G$ et l'on a TP = depl P $\times_M$ P.

Corollaire. Si P et P' sont des fibrés principaux ayant même groupoïde associé, alors depl P = depl P'.

C'est le cas notamment de deux sous-fibrés principaux P et P' d'une même fibré principal se déduisant l'un de l'autre par la translation $z \longrightarrow zs$.

L'espace depl Φ est un algébroïde de Lie au sens de J. Pradines : le faisceau des sections de depl Φ est un faisceau d'algèbres de Lie et le morphisme surjectif de fibrés vectoriels r : depl $\Phi \longrightarrow TM$ induit un morphisme d'algèbres de Lie ; de plus le noyau de r est un fibré en algèbres de Lie. Si Φ est un groupe de Lie, on retrouve son algèbre de Lie au sens usuel. C'est pourquoi A. Kumpera [5] a désigné par algèbre de Lie de Φ le faisceau des sections de depl Φ . J. Pradines [10a] a montré que tout algébroïde de Lie est isomorphe à l'algébroïde d'un groupoïde α-simplement connexe (c'est-à-dire tel que $\alpha^{-1}(x)$ est simplement connexe pour tout $x \in M$).

A tout morphisme de fibrés principaux ou de groupoïdes de Lie correspond un morphisme de l'espace de leurs déplacements infinitésimaux ; en particulier si Φ' est un sous-groupoïde différentiable de Φ, alors depl Φ' est un sous-algébroïde de Lie de depl Φ .

A. Rodrigues [12] a démontré le théorème suivant en utilisant le théorème de Frobenius :

Théorème 4. Si (P,M,π) est un fibré principal de base connexe ; alors pour tout sous algébroïde de Lie E de depl P, il existe un sous-fibré principal connexe P' (connexe en tant que sous-variété de P) tel que depl P' = E' ; tout autre sous-fibré principal connexe P" tel depl P" = E' se déduit de P' par $z \longrightarrow zs$.

La version en termes de groupoïdes de ce théorème devient :

<u>Théorème 5</u>. <u>Soit</u> ϕ <u>un groupoïde de Lie de base connexe M</u> ; <u>si E' est un sous-algé-broïde de Lie</u> de depl ϕ , <u>alors il existe un sous-groupoïde</u> α -<u>connexe unique</u> ϕ' <u>tel que</u> depl ϕ' = E'.

En raison du corollaire du théorème 2 et de la proposition 4, à tout système différentiel d'ordre k dans $\pi^k(M)$ ou $H^k(M)$ correspond un système différentiel d'ordre k dans $J_k T(M)$ (le <u>linéarisé</u> du système précédent) ; l'ensemble des solutions d'un sous-fibré vectoriel R_k de $J_k T(M)$ est appelé un <u>pseudogroupe infinitésimal</u> ([7b]) ; si Γ est un pseudogroupe de Lie, les éléments du pseudogroupe infinitésimal correspondant sont appelés Γ-champs de vecteurs [4b] ; ce sont des champs de vecteurs locaux engendrant des groupes locaux à un paramètre de difféomorphismes appartenant à Γ ; ces questions sont étudiées en détail dans [7b], [4b], [12]. Ces équations, sous le nom d'<u>équations de Lie linéaires</u>,sont l'objet de nombreuses études [3b] , [8], [5].

Si le système d'équations de Lie non linéaires est formellement intégrable, il en est de même de son linéarisé ; inversement on a :

<u>Théorème 6</u>. <u>Soit</u> π_G <u>un sous-groupoïde de Lie de</u> $\pi(M)$, R_1 <u>le sous-fibré vectoriel de</u> $J_1 T(M)$ <u>isomorphe à</u> depl π_G ; <u>si les conditions suivantes sont satisfaites</u> :

1°) π_G <u>est</u> α-<u>connexe</u>

2°) <u>le prolongement</u> $R_2 = J_2 T \cap \bar{R}_2$ <u>de</u> R_1 <u>est un sous-fibré vectoriel de</u> $J_2 T(M)$ <u>et de</u> $\bar{R}_2$.

3°) <u>l'application</u> $R_2 \longrightarrow R_1$ <u>est surjective</u>,

<u>alors le groupoïde</u> π_G <u>est 1-intégrable</u> (c'est-à-dire l'application $\pi^2(M) \cap \bar{\pi}_G^2 \to \pi_G$ <u>est surjective</u> (où $\bar{\pi}_G^2$ est le prolongement semi-holonome de π_G).

Preuve : on est dans les conditions d'application du théorème 5 ; en effet M est connexe, l'application $R_2 \longrightarrow T(M)$ est surjective ; depl $\bar{\pi}_G^2$ est isomorphe à $\bar{R}_2$; par suite depl $\pi^2(M) \cap$ depl $\bar{\pi}_G^2$ est un sous-algèbroïde de Lie de depl $\bar{\pi}^2(M)$ (le crochet de deux sections locales appartient à depl $\pi^2(M)$ et à depl $\bar{\pi}_G^2$). Il existe donc un groupoïde de Lie α-connexe π_G^2 contenu dans $\bar{\pi}_G^2 \cap \pi^2(M)$ tel que depl π_G^2 = depl $\pi^2(M) \cap$ depl $\bar{\pi}_G^2$; d'après l'hypothèse 2, la restriction à π_G^2 de l'application $\varpi : \bar{\pi}_G^2 \longrightarrow \pi_G$ est une submersion et $\varpi(\pi_G^2)$ est ouvert dans π_G ; par action du groupe structural de la fibration principale $\pi_{G,x}^2 \longrightarrow \pi_{G,x}$ (où $x \in M$) on montre que les projections par ϖ de deux variétés intégrables maximales du champ défini par depl $\pi^2(M) \cap$ dep $\bar{\pi}_G^2$ coïncident ou sont disjointes ; $\pi_{G,x}$ étant connexe, l'application $\pi_{G,x}^2 \longrightarrow \pi_{G,x}$ est surjective.

Si le système R_1 est q-intégrable, on démontre par récurrence qu'il en est de même de π_G.

Les conditions d'intégrabilité indiquées par J. Pommaret dans son livre, conditions nécessaires pour que $R_{q+1} \longrightarrow R_q$ soit surjectif, expriment que le tenseur de structure d'ordre q du groupoïde π_G est nul, mais il se place du point du vue local.

Remarque : si l'on considère une surmersion (E,M,π) admettant un parallélisme fibré (cf §II), le théorème 6 ne s'applique pas, même localement, si ce parallèlisme fibré ne vérifie par certaines conditions d'intégrabilité permettant d'appliquer le théorème de Frobenius (voir [7f]).

BIBLIOGRAPHIE

[1] D. BERNARD, Thèse, Ann. Inst. Fourier, 10 (1960), p. 151-270.

[2] C. EHRESMANN

a) "Sur les structures infinitésimales régulières" Congrès. Intern. Math. Amsterdam, 1954 vol. 1, p. 479-480

b) "Connexions Infinitésimales" Colloq. Top. Alg. Bruxelles, 1950, p. 29-55

c) "Structures infinitésimales et pseudogroupes de Lie" Coll. Intern. C.N.R.S. Géom. Diff. Strasbourg 1953 p. 97-110

d) Comptes-rendus Acad. Sc. Paris t. 240 (1954) p. 1762 ; t. 241 (1955) p. 397 et 1755 ; t. 246 (1958) p. 360

e) "Connexions d'ordre supérieur" Atti 5e Congr. dell'Unione Matem. Italiana 1955 ; Cremonese, Roma, 1956, p. 326-328

f) Catégories topologiques et catégories différentiables" Colloq. géom. Diff. Globale Bruxelles, 1958, p. 137-150

g) "Groupoïdes diferenciales" Revista de la Unione Mat. Argentina, XIX, Buenos Aires, 1960, p. 48.

[3] H. GOLDSCHMIDT

a) "Non linear partial differential equations". Journ. of Diff. Geom. 1, 1967, p. 269-307

b) "Sur la structure des équations de Lie" Journ. of Diff. Geometry 6, 1972, p. 357-373; 7, 1972, p. 269-307.

[4] V. GUILLEMIN

a) "The integrability problem for G-structures" Trans. Amer. Math. Soc. 116 (1965) p. 544

and STERNBERG b) "Deformation theory of pseudogroup structures" Mem. Amer. Soc. 64, 1966

c) "The Lewy Counterexample". Journ. of Diff. Geom. 1, 1967, 1967, p. 58-67.

[5] A. KUMPERA and SPENCER "Lie équations" Annals of Math. Studies n° 73 Princeton University Press, Princeton 1972.

[6] M. KURANISHI "Lectures on involutive systems" São Paulo, 1967.

[7] P. LIBERMANN

a) Thèse, Strasbourg 1953. "Sur le problème d'équivalence des structures infinitésimales régulières" Ann. Mat. Pura Appl. 36 (1954) p. 27-120

b) "Pseudogroupes infinitésimaux". Bull. Soc. Math. France 87 (1959) p. 409-425

c) "Connexions d'ordre supérieur et tenseur de structure" Atti. Conv. Internat. Geom. Diff., Bologna, 1967

d) "Sur les prolongements des fibrés principaux et groupoïdes différentiables" Séminaire Analyse Globale, Montréal, 1969, p. 7-108

e) "Groupoïdes différentiables et presque parallélisme" Symposia Mathematica, Vol X (Convegno di Geom. Diff.) Roma 1971, p. 59-93

f) "Parallèlismes" Journ. of Diff. Geometry 8 (1973) p. 511-539

[8] B. MALGRANGE "Equations de Lie" Journ. of Diff. Geometry 6, 1972, p. 503-522 ; 7, 1972, p. 117-141.

[9] P. MOLINO "Sur quelques propriétés des G-structures" Journ. of Diff. Geometry 7, 1972, p. 489-518.

[10] J. PRADINES

a) C.R. Acad. Sc. Paris 263, 1966, p. 907-910 ; 264, 1967, p. 245-248 ; 266, 1968, p. 1194-1196

b) "Fibrés vectoriels doubles et calcul des jets non holonomes". Esquisses Mathématiques, 29 (1977)

[11] N.V. QUE "Nomabelian Spencer cohomology" Journ. of Diff. Geometry 3 (1969) 165-211.

[12] A. ROGRIGUES "G-Structures et pseudogroupes de Lie" Cours Faculté Sciences Grenoble, 1967-68.

[13] D. SPENCER "Overdetermined systems of linear partial differential equations" Bull. Amer. Math. Sco. 75, 1965, p. 1-114.

[14] P. VER EECKE

a) Thèse ; Cahiers Topologie et Geom. Diff. 5, 1963

b) Géométrie Différentielle (São Paulo, 1967) ; Conexiones de orden superior (Zaragoza, 1968)
Traduction anglaise (Cross et Smith), Melbourne, 1978.

INFINITESIMAL DEFORMATIONS

PRESERVING PARALLEL NORMAL VECTOR FIELDS

by

V. I. Oliker

Department of Mathematics
The University of Iowa
Iowa City, Iowa 52242
U.S.A.

The problem of infinitesimal rigidity of isometrically immersed in Euclidean space manifolds is well known and various aspects of it have been studied by many authors. See, for example, papers by Goldstein and Ryan [2], Kahn [3], and the references given there; see, also, Kobayashi and Nomizu [4].

In this work we propose to investigate a similar problem but not for rigidity with respect to the metric, but with respect to certain functions of principal radii of curvature. This problem is not new. It arises naturally when one studies existence and uniqueness in the Minkowski problem and its generalizations, see Cheng and Yau [1], Pogorelov [8], Stoker [9]. It seems to be also of independent interest since certain phenomena occur here, which seemed to be new even in the classical case.

The deformations we consider here, besides being infinitesimal, are also required to preserve a parallel normal vector-field. In codimension one this means that the Gaussian image on the unit hypersphere remains unchanged throughout the deformation. We will show that this condition in fact is not too restrictive. Also, the manifolds we study are assumed to have a globally defined nondegenerate parallel normal vector field.

For deformations of such type we derive the basic differential equation and investigate it in several cases. The results are more complete for convex hypersurfaces, because the equation takes in this case especially simple form and it can be investigated by the methods used in [6] and [9].

§1. Basic notations and formulas.

In this paragraph we follow closely the sections 2 and 3 of [7].

Unless it is stated otherwise the following convention on the ranges of indices is assumed to be in effect throughout the paper:

$$1 \le i,j,k,\ell,r,s \le m, \quad 1 \le \alpha \le n.$$

It is also agreed that repeated lower and upper indices are summed over the respective ranges. It is assumed that all submanifolds and maps are sufficiently smooth, say C^3. We denote by E the Euclidean space of dimension $m+n$ (≥ 3), and fix the origin at some point 0. Consider a smooth connected orientable submanifold M of dimension m (≥ 2) immersed in E, and represented by the position vector field

$$X = X(u^1,\cdots,u^m),$$

where $\{u^i\}$ are the local coordinates on M.

We adopt also the following notations:

$T(M)$, $N(M)$ — the tangent and the normal bundles of M;

$T_x(M)$, $N_x(M)$ — the corresponding restrictions to the fibers at the point $x \in M$;

$\langle\ ,\ \rangle$ — the inner product in E;

$\partial_i = \partial/\partial u^i$, $\partial_{ij} = \partial/\partial u^i \partial u^j$ — the differentiation in E;

$X_i = \partial_i X$, $\xi_i = \partial_i \xi$, $\eta_i = \partial_i \eta$, $X_{ij} = \partial_{ij} X$, $\xi_{ij} = \partial_{ij} \xi$, $\eta_{ij} = \partial_{ij} \eta$, where ξ and η are arbitrary smooth unit vector fields on M with values in $N(M)$;

$G = \langle X_i, X_j \rangle$ — the first fundamental form on M, $G = (G_{ij})$;

$b(\xi) = \langle X_i, \xi_j \rangle$ — the second fundamental form with respect to the unit normal ξ, $b(\xi) = (b_{ij}(\xi))$;

$g(\xi,\eta) = \langle \xi_i, \eta_j \rangle$ — the mixed third fundamental form, $g(\xi,\eta) = (g_{ij}(\xi,\eta))$, $g(\xi) = g(\xi,\xi)$;

$N(1), \cdots, N(n)$ — a smooth field of orthonormal ordered frames on M such that $N(\alpha)\colon M \to N(M)$ for each α.

In the frame $X_1, \cdots, X_m$, $N(1), \cdots, N(m)$ we have for the fields ξ and η as above:

$$\xi_i = b_i^j(\xi) X_j + \sum_\alpha \langle \xi_i, N(\alpha) \rangle N(\alpha) \tag{1}$$

where $b_i^j(\xi) = b_{i\ell}(\xi) G^{\ell j}$, and $G^{\ell j}$ being the inverse of G;

$$g_{ij}(\xi,\eta) = b_i^r(\xi) b_{rj}(\eta) + \sum_\alpha \langle \xi_i, N(\alpha) \rangle \langle \eta_j, N(\alpha) \rangle. \tag{2}$$

If ξ (or η) is parallel, that is, the derivative of ξ in any tangential direction is in $T(M)$, then the second term on the right-hand side of (2) vanishes. Note also, that if in addition, M, ξ, and η are such that $b(\xi)$ and $b(\eta)$ are positive definite then so is $g(\xi,\eta)$. However, $g(\xi)$ is always nonnegative. If $\det b(\xi) \neq 0$ everywhere on M, that is, ξ is nondegenerate, then $g(\xi)$ induces a Riemannian metric on M. We denote by dV_ξ the corresponding volume element. It follows from (1) that if ξ is nondegenerate and parallel, then vectors $\{\xi_i\}$ form a basis in $T_x(M)$, $x \in M$, and according to the Weingarten equation we have:

$$\xi_{ij} = \Gamma^k_{ij}(\xi)\xi_k - \sum_\alpha g_{ij}(\xi,N(\alpha))N(\alpha), \tag{3}$$

where $\Gamma^k_{ij}(\xi)$ denote the Christoffel symbols of the second kind with respect to $g(\xi)$.

Finally, we denote through $h(\xi)$ the support function of M with respect to an arbitrary unit normal field ξ; $h(\xi) = \langle X,\xi\rangle$. If $\xi = N(\alpha)$ we simply write $h(\alpha) \equiv h(N(\alpha))$. Obviously,

$$h(\xi) = \sum_\alpha h(\alpha)\langle \xi,N(\alpha)\rangle.$$

Suppose now that ξ is nondegenerate and parallel. Then we have the representation

$$X = g^{ij}(\xi)h_i(\xi)\xi_j + \sum_\alpha h(\alpha)N(\alpha)$$

$$= \sum_\alpha \{[g^{ij}(\xi)(h_i(\alpha)\langle \xi, N(\alpha)\rangle + h(\alpha)\langle \xi, N_i(\alpha)\rangle)]\xi_j + h(\alpha)N(\alpha)\}, \quad (4)$$

where $g^{ij}(\xi)$ is the matrix inverse to $(g_{ij}(\xi))$.

Put

$$\nabla_{ij} = \partial_{ij} - \Gamma^k_{ij}(\xi)\partial_k.$$

Then, for the ξ as before, we have

$$b_{ij}(\xi) = \nabla_{ij}h(\xi) + \sum_\alpha g_{ij}(\xi, N(\alpha))h(\alpha). \quad (5)$$

The principal radii of curvature associated with ξ (not necessarily parallel now) are denoted by $R_{\xi 1}, \cdots, R_{\xi m}$ and defined at each point of M as the roots of the determinantal equation

$$\det(b(\xi) - Rg(\xi)) = 0.$$

Since for a nondegenerate field ξ $g(\xi)$ is positive definite, the $R_{\xi i}$ are well-defined. Moreover, in this case they don't vanish. Put $|g(\xi)| \equiv \det(g(\xi))$. The elementary symmetric function of order k in $R_{\xi i}$ is defined as

$$S_k(\xi) = \sum_{i_\ell \neq i_r} R_{\xi i_1} \cdots R_{\xi i_k},$$

and it is the coefficient at $(-R)^{m-k}$ of the polynomial

$$\frac{\det(b(\xi)-Rg(\xi))}{|g(\xi)|} = \sum_{k=0}^{m} S_k(\xi)(b(\xi),g(\xi))(-R)^{m-k}, \qquad (6)$$

where $S_0(\xi) = 1$.

§2. Deformation fields.

Let M be a compact submanifold of E. Throughout this section we assume that there exists on M a globally defined nondegenerate parallel unit normal vector field ξ. Let Z be a smooth vector field on M. Consider a map

$$X^t = X + tZ, \quad X^t : [0,\varepsilon] \times M \to E, \quad \varepsilon > 0.$$

Since M is compact, ε can be chosen small enough so that X^t is a family of immersions. Z is called a deformation.

Definition. We say that deformation Z preserves the vector field ξ if ξ is in the normal bundle of X^t for any $t \in [0,\varepsilon]$.

Proposition 2.1. A deformation Z of M preserves a nondegenerate parallel normal vector field ξ if and only if it can be represented in the form

$$Z = g^{ij}(\xi)\langle Z,\xi\rangle_i \xi_j + \sum_\alpha \langle Z,N(\alpha)\rangle N(\alpha). \tag{7}$$

Proof. Let Z be a deformation which preserves ξ. Denote by $f(\alpha) = \langle Z,N(\alpha)\rangle$. Then

$$f(\xi) \equiv \langle Z,\xi\rangle = \sum_\alpha \langle Z,N(\alpha)\rangle\langle \xi,N(\alpha)\rangle = \sum_\alpha f(\alpha)\langle \xi,N(\alpha)\rangle.$$

Since Z preserves ξ, we have for any i, $\langle Z_i,\xi\rangle = 0$, and therefore

$$f_i(\xi) = \langle Z,\xi_i\rangle.$$

Let x be an arbitrary point of M. Vector-fields $\xi_1,\cdots,\xi_m$, $N(1),\cdots,N(m)$ restricted to a neighborhood of x constitute a local basis at x. Let

$$Z = A^j\xi_j + \sum_\alpha B(\alpha)N(\alpha),$$

where A^j and $B(\alpha)$ are the coefficients of expansion of Z in this basis. Obviously, $B(\alpha) = f(\alpha)$, and $A^j = g^{ij}(\xi)f_i(\xi)$.

Conversely, let $f(\alpha)$, $\alpha = 1,\cdots,m$, be m smooth functions defined on M. Define

$$f(\xi) = \sum_\alpha f(\alpha)\langle \xi,N(\alpha)\rangle,$$

and

$$Z = g^{ij}(\xi)f_i(\xi)\xi_j + \sum_\alpha f(\alpha)N(\alpha).$$

We have to show that $\langle Z_k,\xi\rangle = 0$ for any $k = 1,\cdots,m$. In view of (3) and because

$$f_k(\xi) = \sum_\alpha [f_k(\alpha)\langle \xi,N(\alpha)\rangle + f(\alpha)\langle \xi,N_k(\alpha)\rangle],$$

we get

$$\langle Z_{k,\xi}\rangle = g^{ij}(\xi)f_i(\xi)\langle \xi_{jk},\xi\rangle + f_k(\xi)$$

$$= -g^{ij}(\xi)f_i(\xi)g_{jk}(\xi) + f_k(\xi) = 0.$$

Here we also used the fact that $g_{jk}(\xi) = -\langle \xi_{jk},\xi\rangle$. The proposition is proved.

In the following, unless otherwise stated, we always consider deformations which preserve a nondegenerate parallel normal vector field on M.

Let ξ be as above and Z a deformation which preserves ξ. It is easy to see that ξ will remain nondegenerate throughout the deformation and therefore one can consider the k-th elementary symmetric functions of principal radii of curvature with respect to ξ of the immersion X^t. Denote such a function by $S_k^t(\xi)$.

Definition. Let M,ξ,Z be as above. The deformation Z is called an infinitesimal deformation of M with respect to

$S_k(\xi)$ if

$$\frac{\partial}{\partial t} S_k^t(\xi)\Big|_{t=0} = 0 \quad \text{on} \quad M - \partial M, \tag{8}$$

where ∂M denotes the boundary of M. If $\partial M = \emptyset$, then (8) should be satisfied everywhere on M.

Finally, we note that a deformation Z is called trivial if it is a restriction to M of a parallel translation of E.

In this case M is said to be infinitesimally rigid with respect to such deformation.

§3. The main elliptic differential equation.

In this section M and ξ are as before, and Z is an infinitesimal deformation with respect to $S_k(\xi)$ for some k. Of course, we assume that Z preserves ξ. Since the third fundamental form $g(\xi)$ remains the same for the entire family X^t, $t \in [0,\varepsilon]$, the principal radii of curvature are defined as in (6) where $(b_{ij}(\xi))$ is replaced by

$$b^t_{ij}(\xi) = b_{ij}(\xi) + t\langle Z_i, \xi_j\rangle.$$

It follows from here that $S^t_k(\xi)$ is the sum of determinants constructed from $(1/|g(\xi)|)\det(b^t(\xi)-Rg(\xi))$ by taking k columns with $b^t_{ij}(\xi)$ and $m-k$ columns with $g_{ij}(\xi)$ in each of the determinants. If one expands now $S_k^t(\xi)$ in t, then it is not difficult to see that

$$\frac{\partial S_k^t(\xi)}{\partial t}\Big|_{t=0} = \frac{S_k^{ij}(\xi)}{g(\xi)} \langle Z_i, \xi_j \rangle,$$

where $S_k^{ij}(\xi)$ is the sum of the cofactors of the elements $b_{ij}(\xi)$ in the determinants defining $S_k(\xi)$.

Let us compute $\langle Z_i, \xi_j \rangle$. We have $f(\xi) = \langle Z, \xi \rangle$. Since Z preserves ξ, $f_i(\xi) = \langle Z, \xi_i \rangle$. From here

$$f_{ij}(\xi) = \langle Z_j, \xi_i \rangle + \langle Z, \xi_{ij} \rangle,$$

and because of (3)

$$f_{ij}(\xi) = \langle Z_j, \xi_i \rangle + \Gamma_{ij}^s \langle Z, \xi_s \rangle - \sum_\alpha g_{ij}(\xi, N(\alpha)) \langle Z, N(\alpha) \rangle.$$

That is,

$$\langle Z_j, \xi_i \rangle = \nabla_{ij} f(\xi) + \sum_\alpha g_{ij}(\xi, N(\alpha)) f(\alpha),$$

where $f(\alpha) = \langle Z, N(\alpha) \rangle$. Finally, noting that $\langle Z_i, \xi_j \rangle = \langle Z_j, \xi_i \rangle$, we obtain because of (8)

$$\frac{S_k^{ij}(\xi)}{|g(\xi)|} \left(\nabla_{ij} f(\xi) + \sum_\alpha g_{ij}(\xi, N(\alpha)) f(\alpha)\right) = 0 \quad \text{in} \quad M - \partial M. \qquad (9)$$

This is the differential equation of the infinitesimal deformations.

It follows from proposition 3.2 in [7] that

$$\frac{S_k^{ij}(\xi)}{|g(\xi)|}\sum_\alpha g_{ij}(\xi,N(\alpha))f(\alpha) = (m-k+1)S_{k-1}(\xi)f(\xi) + \langle H_k(\xi),Z\rangle ,$$

$$(S_0(\xi) = 1) \qquad (10)$$

where $H_k(\xi)$ is a uniquely defined vector field on M with values in $N(M) \ominus \xi$. If $k = 1$, then $-H_1(\xi)$ is the m times mean curvature vector of the immersion of M into a unit hypersphere of E centered at the origin. The immersion is made via parallel translation of ξ in E to the origin of E. For future reference, let us denote this standard unit hypersphere by Σ.

We note that the coefficients of the equation (9) make sense independent of the fact whether ξ is parallel or not.

Proposition 3.1 [7]. Let $x \in M$ be a point on M and suppose $b(\xi) > 0$ at x. Then the quadratic forms $S_k^{ij}(\xi)\nu_i\nu_j$, $k = 1,\cdots,m$, are positive definite at z. Here $\nu_1,\cdots,\nu_m$ are arbitrary real numbers such that $\nu^2 = \nu_1^2 + \cdots + \nu_m^2 \neq 0$. If M is compact without boundary, ξ is defined on M, and $b(\xi) \neq 0$ everywhere on M, then those quadratic forms are definite everywhere and by selecting a proper orientation of M and E, they can be made positive definite. Also, if ξ is parallel in $N(M)$, then for any c^2 function f on M we have

$$P_{\xi k}(f) \equiv \frac{S_k^{ij}(\xi)}{|g(\xi)|}\nabla_{ij} f = \frac{1}{\sqrt{|g(\xi)|}}\partial_i\left(\frac{S_k^{ij}(\xi)}{\sqrt{|g(\xi)|}}\partial_j f\right).$$

§4. Convex hypersurfaces in E.

Assume for convenience that $n=1$, so that E is $m+1$-dimensional Euclidean space. By a convex hypersurface M in E we understand a smooth compact (with boundary or without), orientable hypersurface whose Gauss map γ is a diffeomorphism onto a closed subset ω of a unit hypersphere Σ centered at the origin, and there exists at least one point on M where the second fundamental form is definite. The last condition is not needed when the boundary ∂M of M is void. Then M is convex because γ is a diffeomorphism onto Σ (see [4], page 41). The unit normal vector field on M in this case is always parallel, and it is nondegenerate since γ is a diffeomorphism. The fact that a deformation preserves this normal vector field means simply that the Gauss image remains unchanged throughout the deformation. The equation (9) of infinitesimal deformations becomes

$$\frac{S_k^{ij}}{|g|}(\nabla_{ij}f+g_{ij}f) = 0 \quad \text{in} \quad M-\partial M. \tag{9'}$$

Since the normal vector field ξ is uniquely defined (up to a change of orientation) we do not indicate it in (9') and further in this section. We also point out that $g = (g_{ij})$ in this case is the standard third fundamental form of M, ∇_{ij} is the operator of second covariant derivative in (g_{ij}) and all these objects as well as equation (9′) are actually defined on the Gauss image $\gamma(M)$ on Σ.

Theorem 1. If the hypersurface M is a closed convex hypersurface, then it does not admit nontrivial infinitesimal deformations with respect to an arbitrary elementary symmetric function of principal radii of curvature S_k, $k = 1, \cdots, m$.

Theorem 2. Any infinitesimal deformation Z with respect to S_k, $k = 1, \cdots, m$, of a convex hypersurface M is trivial if Z is a constant vector field on ∂M and M is a part of a closed convex hypersurface (Z does not have to be a restriction of a deformation of the closed hypersurface).

Theorem 3. An infinitesimal deformation Z with respect to S_k, $k = 1, \cdots, m$, of a convex hypersurface M is trivial if the normal component Z^{ξ} of Z vanishes at the boundary of M and one of the following conditions is satisfied:

(a) the Gauss image $\omega = \gamma(M)$ lies inside an open hemisphere Σ^+ of the hypersphere Σ;

(b) $\omega = \overline{\Sigma}^+$ $(\overline{\Sigma}^+ = \Sigma^+ + \partial\Sigma^+)$;

(c) M is a part of a closed convex hypersurface and either $\omega \subset \overline{\Sigma}^+$ or $\omega \supset \overline{\Sigma}^+$.

<u>Theorem 4</u>. Suppose that M is a convex surface with nonvoid boundary ∂M in a three-dimensional Euclidean space and Z is an infinitesimal deformation of M. Assume that there exists a direction c in E^3 such that $\langle Z,c\rangle = \text{const}$ on ∂M. Then Z is a trivial deformation with respect to S_k, $k = 1,2$.

Theorem 1 is proved by Pogorelov [8], p. 61, for arbitrary k and m. The proof relies on the Aleksandrov-Bonnesen-Fenchel inequality for mixed volumes. When $k = m$ it has been also proved by Cheng and Yau [1]. In both works these results were obtained as a step in a course of proving existence in generalized Minkowski problem. Theorems 2 and 3 can be proved by the methods of our paper [6], with making use of Pogorelov's result on closed convex hypersurfaces. Theorem 4 follows from a theorem due to Stoker [9]. It seems plausible that it is true for any dimension.

The case $k = 1$, that is, when we consider the infinitesimal deformations of the sum of the principal radii of curvature, is special. It is not difficult to see that in this case the equation (9') becomes

$$g^{ij}(\nabla_{ij}f+g_{ij}f) = \Delta_2 f + mf = 0$$

where Δ_2 is the Laplace-Beltrami operator on the unit hypersphere Σ. In this situation one can get results stronger than the previous ones. Namely, if one leaves out the assumption that the hypersurface M is convex, but keeps the assumption that the Gauss map $\gamma : M \to \omega \subset \Sigma$ is a diffeomorphism, then the last equation is still valid and all the theorems 1, 2, 3, 4 hold. Moreover, in this case the requirement in Theorem 2 and part (c) of Theorem 3 that M is a part of a closed hypersurface is no longer needed.

Note also, that in this case the problem from the analytic point of view is equivalent to the Christoffel problem.

Finally, it must be noted that under the boundary condition imposed on infinitesimal deformations Z in Theorem 3, additional conditions are required to assure that Z is trivial. Otherwise, one can construct regions where there exist nontrivial infinitesimal deformations with normal component $Z^{\xi} = 0$ on ∂M. Such examples can be constructed in the same way as it was done in [6] where uniqueness problems were studied.

The following theorem is stronger than Theorem 2 but weaker than Theorem 3. However, because of the examples mentioned above it seems to be of some interest.

Theorem 5. An infinitesimal deformation Z with respect to S_k, $k = 1, \cdots, m$, of a convex hypersurface M is trivial if the normal component Z^{ξ} vanishes at the boundary of M and there exists a point on M at which $Z = 0$. M is assumed to be a part of a closed convex hypersurface.

This theorem can be also proved by the method used in [6].

As it was noted earlier the problem of finding infinitesimal deformations with respect to S_k is crucial for the problem of existence of convex hypersurfaces with a preassigned function S_k as a function of a unit normal. When this problem is solved, then one needs certain a priori estimates of the solution and its derivatives to prove existence. For closed convex hypersurfaces such a priori estimates are known [1],[8]. For open convex surfaces some results in that direction were obtained in [5]. These are the only results for surfaces with boundaries that we are aware of.

The previous theorems can be slightly generalized if instead of infinitesimal deformations with respect to S_k we allow deformations which increase (or decrease) proportionally the function S_k. More precisely, consider a deformation $X^t = X + tZ$ such that

$$\left.\frac{\partial S_k^t}{\partial t}\right|_{t=0} = cS_k \quad \text{in} \quad M-\partial M \tag{11}$$

where c is a constant. Let us show that this case can be reduced to the previous one. Consider first the case when the convex hypersurface M is closed. Then (11) is equivalent to

$$\frac{S_k^{ij}}{|g|} (\nabla_{ij} f + g_{ij} f) = cS_k .$$

But

$$kS_k = \frac{S_k^{ij}}{|g|} (\nabla_{ij} h + g_{ij} h)$$

where h is the support function of M (see [6]). Then

$$\frac{S_k^{ij}}{|g|} [\nabla_{ij}(f - \frac{c}{k}h) + g_{ij}(f - \frac{c}{k}h)] = 0 \quad \text{in} \quad M-\partial M.$$

From Theorem 1 it follows that a deformation whose normal component is $f-(c/k)h$ is a trivial one, that is, $f=(c/k)h$ + support function of a constant vector, and

$$X^t = (1 + \frac{c}{k}t)(g^{ij}h_i\xi_j + h\xi) + tX_0 ,$$

where X_0 is a constant vector. Thus, we have arrived at the following result.

Theorem 6. If a closed convex hypersurface admits a deformation Z such that for some k, $k = 1,\cdots,m$,

$$\left.\frac{\partial S_k^t}{\partial t}\right|_{t=0} = cS_k \quad \text{on} \quad M,$$

where c is constant, then Z is a sum of a parallel translation and a homothetic transformation with coefficient $(1+(c/k)t)$.

Theorems 2, 3, and 4 can be generalized in a similar fashion.

§5. Submanifolds in E.

When M is a submanifold of E and its codimension n is greater than one then equation (9) involves n unknown functions $f(\alpha)$, and obviously one cannot expect infinitesimal rigidity of M if (8) is the only restriction imposed on the deformation. This statement can be put in a more precise form as follows. Let ξ be a nondegenerate parallel unit normal vector field defined globally on M. Choose the vector fields $N(\alpha)$, $\alpha = 1,\cdots,n$, in $N(M)$ so that locally they form a basis and $N(1) = \xi$. Then (9) becomes

$$\frac{S_k^{ij}(\xi)}{|g(\xi)|}(\nabla_{ij}f(\xi)+g_{ij}(\xi)f(\xi)) = -\frac{S_k^{ij}(\xi)}{|g(\xi)|}\sum_{\alpha=2}^{m} g_{ij}(\xi,N(\alpha)f(\alpha). \quad (12)$$

If we set the right hand side of the last equation equal to zero, then from Proposition 3.1 and standard theorems on elliptic partial differential equations it follows that the homogeneous equation has a finite number of linearly independent solutions, and the last equation is solvable for any set of functions $f(\alpha)$, $\alpha = 2,\cdots,m$, such that the right-hand side is orthogonal to the solutions of homogeneous equation. We assume here for simplicity that M does not have a boundary.) Thus, there exist at least $n-1$ degrees of freedom in the way of choosing the deformation vector field Z.

Next we consider a case when the "rigidity" can be established with respect to vector field ξ. Suppose M is actually an imbedded submanifold of the unit hypersphere Σ. It is assumed that Σ is centered at the origin of E. This can be always achieved by performing, if necessary, a parallel translation. Let ξ denote the position vector field of M. Then ξ is also a unit normal vector field, it is nondegenerate, and parallel in the normal bundle of M in E.

Theorem 7. Let M be a compact submanifold isometrically imbedded in Σ and suppose it is totally geodesic in Σ. Assume also that there exists a constant direction c in E such that $\langle c,\xi\rangle \neq 0$ everywhere on M. Then for any k, $k=1,\cdots,n$, an infinitesimal deformation Z of M with respect to $S_k(\xi)$ which preserves ξ and such that $\langle Z,\xi\rangle = 0$ on ∂M, lies completely in the normal bundle of M in Σ. The assertion remains true also if $\partial M = \emptyset$. In this case the boundary condition on $\langle Z,\xi\rangle$ is void.

Proof. Let Z be an infinitesimal deformation of M with respect to some $S_k(\xi)$ and such that it preserves ξ. We choose ξ as one of the vector fields $N(\alpha)$ (say $N(1)$), and denote as before $f(\xi) = \langle Z,\xi\rangle$, $f(\alpha) = \langle Z,N(\alpha)\rangle$. Let $h(\xi) = \langle c,\xi\rangle$. Introduce a function $\tilde{f} : M \to R$ such that

$f(\xi) = \tilde{f}h(\xi)$. Since $h(\xi) \neq 0$ on M, this is always possible. Functions $f(\xi)$ and $f(\alpha)$ have to satisfy the equation (9). But, since M is totally goedesic, the equation (9) assumes a very specific form: namely,

$$\frac{S_k^{ij}(\xi)}{|g(\xi)|}(\nabla_{ij}f(\xi)+g_{ij}(\xi)f(\xi)) = 0 \quad \text{in} \quad M-\partial M. \tag{13}$$

And since $\langle Z,\xi\rangle$ vanishes on ∂M,

$$f(\xi)\Big|_{\partial M} = 0. \tag{14}$$

If M does not have a boundary, then (14) is void.

Let us compute now $\nabla_{ij}f(\xi)$ in terms of $\tilde{f}$ and $h(\xi)$. We have

$$\nabla_{ij}f(\xi) = \tilde{f}\nabla_{ij}h(\xi) + h(\xi)\nabla_{ij}\tilde{f} + \tilde{f}_i h_j(\xi) + \tilde{f}_j h_i(\xi).$$

Substituting it in (13) we obtain

$$h(\xi)\frac{S_k^{ij}(\xi)}{|g(\xi)|}\nabla_{ij}\tilde{f} + \tilde{f}\frac{S_k^{ij}(\xi)}{|g(\xi)|}(\nabla_{ij}h(\xi)+g_{ij}(\xi)h(\xi)) + F = 0,$$

where F denotes the term which contains only the first derivatives of $\tilde{f}$ and $h(\xi)$.

For the function $h(\xi)$ we have

$$\nabla_{ij}h(\xi) = \langle c, \xi_{ij} \rangle - \Gamma^k_{ij}(\xi)\xi_k = -g_{ij}(\xi)h(\xi).$$

Thus, (13) becomes

$$h(\xi)\,\frac{S^{ij}_k(\xi)}{|g(\xi)|}\,\nabla_{ij}\tilde{f} + F = 0.$$

Let us now consider the case when $\partial M \neq \emptyset$. Then, by the maximum principle, which is applicable to the last equation, and in view of (14), we conclude that $\tilde{f} \equiv 0$ since $h(\xi) \neq 0$ on M. The latter implies $f(\xi) = 0$.

The deformation vector field Z preserves ξ and therefore can be represented as

$$Z = g^{ij}(\xi)f_i(\xi)\xi_i + f(\xi)\xi + \sum_{\alpha=2}^{n} f(\alpha)N(\alpha).$$

Obviously, only the last term is actually present.

Now let $\partial M = \emptyset$. From the maximum principle it follows that $\tilde{f}(\xi) = \text{const}$. Denote $A = \tilde{f}(\xi)$. Then $f(\xi) = Ah(\xi)$. Since $f(\xi)$ satisfies (13), we obtain by integration over M with respect to the metric $g(\xi)$:

$$\int_M \frac{S^{ij}_k(\xi)}{|g(\xi)|}\, g_{ij}(\xi)f(\xi)dV_\xi = 0$$

Both matrices $(S^{ij}_k(\xi))$ and $(g_{ij}(\xi))$ are positive definite because ξ is nondegenerate. Then the trace of their product is positive. Thus, there must exist a point on M where $f(\xi) = 0$. It implies that $A = 0$, since $h(\xi) \neq 0$, and, as previously, we conclude that

$$Z = \sum_{\alpha=2}^{n} f(\alpha) N(\alpha).$$

The theorem is proved.

Remark 1. The last theorem can be slightly generalized in the following way. Consider a submanifold of the Euclidean space E. Let ξ be a nondegenerate parallel unit vector field on M. Translating ξ parallel to itself in E to the origin we get a mapping $\gamma_\xi : M \to E$. Since ξ is nondegenerate, γ_ξ is an immersion. Suppose that γ_ξ immerses M as a totally geodesic submanifold of Σ, and assume that there exists a constant direction c in E such that $h(\xi) = \langle c, \xi \rangle \neq 0$ in M. Then the conclusion of the Theorem 7 still holds. The proof of this assertion runs similarly to the proof of Theorem 7.

REFERENCES

1. S.Y. Cheng, S.T. Yau, On the regularity of the solution of n-dimensional Minkowski problem, Comm. Pure Appl. Math. 29(1976), pp. 495-516.

2. R.A. Goldstein, P.J. Ryan, Rigidity and energy, Global Analysis and its Applications, vol. II, Vienna, 1974.

3. E. Kann, A new method for infinitesimal rigidity of surfaces with $k>0$, J. Differential Geometry, 4(1970), pp. 5-12.

4. S. Kobayashi, K. Nomizu, Foundations of Differential Geometry, vol. 2, Interscience, New York, 1969.

5. A.I. Medianik, Existence theorems for convex surfaces with boundaries, I, II. Ukrainski Geom. Sb., 17(1974), pp. 111-120; ibid., 18(1975), pp. 90-98. (In Russian)

6. V.I. Oliker, On certain elliptic differential equations on a hypersphere and their geometric applications, Indiana Univ. Math. J. 28(1979), No. 1, pp. 35-51.

7. ___________, On compact submanifolds with nondegenerate parallel normal vector fields, Pacific J. Math. (to appear).

8. A.V. Pogorelov, The Minkowski multidimensional problem (translated from Russian), J. Wiley and Sons, New York, 1978.

9. J.J. Stoker, Uniqueness theorems for some open and closed surfaces in three-space, Annali della Scula Norm. Sup. di Pisa, (1978), ser. iv, vol. v, 4, pp. 657-677.

DIFFERENTIAL GALOIS THEORY

J.F.POMMARET Collège de France , Paris

Our announcement to that conference is the complete solution of the Galois theory for systems of partial differential equations which will be published in a forthcoming book (17). As it involves a huge amount of differential geometry (16) and algebraic geometry (15) our sketch will be as elementary as possible.

Let us start with some rough ideas.

A) LIE PSEUDOGROUP: It is a group of transformations solutions of a non--linear system of partial differential equations. The following examples show that most of the useful ones are defined by algebraic partial differential equations of any order and that it is not possible in general to know "a priori" or to interpretate the general solution.

$y'' = 0$	$y = ax + b$	$G \times \mathbb{R} \to \mathbb{R}$
$\dfrac{y'''}{y'} - \dfrac{3}{2}\left(\dfrac{y''}{y'}\right)^2 = 0$	$y = \dfrac{ax+b}{cx+d}$	$G \times \mathbb{R} \to \mathbb{R}$
$\dfrac{\partial y^1}{\partial x^1} = 1,\ \dfrac{\partial y^1}{\partial x^2} = 0,\ \dfrac{\partial y^2}{\partial x^2} = 0$	$\begin{cases} y^1 = x^1 + a \\ y^2 = x^2 + f(x^1) \end{cases}$	?
$\dfrac{\partial(y^1, \ldots, y^n)}{\partial(x^1, \ldots, x^n)} = 1$	?	?

B) NEED FOR A NEW THEORY: The first idea is to transform the given equation into a system of a particular kind:

$$y^m - a_1 y^{m-1} + \cdots + (-1)^m a_m = 0 \iff \begin{cases} y^1 + \cdots + y^m = a_1 \\ \cdots\cdots \\ y^1 \cdots y^m = a_m \end{cases}$$

① CLASSICAL GALOIS THEORY:

Systems of algebraic equations $\longleftrightarrow$ finite (permutation) groups

② PICARD-VESSIOT THEORY $\longrightarrow$ KOLCHIN THEORY:

Systems of linear differential equations $\longleftrightarrow$ linear algebraic groups

③ DIFFERENTIAL GALOIS THEORY:

Systems of algebraic p.d.e. $\longleftrightarrow$ algebraic Lie pseudogroups

SUMMARY

I/ BASIC FACTS: A) Systems of p.d.e.

B) Classical Galois theory

II/HISTORICAL SURVEY

III/DIFFERENTIAL INVARIANT THEORY

IV/DIFFERENTIAL GALOIS THEORY

-+-+-+-+-+-+-+-+-+-+-+-+-

I/BASIC FACTS: A)Systems of p.d.e.:(16)

For simplicity we shall be dealing with manifolds and maps of locally constant rank.The notations are standard ones in that field.

$$\begin{cases} \text{fibered manifold} & \pi : \mathcal{E} \to X & (x^i, y^k) \to (x^i) & i = 1, \dots, n & \dim X = n \\ \text{jet bundle} & J_q(\mathcal{E}) \to X & (x^i, y^k_\mu) \quad 0 \leq |\mu| \leq q & \text{or} & (x, y, \frac{\partial y}{\partial x}) \\ \text{vertical bundle} & E = V(\mathcal{E}) & J_q(E) \simeq V(J_q(\mathcal{E})) & & \end{cases}$$

DEFINITION: $\mathcal{R}_q \subset J_q(\mathcal{E})$ $\xleftarrow{\text{projection}}$ $\xrightarrow{\text{prolongation}}$ $\mathcal{R}_{q+r} = J_r(\mathcal{R}_q) \cap J_{q+r}(\mathcal{E})$

$$\phi^\tau\left(x, y, \frac{\partial y}{\partial x}\right) = 0 \qquad\qquad d_\nu \phi^\tau = 0 \quad 0 \leq |\nu| \leq r$$

DEFINITION: $\mathcal{R}_q$ formally integrable $\Leftrightarrow$ $\pi^{q+r+s}_{q+r} : \mathcal{R}_{q+r+s} \to \mathcal{R}_{q+r} \to 0$

DEFINITION: $\mathcal{R}^{(s)}_{q+r} = \pi^{q+r+s}_{q+r}(\mathcal{R}_{q+r+s}) \subseteq \mathcal{R}_{q+r}$

DEFINITION: Symbol $\quad G_q = R_q \cap S_q T^* \otimes E \longrightarrow \mathcal{R}_q$

$$G_q : \quad \frac{\partial \phi}{\partial y^k_\mu} \cdot v^k_\mu = 0 \qquad |\mu| = q$$

prolongation

$$G_{q+r} : \quad \frac{\partial \phi}{\partial y^k_\mu} \cdot v^k_{\mu+\nu} = 0 \qquad |\mu| = q \quad |\nu| = r$$

DEFINITION:Spencer δ-cohomology:

$$\exists\ \delta : \Lambda^s T^* \otimes G_{q+r+1} \longrightarrow \Lambda^{s+1} T^* \otimes G_{q+r} \quad : \quad (\delta\omega)^k_\mu = dx^i \wedge \omega^k_{\mu+1_i}$$

$$I = (i_1, \dots, i_s) \ i_1 < \dots < i_s, \quad dx^I = dx^{i_1} \wedge \dots \wedge dx^{i_s}, \quad \omega^k_\mu = v^k_{\mu, I}\, dx^I$$

$$\delta^2 \equiv 0 \Rightarrow G_q \ s\text{-acyclic} \Leftrightarrow H^1_{q+r} = 0, \dots, H^s_{q+r} = 0$$

THEOREM ①:

$$G_q \text{ 2-acyclic}, \quad \begin{array}{ccc} \mathcal{R}_{q+1} & \longleftarrow & \mathcal{R}_{q+r+1} \\ \downarrow & & \downarrow \\ \mathcal{R}^{(1)}_q & \longleftarrow & \mathcal{R}^{(1)}_{q+r} \\ \cap & & \cap \\ \mathcal{R}_q & \longleftarrow & \mathcal{R}_{q+r} \end{array} \Rightarrow \boxed{\left(\mathcal{R}^{(1)}_q\right)_{+r} = \mathcal{R}^{(1)}_{q+r}}$$

CRITERION: $\left.\begin{array}{l} G_q \text{ 2-acyclic} \\ \mathcal{R}_{q+1} \to \mathcal{R}_q \to 0 \end{array}\right\} \Rightarrow \mathcal{R}_q$ formally integrable

THEOREM ②: $\exists r, s > 0 \Rightarrow \mathcal{R}^{(s)}_{q+r}$ involutive with the same solutions as $\mathcal{R}_q$

EXAMPLE: $\mathcal{R}_2 = \{\partial_{33} y + x^2 \partial_{11} y = 0, \partial_{22} y = 0\} \Rightarrow \mathcal{R}^{(2)}_5$ finite type !!!

B) Classical Galois theory:

$$k \subset k' \subset L \quad \text{fields}$$

DEFINITION: $\Gamma(L/k) = \{\text{automorphisms of } L \text{ fixing } k\}$

$= \text{Galois group of } L \text{ over } k$

EXAMPLES: ① complex numbers:

$$k = \mathbb{R}\ ,\ L = \mathbb{C} = \mathbb{R}(i) = \mathbb{Q}\left(\mathbb{R}[y]/y^2+1\right) \qquad \Gamma(\mathbb{C}/\mathbb{R}) \approx C_2 \quad \begin{matrix} 1 \to 1 \\ i \to \pm i \end{matrix}$$

② cube root of 2 : $\quad x^3 - 2 = 0 \Rightarrow y^3 - 2 \equiv (y - x)(y^2 + xy + x^2)$

$$k = \mathbb{Q}\ ,\ L = \mathbb{Q}(x) = \mathbb{Q}\left(\mathbb{Q}[y]/y^3-2\right) \qquad \Gamma(L/k) = id \quad \begin{matrix} 1 \to 1 \\ 2^{1/3} \to 2^{1/3} \end{matrix}$$

GALOIS CORRESPONDENCE: $\qquad \Gamma(L/k) \supset \Gamma(L/k')$

$$\boxed{\begin{matrix} \Gamma(L/k) & \supset & \Gamma(L/k') & \supset & id \\ \uparrow & & \uparrow & & \uparrow \\ k & \subset & k' & \subset & L \end{matrix}} \qquad \boxed{\begin{matrix} & \Gamma & \supset & \Gamma' & \supset & id \\ & \nearrow\downarrow & & \downarrow & & \downarrow \\ k \subset & L^{\Gamma} & \subset & L^{\Gamma'} & \subset & L \end{matrix}}$$

EQUIVALENT DEFINITIONS:

1) L normal over $k \iff k = L^{\Gamma(L/k)}$

2) Every k-isomorphism of L into a bigger field is a k-automorphism.

3) L is the root field (splitting field) of a polynomial over k.

FUNDAMENTAL THEOREM OF GALOIS THEORY:

Let L be a normal finite extension of k. Then there exists a bijective correspondence between the subgroups of $\Gamma(L/k)$ and the intermediate fields between k and L given by the above Galois correspondence. Moreover, k' is normal over k if and only if $\Gamma(L/k')$ is normal in $\Gamma(L/k)$. In this case $\Gamma(k'/k) = \Gamma(L/k)/\Gamma(L/k')$.

ARTIN POINT OF VIEW: given $\Gamma \subset aut(L) \Rightarrow k = L^{\Gamma}$.

II/ HISTORICAL SURVEY:

1) S. LIE: (1): Discovery of Lie pseudogroups and differential invariants.

2) E. PICARD: (2): Use of Galois theory type methods like the primitive element in order to study linear differential equations by means of an associated linear algebraic group.

3) E. VESSIOT: (1865-1952) (3,4,5,6,7,8): Study of the analogy existing between the classical Galois theory and the Picard-Vessiot theory extended

to other cases. An important result is to bring back the integration to quadratures whenever the linear algebraic group is solvable.

EXAMPLE: A fine problem of constraint is the following:

$$\mathcal{R}_3 \left\{ \begin{array}{l} y^{3'''} - p\, y^{3''} + q\, y^{3'} - r\, y^3 = 0 \\ y^{2'''} - p\, y^{2''} + q\, y^{2'} - r\, y^2 = 0 \\ y^{1'''} - p\, y^{1''} + q\, y^{1'} - r\, y^1 = 0 \\ (y^3)^2 - y^1 \cdot y^2 = 0 \end{array} \right. \qquad \text{find } \mathcal{R}_3^{(5)} \text{ !!!}$$

PROBLEM: What is the condition on the coefficients p, q, r such that there exists a solution with non-zero wronskian?

ANSWER: !!! $9p'' - 18pp' - 27q' + 4p^3 - 18pq + 54r = 0$!!!

4) J.DRACH: (9,10,11) For the first time the idea of differential fields and extensions. However, an enormous mistake in his thesis (I have been able to read private letters between Drach, Vessiot and Painlevé) directed him and many other people in a wrong direction.

5) E.CARTAN: (12): Only one paper giving interesting counterexamples but with a poor understanding of the p.d.e. point of view of Drach and Vessiot.

6) J.F.RITT: (13): Foundation of modern differential algebra and differential algebraic elimination theory. Certainly the best exposition of the work of M.Janet (1921) on systems of p.d.e..

7) I.KAPLANSKY: (14): The best and shortest account of the Picard-Vessiot theory, though in the wrong line.

8) E.KOLCHIN: (15): The bible on that field though using too many notations and nothing more than Janet's work with respect to p.d.e..

9) J.F.POMMARET: (16,17): The second book to appear will be a natural continuation of the preceding one, as it wil use all the differentiable machinery of the formal theory. Applications will be done to analytical mechanics (see above example).

III/ DIFFERENTIAL INVARIANT THEORY:

DEFINITION: A system $\mathcal{A}_q \subset J_q(X,Y)$ is said to be automorphic with respect to the defining equations $\mathcal{R}_q(Y,Y)$ of a pseudogroup Γ of transformations of Y if it is a principal homogenous space under the coresponding groupoïd action, the same property holding also for both prolongations. More precisely,

$$\forall\ \left(x, y, \frac{\partial y}{\partial x}\right),\ \left(x, \bar{y}, \frac{\partial \bar{y}}{\partial x}\right) \in \mathcal{A}_{q+2}$$

$$\Rightarrow \exists \text{ unique } \left(y, \bar{y}, \frac{\partial \bar{y}}{\partial y}\right) \in \mathcal{R}_{q+2}(Y,Y) \quad \text{with} \quad \frac{\partial \bar{y}}{\partial x} = \frac{\partial \bar{y}}{\partial y} \cdot \frac{\partial y}{\partial x}$$

REMARK: In the analytic case, a system $\mathcal{A}_q$ automorphic with respect to $\mathcal{R}_q(Y,Y)$ is also automorphic with respect to Γ, that is to say, whenever $y = f(x)$ is a solution of $\mathcal{A}_q$ both with $\bar{y} = \bar{f}(x)$, then there exists a unique transformation $\bar{y} = \psi(y) \in \Gamma$ such that $\bar{f}(x) \equiv \psi(f(x))$.

IMPORTANT REMARK: The difficulty is in dealing with points and sections of the systems and not with their solutions. The situation is similar to that existing in the study of the deformation cohomology which is also introduced in this background (16).

DEFINITION: A system $\mathcal{A}_q \subset J_q(X,Y)$ is called a natural system if there exist a natural bundle $\mathcal{F}$ (of geometric objects) over X and a natural epimorphism ϕ (invariant in form under any change of source $\bar{x} = \varphi(x)$) together with a section ω of $\mathcal{F}$ giving rise to the following exact sequence:

$$0 \longrightarrow \mathcal{A}_q \longrightarrow J_q(X,Y) \underset{\omega \circ \pi}{\overset{\phi}{\rightrightarrows}} \mathcal{F}$$

THEOREM: Any automorphic system is a natural system and the above construction depends on the choice of a fundamental system of differential invariants of Γ. (The converse may not be true). (7.2.8 of 16)

THEOREM: The integration of any system $\mathcal{B}_q$ invariant under $\mathcal{R}_q(Y,Y)$ is equivalent to that of an automorphic system and a resolvent system.

$$P\left(x, y, \frac{\partial y}{\partial x}\right) = 0 \qquad \Longleftrightarrow \qquad \begin{cases} \Phi\left(y, \frac{\partial y}{\partial x}\right) = u \\ Q\left(x, u, \frac{\partial u}{\partial x}\right) = 0 \end{cases}$$

Moreover, in the analytic case, the Cauchy data of the given system are just made of the sum of the Cauchy data of the automorphic system and the Cauchy data of the resolvent system.

KEY PICTURE:

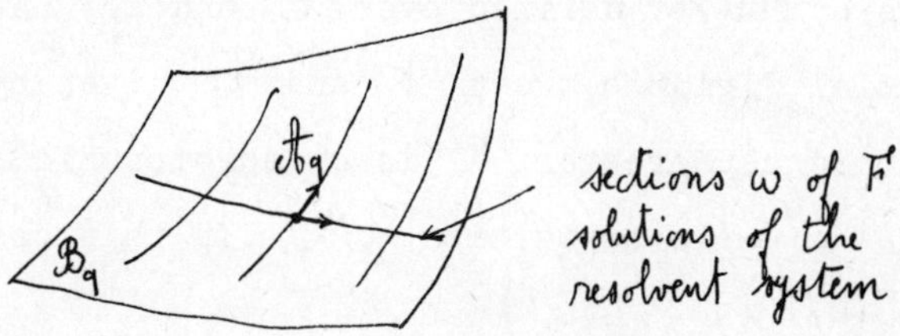

EXAMPLE:

$$y'' + a(x)\, y = 0 \qquad \Longleftrightarrow \qquad \begin{cases} \frac{y'}{y} = u \\ u' + u^2 + a(x) = 0 \quad \text{(Riccati)} \end{cases}$$

IV/ DIFFERENTIAL GALOIS THEORY:

Let $K \subset K' \subset L$ be differential fields with derivations $\partial_1, \ldots, \partial_n$.

EXAMPLE: Differential field of rational (meromorphic) functions.

STANDARD NOTATIONS: (15)

$K\{y\}$ differential ring of differential polynomial functions (see $K[y]$)

$K\langle y\rangle$ differential field of differential rational functions (see $K(y)$)

DEFINITION: L is said to be a finitely generated automorphic (natural) extension of K whenever $L = K\langle\eta\rangle$ where η is the generic solution of an irreducible automorphic (natural) system defined over K.

THEOREM: In the general case, the differential fields K and L have the same field of constants C.

DEFINITION: A differential isomorphism of L is a differential monomorphism of L into a bigger differential field.

THEOREM: The set $\Gamma(L/K)$ of differential isomorphisms of L fixing K has a natural structure of algebraic Lie pseudogroup defined over C.

CARE: Any normal extension is an automorphic extension but the converse is not true in general.

<u>EXAMPLE</u>:(see I/B ex 2)

$$k = \mathbb{R}(x) = \mathbb{R}\langle x\rangle \subset L = \mathbb{Q}\left(k\{y\}/y^3 - x\right) \Rightarrow \Gamma(L/k) = \{\bar{y}^3 = y^3\}, \ C = \mathbb{R}$$

<u>FUNDAMENTAL THEOREM OF DIFFERENTIAL GALOIS THEORY</u>:

Let L be a finitely generated automorphic extension of k. Then there is a bijective correspondence between the algebraic Lie subpseudogroups of $\Gamma(L/k)$ defined over C and the natural intermediate differential fields between k and L given by the preceding Galois correspondence.Moreover k' is an automorphic extension of k if and only if $\Gamma(L/k')$ is normal in $\Gamma(L/k)$ .In this case we have

$$\Gamma(k'/k) = \Gamma(L/k) / \Gamma(L/k') .$$

<u>COROLLARY</u>:If k° is the algebraic closure of k in L and $\Gamma^\circ(L/k)$ is the component of the identity of $\Gamma(L/k)$ then $\Gamma^\circ(L/k) = \Gamma(L/k^\circ)$.

<u>REMARK</u>:If L is of <u>finite type</u> over k (the definition is coming from any system defining the generators) then every intermediate differential field is natural.This is absolutely not evident a priori (!!!) and this kind of situation has never been considered by Kolchin. Drach and Vessiot did not notice this fact,though they were very close to the solution.

<u>EXAMPLE</u> (see III/) $k = \mathbb{Q}\langle a(x)\rangle \subset k' = \mathbb{Q}\left(k\{u\}/u' + u^2 + a(x)\right) \subset L = \mathbb{Q}\left(k\{y\}/y'' + a(x)y\right)$

$L = k\langle\eta\rangle$ is automorphic over $k' = k\langle \eta'/\eta\rangle$ and not over k .

Main historical BIBLIOGRAPHY

1)<u>S.LIE</u>:Leipziger berichte (collected work,1895)

2)<u>E.PICARD</u>:Traité d'analyse,t 3,1910

3)<u>E.VESSIOT</u>:Thesis (1892)

4) " :Groupes infinis (Ann.Ecole Normale Supérieure,1903)

5) " :Sur l'intégration des systèmes (Acta Mathematica,1904)

6) " :Sur lathéorie de Galois (Ann.ENS,1904)

7) " :Réductibilité des systèmes complets (Ann.ENS,1912)

8) " :Réductibilité des e.d.p.du premier ordre (Ann.ENS,1915)

9)J.DRACH:Essai sur une théorie d'intégration (Ann.ENS,1898)

10) " :Problème logique de l'intégration (Ann.Fac.Toulouse,1908)

1912

11) " :Integration logique (Int.Congress of math.,1912,1920,1924)

12)E.CARTAN:La théorie de Galois et ses généralisations (Oeuvres,1938)

13)J.F.RITT:Differential algebra (Dover,1960)

14)I.KAPLANSKY:Introduction to differential algebra (Hermann,1976)

15)E.KOLCHIN:Differential algebra and algebraic groups (Acad.Press, 1973)

16)J.F.POMMARET:Systems of partial differential equations and Lie pseudogroups (Gordon and Breach,1978)

17) " :Differential Galois theory (To appear)

COUNTEREXAMPLES TO A CONJECTURE OF RENE THOM

by

George M.Rassias
Department of Mathematics,University of California
Berkeley,California 94720/U.S.A.

René Thom made the following conjecture which appeared in Saunders Maclane{3}.The purpose of the conjecture was to find information about a degenerate critical point of a real-valued function, by perturbing it to a Morse function and then using Morse theory to study the properties of the degenerate critical point.

Conjecture (R.Thom)

Let F be a C^2 differentiable function on a closed differentiable manifold M, $F: M \to R$, having one degenerate critical point p. Perturb F to get a new function $G = F + \delta F$, which may be assumed it is a Morse function. Thus, the new function G has a finite number of non-degenerate critical points q_i corresponding to the original degenerate critical point p. Suppose also that the perturbation is such that the stable and unstable manifolds of the critical points meet transversally. The gradient trajectories between two of the relevant critical points form a set Σ. The conjecture asserts; (i) that Σ is contractible in itself, and (ii) that the degenerate case arises by collapsing Σ (i.e. that the sets Γ_p^- and Γ_p^+ of trajectories entering and leaving p are determined by the sets $\Gamma_{q_i}^-$ and $\Gamma_{q_i}^+$ for the q_i, perhaps by taking the disjoint union of the $\Gamma_{q_i}^+$ and identifying the points q_i. Then, degenerate critical points could be characterized by the number of points and their indices in the nondegenerate approximation.

Definitions

Let $F : M \to R$ be a Morse function on a closed differentiable manifold M, and Φ the flow of grad (F). Consider z to be a critical point of F.
Then, the stable manifold of z is

$$W^s(z) = \{ x \in M : \lim_{n \to +\infty} \Phi_{t_n}(x) = z \}$$

and the unstable manifold of z is

$$W^u(z) = \{ x \in M : \lim_{n \to -\infty} \Phi_{t_n}(x) = z \}$$

More information on these manifolds can be found in M.W. Hirsch {2}.

Key words and phrases: Morse function, stable manifold, transversality, gradient vector field, contractibility.
A.M.S. (MOS) Subject classification (1970) Primary 58E05, 58F10

Counterexamples (Description)

The first counterexample is a Morse function on the 3-dimensional sphere S^3, having two (non-degenerate) critical points such that the set of gradient trajectories between these two critical points, is not contractible in itself.

The second counterexample, is a Morse function on $\mathbb{R}P^3 \approx L(2, 1)$ and the Lens spaces $L(p,1)$, $p > 2$, such that the set of gradient trajectories between two of the critical points of the Morse function, is not contractible in itself.

Representation of the 3-sphere S^3 as the union of two solid tori, by a homeomorphism on their boundaries.

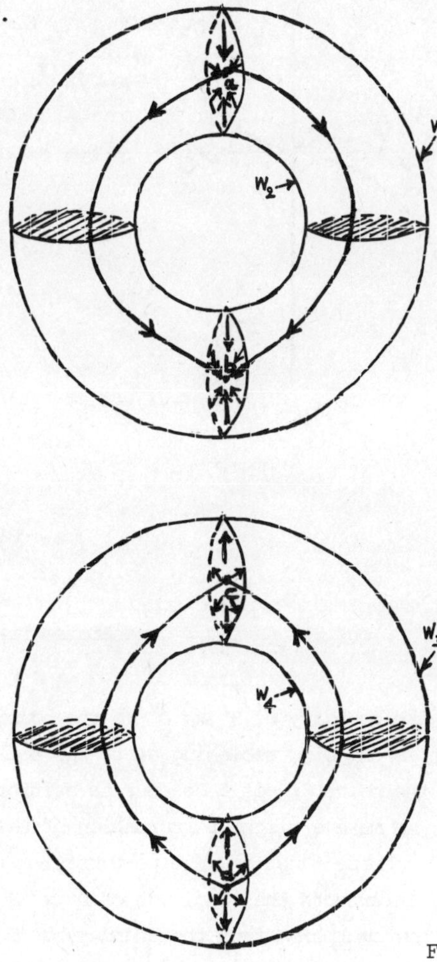

ind(a) = 2

← S^1 x D^2

ind(b) = 3

ind(c) = 1

← D^2 x S^1

ind(d) = 0

Figure 1

Counterexample 1st

The function $y = x^3$ has a degenerate critical point at zero and a slight perturbation of y to $\bar{y} = x^3 - x$ gives a Morse function with two (non-degenerate) critical points $(\pm\ \sqrt{3}/_3)$ as the following figure indicates.

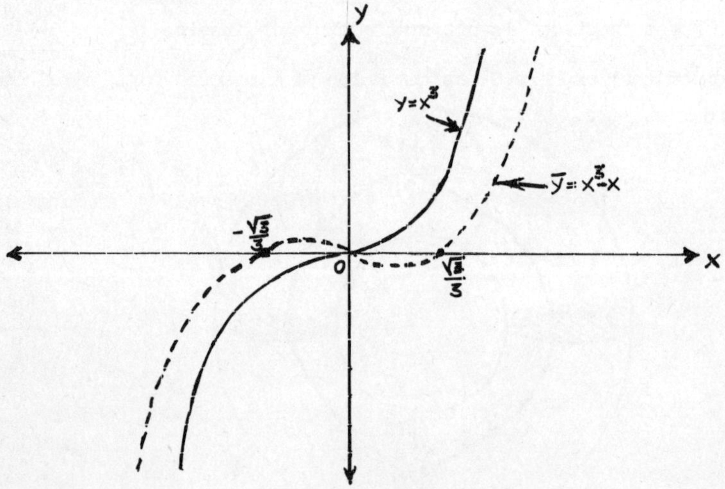

Figure 2

So, we start with a function f on S^3 having a degenerate critical point in the following way.

Define a homeomorphism from the boundary of $S^1 \times D^2$ to the boundary of $D^2 \times S^1$, which sends the longitude of $S^1 \times D^2$, enclosing a, to the following kind of meridian of $D^2 \times S^1$, such that the point x is a degenerate critical point as in the above graph of $y = x^3$. Then, we make a slight perturbation of the function f such that the new function g is a Morse function. The perturbation is made such that the meridian shown in figure {3} intersects the longitude of $D^2 \times S^1$ as in figure {2}. In other words, we define a homeomorphism from the boundary of $S^1 \times D^2$ to the boundary of $D^2 \times S^1$, sending the longitude of $S^1 \times D^2$, endlosing a, to the following kind of meridian of $D^2 \times S^1$, intersecting the longitude of $D^2 \times S^1$ enclosing c, three times.
Also, this homeomorphism maps the longitude of $S^1 \times D^2$ enclosing b to the outside circle of the boundary of $D^2 \times S^1$, and it maps W_1 to W_4.
Identifying the two solid tori with the above homeomorphism we get a Morse function

on S^3 with two (non-degenerate) critical points, e.g., the height function on S^3. Because of the assumption that the stable and

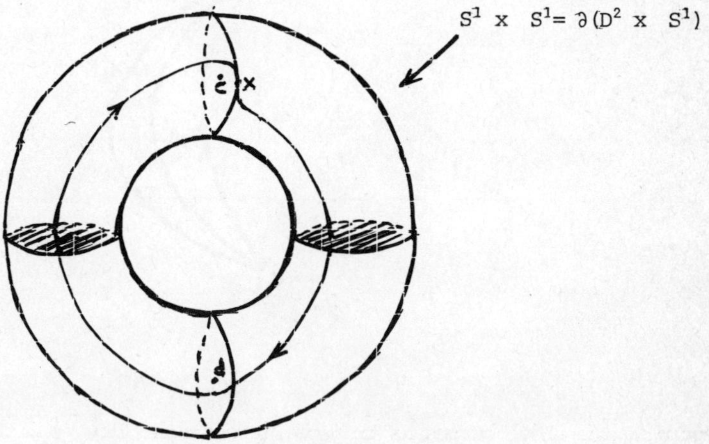

Figure 3

unstable manifolds of the critical points intersect transversally, it follows that in the above homeomorphism we have an odd number of intersection points.

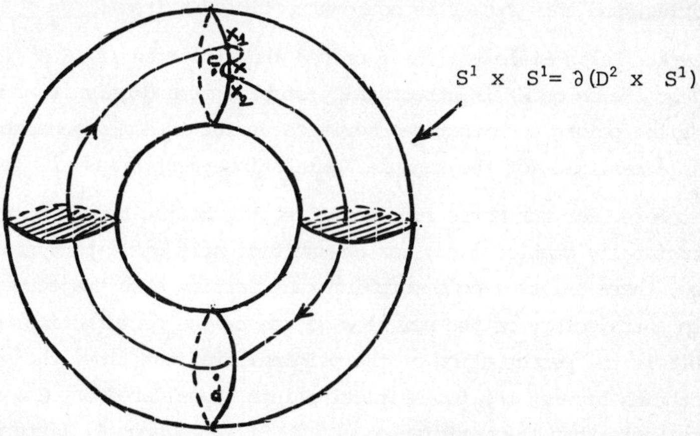

Figure 4

Now, join the critical point c with the intersection points x, x_1, x_2 and extend these trajectories on the boundary of $D^2 \times S^1$ so that to meet the critical point d in the following way.

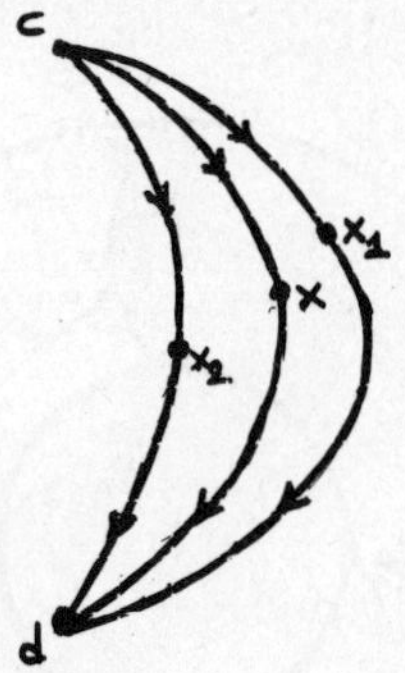

Figure 5

Regarding the above curves as trajectories of some flow Φ_t, then Φ_t is a gradient flow, i.e., the 1-parameter group of transformations generated by some gradient vector field X. This is true because of Smale {6} asserting that, given a vector field X and assuming that the zeros of X are transverse, each a- and ω-limit set is a zero, and the stable and unstable manifolds of the zeros meet transversally, then X is a gradient vector field. The equivalence of the first condition with that given by Smale, has been shown by Wall {8}.

So, after the identification of the boundaries of $S^1 \times D^2$ and $D^2 \times S^1$, we obtain S^3 with three gradient trajectories joining the maximum and minimum of the Morse function (height function). Similarly, we could obtain an odd number of gradient trajectories, and this set Σ is not contractible in itself.

Remark. Palis-Smale {5} have proved that the grad (f) $\in X^r$ (M) (= the space of C^r vector fields on M) is structurally stable, if and only if, f has a finite number of critical points such that each critical point is non-degenerate and the stable, unstable manifolds of the singular points intersect transversally.

Therefore, for our Morse function (i.e. the height function f on S^3) the grad (f) is structurally stable, i.e., for any sufficiently small perturbation of the gradient flow, there exists a homeomorphism that carries each trajectory of the original flow onto a trajectory of the new flow gotten by the perturbation. Hence, no matter how small is the perturbation of the original gradient flow, the set Σ of gradient trajectories between the two critical points, considered on S^3, is not contractible. Also, note that the meridian of $D^2 \times S^1$ in figure {4} (intersecting the longitude of $D^2 \times S^1$ three times) cannot be deformed through some regular homotopy to a meridian of $D^2 \times S^1$ intersecting the longitude of $D^2 \times S^1$ just once, because for some $t > 0$, the meridian of $D^2 \times S^1$ will be tangent to the longitude of $D^2 \times S^1$, which is not permissible because of the transversality property which must be preserved for all t according to the hypothesis.

Counterexample 2nd

We choose a homeomorphism from the boundary of $S^1 \times D^2$ to the boundary of $D^2 \times S^1$, such that the resulting closed 3-dimensional manifold is $\mathbb{R}P^3 \approx L(2,1)$.

If we modify appropriately this homeomorphism we obtain the Lens spaces $L(p,1)$, $p > 2$.

N o t e,

$$L\,(p,\,1) \approx S^3/_{(z_1,\,z_2)} = \left\{ e^{2\pi i/p} .\, z_1 \,,\; e^{2\pi i/p} . z_2 \right\} , \quad p \geq 1$$

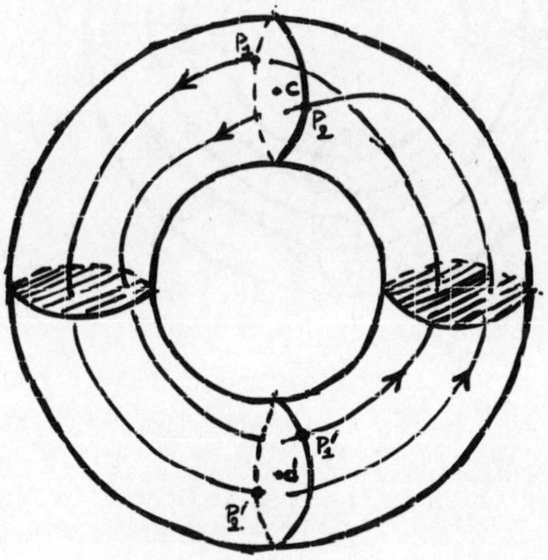

Figure 6

This homeomorphism maps the longitude of $S^1 \times D^2$ enlosing a, to the following kind of meridian of $D^2 \times S^1$ wrapping around the boundary of $D^2 \times S^1$ and intersecting the longitude of $D^2 \times S^1$ enclosing c, two times. Of course, it may intersect the longitude as many times as we want. We join the critical point d with the points p_1', p_2' and then extend these trajectories so that to meet the points p_1, p_2 and then to end up at the critical point c, in the following way.

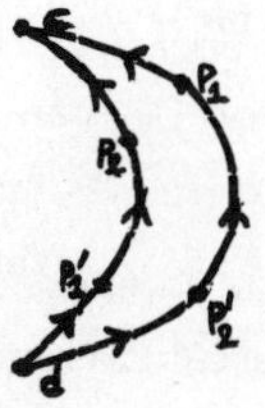

Figure 7

So, after the identification we obtain $\mathbb{RP}^3$ with two gradient trajectories joining two of the critical points of some Morse function on $\mathbb{RP}^3$, like in figure {7}, and of course this set Σ is not contractible in itself, similarly, we can have a set of $p > 2$ gradient trajectories on $L(p, 1)$, which is not contractible in itself.

The homeomorphism gives us $L(3, 1)$, indicated in figure {8}.

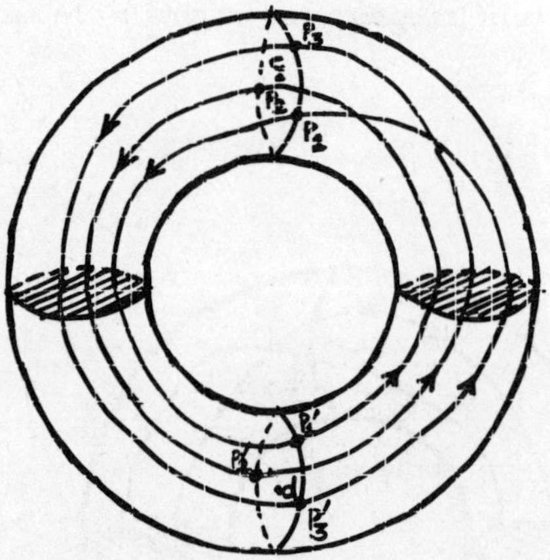

Figure 8

The set Σ of gradient trajectories is the one, indicated in figure {9}

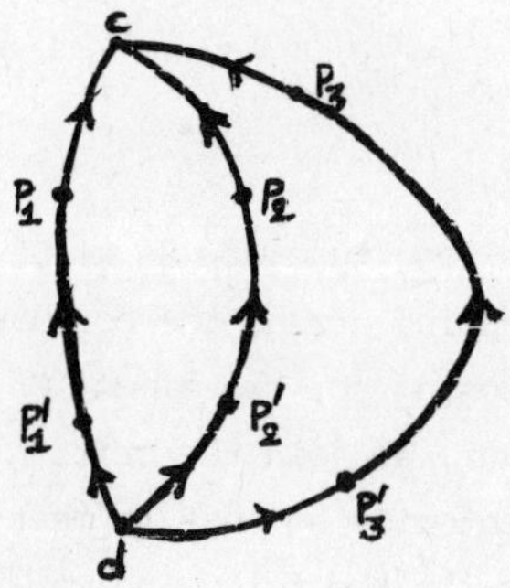

Figure 9

which is not contractible in itself.

Remark. An example in J.Guckenheimer {1} can be shown to be another counterexample to this conjecture.

Acknowledgements. It is my pleasure to express my warm thanks to Professors S.Smale and R.F.Williams for helpful conversations and the refereé for informing me about

J.Guckenheimer's work.

REFERENCES

{1} J.GUCKEINHEIMER, Bifurcation and catastrophe, Proc. on Dynamical Systems, edited by M.M.Peixoto,1973, N.Y. Academic Press, pp. 95-111.

{2} M.W.HIRSCH, Differential Topology, Grad. Texts of Math., 1975.

{3} S.MACLANE, Geometrical Mechanics, Part II, Lect. Notes, Dept. of Math., Univ. of Chicago, 1968, pp. 63.

{4} S.MACLANE, Hamiltonian Mechanics and Geometry, Amer. Math. Monthly, June-Dec., 1970.

{5} J.PALIS and S.SMALE, Structural Stability theorems, Proc. of the Symposium in Pure Math., XIV, edited by S.S. Chern and S.Smale, 1970, Amer. Math. Soc.

{6} S.SMALE, On gradient dynamical systems, Ann. of Math., Vol. 74 (1961) 199 - 206.

{7} S.SMALE, A survey of some recent developments in differential topology, Bull. A.M.S., vol. 69 (1963) 131 - 146.

{8} C.T.C. WALL, Reflections of gradient vector fields, Proc. of Liverpool Singularities, Symposium II, Lect. Notes in Math., Springer - Verlag. No. 209, 1971.

Conformal invariants on almost Hermitian manifolds

by

F. Tricerri and L. Vanhecke

1. INTRODUCTION

Let (V,g) be an n-dimensional real vector space with positive definite inner product g and denote by $\mathcal{R}(V)$ the subspace of $V^* \otimes V^* \otimes V^* \otimes V^*$ consisting of all tensors having the same symmetries as the curvature tensor of a Riemannian manifold (including the first Bianchi identity). In a well known paper [9] Singer and Thorpe considered $\mathcal{R}(V)$ (in particular for $n=4$) and gave a geometrically useful description of the splitting of $\mathcal{R}(V)$ under the action of $\mathcal{O}(n)$ into three components. This was also studied by Nomizu [8] for generalized curvature tensor fields.

An analogous decomposition was given [7], [10] when V is a 2n-dimensional real vector space endowed with a complex structure J compatible with a positive definite inner product g and for the subspace $\mathcal{K}(V)$ of $\mathcal{R}(V)$ consisting of tensors satisfying the Kähler identity. In this case the splitting of $\mathcal{K}(V)$ is treated for the action of $\mathcal{U}(n)$.

Of course all these decompositions are in principle consequences of general theorems on group representations. (For other decompositions see [3], [5].) On the other hand these decompositions do provide insight in some problems of differential geometry. For example, in the splitting of $\mathcal{R}(V)$, one of the projection operators gives the Weyl conformal tensor and in the splitting of $\mathcal{K}(V)$ one obtains in the same way the Bochner tensor.

Several attempts have been made [13], [14], [15] to determine the splitting of $\mathcal{R}(V)$ under the action of $\mathcal{U}(n)$. One of the main purposes of these papers was to obtain a generalization of the Bochner tensor and to understand better its geometrical meaning.

In this note we will give the complete decomposition of $\mathcal{R}(V)$ under the action of $\mathcal{U}(n)$. One of the main purposes is to decompose the Riemann curvature tensor on a general almost Hermitian manifold and to consider conformal changes of the metric. This will provide several conformal invariants including a Bochner conformal tensor which in the case of a Kähler manifold reduces to the usual one.

Details, proofs and other results will be given in a more extended forthcoming paper [11].

Finally we wish to thank Alfred Gray for his interest in our work and for several useful discussions.

2. THE DECOMPOSITION

Let V be an n-dimensional real vector space with positive definite inner product g. A tensor R of type (1,3) over V is a bilinear mapping $R : V \times V \to \mathrm{Hom}(V,V) : (x,y) \longmapsto R(x,y)$. R is called a *curvature tensor* over V if it has the following properties for all $x,y,z,w \in V$:

i) $R(x,y) = -R(y,x)$;

ii) $R(x,y)$ is a skew symmetric endomorphism of V, i.e. $R(x,y,z,w) + R(x,y,w,z) = 0$ where $R(x,y,z,w) = g(R(x,y)z,w)$;

iii) $\mathfrak{S}\, R(x,y)z = 0$, where $\mathfrak{S}$ denotes the cyclic sum over x, y and z. This is the first Bianchi identity.

The *Ricci tensor* $\rho(R)$ of type (0,2) associated with R is a symmetric bilinear function on $V \times V$ defined by

$$\rho(R)(x,y) = \mathrm{trace}(z \in V \longmapsto R(x,z)y \in V).$$

Then, the Ricci tensor $Q = Q(R)$ of type $(1,1)$ is given by $\rho(R)(x,y) = g(Qx,y)$ and the trace of Q is called the *scalar curvature* $\tau = \tau(R)$ of R.

Next let V be a 2n-dimensional real vector space with a complex structure J and a Hermitian product g, i.e.

$$J^2 = -I, \quad g(Jx,Jy) = g(x,y)$$

for all $x,y \in V$, and I denoting the identity transformation of V. The *Ricci* $*$ *tensor* $\rho^*(R)$ of type $(0,2)$ resp. $Q^* = Q^*(R)$ of type $(1,1)$ associated with a curvature tensor R is defined by

$$\rho^*(R)(x,y) = g(Q^*x,y) = \mathrm{trace}(z \in V \longmapsto R(Jz,x)Jy \in V)$$
$$= \sum_{i=1}^{2n} R(x,e_i,Jy,Je_i)$$

where $\{e_i, e_{n+i} = Je_i \,;\, i = 1,\dots,n\}$ is an adapted orthonormal basis of V. Then the trace of Q^* is called the $*$ *scalar curvature* $\tau^* = \tau^*(R)$ of R.

Let $\mathcal{R}(V)$ denote the vector space of all curvature tensors over V. This space has a natural inner product induced from that on V:

$$\langle R,\widetilde{R}\rangle = \sum_{i,j,k=1}^{2n} g(R(e_i,e_j)e_k,\widetilde{R}(e_i,e_j)e_k),$$

$\{e_i\}$ being an arbitrary orthonormal basis of V. Further, let ρ be the standard representation of $\mathcal{U}(n)$ in V. Then there is a natural induced representation $\widetilde{\rho}$ of $\mathcal{U}(n)$ in $\mathcal{R}(V)$ given by

$$\widetilde{\rho}(a)(R)(x,y,z,w) = R(\rho(a^{-1})x,\rho(a^{-1})y,\rho(a^{-1})z,\rho(a^{-1})w)$$

for all $x,y,z,w \in V$, $R \in \mathcal{R}(V)$ and $a \in \mathcal{U}(n)$.

Our aim is to give a complete decomposition of $\mathcal{R}(V)$ into orthogonal irreducible factors. Before doing this we introduce three particular subspaces $\mathcal{R}_i(V)$, $i = 1,2,3$, of $\mathcal{R}(V)$:

$$\mathcal{R}_1(V) = \{R \in \mathcal{R}(V) \,|\, R(x,y,z,w) = R(x,y,Jz,Jw)\},$$

$$\mathcal{R}_2(V) = \{R \in \mathcal{R}(V) \,|\, R(x,y,z,w) = R(Jx,Jy,z,w) + R(Jx,y,Jz,w) + R(Jx,y,z,Jw)\},$$

$\mathcal{R}_3(V) = \{R \in \mathcal{R}(V) \mid R(x,y,z,w) = R(Jx,Jy,Jz,Jw)\}$.

These subspaces are invariant under the action of $\mathcal{U}(n)$ and $\mathcal{R}_1(V) \subset \mathcal{R}_2(V) \subset \mathcal{R}_3(V)$ (see for example [2], [4]). Further we put

$$\mathcal{R}_1^{\perp}(V) = \text{orthogonal complement of } \mathcal{R}_1(V) \text{ in } \mathcal{R}_2(V),$$

$$\mathcal{R}_2^{\perp}(V) = \text{orthogonal complement of } \mathcal{R}_2(V) \text{ in } \mathcal{R}_3(V),$$

$$\mathcal{R}_3^{\perp}(V) = \text{orthogonal complement of } \mathcal{R}_3(V) \text{ in } \mathcal{R}(V).$$

Hence we have already the following decomposition of $\mathcal{R}(V)$ into orthogonal invariant subspaces:

$$\mathcal{R}(V) = \mathcal{R}_1(V) \oplus \mathcal{R}_1^{\perp}(V) \oplus \mathcal{R}_2^{\perp}(V) \oplus \mathcal{R}_3^{\perp}(V). \tag{2.1}$$

Of course these factors are not all irreducible. For example, a complete decomposition of $\mathcal{R}_1(V)$ into three irreducible factors is given in [7], [10].

In order to be able to give a complete decomposition we introduce two other curvature tensors π_1 and π_2 by

$$\pi_1(x,y,z,w) = g(x,z)g(y,w) - g(x,w)g(y,z),$$

$$\pi_2(x,y,z,w) = 2g(x,Jy)g(z,Jw) + g(x,Jz)g(y,Jw) - g(x,Jw)g(y,Jz).$$

The following theorem is a consequence of the invariance theory (see [1], [6], [16]).

THEOREM 2.1. The curvature tensors π_1 and π_2 form a basis for the subspace of invariant curvature tensors (i.e. $\tilde{\rho}(a)R = R$ for all $a \in \mathcal{U}(n)$).

Note that π_1 is also an invariant tensor for $\mathcal{O}(n)$. This curvature tensor generates the 1-dimensional space of invariant curvature tensors for the action of $\mathcal{O}(n)$ and this gives one of the factors in the splitting of $\mathcal{R}(V)$ under the action of $\mathcal{O}(n)$.

Now we give the decomposition theorems 2.2 and 2.3.

THEOREM 2.2. We have for $n \geqslant 4$:

$$\mathcal{R}(V) = \mathcal{W}_1 \oplus \dots \oplus \mathcal{W}_{10} \qquad \text{(orthogonal)}$$

where

$$\mathcal{R}_1(V) = \mathcal{W}_1 \oplus \mathcal{W}_2 \oplus \mathcal{W}_3,$$

$$\mathcal{R}_1^{\perp}(V) = \mathcal{W}_4 \oplus \mathcal{W}_5 \oplus \mathcal{W}_6,$$

$$\mathcal{R}_2^{\perp}(V) = \mathcal{W}_7,$$

$$\mathcal{R}_3^{\perp}(V) = \mathcal{W}_8 \oplus \mathcal{W}_9 \oplus \mathcal{W}_{10}$$

and the irreducible factors $\mathcal{W}_i$ are given by

$$\mathcal{W}_1 = \mathcal{L}(\pi_1 + \pi_2) = \text{vector space generated by } \pi_1 + \pi_2,$$

$$\mathcal{W}_3 = \{R \in \mathcal{R}_1(V) \mid \rho(R) = 0\},$$

$$\mathcal{W}_2 = \text{orthogonal complement of } \mathcal{W}_1 \oplus \mathcal{W}_3 \text{ in } \mathcal{R}_1(V),$$

$$\mathcal{W}_4 = \mathcal{L}(3\pi_1 - \pi_2) = \text{vector space generated by } 3\pi_1 - \pi_2,$$

$$\mathcal{W}_6 = \{R \in \mathcal{R}_1^{\perp}(V) \mid \rho(R) = 0\},$$

$$\mathcal{W}_5 = \text{orthogonal complement of } \mathcal{W}_4 \oplus \mathcal{W}_6 \text{ in } \mathcal{R}_1^{\perp}(V),$$

$$\mathcal{W}_{10} = \{R \in \mathcal{R}_3^{\perp}(V) \mid \rho(R) = \rho^*(R) = 0\},$$

$$\mathcal{W}_8 \oplus \mathcal{W}_9 = \text{orthogonal complement of } \mathcal{W}_{10} \text{ in } \mathcal{R}_3^{\perp}(V),$$

$$\mathcal{W}_8 = \{R \in \mathcal{W}_8 \oplus \mathcal{W}_9 \mid \rho^*(R) = 0\},$$

$$\mathcal{W}_9 = \{R \in \mathcal{W}_8 \oplus \mathcal{W}_9 \mid \rho(R) = 0\}.$$

THEOREM 2.3. We have $\mathcal{W}_6 = \{0\}$ for $n = 3$ and $\mathcal{W}_5 = \mathcal{W}_6 = \mathcal{W}_{10} = \{0\}$ for $n = 2$.

REMARKS

A. The projections of $R \in \mathcal{R}(V)$ on each $\mathcal{W}_i$, the length of these projections and the dimensions of the $\mathcal{W}_i$ will be given in [11] where also considerations about the quadratic invariants will be included.

B. The spaces considered in (2.1) may be defined by introducing three operators L_i, $i = 1,2,3$, as follows.

Let

$$L_3(R)(x,y,z,w) = R(Jx,Jy,Jz,Jw)$$

for $R \in \mathcal{R}(V)$ and all $x,y,z,w \in V$. It is easy to see that L_3 is an involutive isometry of $\mathcal{R}(V)$ which commute with all $\tilde{\rho}(a)$, $a \in \mathcal{U}(n)$. Then

$$\mathcal{R}_3(V) = (+1)\text{-eigenspace of } L_3,$$
$$\mathcal{R}_3^{\perp}(V) = (-1)\text{-eigenspace of } L_3.$$

Further let

$$L_2(R)(x,y,z,w) = \tfrac{1}{2}\{R(x,y,z,w) + R(Jx,Jy,z,w) + R(Jx,y,Jz,w) + R(Jx,y,z,Jw)\},$$

$$L_1(R)(x,y,z,w) = \tfrac{1}{4}\{3R(Jx,Jy,z,w) - R(x,y,z,w) + R(Jz,x,Jy,w) - R(Jz,y,Jx,w) + R(Jz,Jy,x,w) - R(Jz,Jx,y,w)\}$$

for $R \in \mathcal{R}_3(V)$ and all $x,y,z,w \in V$. L_1 and L_2 are both involutive isometries of $\mathcal{R}_3(V)$ commuting with all $\tilde{\rho}(a)$, $a \in \mathcal{U}(n)$. Further $L_1 L_2 = L_2 L_1 = L_1 - L_2 + I$, where I is the identity transformation of $\mathcal{R}(V)$. We have

$$\mathcal{R}_2(V) = (+1)\text{-eigenspace of } L_2,$$
$$\mathcal{R}_2^{\perp}(V) = (-1)\text{-eigenspace of } L_2,$$
$$\mathcal{R}_1(V) = (+1)\text{-eigenspace of } L_1,$$
$$\mathcal{R}_1^{\perp}(V) = (-1)\text{-eigenspace of } L_1.$$

C. The spaces $\mathcal{W}_2$, $\mathcal{W}_5$, $\mathcal{W}_8$ and $\mathcal{W}_9$ may be defined as follows. Let S be an arbitrary element of $V^* \otimes V^*$ and put

$$\varphi(S)(x,y,z,w) = g(x,y)S(y,w) + g(y,w)S(x,z) - g(x,w)S(y,z) - g(y,z)S(x,w),$$

$$\psi(S)(x,y,z,w) = 2g(x,Jy)S(z,Jw) + 2g(z,Jw)S(x,Jy) + g(x,Jz)S(y,Jw) + g(y,Jw)S(x,Jz) - g(x,Jw)S(y,Jz) - g(y,Jz)S(x,Jw)$$

for all $x,y,z,w \in V$. Then $\varphi(S) \in \mathcal{R}(V)$ if and only if S is symmetric and $\psi(S) \in \mathcal{R}(V)$ if and only if $S(x,Jy) + S(y,Jx) = 0$. Further we put

$$\mathcal{S}_2^{0+} = \{S \in V^* \otimes V^* \mid S(x,y) = S(y,x), S(Jx,Jy) = S(x,y), \text{trace } S = 0\},$$

$$\mathcal{S}_2^{-} = \{S \in V^* \otimes V^* \mid S(x,y) = S(y,x), S(Jx,Jy) = -S(x,y)\},$$

$$\mathcal{A}_2^{-} = \{S \in V^* \otimes V^* \mid S(x,y) = -S(y,x), S(Jx,Jy) = -S(x,y)\}.$$

We have

$$\mathcal{W}_2 = (\varphi + \psi)(\mathcal{S}_2^{0+}) \ , \quad \mathcal{W}_5 = (3\varphi - \psi)(\mathcal{S}_2^{0+}),$$
$$\mathcal{W}_8 = \varphi(\mathcal{S}_2^{-}) \ , \quad \mathcal{W}_9 = \psi(\mathcal{A}_2^{-}).$$

We note that $\psi(S) = 3\varphi(S)$ for $n = 2$ and arbitrary $S \in V^* \otimes V^*$. This implies $\mathcal{W}_5 = \{0\}$ for $n = 2$. Further we have in general $2\pi_1 = \varphi(g)$, $2\pi_2 = \psi(g)$, g being the inner product on V.

D. Let $\mathcal{R}_W(V)$ be the subspace of $\mathcal{R}(V)$ defined by $\rho(R) = 0$. It is well known that the projection of a curvature tensor R on $\mathcal{R}_W(V)$ gives the *Weyl conformal tensor* $C(R)$ associated with R. For real dimension $n > 3$ of V we have

$$C(R)(x,y)z = R(x,y)z - \frac{1}{n-2}\{g(x,z)Qy - g(y,z)Qx + g(Qx,z)y - g(Qy,z)x\} + \frac{\tau}{(n-1)(n-2)}\{g(x,z)y - g(y,z)x\}$$

and $C(R) = 0$ for $n = 3$. The fact that $\mathcal{R}_W(V)$ is defined by $\rho(R) = 0$ implies that this space is the orthogonal complement of the subspace of $\mathcal{R}(V)$ generated by all curvature tensors $\varphi(S)$, S symmetric.

Under the action of $\mathcal{U}(n)$ this space is reducible. We have the following orthogonal splitting into irreducible factors:

$$\mathcal{R}_W(V) = \mathcal{L}(3\pi_1 - (2n-1)\pi_2) \oplus (3\varphi - (n-1)\psi)(\mathcal{S}_2^{0+}) \oplus \mathcal{W}_3 \oplus \mathcal{W}_6 \oplus \mathcal{W}_7 \oplus \mathcal{W}_9 \oplus \mathcal{W}_{10}$$

and further

$$\mathcal{R}_W^{\perp}(V) = \mathcal{L}(\pi_1) \oplus \varphi(\mathcal{S}_2^{0+}) \oplus \varphi(\mathcal{S}_2^{-}).$$

For $n=2$ we have only four factors since $\mathcal{W}_6 = \mathcal{W}_{10} = \{0\}$ and $(3\varphi-\psi)(S_2^{0+})=0$. Finally for $n=3$ there are six factors since only $\mathcal{W}_6 = \{0\}$.

E. Let $\mathcal{R}_B(V)$ be the subspace of $\mathcal{R}(V)$ defined by $\rho(R) = \rho^*(R) = 0$ for $R \in \mathcal{R}(V)$. Then

$$\mathcal{R}_B(V) = \mathcal{W}_3 \oplus \mathcal{W}_6 \oplus \mathcal{W}_7 \oplus \mathcal{W}_{10}.$$

A tensor $B \in \mathcal{R}_B(V)$ is called a Bochner tensor and the projection $B(R)$ of $R \in \mathcal{R}(V)$ on $\mathcal{R}_B(V)$ is called the *Bochner conformal tensor* associated with R.

For $n \geqslant 3$ we have

$$\begin{aligned} B(R) = R &- \frac{1}{16(n+2)}(\varphi+\psi)(\rho+3\rho^*)(R+L_3R) \\ &- \frac{1}{16(n-2)}(3\varphi-\psi)(\rho-\rho^*)(R+L_3R) \\ &- \left\{\frac{1}{4(n+1)}\psi(\rho^*) + \frac{1}{4(n-1)}\varphi(\rho)\right\}(R-L_3R) \\ &+ \frac{1}{16(n+1)(n+2)}(\tau+3\tau^*)(R)(\pi_1+\pi_2) + \frac{1}{16(n-1)(n-2)}(\tau-\tau^*)(R)(3\pi_1-\pi_2) \end{aligned}$$

and for $n=2$:

$$\begin{aligned} B(R) = R &- \frac{1}{2}\varphi(\rho - \frac{\tau}{4}g)(R) - \frac{1}{2}\psi(\rho^*)(R-L_3R) \\ &- \frac{1}{96}(\tau+3\tau^*)(R)(\pi_1+\pi_2) - \frac{1}{32}(\tau-\tau^*)(R)(3\pi_1-\pi_2). \end{aligned}$$

3. APPLICATIONS FOR ALMOST HERMITIAN MANIFOLDS

Let (M,J,g) be an almost Hermitian manifold and denote by $\mathcal{R}(M)$ the vector bundle with fiber $\mathcal{R}(T_mM)$, $m \in M$. The decomposition given in theorem 2.2 induces a decomposition of $\mathcal{R}(M)$. We still denote the factors (subbundles of $\mathcal{R}(M)$) by $\mathcal{W}_i$, $i = 1,\ldots,10$.

Further, let ∇ resp. R denote the Riemannian connection resp. Riemann curvature tensor on M, where $R(X,Y) = \nabla_{[X,Y]} - [\nabla_X,\nabla_Y]$ for all $X,Y \in \mathfrak{X}(M)$, the algebra of C^∞ vector fields on M.

We consider now the conformal change of metric defined by

$$\tilde{g} = e^{\sigma} g$$

where σ is a C^{∞} function on M. Then we have the following well known relation between $\tilde{R}$ and R:

$$e^{-\sigma}\tilde{R}(X,Y,Z,W) = R(X,Y,Z,W) - \frac{1}{2}\{L(X,Z)g(Y,W) + L(Y,W)g(X,Z) - L(X,W)g(Y,Z) - L(Y,Z)g(X,W)\} - \frac{\|\omega\|^2}{4}\{g(X,Z)g(Y,W) - g(Y,Z)g(X,W)\}$$

where

$$\omega = d\sigma\ ,\quad L(X,Y) = (\nabla_X\omega)(Y) - \frac{1}{2}\omega(X)\omega(Y).$$

L is symmetric and using the notations of section 2 we have:

$$e^{-\sigma}\tilde{R} = R + T_{\sigma}$$

where

$$T_{\sigma} = -\frac{1}{2}\varphi(L) - \frac{\|\omega\|^2}{4}\pi_1 .$$

The curvature tensor T_{σ} is orthogonal to each element of $\mathcal{R}_W(M)$ and hence the associated Weyl tensor $C(T_{\sigma})$ vanishes.

Now we put

$$\mathcal{C}_1 = \mathcal{L}(3\pi_1 - (2n-1)\pi_2),$$
$$\mathcal{C}_2 = (3\varphi - (n-1)\psi)(\mathcal{S}_2^{0+}),$$
$$\mathcal{C}_3 = \mathcal{W}_3\ ,\ \mathcal{C}_4 = \mathcal{W}_6\ ,\ \mathcal{C}_5 = \mathcal{W}_7\ ,\ \mathcal{C}_6 = \mathcal{W}_9\ ,\ \mathcal{C}_7 = \mathcal{W}_{10}.$$

Then we have

$$\mathcal{R}_W(M) = \mathcal{C}_1 \oplus \dots \oplus \mathcal{C}_7 .$$

Let $C_i(R)$ denote the projection of R on $\mathcal{C}_i$. Then we have for the Weyl conformal tensor

(3.1) $$C(R) = \sum_{i=1}^{7} C_i(R)$$

and for the Bochner conformal tensor

$$B(R) = C_3(R) + C_4(R) + C_5(R) + C_7(R).$$

The following theorem shows why we call $B(R)$ a conformal tensor.

THEOREM 3.1. The curvature tensors $C_i(R)$ and consequently the Bochner conformal tensor are conformally invariant.

Further we have

THEOREM 3.2. Let R be a curvature tensor field on M. Then:

$$C_1(R) = \frac{1}{8n(n^2-1)(2n-1)} \{\tau - (2n-1)\tau^*\}(R)\{3\pi_1 - (2n-1)\pi_2\},$$

$$C_2(R) = \frac{1}{n-1} \{3\varphi - (n-1)\psi\}(S_2 - S_1),$$

$$C_3(R) = \frac{1}{4} (I + L_1)(R + L_3R) - (\varphi + \psi)(S_1) - \frac{1}{16n(n+1)} (\tau + 3\tau^*)(R)(\pi_1 + \pi_2),$$

$$C_4(R) = \frac{1}{4} (L_2 - L_1)(R + L_3R) - (3\varphi - \psi)(S_2) - \frac{1}{16n(n-1)} (\tau - \tau^*)(R)(3\pi_1 - \pi_2),$$

$$C_5(R) = \frac{1}{4} (I - L_2)(R + L_3R),$$

$$C_6(R) = \frac{1}{4(n+1)} \psi(\rho^*)(R - L_3R),$$

$$C_7(R) = \frac{1}{2} (R - L_3R) - \{\frac{1}{4(n+1)} \psi(\rho^*) + \frac{1}{4(n-1)} \varphi(\rho)\}(R - L_3R),$$

where

$$16(n+2)S_1 = (\rho + 3\rho^*)(R + L_3R) - \frac{1}{n} (\tau + 3\tau^*)(R)g,$$

$$16(n-2)S_2 = (\rho - \rho^*)(R + L_3R) - \frac{1}{n} (\tau - \tau^*)(R)g$$

for $n \geqslant 3$ and $S_2 = 0$ for $n = 2$.

Some of the $C_i(R)$ vanish on manifolds belonging to special classes of almost Hermitian manifolds. We give some examples.

THEOREM 3.3. Let R be an arbitrary curvature tensor field on a general almost Hermitian manifold of complex dimension n. We have

i) $C_2(R) = C_4(R) = C_7(R) = 0$ for $n = 2$;

ii) $C_4(R) = 0$ for $n = 3$.

Next we consider a Hermitian manifold M. A. Gray proved [2] that the Riemann curvature tensor of M satisfies the following identity:

$$R(X,Y,Z,W) + R(JX,JY,JZ,JW) = R(JX,JY,Z,W) + R(X,Y,JZ,JW) + R(JX,Y,JZ,W) + R(X,JY,Z,JW) + R(JX,Y,Z,JW) + R(X,JY,JZ,W).$$

This is equivalent with the condition $(I - L_2)(R + L_3R) = 0$. Hence

THEOREM 3.4. Let R be the Riemann curvature tensor on a Hermitian manifold. Then $C_5(R) = 0$.

Now let M be a Hermitian manifold which is locally conformal to a Kähler manifold (see [3], [12]). Since $C_4(R) = C_5(R) = C_6(R) = C_7(R) = 0$ on a Kähler manifold, we deduce from theorem 3.1 the following result.

THEOREM 3.5. Let M be a locally conformal Kähler manifold. Then $C_i(R) = 0$ for $i = 4,5,6,7$.

These conditions are only necessary conditions. Indeed, consider the manifold $M = M^2 \times \mathbb{R}^4$ where M^2 is a minimal surface in $\mathbb{R}^3$. It is proved in [2] that M is a Hermitian manifold such that $L_3R = R$. This implies $C_5 = C_6 = C_7 = 0$ and since $\dim M = 6$ we have also $C_4 = 0$. On the other hand M is not a locally conformal Kähler manifold [3].

From (3.1) and (3.2) we have further

THEOREM 3.6. Any conformally flat almost Hermitian manifold is Bochner flat.

Examples of such manifolds are given by the Hopf manifolds $S^1 \times S^{2q+1}$ $(q \geqslant 0)$. For the other Calabi-Eckmann manifolds $S^{2p+1} \times S^{2q+1}$ $(p \geqslant 1, q \geqslant 1)$ we have $C_5(R) = C_6(R) = 0$ but $C_i(R) \neq 0$ for $i = 1,2,3,4$ and 7. Examples of

Bochner flat manifolds which are not conformally flat are provided by taking conformal transformations of Bochner flat Kähler manifolds.

REFERENCES

[1] J.A. DIEUDONNE & J.B. CARRELL, Invariant Theory, Old and New, Advances in Math. 4 (1970), 1-80.

[2] A. GRAY, Curvature identities for Hermitian and almost Hermitian manifolds, Tôhoku Math. J., 28 (1976), 601-612.

[3] A. GRAY & L.M. HERVELLA, The sixteen classes of almost Hermitian manifolds and their linear invariants, to appear in Ann. Math. Pura Appl..

[4] A. GRAY & L. VANHECKE, Almost Hermitian manifolds with constant holomorphic sectional curvature, to appear in Czechoslovak Math. J..

[5] A. GRAY & L. VANHECKE, Decomposition of the space of covariant derivatives of curvature operators, to appear.

[6] N. IWAHORI, Some remarks on tensor invariants of O(n), U(n), Sp(n), J. Math. Soc. Japan, 10 (1958), 145-160.

[7] H. MORI, On the decomposition of generalized K-curvature tensor fields, Tôhoku Math. J. 25 (1973), 225-235.

[8] K. NOMIZU, On the decomposition of generalized curvature tensor fields, Differential Geometry (in honor of K. Yano), Kinokuniya, Tokyo, 1972, 335-345.

[9] I.M. SINGER & J.A. THORPE, The curvature of 4-dimensional Einstein spaces, Global Analyşis (paper in honor of K. Kodaira), University of Tokyo Press, Tokyo, 1969, 355-365.

[10] M. SITARAMAYYA, Curvature tensors in Kaehler manifolds, Trans. Amer. Math. Soc. 183 (1973), 341-351.

[11] F. TRICERRI & L. VANHECKE, Decomposition of the space of curvature operators, to appear.

[12] I. VAISMAN, On locally conformal almost Kähler manifolds, Israel J. Math. 24 (1976), 338-351.

[13] L. VANHECKE, On the decomposition of curvature tensor fields on almost

Hermitian manifolds, Proc. Conference on Differential Geometry, Michigan State University, East Lansing, 1976, 16-33.

[14] L. VANHECKE, The Bochner curvature tensor on almost Hermitian manifolds, Rend. Sem. Mat. Univ. e Politec. Torino 34 (1975-76), 21-38.

[15] L. VANHECKE & D. JANSSENS, The Bochner curvature tensor on almost Hermitian manifolds, Hokkaido Math. J. 7 (1978), 252-258.

[16] H. WEYL, Classical Groups, their Invariants and Representations, Princeton University Press, 1946.

Istituto di Geometria
Università di Torino
Via Principe Amedeo 8
10123 TORINO (Italia)

Departement Wiskunde
Katholieke Universiteit Leuven
Celestijnenlaan 200B
B-3030 LEUVEN (Belgium)

CONFORMAL CHANGES OF ALMOST CONTACT METRIC STRUCTURES

Izu Vaisman

Department of Mathematics
University of Haifa, Israel

Consider a differentiable manifold M^{2n+1} endowed with an almost contact metric structure (φ,ξ,η,g) [2]. A conformal change of the metric g leads to a metric which is no more compatible with the almost contact structure (φ,ξ,η). This can be corrected by a convenient change of ξ and η, which implies rather strong restrictions.

Using such a definition for the conformal change of an almost contact metric structure we shall characterize new types of almost contact manifolds and we shall discuss some examples. As an application we get the interesting result that, under specified conditions, a product of two locally conformal Kähler manifolds has a foliation of codimension two, whose leaves are again locally conformal Kähler manifolds.

1. We proceed with the use of the above notation and we recall the defining properties of an almost contact metric (a.ct.m.) structure

$$\begin{aligned} &\varphi^2 = -I + \eta \otimes \xi \quad , \quad \eta(\xi) = 1 \, , \\ &g(\varphi X,\varphi Y) = g(X,Y) - \eta(X)\eta(Y) \, , \end{aligned} \tag{1.1}$$

where I is the identity endomorphism of the tangent bundle of M and $X,Y,..$ denote, like usually, vector fields on M. Recall also that (1.1) imply

$$\varphi(\xi) = 0 \, , \quad \eta\circ\varphi = 0 \, , \quad \eta(X) = g(\xi,X) \, . \tag{1.2}$$

Since we want to preserve the relations (1.1), we are led to give

DEFINITION 1.1. A *conformal change* of the a.ct.m. structure on M is a change of the form

$$\varphi' = \varphi \, , \quad \xi' = e^{\sigma}\xi \, , \quad \eta' = e^{-\sigma}\eta \, , \quad g' = e^{-2\sigma}g \, , \tag{1.3}$$

where σ is a differentiable function on M.

Correspondingly, a diffeomorphism $F : M \to M'$ between two a.ct.m. manifolds will be called a *conformal transformation* if it induces a conformal change of the two structures, and an infinitesimal transformation X on M will be called *conformal* if the corresponding local one-parameter group consists of conformal transformations. We devote this section to some properties of this last notion.

It is rather classical that, for a conformal infinitesimal transformation (c.i.t.) u, we shall have

$$L_u\varphi = 0 \, , \quad L_u\xi = \lambda\xi \, , \quad L_u\eta = \mu\eta \, , \; L_ug = \rho g \, , \tag{1.4}$$

where L_u denotes the Lie derivative.

Moreover, if we apply the operator L_u to the second and the third relations of (1.1) and if we use in our computation (1.1), it is easy to obtain

$$\mu = -\lambda \quad , \quad \rho = 2\mu \quad . \tag{1.5}$$

It follows that the vector field u is a c.i.t. iff it satisfies

$$L_u\varphi = 0 \ , \ L_u\xi = \lambda\xi \ , \ L_u\eta = -\lambda\eta \ , \ L_u g = -2\lambda g \ . \tag{1.6}$$

But we can prove

PROPOSITION 1.1. *The vector field u is a c.i.t. iff*

$$L_u\varphi = 0 \ , \ L_u g = -2\lambda g \ . \tag{1.7}$$

PROOF. Using the general formula

$$(L_u\varphi)(X) = L_u(\varphi X) - \varphi(L_u X) \ , \tag{1.8}$$

we compute $(L_u\varphi)(\xi)$ and get $\varphi(L_u\xi) = 0$, whence, since rank $\varphi = 2n$, $L_u\xi = \rho\xi$.

Similarly, we get $(L_u\eta)\circ\varphi = 0$ which implies $L_u\eta = \sigma\eta$.

Finally, using $\eta(\xi) = 1$ and the last relation (1.2), one derives $\sigma = -\rho = \lambda$. Q.e.d.

Let us also consider the fundamental 2-form of the structure, which is defined by [2]

$$\Phi(X,Y) = g(X,\varphi Y) \ . \tag{1.9}$$

Then we have (compare with [6])

PROPOSITION 1.2. *The vector field u is a c.i.t. iff there are two functions μ and λ such that*

$$L_u g = -2\lambda g \ , \ L_u\Phi = 2\mu\Phi \tag{1.10}$$

and then $\mu = -\lambda$.

PROOF. Using the definition of the Lie derivative we get first

$$(L_u\Phi)(X,Y) = (L_u g)(X,\varphi Y) + g\big(X,(L_u\varphi)Y\big) \ , \tag{1.11}$$

which shows that (1.7) imply (1.10) with $\mu = -\lambda$.

Conversely, from (1.10) and (1.11) we get

$$L_u\varphi = 2(\lambda+\mu)\varphi. \tag{1.12}$$

Now it is easy to establish the following auxiliary formula

$$L_u(\varphi^3) = (L_u\varphi)\circ\varphi^2 + \varphi\circ L_u\varphi\circ\varphi + \varphi^2\circ(L_u\varphi) \ . \tag{1.13}$$

Using the known fact [2] that

$$\varphi^3 + \varphi = 0 \ , \tag{1.14}$$

and using (1.12), formula (1.13) yields

$$\lambda + \mu = 0 \ , \tag{1.14}$$

which completes the proof of Proposition 1.2.

One can see that the conditions for u to be a c.i.t. are rather restrictive, a fact which is also clear from the results of Tanno [5, 6].

To illustrate this thing here, we prove

PROPOSITION 1.3. (i) [5]. *Every c.i.t. of a contact metric manifold* $M^{2n+1}(n \geq 1)$ *is an infinitesimal automorphism.* (ii) *Every c.i.t. of a compact metric cosymplectic manifold* $M^{2n+1}(n \geq 1)$ *is an infinitesimal automorphism.* (iii) *If u is a c.i.t. of a compact a.ct.m. manifold and if $g(\xi,[\xi,u])$ has a fixed sign, then u is an infinitesimal automorphism.*

PROOF. (i) In the mentioned case we have $\Phi = d\eta$, whence, by (1.10)

$$L_u d\eta = -2\lambda d\eta \ .$$

But since L_u and d commute, this implies by (1.6)

$$d\lambda \wedge \eta = \lambda d\eta \ ,$$

which contradicts $\eta \wedge (d\eta)^n \neq 0$ if $\lambda \neq 0$ in at least one point. Hence $\lambda = 0$, q.e.d.

(ii) By a metric cosymplectic manifold we understand an a.ct.m. manifold for which the forms η and Φ are closed.

Then, by differentiating $L_u\eta = -\lambda\eta$, $L_u\Phi = -2\lambda\Phi$, we get

$$d\lambda \wedge \eta = 0 \ , \quad d\lambda \wedge \Phi = 0 \ ,$$

whence, if $d\lambda \neq 0$, we get

$$\eta \wedge \Phi = 0$$

in contradiction with $\eta \wedge \Phi^n \neq 0$.

It follows therefore $\lambda =$ const., and by using the classical formula

$$L_u = i(u)d + di(u)$$

we deduce that if $\lambda \neq 0$, η and Φ are, actually, exact forms.

But in this case the volume element $\eta \wedge \Phi^n$ would be exact which is impossible for a compact M.

Therefore we must have $\lambda = 0$, q.e.d.

(iii) We start with the relations

$$L_u\eta = -\lambda\eta \ , \ L_u g = -2\lambda g \ .$$

The first yields

$$\lambda = -(L_u\eta)(\xi) \ , \tag{1.15}$$

while the second gives [4]

$$\lambda = -\frac{1}{2n+1}\,\delta U \ , \tag{1.16}$$

where $U(X) = g(u,X)$.

From these two relations we get

$$\delta U = (2n+1)g(\xi,[\xi,u]). \tag{1.17}$$

The stated result follows by integrating over M and using the known relation [4] $\int_M \delta U = 0$.

2. We come back now to the conformal changes of the a.ct.m. structures, defined by (1.3) and we shall be interested in considering a.ct.m. manifolds M^{2n+1} which are *locally conformal* (l.c.) to different important types of manifolds.

We begin with the *l.c. contact metric manifolds*. The a.ct.m. structure (φ,ξ,η,g) is a contact structure iff

$$\Phi = d\eta \tag{2.1}$$

Hence (φ,ξ,η,g) is l.c. contact iff M has an open covering $\{U_a\}$ endowed with differentiable functions $\sigma_a : U_a \to R$ such that over each U_a one has

$$e^{-2\sigma_a}\,\Phi = d(e^{-\sigma_a}\,\eta) \ , \tag{2.2}$$

or, equivalently

$$e^{-\sigma_a}\,\Phi = d\eta - d\sigma_a \wedge \eta \ . \tag{2.3}$$

PROPOSITION 2.1. *A l.c. contact metric structure is necessarily globally conformal contact and its form* η *is a contact form on* M.

PROOF. The characteristic relation (2.3) implies

$$e^{-n\sigma_a}\,\eta \wedge \Phi^n = \eta \wedge (d\eta)^n \ , \tag{2.4}$$

which proves that η is a contact form. (2.4) also gives $\sigma_a = \sigma_\beta$ over every $U_a \cap U_\beta$. Q.e.d.

COROLLARY 2.2. *Any l.c. contact K-structure is a globally conformal K-structure. Any l.c. Sasakian structure is globally conformal Sasakian.*

(See [2] for the corresponding definitions).

We shall consider now the normality condition which will be more interesting. This condition is [2]*

$$\varphi^2[X,Y] + [\varphi X,\varphi Y] - \varphi[\varphi X,Y] - \varphi[X,\varphi Y] + d\eta(X,Y)\xi = 0 . \qquad (2.5)$$

With the notation of (2.3) we get then that M is a *l.c. normal a.ct.m. manifold* iff one has for some system $\{\sigma_a\}$

$$[\varphi,\varphi] + \xi \otimes (d\eta - d\sigma_a \wedge \eta) = 0 \quad , \qquad (2.6)$$

where $[\varphi,\varphi]$ denotes the Nijenhuis tensor given by the first four terms of (2.5).

PROPOSITION 2.3. *An a.ct.m. structure for which* $\eta \wedge d\eta \neq 0$ *is l.c. normal iff there is a closed Pfaff form* ω *on* M *such that*

$$[\varphi,\varphi] + \xi \otimes d\eta = \xi \otimes (\omega \wedge \eta) . \qquad (2.7)$$

Moreover, the structure is globally conformal normal iff ω *is an exact form.*

PROOF. Comparing (2.6) for two neighbourhoods U_a and U_β, we get

$$d\sigma_a \wedge \eta = d\sigma_\beta \wedge \eta \quad , \qquad (2.8)$$

$$d\sigma_a \wedge d\eta = d\sigma_\beta \wedge d\eta \quad . \qquad (2.9)$$

Now, (2.8) implies $d\sigma_a = d\sigma_\beta + \lambda_{a\beta}\eta$, which together with (2.9) and with $\eta \wedge d\eta \neq 0$ yields $\lambda_{a\beta} = 0$, whence $d\sigma_a = d\sigma_\beta$. Therefore the local forms $d\sigma_a$ define a global closed Pfaff form ω and (2.6) becomes (2.7).

Conversely, if ω exists we can put locally $\omega = d\sigma_a$ and get thereby the local conformal changes to a normal structure.

The last assertion of Proposition 2.3 is trivial. Q.e.d.

COROLLARY 2.4. *An a.ct.m. structure for which* $\eta \wedge d\eta \neq 0$ *is l.c. quasi-Sasakian iff it is l.c. normal and satisfies the supplementary condition*

$$d\Phi = 2\omega \wedge \Phi \qquad (2.10)$$

where ω *is the form of (2.7).*

PROOF. The quasi-Sasakian manifolds were studied in [1] and they are characterized by normality and the closedness of their fundamental form. Hence, in our case we must have $d(e^{-2\sigma_a}\Phi) = 0$, and this is equivalent to (2.10).

REMARK. Let us call *almost quasi-Sasakian* a structure which is a.ct.m. and has closed fundamental form (but it is not necessarily normal). Then (2.10) is the condition which characterizes l.c. almost quasi-Sasakian manifolds of dimension ≥ 5. Moreover, to get the characterization of the l.c. metric cosymplectic manifolds we

---

* Our $d\eta$ differs by a factor 2 from that in [2]. Hence, also, our contact metric case is (inessentially) different. (φ,ξ,η,g) is contact metric here iff $(\varphi,2\xi,\frac{1}{2}\eta,\frac{1}{4}g)$ is such in [2].

have to add

$$d\eta = \omega \wedge \eta \ . \tag{2.11}$$

Note also the intermediate case when Φ is closed and η integrable, when we shall say that the structure is *special almost quasi-Sasakian*. The corresponding l.c. case is characterized by (2.10) and

$$\eta \wedge d\eta = 0 \ , \tag{2.12}$$

i.e., η itself must be completely integrable.

3. Examples of the situations in Section 2 can be obtained by considering hypersurfaces in complex manifolds and we begin by

PROPOSITION 3.1. *Let M^{2n} be a Hermitian manifold and τ a closed and nowhere vanishing Pfaff form on M. Denote by ζ the unit vector field normal to the foliation $\mathcal{F}$ defined by $\tau = 0$. Then, if ζ is an l.c. holomorphic vector field, the leaves of $\mathcal{F}$ carry an induced l.c. normal a.ct.m. structure.*

PROOF. Denote by J the complex structure and by g the metric of M. We get a.ct. m. structures on the leaves of $\mathcal{F}$ as in the known case of a single hypersurface [2]. Namely, for every X on M we have a decomposition

$$JX = \varphi X + \eta(X)\zeta \ , \tag{3.1}$$

where φX is tangent to the leaves of $\mathcal{F}$. By an abuse of notation we denote also by φ, η the induced tensors on the leaves and we do the same for the vector field

$$\xi = -J\zeta \ . \tag{3.2}$$

Then we see by an easy verification that (φ,ξ,η,g) yields an a.ct.m. structure on every leaf of $\mathcal{F}$.

In order to discuss normality, we calculate the Nijenhuis tensor $[\varphi,\varphi]\,(X,Y)$ for a couple X,Y of vector fields on M which are tangent to the leaves of $\mathcal{F}$, i.e., satisfy

$$\tau(X) = \tau(Y) = 0 \ . \tag{3.3}$$

By a lengthy but technical computation and using $[J,J] = 0$, we get

$$\begin{aligned}[\varphi,\varphi](X,Y) = & -d\eta(X,Y)\xi + \eta(Y)(L_\xi\varphi)(X) - \eta(X)(L_\xi\varphi)(Y) + \\ & + \Big\{\varphi Y(\eta(X)) - \varphi X(\eta(Y)) + \eta(X)\eta([\zeta,Y]) + \eta(Y)\eta([X,\zeta]) - \\ & - \eta(X)\zeta(\eta(Y)) + \eta(Y)\zeta(\eta(X)) + \eta([\varphi X,Y]) + \eta([X,\varphi Y])\Big\}\zeta \ .\end{aligned} \tag{3.4}$$

Note now that any vector field V on M can be decomposed as $V = V' + \alpha(V)\zeta$, where V' is tangent to $\mathcal{F}$ and

$$\alpha(V) = \frac{\tau(V)}{|\tau|} \ .$$

Using this decomposition for $(L_\xi\varphi)(X)$ and $(L_\xi\varphi)(Y)$ and separating the tangent part in (3.4) we get

$$[\varphi,\varphi](X,Y) + d\eta(X,Y)\xi = \eta(Y)(L_\xi\varphi)(X) - \eta(X)(L_\xi\varphi)(Y) +$$

$$+ \{\eta(X)(\varphi Y)(\ln|\tau|) - \eta(Y)(\varphi X)(\ln|\tau|)\}\zeta \ . \tag{3.5}$$

(We also used $d\tau = 0$.)

Now we have to use the hypothesis that ζ is l.c. holomorphic, i.e., [8], is locally proportional to holomorphic fields. As we proved in [8], this is equivalent with the fact that the Pfaff form

$$\alpha_\xi(X) = g(J\zeta,[\zeta,JX]) - g(\zeta,[\zeta,X]) \tag{3.6}$$

is closed and satisfies the relation

$$(L_\xi J)(X) = \alpha_\xi(X)J\zeta - \alpha_\xi(JX)\zeta \ . \tag{3.7}$$

But (3.1) yields

$$(L_\xi J)(X) = (L_\xi\varphi)(X) + (L_\xi\eta)(X) \ . \tag{3.8}$$

Let us note, passing by, that (3.6) and (3.8) give us

$$\alpha_\xi = -\eta\circ(L_\xi\varphi) \ . \tag{3.9}$$

Now, (3.7) and (3.8) provide the relation

$$(L_\xi\varphi)(X) = -\alpha_\xi(X)\xi - \{\alpha_\xi(JX) + (L_\xi\eta)(X)\}\zeta \ , \tag{3.10}$$

which we shall introduce in (3.5). By a new separation of the tangent part, it remains that, for any two fields X,Y which are tangent to the foliation F, the following relation holds good

$$[\varphi,\varphi](X,Y) + d\eta(X,Y)\xi = -(\alpha_\xi \wedge \eta)(X,Y)\xi \ . \tag{3.11}$$

In view of (2.6), (2.7), this ends the proof of Proposition 3.1. Moreover, if α_ξ is not exact (i.e., ζ is not globally conformal holomorphic), then the structure of the leaves is not globally conformal normal.

COROLLARY 3.2. *If, for (M,τ,ζ) of Proposition 3.1, ζ is a holomorphic field, the a.ct.m. structure induced on the leaves of F is normal.*

The next result which we can prove is

PROPOSITION 3.3. *Let M^{2n} be a locally conformal Kähler (l.c.K. [7]) manifold and τ a closed and nowhere vanishing Pfaff form on M. Denote by F the foliation $\tau = 0$.*

Then the leaves of F *carry an induced l.c. almost quasi-Sasakian structure, which is special if* $\tau \circ J$ *is completely integrable.*

PROOF. We get on the leaves of F the a.ct.m. structure already considered in Proposition 3.1 and for which we shall use the same notation. It is trivial that, for vector fields X,Y,Z tangent to F, the following relations hold

$$\Phi(X,Y) = \Omega(X,Y), \quad d\Phi(X,Y,Z) = d\Omega(X,Y,Z) . \tag{3.12}$$

Now, since l.c.K. means

$$d\Omega = \omega \wedge \Omega \quad , \tag{3.13}$$

where ω is the Lee form of M [7], (3.13) implies (2.10) and we are done. As for the last statement, it is true since by (3.1) we get

$$\eta = \frac{\tau \circ J}{|\tau|} \quad . \tag{3.14}$$

Proposition 3.3 admits the following converse.

PROPOSITION 3.4. *Let* M^{2n+1} *be a normal l.c. special almost quasi-Sasakian manifold with the structure* (φ,ξ,η,g). *Then,* $\eta = 0$ *defines a foliation* F *on* M *whose leaves carry an induced l.c.K. structure.*

PROOF. It is known from [3, Theorem 3.4] that the leaves of F carry an induced complex structure whose fundamental form Ω satisfies

$$d\Omega(X,Y,Z) = d\Phi(X,Y,Z) \tag{3.15}$$

for X,Y,Z tangent to F. Hence, since the given structure on M is l.c. almost-quasi-Sasakian, the induced structure on the leaves is clearly l.c.K. Q.e.d.

Now, the results of this Section can be combined to give the following interesting result

PROPOSITION 3.5. *Let* M_i $(i = 1,2)$ *be two l.c.K. manifolds with Lee forms* ω_i , *and put* $\zeta_i = (1/|\omega_i|) \# \omega_i$. *Suppose that* ζ_i *are analytic vector fields and that the Pfaff forms* $\omega_i \circ J$ *are completely integrable. Denote by* F *the foliation of codimension* 2 *defined on* $M_1 \times M_2$ *by the equations*

$$\omega_1 - \omega_2 = 0 , \quad \omega_1 \circ J - \omega_2 \circ J = 0 . \tag{3.16}$$

Then, every leaf of F *carries an induced l.c.K. structure.*

PROOF. Consider first the foliation F' defined on $M_1 \times M_2$ by $\omega_1 - \omega_2 = 0$. From the analyticity of ζ_i (# is the well known musical isomorphism) we easily deduce that the hypotheses of Corollary 3.2 are satisfied, whence the leaves of F' have an induced normal a.ct.m. structure. This structure is seen to be a special l.c. almost quasi-Sasakian structure just like in the Proof of Proposition 3.3. Indeed, the only difference is that, instead of (3.13), we have

Vol. 640: J. L. Dupont, Curvature and Characteristic Classes. X, 175 pages. 1978.

Vol. 641: Séminaire d'Algèbre Paul Dubreil, Proceedings Paris 1976–1977. Edité par M. P. Malliavin. IV, 367 pages. 1978.

Vol. 642: Theory and Applications of Graphs, Proceedings, Michigan 1976. Edited by Y. Alavi and D. R. Lick. XIV, 635 pages. 1978.

Vol. 643: M. Davis, Multiaxial Actions on Manifolds. VI, 141 pages. 1978.

Vol. 644: Vector Space Measures and Applications I, Proceedings 1977. Edited by R. M. Aron and S. Dineen. VIII, 451 pages. 1978.

Vol. 645: Vector Space Measures and Applications II, Proceedings 1977. Edited by R. M. Aron and S. Dineen. VIII, 218 pages. 1978.

Vol. 646: O. Tammi, Extremum Problems for Bounded Univalent Functions. VIII, 313 pages. 1978.

Vol. 647: L. J. Ratliff, Jr., Chain Conjectures in Ring Theory. VIII, 133 pages. 1978.

Vol. 648: Nonlinear Partial Differential Equations and Applications, Proceedings, Indiana 1976–1977. Edited by J. M. Chadam. VI, 206 pages. 1978.

Vol. 649: Séminaire de Probabilités XII, Proceedings, Strasbourg, 1976–1977. Edité par C. Dellacherie, P. A. Meyer et M. Weil. VIII, 805 pages. 1978.

Vol. 650: C*-Algebras and Applications to Physics. Proceedings 1977. Edited by H. Araki and R. V. Kadison. V, 192 pages. 1978.

Vol. 651: P. W. Michor, Functors and Categories of Banach Spaces. VI, 99 pages. 1978.

Vol. 652: Differential Topology, Foliations and Gelfand-Fuks-Cohomology, Proceedings 1976. Edited by P. A. Schweitzer. XIV, 252 pages. 1978.

Vol. 653: Locally Interacting Systems and Their Application in Biology. Proceedings, 1976. Edited by R. L. Dobrushin, V. I. Kryukov and A. L. Toom. XI, 202 pages. 1978.

Vol. 654: J. P. Buhler, Icosahedral Golois Representations. III, 143 pages. 1978.

Vol. 655: R. Baeza, Quadratic Forms Over Semilocal Rings. VI, 199 pages. 1978.

Vol. 656: Probability Theory on Vector Spaces. Proceedings, 1977. Edited by A. Weron. VIII, 274 pages. 1978.

Vol. 657: Geometric Applications of Homotopy Theory I, Proceedings 1977. Edited by M. G. Barratt and M. E. Mahowald. VIII, 459 pages. 1978.

Vol. 658: Geometric Applications of Homotopy Theory II, Proceedings 1977. Edited by M. G. Barratt and M. E. Mahowald. VIII, 487 pages. 1978.

Vol. 659: Bruckner, Differentiation of Real Functions. X, 247 pages. 1978.

Vol. 660: Equations aux Dérivée Partielles. Proceedings, 1977. Edité par Pham The Lai. VI, 216 pages. 1978.

Vol. 661: P. T. Johnstone, R. Paré, R. D. Rosebrugh, D. Schumacher, R. J. Wood, and G. C. Wraith, Indexed Categories and Their Applications. VII, 260 pages. 1978.

Vol. 662: Akin, The Metric Theory of Banach Manifolds. XIX, 306 pages. 1978.

Vol. 663: J. F. Berglund, H. D. Junghenn, P. Milnes, Compact Right Topological Semigroups and Generalizations of Almost Periodicity. X, 243 pages. 1978.

Vol. 664: Algebraic and Geometric Topology, Proceedings, 1977. Edited by K. C. Millett. XI, 240 pages. 1978.

Vol. 665: Journées d'Analyse Non Linéaire. Proceedings, 1977. Edité par P. Bénilan et J. Robert. VIII, 256 pages. 1978.

Vol. 666: B. Beauzamy, Espaces d'Interpolation Réels: Topologie et Géometrie. X, 104 pages. 1978.

Vol. 667: J. Gilewicz, Approximants de Padé. XIV, 511 pages. 1978.

Vol. 668: The Structure of Attractors in Dynamical Systems. Proceedings, 1977. Edited by J. C. Martin, N. G. Markley and W. Perrizo. VI, 264 pages. 1978.

Vol. 669: Higher Set Theory. Proceedings, 1977. Edited by G. H. Müller and D. S. Scott. XII, 476 pages. 1978.

Vol. 670: Fonctions de Plusieurs Variables Complexes III, Proceedings, 1977. Edité par F. Norguet. XII, 394 pages. 1978.

Vol. 671: R. T. Smythe and J. C. Wierman, First-Passage Perculation on the Square Lattice. VIII, 196 pages. 1978.

Vol. 672: R. L. Taylor, Stochastic Convergence of Weighted Sums of Random Elements in Linear Spaces. VII, 216 pages. 1978.

Vol. 673: Algebraic Topology, Proceedings 1977. Edited by P. Hoffman, R. Piccinini and D. Sjerve. VI, 278 pages. 1978.

Vol. 674: Z. Fiedorowicz and S. Priddy, Homology of Classical Groups Over Finite Fields and Their Associated Infinite Loop Spaces. VI, 434 pages. 1978.

Vol. 675: J. Galambos and S. Kotz, Characterizations of Probability Distributions. VIII, 169 pages. 1978.

Vol. 676: Differential Geometrical Methods in Mathematical Physics II, Proceedings, 1977. Edited by K. Bleuler, H. R. Petry and A. Reetz. VI, 626 pages. 1978.

Vol. 677: Séminaire Bourbaki, vol. 1976/77, Exposés 489–506. IV, 264 pages. 1978.

Vol. 678: D. Dacunha-Castelle, H. Heyer et B. Roynette. Ecole d'Eté de Probabilités de Saint-Flour. VII-1977. Edité par P. L. Hennequin. IX, 379 pages. 1978.

Vol. 679: Numerical Treatment of Differential Equations in Applications, Proceedings, 1977. Edited by R. Ansorge and W. Törnig. IX, 163 pages. 1978.

Vol. 680: Mathematical Control Theory, Proceedings, 1977. Edited by W. A. Coppel. IX, 257 pages. 1978.

Vol. 681: Séminaire de Théorie du Potentiel Paris, No. 3, Directeurs: M. Brelot, G. Choquet et J. Deny. Rédacteurs: F. Hirsch et G. Mokobodzki. VII, 294 pages. 1978.

Vol. 682: G. D. James, The Representation Theory of the Symmetric Groups. V, 156 pages. 1978.

Vol. 683: Variétés Analytiques Compactes, Proceedings, 1977. Edité par Y. Hervier et A. Hirschowitz. V, 248 pages. 1978.

Vol. 684: E. E. Rosinger, Distributions and Nonlinear Partial Differential Equations. XI, 146 pages. 1978.

Vol. 685: Knot Theory, Proceedings, 1977. Edited by J. C. Hausmann. VII, 311 pages. 1978.

Vol. 686: Combinatorial Mathematics, Proceedings, 1977. Edited by D. A. Holton and J. Seberry. IX, 353 pages. 1978.

Vol. 687: Algebraic Geometry, Proceedings, 1977. Edited by L. D. Olson. V, 244 pages. 1978.

Vol. 688: J. Dydak and J. Segal, Shape Theory. VI, 150 pages. 1978.

Vol. 689: Cabal Seminar 76–77, Proceedings, 1976–77. Edited by A.S. Kechris and Y. N. Moschovakis. V, 282 pages. 1978.

Vol. 690: W. J. J. Rey, Robust Statistical Methods. VI, 128 pages. 1978.

Vol. 691: G. Viennot, Algèbres de Lie Libres et Monoïdes Libres. III, 124 pages. 1978.

Vol. 692: T. Husain and S. M. Khaleelulla, Barrelledness in Topological and Ordered Vector Spaces. IX, 258 pages. 1978.

Vol. 693: Hilbert Space Operators, Proceedings, 1977. Edited by J. M. Bachar Jr. and D. W. Hadwin. VIII, 184 pages. 1978.

Vol. 694: Séminaire Pierre Lelong – Henri Skoda (Analyse) Année 1976/77. VII, 334 pages. 1978.

Vol. 695: Measure Theory Applications to Stochastic Analysis, Proceedings, 1977. Edited by G. Kallianpur and D. Kölzow. XII, 261 pages. 1978.

Vol. 696: P. J. Feinsilver, Special Functions, Probability Semigroups, and Hamiltonian Flows. VI, 112 pages. 1978.

Vol. 697: Topics in Algebra, Proceedings, 1978. Edited by M. F. Newman. XI, 229 pages. 1978.

Vol. 698: E. Grosswald, Bessel Polynomials. XIV, 182 pages. 1978.

Vol. 699: R. E. Greene and H.-H. Wu, Function Theory on Manifolds Which Possess a Pole. III, 215 pages. 1979.

Vol. 700: Module Theory, Proceedings, 1977. Edited by C. Faith and S. Wiegand. X, 239 pages. 1979.

Vol. 701: Functional Analysis Methods in Numerical Analysis, Proceedings, 1977. Edited by M. Zuhair Nashed. VII, 333 pages. 1979.

Vol. 702: Yuri N. Bibikov, Local Theory of Nonlinear Analytic Ordinary Differential Equations. IX, 147 pages. 1979.

Vol. 703: Equadiff IV, Proceedings, 1977. Edited by J. Fábera. XIX, 441 pages. 1979.

Vol. 704: Computing Methods in Applied Sciences and Engineering, 1977, I. Proceedings, 1977. Edited by R. Glowinski and J. L. Lions. VI, 391 pages. 1979.

Vol. 705: O. Forster und K. Knorr, Konstruktion verseller Familien kompakter komplexer Räume. VII, 141 Seiten. 1979.

Vol. 706: Probability Measures on Groups, Proceedings, 1978. Edited by H. Heyer. XIII, 348 pages. 1979.

Vol. 707: R. Zielke, Discontinuous Čebyšev Systems. VI, 111 pages. 1979.

Vol. 708: J. P. Jouanolou, Equations de Pfaff algébriques. V, 255 pages. 1979.

Vol. 709: Probability in Banach Spaces II. Proceedings, 1978. Edited by A. Beck. V, 205 pages. 1979.

Vol. 710: Séminaire Bourbaki vol. 1977/78, Exposés 507–524. IV, 328 pages. 1979.

Vol. 711: Asymptotic Analysis. Edited by F. Verhulst. V, 240 pages. 1979.

Vol. 712: Equations Différentielles et Systèmes de Pfaff dans le Champ Complexe. Edité par R. Gérard et J.-P. Ramis. V, 364 pages. 1979.

Vol. 713: Séminaire de Théorie du Potentiel, Paris No. 4. Edité par F. Hirsch et G. Mokobodzki. VII, 281 pages. 1979.

Vol. 714: J. Jacod, Calcul Stochastique et Problèmes de Martingales. X, 539 pages. 1979.

Vol. 715: Inder Bir S. Passi, Group Rings and Their Augmentation Ideals. VI, 137 pages. 1979.

Vol. 716: M. A. Scheunert, The Theory of Lie Superalgebras. X, 271 pages. 1979.

Vol. 717: Grosser, Bidualräume und Vervollständigungen von Banachmoduln. III, 209 pages. 1979.

Vol. 718: J. Ferrante and C. W. Rackoff, The Computational Complexity of Logical Theories. X, 243 pages. 1979.

Vol. 719: Categorial Topology, Proceedings, 1978. Edited by H. Herrlich and G. Preuß. XII, 420 pages. 1979.

Vol. 720: E. Dubinsky, The Structure of Nuclear Fréchet Spaces. V, 187 pages. 1979.

Vol. 721: Séminaire de Probabilités XIII. Proceedings, Strasbourg, 1977/78. Edité par C. Dellacherie, P. A. Meyer et M. Weil. VII, 647 pages. 1979.

Vol. 722: Topology of Low-Dimensional Manifolds. Proceedings, 1977. Edited by R. Fenn. VI, 154 pages. 1979.

Vol. 723: W. Brandal, Commutative Rings whose Finitely Generated Modules Decompose. II, 116 pages. 1979.

Vol. 724: D. Griffeath, Additive and Cancellative Interacting Particle Systems. V, 108 pages. 1979.

Vol. 725: Algèbres d'Opérateurs. Proceedings, 1978. Edité par P. de la Harpe. VII, 309 pages. 1979.

Vol. 726: Y.-C. Wong, Schwartz Spaces, Nuclear Spaces and Tensor Products. VI, 418 pages. 1979.

Vol. 727: Y. Saito, Spectral Representations for Schrödinger Operators With Long-Range Potentials. V, 149 pages. 1979.

Vol. 728: Non-Commutative Harmonic Analysis. Proceedings, 1978. Edited by J. Carmona and M. Vergne. V, 244 pages. 1979.

Vol. 729: Ergodic Theory. Proceedings, 1978. Edited by M. Denker and K. Jacobs. XII, 209 pages. 1979.

Vol. 730: Functional Differential Equations and Approximation of Fixed Points. Proceedings, 1978. Edited by H.-O. Peitgen and H.-O. Walther. XV, 503 pages. 1979.

Vol. 731: Y. Nakagami and M. Takesaki, Duality for Crossed Products of von Neumann Algebras. IX, 139 pages. 1979.

Vol. 732: Algebraic Geometry. Proceedings, 1978. Edited by K. Lønsted. IV, 658 pages. 1979.

Vol. 733: F. Bloom, Modern Differential Geometric Techniques in the Theory of Continuous Distributions of Dislocations. XII, 206 pages. 1979.

Vol. 734: Ring Theory, Waterloo, 1978. Proceedings, 1978. Edited by D. Handelman and J. Lawrence. XI, 352 pages. 1979.

Vol. 735: B. Aupetit, Propriétés Spectrales des Algèbres de Banach. XII, 192 pages. 1979.

Vol. 736: E. Behrends, M-Structure and the Banach-Stone Theorem. X, 217 pages. 1979.

Vol. 737: Volterra Equations. Proceedings 1978. Edited by S.-O. Londen and O. J. Staffans. VIII, 314 pages. 1979.

Vol. 738: P. E. Conner, Differentiable Periodic Maps. 2nd edition, IV, 181 pages. 1979.

Vol. 739: Analyse Harmonique sur les Groupes de Lie II. Proceedings, 1976–78. Edited by P. Eymard et al. VI, 646 pages. 1979.

Vol. 740: Séminaire d'Algèbre Paul Dubreil. Proceedings, 1977–78. Edited by M.-P. Malliavin. V, 456 pages. 1979.

Vol. 741: Algebraic Topology, Waterloo 1978. Proceedings. Edited by P. Hoffman and V. Snaith. XI, 655 pages. 1979.

Vol. 742: K. Clancey, Seminormal Operators. VII, 125 pages. 1979.

Vol. 743: Romanian-Finnish Seminar on Complex Analysis. Proceedings, 1976. Edited by C. Andreian Cazacu et al. XVI, 713 pages. 1979.

Vol. 744: I. Reiner and K. W. Roggenkamp, Integral Representations. VIII, 275 pages. 1979.

Vol. 745: D. K. Haley, Equational Compactness in Rings. III, 167 pages. 1979.

Vol. 746: P. Hoffman, τ-Rings and Wreath Product Representations. V, 148 pages. 1979.

Vol. 747: Complex Analysis, Joensuu 1978. Proceedings, 1978. Edited by I. Laine, O. Lehto and T. Sorvali. XV, 450 pages. 1979.

Vol. 748: Combinatorial Mathematics VI. Proceedings, 1978. Edited by A. F. Horadam and W. D. Wallis. IX, 206 pages. 1979.

Vol. 749: V. Girault and P.-A. Raviart, Finite Element Approximation of the Navier-Stokes Equations. VII, 200 pages. 1979.

Vol. 750: J. C. Jantzen, Moduln mit einem höchsten Gewicht. III, 195 Seiten. 1979.

Vol. 751: Number Theory, Carbondale 1979. Proceedings. Edited by M. B. Nathanson. V, 342 pages. 1979.

Vol. 752: M. Barr, *-Autonomous Categories. VI, 140 pages. 1979.

Vol. 753: Applications of Sheaves. Proceedings, 1977. Edited by M. Fourman, C. Mulvey and D. Scott. XIV, 779 pages. 1979.

Vol. 754: O. A. Laudal, Formal Moduli of Algebraic Structures. III, 161 pages. 1979.

Vol. 755: Global Analysis. Proceedings, 1978. Edited by M. Grmela and J. E. Marsden. VII, 377 pages. 1979.

Vol. 756: H. O. Cordes, Elliptic Pseudo-Differential Operators – An Abstract Theory. IX, 331 pages. 1979.

Vol. 757: Smoothing Techniques for Curve Estimation. Proceedings, 1979. Edited by Th. Gasser and M. Rosenblatt. V, 245 pages. 1979.

Vol. 758: C. Năstăsescu and F. Van Oystaeyen; Graded and Filtered Rings and Modules. X, 148 pages. 1979.

$$d\Omega = \omega_1 \wedge \Omega_1 + \omega_2 \wedge \Omega_2$$

where Ω_i are the fundamental forms of the manifolds M_i $(i = 1,2)$ and $\Omega = \Omega_1 + \Omega_2$ is the fundamental form of $M_1 \times M_2$. But the result follows since on the leaves of F' we have $\omega_1 = \omega_2$.

Finally, the leaves of F can be considered as leaves of the leaves of F' satisfying the hypotheses of Proposition 3.4. Hence the stated result follows by using Proposition 3.4. Q.e.d.

REFERENCES

1. D.E. Blair, The Theory of Quasi-Sasakian Structures, *J. Diff. Geom. 1* (1967), 331-345.
2. D.E. Blair, Contact Manifolds in Riemannian Geometry, *Lect. Notes in Math., 509*, Springer-Verlag, New York (1976).
3. D.E. Blair and G.D. Ludden, Hypersurfaces in Almost Contact Manifolds, *Tôhoku Math. J. 21* (1969), 354-362.
4. S.I. Goldberg, *Curvature and Homology*, Academic Press, New York (1962).
5. S. Tanno, Note on Infinitesimal Transformations over Contact Manifolds, *Tôhoku Math. J., 14* (1962), 416-430.
6. S. Tanno, Some Transformations on Manifolds with Almost Contact and Contact Metric Structures, I., II., *Tôhoku Math. J., 15* (1963), 140-147, 322-331.
7. I. Vaisman, On Locally Conformal Almost Kähler Manifolds, *Israel J. of Math., 24* (1976), 338-351.
8. I. Vaisman, Holomorphic Vector Fields on Locally Conformal Kähler Manifolds, *Anal. St. Univ. Iasi, 24* (1978), 357-362.